THIRD EDITION

MEASUREMENT AND DETECTION OF RADIATION

THIRD EDITION

MEASUREMENT AND DETECTION OF RADIATION

GEORGE GREEN LIBRARY OF
SCIENCE AND ENGINEERING

NICHOLAS TSOULFANIDIS

SHELDON LANDSBERGER

CRC Press
Taylor & Francis Group
Boca Raton   London   New York

CRC Press is an imprint of the
Taylor & Francis Group, an **informa** business

CRC Press
Taylor & Francis Group
6000 Broken Sound Parkway NW, Suite 300
Boca Raton, FL 33487-2742

© 2011 by Taylor and Francis Group, LLC
CRC Press is an imprint of Taylor & Francis Group, an Informa business

No claim to original U.S. Government works

Printed in the United States of America on acid-free paper
10 9 8 7 6 5 4 3 2 1

International Standard Book Number: 978-1-4200-9185-4 (Hardback)

Library of Congress Cataloging-in-Publication Data

Tsoulfanidis, Nicholas, 1938-
  Measurement and detection of radiation. -- 3rd ed. / Nicholas Tsoulfanidis, Sheldon Landsberger.
    p. cm.
  Summary: "The research and applications of nuclear instrumentation have grown substantially since publication of the previous editions. With the miniaturization of equipment, increased speed of electronic components, and more sophisticated software, radiation detection systems are now more productively used in many disciplines, including nuclear nonproliferation, homeland security, and nuclear medicine. Continuing in the tradition of its bestselling predecessors, measurement and detection of radiation, third edition illustrates the fundamentals of nuclear interactions and radiation detection with a multitude of examples and problems. It offers a clearly written, accessible introduction to nuclear instrumentation concepts. New to the third edition. A new chapter on the latest applications of radiation detection, covering nuclear medicine, dosimetry, health physics, nonproliferation, and homeland security. Updates to all chapters and subtopics within chapters, as needed. Many new references and a completely updated bibliography. This third edition of a classic textbook continues to serve new students entering the nuclear science and engineering fields. It enables them to select the proper detector, analyze the results of counting experiments, and perform radiation measurements that follow proper health physics procedures. A solutions manual is available with qualifying course adoption"-- Provided by publisher.
  Summary: "The field of nuclear instrumentation has greatly increased both in basic research and applications, since the appearance of the 2nd edition of this book. With the miniaturization of equipment and increased speed of electronic components, radiation detection systems are now more productively used in many disciplines. In particular, areas of nuclear non-proliferation, homeland security, and nuclear medicine all have benefitted from these technological advances. With the emergence of the internet in the 1990's the dissemination of information is also much readily available. While the number of nuclear engineering programs has remained more or less constant in the United States, over the last 20 years, there are now many new academic programs that have nuclear and science and engineering as part of their undergraduate curriculum"-- Provided by publisher.
  Includes bibliographical references and index.
  ISBN 978-1-4200-9185-4 (hardback)
  1. Radiation--Measurement. 2. Nuclear counters. I. Landsberger, Sheldon. II. Title.

QC795.42.T78 2010
539.7'7--dc22   2010036515

Visit the Taylor & Francis Web site at
http://www.taylorandfrancis.com

and the CRC Press Web site at
http://www.crcpress.com

To my wife Zizeta from Nicholas
To my wife Marsha from Sheldon

Contents

Preface to the First Edition

The material in this book, which is the result of a 10-year experience obtained in teaching courses related to radiation measurements at the University of Missouri-Rolla, is intended to provide an introductory text on the subject. It includes not only what I believe the beginner ought to be taught but also some of the background material that people involved in radiation measurements should have. The subject matter is addressed to upper-level undergraduates and first-year graduate students. It is assumed that the students have had courses in calculus and differential equations and in basic atomic and nuclear physics. The book should be useful to students in nuclear, mechanical, and electrical engineering, physics, chemistry (for radiochemistry), nuclear medicine, and health physics; to engineers and scientists in laboratories using radiation sources; and to personnel in nuclear power plants.

The structure and the contents of the book are such that the person who masters the material will be able to

1. Select the proper detector given the energy and type of particle to be counted and the purpose of the measurement.

2. Analyze the results of counting experiments, that is, calculate errors, smooth results, unfold energy spectra, fit results with a function, etc.

3. Perform radiation measurements following proper health physics procedures.

Chapter 1 defines the energy range of the different types of radiation for which instruments and methods of measurement are considered; it gives a brief discussion of errors that emphasizes their importance; and, finally, it presents a very general description of the components of a counting system. This last part of the chapter is necessary because a course on radiation measurements involves laboratory work, and for this reason the students should be familiar from the very beginning with the general features and functions of radiation instruments.

Chapter 2 addresses the very important subject of errors. Since all experimental results have errors, and results reported without their corresponding errors are meaningless, this chapter is fundamental for a book such as this one. Further discussion of errors caused by the analysis of the results is presented in Chapter 11.

Chapters 3 and 4 constitute a quick review of material that should have been covered in previous courses. My experience has been that students need this review of atomic and nuclear physics and of penetration of radiation through matter. These two chapters can be omitted if the instructor feels that the students know the subject.

Chapters 5–7 describe the different types of radiation detectors. Full chapters have been devoted to gas-filled counters, scintillation detectors, and semiconductor detectors. Detectors with "special" functions are discussed in Chapter 17.

The subject of relative and absolute measurements is presented in Chapter 8. The solid angle (geometry factor) between source and detector and effects due to the source and the detector, such as efficiency, backscattering, and source self-absorption are all discussed in detail.

Chapter 9 is an introduction to spectroscopy. It introduces and defines the concepts used in the next four chapters. Chapter 10 discusses the features of the electronic components of a counting system that are important in spectroscopy. Its objective is not to make the reader an expert in electronics but to show how the characteristics of the instruments may influence the measurements.

Chapter 11 presents methods of analysis of experimental data. Methods of curve fitting, of interpolation, and of least-squares fitting are discussed concisely but clearly. A general discussion of folding, unfolding, and data smoothing, which are necessary tools in analysis of spectroscopic measurements, occupies the second half of this chapter. Special methods of unfolding for photons, charged particles, and neutrons are further discussed in Chapters 12 through 14, which also cover spectroscopy. Individual chapters are devoted to photons, charged particles, and neutrons. All the factors that affect spectroscopic measurements and the methods of analysis of the results are discussed in detail.

Chapter 15 is devoted to activation analysis, a field with wide-ranging applications. Health physics is discussed in Chapter 16. I feel that every person who handles radiation should know at least something about the effects of radiation, radiation units, and regulations related to radiation protection. This chapter may be omitted if the reader has already studied the subject.

Chapter 17 deals with special detectors and spectrometers that have found applications in many different fields but do not fit in any of the previous chapters. Examples are the self-powered detectors, which may be gamma or neutron detectors, fission track detectors, thermoluminescent dosimeters, photographic emulsions, and others.

The problems at the end of each chapter should help the student understand the concepts presented in the text. They are arranged not according to difficulty but in the order of presentation of the material needed for their solution.

The appendixes at the end of the book provide useful information to the reader.

I use the SI (metric) units with the exception of some well-established nonmetric units, which, it seems, are here to stay. Examples are MeV, keV, and eV for energy; the barn for cross sections; the curie; and the rem. These units are given in parentheses along with their SI counterparts.

Writing a book is a tremendous undertaking, a task too big for any single person. I was fortunate to have been helped by many individuals, and it gives me great pleasure to recognize them here. First and foremost, I thank all the former students who struggled through my typed notes when they took the radiation measurements course at the University of Missouri-Rolla. Their numerous critical comments are deeply appreciated. I thank my colleagues, Dr. D. Ray Edwards for his continuous support, Dr. G. E. Mueller for his many useful suggestions, and Drs. A. E. Bolon and T. J. Dolan for many helpful discussions over the last 10 years. I also thank Dr. R. H. Johnson of Purdue University for reviewing certain chapters. I especially thank my dear friend Professor B. W. Wehring of the University of Illinois for numerous lengthy discussions following his detailed critical review of most of the chapters. I am grateful to Mrs. Susan Elizagary for expertly typing most of the manuscript and to Mrs. Betty Volosin for helping in the final stages of typing.

No single word or expression of appreciation can adequately reflect my gratitude to my wife Zizeta for her moral support and understanding during the last three painstaking years, and to my children Steve and Lena for providing pleasant and comforting distraction.

Nicholas Tsoulfanidis

Preface to the Second Edition

For an author it is very gratifying to discover that a technical book is still relevant more than ten years after it was first published. This is the case with this book because it addresses the fundamentals of nuclear radiation counting, which have not significantly changed during that period of time. Like the first edition, this book is written for persons who have no prior knowledge of radiation counting. These include undergraduate students in nuclear science and engineering; first-year graduate students who enter this field from another discipline; health physicists and health physics technicians; nuclear medicine technical personnel; and scientists, engineers, and technicians in laboratories where atomic and nuclear radiation are used. In addition, according to comments from former students and colleagues, the book has proven to be an excellent reference.

The second edition follows the same guidelines as the first—namely simplicity in writing and use of many examples. The main structural change is the elimination of Chapter 17 (Special Detectors and Spectrometers) and the relocation of the material in appropriate chapters. For example, rate meters and gas-filled detectors are now discussed in Chapter 5. Self-powered detectors are now included in Chapter 14 along with other neutron detectors. Chapter 16 deals with solid-state track recorders and thermoluminescent dosimeters.

As should be expected, all chapters have been corrected for errors, revised for clarification, and new examples have been added as needed. The more substantive revisions were made in the following chapters: In Chapter 2, there is now a better explanation of the x^2 procedure and the minimum detectable activity (MDA). In Chapter 4, relative to the stopping power of charged particles, there is a more detailed discussion and presentation of the latest formulas of gamma-ray build-up factors. The Long Range Alpha Detector (LRAD), a clever new counter of alpha radiation, is introduced in Chapter 5. In Chapter 7, pure germanium detectors, which are prominent devices for the detection of gamma rays, are introduced. In Chapter 12 the latest information about Ge detectors is presented. Magnetic and electrostatic spectrometers and the position-sensitive detectors are included in Chapter 13. In Chapter 14, the LSL-M2 unfolding code is introduced as well as compensated ion chambers and self-powered neutron detectors. Chapter 16 is almost completely rewritten. There is an improved presentation in the dose rate calculation, detailed discussion of the new protection guides and exposure limits, and an expanded list of dosimeters.

I am grateful to Dr. Eiji Sakai who translated the First Edition into Japanese and in doing so discovered several typos and, more importantly, offered many suggestions that are incorporated into the Second Edition and make it better.

Nicholas Tsoulfanidis

Preface to the Third Edition

The field of nuclear instrumentation has greatly increased both in basic research and applications, since the appearance of the second edition of this book. With the miniaturization of equipment and increased speed of electronic components, radiation detection systems are now more productively used in many disciplines. In particular, areas of nuclear nonproliferation, homeland security, and nuclear medicine all have benefited from these technological advances. With the emergence of the internet in the 1990s the dissemination of information is also much readily available. While the number of nuclear engineering programs has remained more or less constant in the United States over the last 20 years, there are now many new academic programs that have nuclear science and engineering as part of their undergraduate curriculum. These programs reside in engineering, chemistry, and general science departments; in Canada and Europe new complete nuclear engineering programs have emerged. While there are also new research reactors, many newer countries also have fully developed radiation protection programs. Currently, there are many countries in the developing world that have shown an interest in pursuing nuclear power. For example, China and India are on the road to an aggressive nuclear power expansion. All these nuclear-related programs and activities require at least one course in radiation detection; thus the need for a textbook in this field is still of paramount importance in an educational and training setting.

While several recent books have appeared on the market in radiation detection and measurements, they are much better suited for the advanced student or as a reference book. This third edition remains fundamentally as an undergraduate or first year graduate textbook and will serve very well the many new students entering the nuclear field from the various disciplines of nuclear science and engineering.

In the third edition Sheldon Landsberger was invited to become a coauthor; we have followed the same guidelines as the first two—appreciation of the fundamentals of nuclear interactions and radiation detection with a multitude of examples and problems. The main changes in this third edition include the introduction of a new chapter called Latest Applications of Radiation Detection with topics such as nuclear medicine, dosimetry, health physics, nonproliferation, and homeland security; several sections in the second edition have been eliminated to be more consistent with today's state of radiation measurements. All the references have been checked for relevance; some were kept, and many new ones have been added. The bibliography was completely updated, and web site addresses are provided in many chapters.

We are grateful for the comments expressed by the reviewers of our proposal for the third edition, especially suggestions for improvements. Both of us are especially appreciative of the encouragement we have received from our peers to update this book.

Nicholas Tsoulfanidis
Sheldon Landsberger

Authors

Nicholas Tsoulfanidis is nuclear engineering professor emeritus of the Missouri University of Science & Technology and adjunct professor at the University of Nevada–Reno. Since June 1997, he has served as the editor of *Nuclear Technology*, an international journal published by the American Nuclear Society. During his professional career, he has performed research in radiation transport, radiation protection, and the nuclear fuel cycle, and he published numerous technical papers on these topics in addition to teaching. He is also the coauthor of the book *The Nuclear Fuel Cycle: Analysis and Management.*

Sheldon Landsberger has been the coordinator of the nuclear and radiation engineering program at the University of Texas at Austin since 1997. From 1987 to 1997 he was a faculty member in the nuclear engineering department at the University of Illinois at Urbana–Champaign. During his professional career he has been involved in a wide variety of experimental research projects, including fundamental nuclear physics and applied nuclear analytical techniques in environmental research. He holds two honors for teaching: the Glenn Murphy award from the American Society of Engineering Education, Nuclear and Radiological Division, and the Holly Compton award from the American Nuclear Society.

1

Introduction to Radiation Measurements

1.1 WHAT IS MEANT BY RADIATION?

The word *radiation* was used until about 1900 to describe electromagnetic waves. Around the turn of the century, electrons, X-rays, and natural radioactivity were discovered and were included under the umbrella of the term radiation. The newly discovered radiation showed characteristics of particles, in contrast to the electromagnetic radiation, which was treated as a wave. In the 1920s, DeBroglie developed his theory of the duality of matter, which was soon proved correct by electron diffraction experiments, and the distinction between particles and waves ceased to be important. Today, radiation refers to the whole electromagnetic spectrum as well as to all the atomic and subatomic particles that have been discovered.

One of the many ways in which different types of radiation are grouped together is in terms of ionizing and nonionizing radiation. The word *ionizing* refers to the ability of the radiation to ionize an atom or a molecule of the medium it traverses.

Nonionizing radiation is electromagnetic radiation with wavelength λ of about 10 nm or longer. That part of the electromagnetic spectrum includes radio waves, microwaves, visible light ($\lambda = 770\text{–}390$ nm), and ultraviolet light ($\lambda = 390\text{–}10$ nm).

Ionizing radiation includes the rest of the electromagnetic spectrum (X-rays, $\lambda \approx 0.01\text{–}10$ nm) and γ-rays with wavelength shorter than that of X-rays. It also includes all the atomic and subatomic particles, such as electrons, positrons, protons, alphas, neutrons, heavy ions, and mesons.

The material in this text refers only to ionizing radiation. Specifically, it deals with detection instruments and methods, experimental techniques, and analysis of results for radiation in the energy range shown in Table 1.1. Particles with energies listed in Table 1.1 are encountered around nuclear reactors and low-energy accelerators, around installations involving production or use of natural or manufactured radioisotopes, in medical and nuclear medicine isotopes, biological, biochemical, geological and environmental research involving radioactive tracers, and in naturally occurring radioactive materials (NORMs). Not included in Table 1.1 are cosmic rays and particles produced by high-energy accelerators (GeV energy range).

1.2 STATISTICAL NATURE OF RADIATION EMISSION

Radiation emission is nothing more than release of energy by a system as it moves from one state to another. According to classical physics, exchange or release of energy takes place on a continuous basis; that is, any amount of energy, no matter how small, may be exchanged as long as the exchange is consistent with conservation laws. The fate of a system is exactly determined if initial conditions and forces acting upon it are given. One may say that classical physics prescribed a "deterministic" view of the world.

Quantum theory changed all that. According to quantum theory, energy can be exchanged only in discrete amounts when a system moves from one state to another. The fact that conservation laws are satisfied is a necessary but not a sufficient condition for the change of a system. The fate of the system is not determined exactly if initial conditions and forces are known. One can only talk about the probability that the system will do something or do nothing. Thus, with the introduction of quantum theory, the study of the physical world changed from "deterministic" to "probabilistic."

The emission of atomic and nuclear radiation obeys the rules of quantum theory. Thus, one can only talk about the probability that a reaction will take place or that a particle will be emitted. If one attempts to measure the number of particles emitted by a nuclear reaction, that number

Table 1.1
Maximum Energy Considered

Particle	Energy (MeV)
α	20
β	10
γ	20
n	20
Heavy ions	100

is not constant in time; it has a statistical uncertainty because of the probabilistic nature of the phenomenon under study.

Consider a radioactive source emitting electrons and assume that one attempts to measure the number of electrons per unit time emitted by the source. For every atom of the source there is a probability, not a certainty, that an electron will be emitted during the next unit of time. One can never measure the "exact" number. The number of particles emitted per unit time is different for successive units of time. Therefore, one can only determine the average number of particles emitted. That average, like any average, carries with it an uncertainty, an error. The determination of this error is an integral part of any radiation measurement.

1.3 THE ERRORS AND ACCURACY AND PRECISION OF MEASUREMENTS

A measurement is an attempt to determine the value of a certain parameter or quantity. Anyone attempting a measurement should keep in mind the following two axioms regarding the result of the measurement:

Axiom 1 No measurement yields a result without an error.

Axiom 2 The result of a measurement is almost worthless unless the error associated with that result is also reported.

The term *error* is used to define the following concept:

$$\text{Error} = (\text{measured or computed value of quantity } Q) - (\text{true value of } Q)$$

or

$$\text{Error} = \text{estimated uncertainty of the measured or computed value of } Q$$

Related to the error of a measurement are the terms *accuracy* and *precision*. The dictionary gives essentially the same meaning for both accuracy and precision, but in experimental work they have different meanings.

The accuracy of an experiment tells us how close the result of the measurement is to the true value of the measured quantity. The precision of an experiment is a measure of the exactness of the result. As an example, consider the measurement of the speed of light, which is known from measurements, to be equal to 2.997930×10^8 m/s.

Assume that a measurement gave the result 2.9998×10^8 m/s. The difference between these two numbers is an estimate of the accuracy of the measurement. On the other hand, the precision of the measurement is related to the number of significant figures* representing the result. The number 2.9998×10^8 indicates that the result has been determined to be between 2.9997 and 2.9999 or, equivalently, that it is known to 1 part in 30,000 (1/29998).

If the measurement is repeated and the new result is 2.9999×10^8 m/s, the accuracy has changed but not the precision. If, on the other hand, the result of the measurement is 2.99985×10^8 m/s, both precision and accuracy have changed.

Another way to look at the accuracy and precision of a measurement is in terms of the distribution of the data obtained (Figure 1.1). To improve the error of a measurement, the process is repeated many times, if practical. The results recorded, after repeated identical tries, are not identical. Instead, the data follow a distribution, almost Gaussian in most cases (see Chapter 2 for more details), and the measured value reported is an average based on the shape of the distribution of data. The width of the distribution of individual results is a measure of the precision of the measurement; the distance of the average of the distribution from the true value is a measure of the accuracy of the measurement.

Every experimenter should consider accuracy and precision simultaneously. It would be a waste of effort to try to improve the precision of a measurement if it is known that the result is inaccurate. On the other hand, it is almost useless to try to achieve very high accuracy if the precision of the measurement is low.

Limitations in the accuracy and precision of measurements result from many causes. Among the most important are the following:

1. Incorrectly calibrated instruments.

2. Algebraic or reading errors of the observer.

---

* As an example of the number of significant figures, each of the following numbers has five significant figures: 2.9998, 29998, 20009, .0029998, 2.9880×10^8.

3. Uncontrolled changes in environmental conditions, such as temperature, pressure, and humidity.

4. Inability to construct arbitrarily small measuring meter-sticks, rods, pointers, clocks, apertures, lenses, and so on.

5. A natural limit of sensitivity for any real measuring instrument detecting individual effects of atoms, electrons, molecules, and protons.

6. Imperfect method of measurement in most cases.

7. Unknown exact initial state of the system. Or, even if the initial state is known, it is impossible to follow the evolution of the system. For example, to determine the state of a gas in a container, one should know the exact position and velocity of every molecule at $t = 0$. Even if this is known, how practical is it to follow 10^{20} atoms or molecules moving in a box?

8. Statistical nature of some processes, for example, radioactive decay. There is a probability that an atom of a radioactive isotope will decay in the next 10 s, and this is as much information as one can report on this matter. The probability can be calculated, but it is still a probability never a certainty.

In addition to true measurements, scientists and engineers perform numerical computations that provide a result with some degree of uncertainty or error. This uncertainty or error is the result of the imperfect numerical model used or of the limitations of the computational tool (computer or computer code used). Although there are ways to estimate such errors, the process is not trivial. Such methods will not be discussed in this book. The only reason for this note is to make the user aware of the limitations, in terms of uncertainties/errors, of numerical computations. Examples of such errors in nuclear instrumentation included peak fitting routines and incorrect pileup and dead-time corrections.

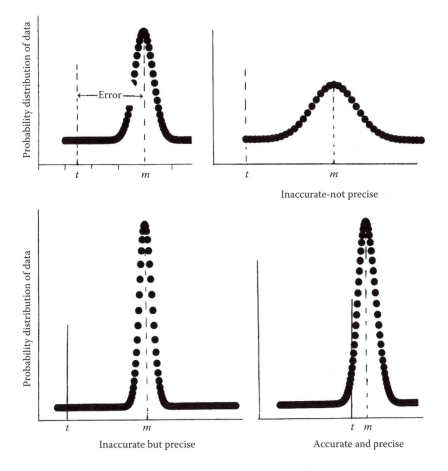

Figure 1.1 Accuracy and precision of measurements; t, true value; m, measured value.

1.4 TYPES OF ERRORS

There are many types of errors, but they are usually grouped into two broad categories namely systematic and random.

Systematic (or determinate) errors are those that affect all the results in the same way. Examples of systematic errors are given below:

1. Errors from badly calibrated instruments.

2. Personal errors (algebraic, wrong readings, and so on).

3. Imperfect technique.

Systematic errors introduce uncertainties that do not obey a particular law and cannot be estimated by repeating the measurement. The experimenter should make every reasonable effort to minimize or, better yet, eliminate systematic errors. Once a systematic error is identified, all results are corrected appropriately. For example, if a measurement of temperature is made and it is discovered that the thermocouple used overestimates the temperature by 10%, all temperatures measured are decreased by 10%.

Random (or statistical) errors can either decrease or increase the results of a measurement, but in a nonreproducible way. Most of the random errors cannot be eliminated. They can be reduced, however, by improving the experimental apparatus, improving the technique, and/or repeating the experiment many times. Examples of random errors include the following:

1. Errors resulting from experimental apparatus (reading of instruments, electronic noise, and so on).

2. Errors from uncontrolled change in condition such as voltage, temperature, or pressure.

3. Probabilistic nature of the phenomenon under study.

The determination of error associated with the measurement is a very important task. It is probably as important as the measurement. Technical journals and scientific reports never report results of experiments without the error corresponding to these results. A measurement reported without an error is almost worthless. For this reason, the study of errors is a topic of great importance for scientists and engineers.

This text does not give a complete theory of error. Only the fundamentals needed for a basic understanding of the statistical analysis of errors are presented. The objective is to present methods that provide an estimate of the error of a certain measurement or a series of measurements and procedures that minimize the error.

Only random errors are discussed from here on. In every measurement, systematic and random errors should be treated separately. Systematic and random errors should never be combined using the methods discussed in Chapter 2. Those methods apply to random errors only.

1.5 NUCLEAR INSTRUMENTATION

1.5.1 Introduction

This section is addressed to the person who has not seen or used radiation instruments.* Its purpose is to present a general description of the physical appearance and operation of the basic components of a radiation counting system. Every component is treated like a "black box," that is, input and output are discussed without any details about how the output is obtained. Details about the construction and operation of individual units are given in later chapters.

Detectors are discussed in Chapters 5 through 7 and the rest of the electronics are discussed in Chapter 10.

Counting systems are classified into two types, according to the method of operation as given below:

1. *Pulse type systems*: The output consists of voltage pulses, one pulse per particle detected.

2. *Current type systems*: The output is an average value, resulting from the detection of many particles.

---

* The term *radiation instruments* refers to instruments used for the detection of ionizing radiation as explained in Section 1.1.

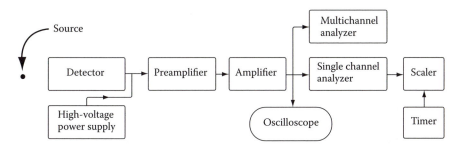

Figure 1.2 A basic pulse type detection system.

A basic pulse type system consists of the instruments shown in Figure 1.2. The function of each component is discussed in later sections of this chapter.

A current type system (e.g., an electrometer or a rate meter) is simpler than the pulse type system. Such systems are discussed in Chapter 5. The remainder of this chapter concerns only pulse type counting systems.

1.5.2 The Detector

The function of the detector is to produce a signal for every particle entering into it. Every detector works by using some interaction of radiation with matter. Following is a list of the most common detector types:

1. Gas-filled detectors (ionization, proportional, Geiger–Muller detectors)

2. Scintillation detectors

3. Semiconductor detectors

4. Spark chambers (used with high energy particles)

5. Charged particle alpha detectors

6. Bubble chambers

7. Photographic emulsions

8. Thermoluminescent dosimeters (TLDs)

9. Cerenkov counters

10. Neutron detectors

11. Whole body counters

12. Electronic dosimeters

13. Homeland security portal monitors

14. Continuous air monitors

15. Smoke detectors

16. Nuclear medicine detectors.

The signal at the output of most detectors is a voltage pulse, such as the one shown in Figure 1.3. For others, the signal may be a change in color (emulsions) or some trace that can be photographed (bubble or spark chambers).

The ideal pulse type counter should satisfy the following requirements:

1. Every particle entering the detector should produce a pulse at the exit of the counter, which is higher than the electronic noise* level of the unit that accepts it (usually this unit is the

---

* Electronic noise is any type of interference that tends to "mask" the quantity to be observed. It is usually the result of the thermal motion of charge carriers in the components of the detection system (cables, resistors, the detector itself, etc.) and manifests itself as a large number of low voltage pulses. Electronic noise should be distinguished from background pulses resulting from radiation sources that are always present, for example, cosmic rays.

preamplifier). In such a case, every particle entering the detector will be detected, and the detector efficiency, defined as the ratio of the number of particles detected to the number of particles entering the counter, will be equal to 100% (for more details on efficiency, see Chapter 8).

2. The duration of the pulse should be short, so that particles coming in one after the other in quick succession produce separate pulses. The duration of the pulse is a measure of the dead time of the counter (see Section 2.21) and may result in loss of counts in the case of high counting rates.

3. If the energy of the particle is to be measured, the height of the pulse should have some known fixed relationship to the energy of the particle. To achieve this, it is important that the size of the counter is such that the particle deposits all its energy (or a known fraction) in it.

4. If two or more particles deposit the same energy in the detector, the corresponding pulses should have the same height. This requirement is expressed in terms of the energy resolution of the detector (see Chapter 9). Good energy resolution is extremely important if the radiation field consists of particles with different energies and the objective of the measurement is to identify (resolve) these energies. Figure 1.4 shows an example of good and bad energy resolution.

There is no detector that satisfies all these requirements. Few detectors have 100% efficiency. In practice, it is not feasible for gamma and neutron detectors to have all the energy of the particle deposited in the counter. Because of statistical effects, there is no detector with ideal energy resolution. What should one do?

In practice, the experimenter selects a detector that satisfies as many of these properties as possible to the highest degree possible and, depending on the objective of the measurement, applies appropriate corrections to the measured data.

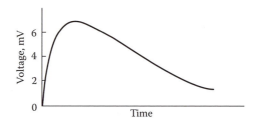

Figure 1.3 A typical pulse type detector signal.

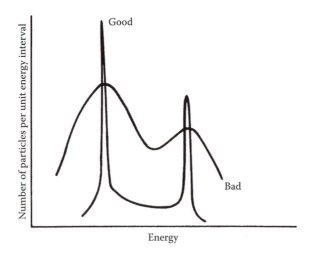

Figure 1.4 Good and bad energy resolution.

6

1.5.3 The Nuclear Instrument Module Concept

While most of the commercially available instruments that are used in radiation measurements conform to the standards on nuclear instrument modules (NIM)[1–3] developed by the U.S. Atomic Energy Commission, there are now several companies involved in developing data acquisition systems that utilize very small digital signal processing units, thus negating the need for NIMs.

The objective of the NIM standard was the design of commercial modules that are interchangeable physically and electrically. The electrical interchangeability is confined to the supply of power to the modules and in general does not cover the design of the internal circuits.

Multiple-width NIMs are also made. The standard NIM bin will accommodate 12 single-width NIMs or any combination of them having the same total equivalent width. Figure 1.5 is a photograph of the front and back sides of a commercial standard bin. Figure 1.6 is a photograph of the bin filled with NIMs of different widths, made by different manufacturers.

1.5.4 The High Voltage Power Supply

The *high voltage power supply* (HVPS) provides a positive or negative voltage necessary for the operation of the detector. Most detectors need positive high voltage (HV). Typical HVs for common detectors are given in Table 1.2. The HVPS is constructed in such a way that the HV at the output changes very little even though the input voltage (110 V, AC) may fluctuate.

A typical commercial HVPS is shown in Figure 1.7. The front panel has an indicator light that shows whether the unit is on or off and, if it is on, whether the output is positive or negative voltage. Also shown is an amplifier, single channel analyzer, pulser, and timer counter.

1.5.5 The Preamplifier

The primary purpose of the preamplifier is to provide an optimized coupling between the output of the detector and the rest of the counting system. The preamplifier is also necessary to minimize any sources of noise that may change the signal.

The signal that comes out of the detector is very weak, in the millivolt (mV) range (Figure 1.3). Before it can be recorded, it will have to be amplified by a factor of a thousand or more. To achieve this, the signal will have to be transmitted through a cable to the next instrument of the counting system, which is the amplifier. Transmission of any signal through a cable attenuates it to a

Figure 1.5 Photograph of a commercial NIM bin. (From Advanced Measurement Technology—ORTEC.)

Figure 1.6 A typical bin filled with a combination of NIMs.[1] (From Advanced Measurement Technology—ORTEC.)

Table 1.2
High Voltage Needed for Certain Common Detectors

Detector	High Voltage (V)
Ionization counters	HV < 1000
Proportional counters	500 < HV < 1500
GM counters	500 < HV < 1500
Scintillators	750 < HV < 1250
Neutron detectors	1000 < HV < 2000
Semiconductor detectors	
Surface barrier	HV < 100
Hyper-pure germanium	2500 < HV < 4500

certain extent. If it is weak at the output of the detector, it might be lost in the electronic noise that accompanies the transmission. This is avoided by placing the preamplifier as close to the detector as possible. The preamplifier shapes the signal and reduces its attenuation by matching the impedance of the detector with that of the amplifier. After going through the preamplifier, the signal may be safely transmitted to the amplifier, which may be located at a considerable distance away. Although some preamplifiers amplify the signal slightly, their primary function is that of providing electronic matching between the output of the detector and the input of the amplifier.

There are many types of commercial preamplifiers, two of which are shown in Figure 1.8. In most cases, the HV is fed to the detector through the preamplifier.

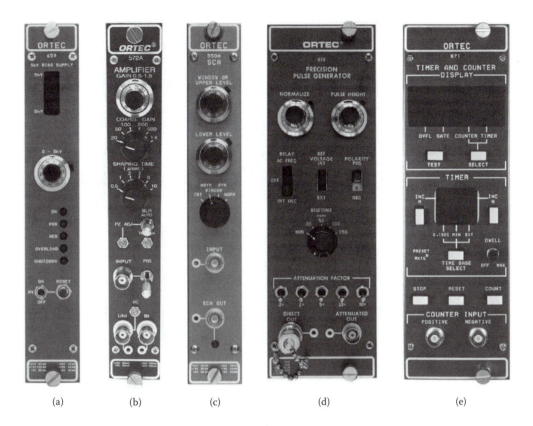

(a) (b) (c) (d) (e)

Figure 1.7 Typical: (a) HV power supply, (b) amplifier, (c) single channel analyzer, (d) pulser, (e) timer counter. (From Advanced Measurement Technology—ORTEC.)

1.5.6 The Amplifier

The main amplification unit is the amplifier. It increases the signal by as many as 1000 times or more. Modern commercial amplifiers produce a maximum signal of 10 V, regardless of the input and the amplification. For example, consider a preamplifier that gives at its output three pulses with heights 50, 100, and 150 mV. Assume that the amplifier is set to amplification of 100. At the output of the unit, the three pulses will be

$$50 \times 10^{-3} \times 100 = 5 \text{ V}$$
$$100 \times 10^{-3} \times 100 = 10 \text{ V}$$
$$150 \times 10^{-3} \times 100 = 10 \text{ V}$$

Note that the third value should be 15 V, but since the amplifier produces a maximum signal of 10 V, the three different input pulses will show, erroneously, as two different pulses at the output. If only the number of particles is measured, there is no error introduced—but if the energy of the particles is measured, then the error is very serious. In the example given above, if gammas of three different energies produce the pulses at the output of the preamplifier, the pulses at the output of the amplifier will be attributed erroneously to gammas of two different energies. To avoid such an error, an observer should follow this rule:

Before any measurement of particle energy, make certain that the highest pulse of the spectrum to be measured is less than 10 V at the output of the amplifier.

In addition to signal amplification, an equally important function of the amplifier is to convert the signal at the output of the preamplifier into a form suitable for the measurement desired. More details on this subject are given in Chapter 10. The front panel of a typical commercial amplifier is shown in Figure 1.7.

(a)

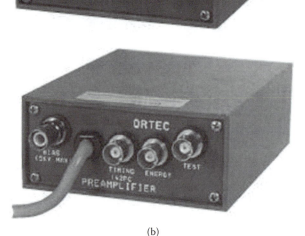

(b)

Figure 1.8 Typical preamplifier: (a) used with a photomultiplier tube showing back and front views and (b) used with semiconductor detectors showing back and front views. (From Advanced Measurement Technology—ORTEC.)

Commercial amplifiers have two dials for adjusting the amplification as given below:

1. *Coarse gain*: This dial adjusts the amplification in *steps*. Each step is a fraction of the maximum amplification. For example, the dial may show the numbers 1, 2, 4, 8, 16. If the maximum amplification is 100, then the coarse gain on 16 will give a maximum of 100, the coarse gain on 8 will give 50, etc. Some amplifiers have the numbers 1/6, 1/8, 1/4, 1/2, 1, and some newer ones have 1, 10, 100, 1000, and so on.

2. *Fine gain*: This dial adjusts the amplification *continuously* within each step of the coarse gain. The numbers, in most units, go from 0 to 10. The highest number provides the maximum amplification indicated by the coarse gain. As an example, consider the maximum amplification to be 100. If the coarse gain is 8 (highest number 16) and the fine gain 5 (highest number 10), the amplification will be $100 \times (1/2)(\text{coarse gain}) \times (1/2)(\text{fine gain}) = 25$.

Most commercial amplifiers provide at the output two types of pulses, called unipolar and bipolar (Figure 1.9).

It is now more common that systems have built in amplifiers designed such that coarse and fine gains can be computer controlled.

1.5.7 The Oscilloscope

The oscilloscope is an instrument that permits the study of rapidly changing phenomena, such as a sinusoidal voltage or the pulse of a counter. The phenomenon is observed on a fluorescent screen as shown in Figure 1.10. The horizontal axis of the screen measures time. The vertical axis gives volts.

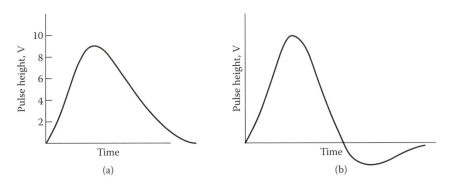

Figure 1.9   The pulse at the output of the amplifier: (a) unipolar pulse and (b) bipolar pulse.

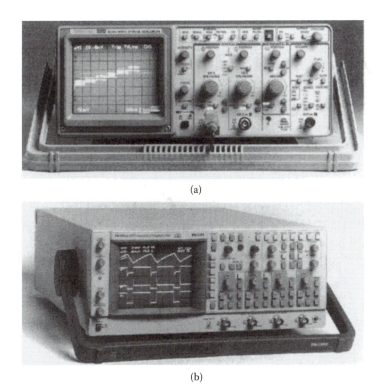

Figure 1.10   Two commercial oscilloscopes: (a) a Tektronix 2212 oscilloscope (Copyright © 1994 by Tektronix, Inc. All rights reserved. Reproduced by permission); and (b) a Philips PM3394A auto ranging combiscope (reproduced with permission).

In radiation measurements the oscilloscope is used to check the quality of the signal as well as the level and type of the electronic noise. It is always a good practice before any measurement is attempted to examine the signal at the output of the amplifier. A few examples of good and bad pulses are shown in Figure 1.11. In Figure 1.11, labels a and b represent good pulses and label c is probably an electrical discharge not good for counting. Label d in Figure 1.11 is no good either, because a high frequency signal is "riding" on the output of the preamplifier. If the pulse is not good, the observer should not proceed with the measurement unless the source of noise is identified and eliminated.

Modern oscilloscopes provide analog as well as digital signals and are capable of having various outputs including ones for printing and picture taking.

1.5.8 The Discriminator or Single-Channel Analyzer

The single channel analyzer (SCA) is used to eliminate the electronic noise and, in general, to reject unwanted pulses. When a pulse is amplified, the electronic noise that is always present in a circuit is also amplified. If one attempts to count all the pulses present, the counting rate may be exceedingly high. But electronic noise is a nuisance and it should not be counted.

In some cases, one may want to count only pulses above a certain height, that is, particles with energy above certain threshold energy. Pulses lower than that height should be rejected. The discriminator or SCA is the unit that can make the selection of the desired pulses. Figure 1.7c shows the front panel of a typical commercial SCA. Modern SCAs work in the following way.

There are two dials on the front panel of the unit. One is marked E, for energy, or LLD, for lower level dial; the other is marked ΔE or ULD/ΔE, for upper level dial/ΔE. There is also a two-position switch with integral (INT) and differential (DIFF) positions. In the INT position, only the E dial operates and the unit functions as a *discriminator*. In the DIFF position, both E and ΔE operate and the unit is then a *single channel analyzer*.

In some other commercial models, instead of INT and DIFF positions, the instrument has special connectors for the desired output.

The discriminator (switch position: INT). The dial E (for energy) may be changed continuously from 0 to 100. Of course, the discriminator works with voltage pulses, but there is a one-to-one correspondence between a pulse height and the energy of a particle. Assume that the discriminator is set to $E = 2.00$ V (the 2 V may also correspond to 2 MeV of energy). Only pulses with height greater than 2 V will pass through the discriminator. Pulses lower than 2 V will be

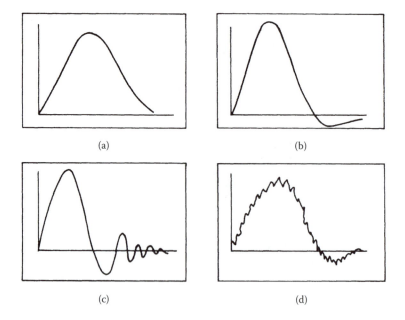

(a)

(b)

(c)

(d)

Figure 1.11 Samples of good (a and b) and bad (c and d) pulses as seen on the screen of the oscilloscope.

rejected. For every pulse that is larger than 2 V, the discriminator will provide at the output a rectangular pulse with height equal to 10 V (Figure 1.12) regardless of the actual height of the input pulse. The output pulse of the discriminator is a pulse that triggers the unit (scaler), which counts individual pulses and tells it, "a pulse with height bigger than 2 V has arrived; count 1." Thus, the discriminator eliminates all pulses below E and allows only pulses that are higher than E to be counted.

The single channel analyzer (switch position: DIFF). Both E and ΔE dials operate. Only pulses with heights between E and $E + \Delta E$ are counted (Figure 1.13). The two dials form a "channel";

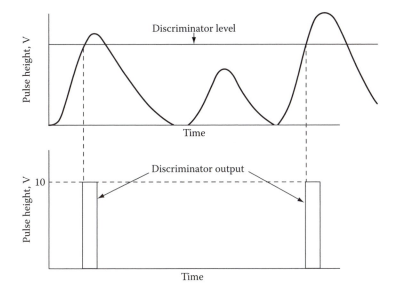

Figure 1.12 The pulse at the output of a discriminator.

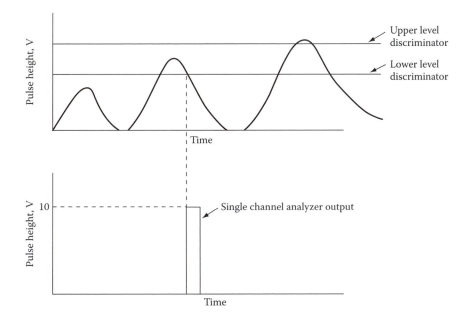

Figure 1.13 The operation of a single channel analyzer.

hence the name single channel analyzer. If the E dial is changed to E_1, then pulses with heights between E_1 and $E_1 + \Delta E$ will be counted. In other words, the width ΔE, or window, of the channel is always added to E.

As with the amplifiers it is now more common that systems have built in single channel analyzers have been digitally designed that can be computer controlled.

1.5.9 The Scaler

The scaler is a recorder of pulses. For every pulse entering the scaler, a count of 1 is added to the previous total. At the end of the counting period, the total number of pulses recorded is displayed. Figure l.7d shows the front panel of a typical commercial scaler.

1.5.10 The Timer

The timer is connected to the scaler and its purpose is to start and stop the scaler at desired counting time intervals. The front panel of a typical timer is shown in Figure 1.7e. Some models combine the timer with the scaler in one module.

1.5.11 The Multichannel Analyzer

The multichannel analyzer (MCA) records and stores pulses according to their height. Each storage unit is called a channel.

The height of the pulse has some known relationship—usually proportional—to the energy of the particle that enters into the detector. Each pulse is in turn stored in a particular channel corresponding to a certain energy. The distribution of pulses in the channels is an image of the distribution of the energies of the particles. At the end of a counting period, the spectrum that was recorded may be displayed on the screen of the MCA (Figure 1.14). The horizontal axis is a channel number, or particle energy. The vertical axis is a number of particles recorded per channel. More details about the MCA and its use are given in Chapters 9 and 10.

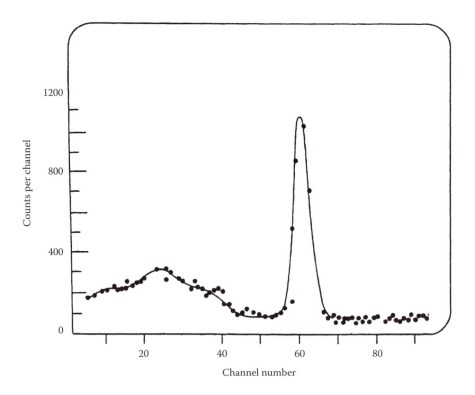

Figure 1.14 An energy spectrum shown on the screen of an MCA.

BIBLIOGRAPHY

Lyons, L., *Statistics for Nuclear and Particle Physicists*, Cambridge University Press, Cambridge, 1986.

Rabinovich, S. G., *Measurement Errors and Uncertainities*, 3rd ed., Springer Verlag, New York, 2005.

Taylor, J. R., *An Introduction to Error Analysis: The Study of Uncertainties in Physical Measurements*, University Science Books, Sausalito, CA, 1997.

REFERENCES

1. http://www.canberra.com/

2. http://www.ortec-online.com/

3. http://www.pgt.com/

2

Statistical Errors of Radiation Counting

2.1 INTRODUCTION

This chapter discusses statistics at the level needed for radiation measurements and analysis of their results. People who perform experiments need statistics for analysis of experiments that are statistical in nature, treatment of errors, and fitting a function to the experimental data. The first two uses are presented in this chapter. Data management and spectral fitting are presented in Chapter 11.

2.2 DEFINITION OF PROBABILITY

Assume that one repeats an experiment many times and observes whether or not a certain event x is the outcome. The event is a certain observable result defined by the experimenter. If the experiment was performed N times, and n results were of type x, then the probability $P(x)$ that any single event will be of type x is equal to

$$P(x) = \lim_{N \to \infty} \frac{n}{N} \tag{2.1}$$

The ratio n/N is sometimes called the relative frequency of occurrence of x in the first N trials.

There is an obvious difficulty with the definition given by Eq. 2.1—the requirement of an infinite number of trials. Clearly, it is impossible to perform an infinite number of experiments. Instead, the experiment is repeated N times, and if the event x occurs n times out of N, the probability $P(x)$ is

$$P(x) = \frac{n}{N} \tag{2.2}$$

Equation 2.2 will not make a mathematician happy, but it is extensively used in practice because it is in accord with the idea behind Eq. 2.1 and gives useful results.

As an illustration of the use of Eq. 2.2, consider the experiment of tossing a coin 100 times and recording how many times the result is "heads" and how many it is "tails." Assume that the result is as follows:

Heads: 48 times

Tails: 52 times

Based on Eq. 2.2, the probability of getting heads or tails if the coin is tossed once more is as follows:

$$P(\text{heads}) = \frac{48}{100} = 0.48$$

$$P(\text{tails}) = \frac{52}{100} = 0.52$$

For this simple experiment, the correct result is known to be

P(tails) = P(heads) = 0.5

and one expects to approach the correct result as the number of trials increases. That is, Eq. 2.2 does not give the correct probability, but as $N \to \infty$, Eq. 2.2 approaches Eq. 2.1.

Since both n and N are positive numbers, $0 \le n/N \le 1$ therefore,

$$0 \le P(x) \le 1$$

that is, the probability is measured on a scale from 0 to 1.

If the event x occurs every time the experiment is performed, then $n = N$ and $P(x) = 1$. Thus, the probability of a certain (sure) event to occur is equal to 1.

If the event x never occurs, then $n = 0$ and $P(x) = 0$. In this case, the probability of an impossible event to occur is 0.

If the result of a measurement has N possible outcomes, each having equal probability, then the probability for the individual event x_i to occur is

$$P(x_i) = \frac{1}{N}, \quad i = 1, \ldots, N$$

For example, in the case of coin tossing there are two events of equal probability; therefore

$$P(\text{heads}) = P(\text{tails}) = \frac{1}{2}$$

2.3 BASIC PROBABILITY THEOREMS

In the language of probability, an "event" is an outcome of one or more experiments or trials and is defined by the experimenter. Some examples of events are as follows:

1. Tossing a coin once.

2. Tossing a coin twice and getting heads both times.

3. Tossing a coin 10 times and getting heads for the first five times and tails for the other five.

4. Picking up one card from a deck of cards and that card being red.

5. Picking up 10 cards from a deck and all of them being hearts.

6. Watching the street for 10 min and observing two cyclists pass by.

7. Counting a radioactive sample for 10 s and recording 100 counts.

8. Inspecting all the fuel rods in a nuclear reactor and finding faults in two of them.

Given enough information, one can calculate the probability that any one of these events will occur. In some cases, an event may consist of simpler components and one would like to know how to calculate the probability of the complex event from the probabilities of its components.

Consider two events x and y and a series of N trials. The result of each trial will be only one of the following four possibilities:

1. x occurred but not y.

2. y occurred but not x.

3. Both x and y occurred.

4. Neither x nor y occurred.

Let n_1, n_2, n_3, n_4 be the number of times in the N observations that the respective possibilities occurred. Then,

$$n_1 + n_2 + n_3 + n_4 = N \tag{2.3}$$

The following probabilities are defined with respect to the events x and y:

$P(x)$ = probability that x occurred
$P(y)$ = probability that y occurred
$P(x + y)$ = probability that either x or y occurred
$P(xy)$ = probability that both x and y occurred
$P(x|y)$ = conditional probability of x given y
= probability of x occurring given that y has occurred
$P(y|x)$ = conditional probability of y given x
= probability of y occurring given that x has occurred.

Using Eq. 2.2, these probabilities are

$$P(x) = \frac{n_1 + n_3}{N} \tag{2.4}$$

$$P(y) = \frac{n_2 + n_3}{N} \tag{2.5}$$

$$P(x+y) = \frac{n_1 + n_2 + n_3}{N} \tag{2.6}$$

$$P(xy) = \frac{n_3}{N} \tag{2.7}$$

$$P(x|y) = \frac{n_3}{n_2 + n_3} \tag{2.8}$$

$$P(y|x) = \frac{n_3}{n_1 + n_3} \tag{2.9}$$

For the six probabilities given by Eqs. 2.4–2.9, the following two relations hold:

$$P(x+y) = P(x) + P(y) - P(xy) \tag{2.10}$$

$$P(xy) = P(x)P(y|x) = P(y)P(x|y) \tag{2.11}$$

Equation 2.10 is called the addition law of probability. Equation 2.11 is called the multiplication law of probability.

Example 2.1   Consider two well-shuffled decks of cards. What is the probability of drawing one card from each deck with both of them being the ace of spades?
Answer   The events of interest are as follows:

Event x = event y = (drawing one card and that card being ace of spades)

Since each deck has only one ace of spades,

$$P(x) = P(y) = P(\text{ace of spades}) = \frac{1}{52}$$

The conditional probability is

$$P(x|y) = P \text{ (First card ace of spades when second card is ace of spades)} = \frac{1}{52}$$

In this case, $P(x|y) = P(x)$ because the two events are independent. The fact that the first card from the first deck is the ace of spades has no influence on what the first card from the second deck is going to be. Similarly, $P(y|x) = P(y)$.

Therefore, using Eq. 2.11, one has

$$P(xy) = P(x)P(y) = \left(\frac{1}{52}\right)\left(\frac{1}{52}\right) = 0.00037$$

Example 2.2   Consider two well-shuffled decks of cards and assume one card is drawn from each of them. What is the probability of one of the two cards being the ace of spades?
Answer   Using Eq. 2.10,

$$P(x+y) = \frac{1}{52} + \frac{1}{52} - \left(\frac{1}{52}\right)\left(\frac{1}{52}\right) = 0.038$$

Under certain conditions, the addition and multiplication laws expressed by Eqs. 2.10 and 2.11 are simplified.

If the events x and y are mutually exclusive—that is, they cannot occur simultaneously—then $P(xy) = 0$ and the addition law becomes

$$P(x+y) = P(x) + P(y) \tag{2.12}$$

If the probability that x occurs is independent of whether or not y occurs, and vice versa, then as shown in Example 2.1,

$$P(y|x) = P(y)$$
$$P(x|y) = P(x)$$

In that case, the events x and y are called stochastically independent and the multiplication law takes the form

$$P(xy) = P(x)P(y) \tag{2.13}$$

Equations 2.12 and 2.13 are also known as the addition and multiplication laws of probability, but the reader should keep in mind that Eqs. 2.12 and 2.13 are special cases of Eqs. 2.10 and 2.11.

Example 2.3   What is the probability that a single throw of a die will result in either 2 or 5?
Answer

$$P(2) = \frac{1}{6} \quad P(5) = \frac{1}{6}$$
$$P(2+5) = P(2) + P(5) = \frac{1}{6} + \frac{1}{6} = \frac{1}{3}$$

Example 2.4   Consider two well-shuffled decks of cards and assume one card is drawn from each deck. What is the probability of both cards being spades?
Answer

$$P(\text{one spade}) = \frac{13}{52}$$
$$P[(\text{spade})(\text{spade})] = \left(\frac{13}{52}\right)\left(\frac{13}{52}\right) = \frac{1}{16}$$

Equations 2.12 and 2.13 hold for any number of events provided the events are mutually exclusive or stochastically independent. Thus, if we have N such events $x_{n|n=1,\dots,N}$

$$P(x_1 + x_2 + \cdots + x_N) = P(x_1) + P(x_2) + \cdots + P(x_N) \tag{2.14}$$

$$P(x_1 x_2 \cdots x_N) = P(x_1) P(x_2) \cdots P(x_N) \tag{2.15}$$

2.4 PROBABILITY DISTRIBUTIONS AND RANDOM VARIABLES

When an experiment is repeated many times under identical conditions, the results of the measurement will not necessarily be identical. In fact, as a rule rather than as an exception, the results will be different. Therefore, it is very desirable to know if there is a law that governs the individual outcomes of the experiment. Such a law, if it exists and is known, would be helpful in two ways. First, from a small number of measurements, the experimenter may obtain information about expected results of subsequent measurements. Second, a series of measurements may be checked for faults. If it is known that the results of an experiment obey a certain law and a given series of outcomes of such an experiment does not follow that law, then that series of outcomes is suspect and should be thoroughly investigated before it becomes acceptable.

There are many such laws governing different types of measurements. The three most frequently used will be discussed in later sections of this chapter, but first some general definitions and the concept of the random variable are introduced.

A quantity x that can be determined quantitatively and that in successive but similar experiments can assume different values is called a random variable. Examples of random variables are the result of drawing one card from a deck of cards, the result of the throw of a die, the result of measuring the length of a nuclear fuel rod, and the result of counting the radioactivity of a sample. There are two types of random variables, discrete and continuous.

A discrete random variable takes one of a set of discrete values. Discrete random variables are especially useful in representing results that take integer values—for example, number of persons, number of defective batteries, or number of counts recorded in a scaler.

A continuous random variable can take any value within a certain interval—for example, weight or height of people, the length of a rod, or the temperature of the water coming out of a reactor.

For every random variable x one may define a function $f(x)$ as follows:

Discrete random variables

$f(x_i)$ = probability that the value of the random variable is x_i, $i = 1, 2, \ldots, N$ where N is the number of possible (discrete) values of x. Since x takes only one value at a time, the events represented by the probabilities $f(x_i)$ are mutually exclusive; therefore, using Eq. 2.14,

$$\sum_{i=1}^{N} f(x_i) = 1 \qquad (2.16)$$

Continuous random variables

Assume that a random variable may take any value between a and $b\,(a \leq x \leq b)$. Then,

$f(x)\,dx$ = probability that the value of x lies between x and $x + dx$.

One should notice that for a continuous variable what is important is not the probability that x will take a specific value, but only the probability that x falls within an interval defined by two values of x. The equation corresponding to Eq. 2.16 is now

$$\int_{a}^{b} f(x)\,dx = 1 \qquad (2.17)$$

Equations 2.16 and 2.17 give the probability of a sure event, because x will certainly have one of the values $x_1, x_2, \ldots, x_N$ and will certainly have a value between a and b.

The function $f(x)$ is called the probability density function (pdf).*

Consider now the following function:

$$F(x_j) = \int_{a}^{x_j} f(x')\,dx' \qquad (2.18)$$

For a discrete variable,

$$F(x_j) = \sum_{i=1}^{j} f(x_i) \qquad (2.19)$$

Thus, $F(x_j)$ = probability that the value of x is less than or equal to x_j.

The function $F(x)$ is called the cumulative distribution function (cdf).† The cdf has the following properties:

$$F(x_N) = 1 \qquad (2.20)$$

$$F(a) = 0 \qquad (2.21)$$

$$F(b) = 1 \qquad (2.22)$$

---

* It has also been called the frequency function.
† It has also been called the integral or total distribution function.

The cdf is a positive monotonously increasing function, that is, $F(b) > F(a)$, if $b > a$. There is a relationship between the cdf and the pdf obtained from Eq. 2.18, namely,

$$f(x) = \frac{\mathrm{d}F(x)}{\mathrm{d}x}$$

(2.23)

2.5 LOCATION INDEXES (MODE, MEDIAN, MEAN)

If the distribution function $F(x)$ or $f(x)$ is known, a great deal of information can be obtained about the values of the random variable x. Conversely, if $F(x)$ or $f(x)$ is not completely known, certain values of x provide valuable information about the distribution functions. In most practical applications, the important values of x are clustered within a relatively narrow interval. To obtain an idea about the whole distribution, it is often adequate to indicate the position of this interval by "location indexes" providing typical values of x.

In theory, an infinite number of location indexes* may be constructed, but in practice, the following three are most frequently used: the mode, the median, and the mean of a distribution. Their definitions and physical meanings will be presented with the help of an example.

Consider the continuous pdf shown in Figure 2.1. The function $f(x)$ satisfies Eq. 2.17, that is, the total area under the curve of Figure 2.1 is equal to 1, with $a = -\infty$ and $a = +\infty$.

The *mode* is defined as the most probable value of x. Therefore, the mode x_1 is that x for which $f(x)$ is maximum and is obtained from

$$\frac{\mathrm{d}f(x)}{\mathrm{d}x} = 0$$

(2.24)

The *median* is the value x_2 for which

$$F(x_2) = \int_{-\infty}^{x_2} f(x)\,\mathrm{d}x = \frac{1}{2}$$

(2.25)

that is, the probability of x taking a value less than x_2 is equal to the probability of x taking a value greater than x_2.

The *mean*, also known as the "average" or the "expectation value" of x, is defined by the equation

$$\bar{x} = m = \int_{-\infty}^{\infty} xf(x)\,\mathrm{d}x$$

(2.26)

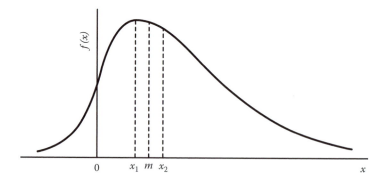

Figure 2.1 The mode (x_1), the median (x_2) and the mean (m) for a continuous probability distribution function.

---

* Measure of location is another name for location indexes.

An expression more general than Eq. 2.26 that gives the mean or average of any function $g(x)$, regardless of whether or not $f(x)$ satisfies Eq. 2.17, is as follows:

$$\overline{g(x)} = \frac{\int_{-\infty}^{\infty} g(x)f(x)\,dx}{\int_{+\infty}^{\infty} f(x)\,dx} \tag{2.27}$$

For a discrete pdf, the location indexes are defined in a similar way. If the pdf satisfies Eq. 2.16, the mean is given by

$$m = \overline{x} = \sum_{i=1}^{N} x_i f(x_i) \tag{2.28}$$

Equation 2.28 is an approximation because the true mean can only be determined with an infinite number of measurements. However, in practice, it is always a finite number of measurements that is available, and the average $\overline{x}$ instead of the true m is determined. Equation 2.28 is analogous to Eq. 2.2, which defines the probability based on a finite number of events.

The general expression for the average of a discrete pdf, equivalent to Eq. 2.27, is

$$\overline{g(x)} = \frac{\sum_{i=1}^{N} g(x_i)f(x_i)}{\sum_{i=1}^{N} f(x_i)} \tag{2.29}$$

Which of these or some other location indexes one uses is a matter of personal choice and convenience, depending on the type of problem studied. The mean is by far the most frequently used index and for this reason, only the mean will be discussed further.

Some elementary but useful properties of the mean that can be easily proven using Eqs. 2.26 or 2.28 are

$$\overline{ax} = a\overline{x} = am, \quad a = \text{constant}$$
$$\overline{a+x} = a + \overline{x} = a + m \tag{2.30}$$
$$\overline{g_1(x) + g_2(x) + \cdots + g_i(x)} = \overline{g_1(x)} + \overline{g_2(x)} + \cdots + \overline{g_i(x)}$$

Example 2.5   Calculation of the mean. The probability that a radioactive nucleus will not decay for time t is equal to

$$f(t) = \lambda e^{-\lambda t}$$

where λ is a constant. What is the mean life of such a nucleus?
Answer   Using Eq. 2.26, the mean life $\overline{t}$ is

$$\overline{t} = \int_0^{\infty} t\lambda e^{-\lambda t}\,dt = \frac{1}{\lambda}$$

Example 2.6   Consider the throw of a die. The probability of getting any number between 1 and 6 is $1/6$. What is the average number?
Answer   Using Eq. 2.29,

$$m = \overline{x} = \sum_{i=1}^{6} (i)\frac{1}{6} = \frac{1}{6}\sum_{i=1}^{6} i = \left(\frac{1}{6}\right)\frac{6(6+1)}{2} = 3.5$$

Example 2.7   Consider an experiment repeated N times giving the results $x_i|_{i=1,...,N}$. What is the average of the results?

Answer   Since the experiments were identical, all the results have the same probability of occurring, a probability that is equal to $1/N$. Therefore, the mean is

$$\bar{x} = \bar{m} = \sum_{i=1}^{N} x_i \left(\frac{1}{N}\right) \tag{2.31}$$

Equation 2.31 defines the so-called arithmetic mean of a series of N random variables. It is used extensively when the results of several measurements of the same variable are combined.

An extension of Eq. 2.31 is the calculation of the "means of means." Assume that one has obtained the averages $\bar{x}_1, \bar{x}_2, ..., \bar{x}_M$ by performing a series of M measurements, each involving $N_1, N_2, ..., N_M$ events, respectively.

The arithmetic mean of all the measurements, $\bar{X}$ is

$$\bar{X} = \frac{\bar{x}_1 + \bar{x}_2 + \cdots + \bar{x}_M}{M} \tag{2.32}$$

where

$$\bar{x}_j = \sum_{i=1}^{N_j} \frac{x_{ji}}{N_j}, \quad j = 1, ..., M$$

2.6 DISPERSION INDEXES, VARIANCE, AND STANDARD DEVIATION

A pdf or cdf is determined only approximately by any location index. For practical purposes it is sufficient to know the value of one location index—for example, the mean—together with a measure indicating how the probability density is distributed around the chosen location index. There are several such measures called *dispersion indexes*. The dispersion index most commonly used and the only one to be discussed here is the variance $V(x)$ and its square root, which is called the standard deviation σ.

The variance of a pdf is defined as shown by Eqs. 2.33 and 2.34. For continuous distributions,

$$V(x) = \sigma^2 = \int_{-\infty}^{\infty} (x-m)^2 f(x)\, dx \tag{2.33}$$

For discrete distributions,

$$V(x) = \sigma^2 = \sum_{i=1}^{N} (x_i - m)^2 f(x_i) \tag{2.34}$$

It is assumed that $f(x)$ satisfies Eq. 2.16 or 2.17 and N is a large number. It is worth noting that the variance is nothing more than the average of $(x-m)^2$. The variance of a linear function of x, $a + bx$, is

$$V(a+bx) = b^2 V(x) \tag{2.35}$$

where a and b are constants.

2.7 COVARIANCE AND CORRELATION

Consider the random variables $X_1, X_2, ..., X_M$ with means $m_1, m_2, ..., m_M$ and variances $\sigma_1^2, \sigma_2^2, ..., \sigma_M^2$. A question that arises frequently is, what is the average and the variance of the linear function

$$Q = a_1 X_1 + a_2 X_2 + \cdots + a_M X_M \tag{2.36a}$$

where the values of $a_i|_{i=1,...,M}$ are constants?

The average is simply (using Eq. 2.28)

$$\bar{Q} = a_1 m_1 + a_2 m_2 + \cdots + a_M m_M = \sum_{i=1}^{M} a_i m_i \tag{2.36b}$$

The variance is

$$V(Q) = \sigma^2 \overline{(Q - \bar{Q})^2} = \left[\sum_{i=1}^{M} a_i (X_i - m_i) \right]^2$$
$$= \sum_{i=1}^{M} a_i^2 \sigma_i^2 + 2 \sum_{j>i}^{M} \sum_{i}^{M} a_i a_j \overline{(X_i - m_i)(X_j - m_j)} \tag{2.37}$$

The quantity $\overline{(X_i - m_i)(X_j - m_j)}$ is called the "covariance" between X_i and X_j:

$$\mathrm{cov}(X_i, X_j) = \overline{(X_i - m_i)(X_j - m_j)} \tag{2.38}$$

The covariance, as defined by Eq. 2.38, suffers from the serious drawback that its value changes with the units used for the measurement of X_i, X_j. To eliminate this effect, the covariance is divided by the product of the standard deviations σ_i, σ_j, and the resulting ratio is called the correlation coefficient $\rho(X_i, X_j)$. Thus,

$$\rho_{ij} = \rho(X_i, X_j) = \frac{\mathrm{cov}(X_i, X_j)}{\sigma_i \sigma_j} \tag{2.39}$$

Using Eq. 2.39, the variance of Q becomes

$$\sigma^2 = V(Q) = V \left(\sum_{i=1}^{M} a_i X_i \right) = \sum_{i=1}^{M} a_i^2 \sigma_i^2 + 2 \sum_{j>i}^{M} \sum_{i}^{M} a_i a_j \rho_{ij} \sigma_i \sigma_j \tag{2.40}$$

Random variables for which $\rho_{ij} = 0$ are said to be uncorrelated.
If the X_i's are mutually uncorrelated, Eq. 2.40 takes the simpler form

$$\sigma^2 = V \left(\sum_{i=1}^{M} a_i X_i \right) = \sum_{i=1}^{M} a_i^2 \sigma_i^2 \tag{2.41}$$

Consider now a second linear function of the variables $X_1, X_2, X_3, \ldots, X_M$, namely, $R = b_1 X_1 + \cdots + b_M X_M$. The average of R is

$$\bar{R} = b_1 m_1 + b_2 m_2 + \cdots + b_M m_M = \sum_{i=1}^{M} b_i m_i$$

The covariance of Q, R is

$$\mathrm{cov}(Q, R) = \left[\sum_{i=1}^{M} a_i (X_i - m_i) \right] \left[\sum_{j=1}^{M} b_j (X_j - m_j) \right]$$
$$= \sum_{i=1}^{M} a_i b_i \sigma_i^2 + \sum_{i \neq j}^{M} \sum_{j}^{M} a_j b_j \rho_{ij} \sigma_i \sigma_j \tag{2.42}$$

If all the X's are mutually uncorrelated, then $\rho_{ij=0}$ and

$$\mathrm{cov}(Q, R) = \sum_{i=1}^{M} a_i b_i \sigma_i^2 \tag{2.43}$$

If all the X's have the same variance σ^2,

$$\text{cov}(Q, R) = \sigma^2 \sum_{i=1}^{M} a_i b_i \tag{2.44}$$

Equations 2.40–2.44 will be applied in Section 2.15 for the calculation of the *propagation of errors*.

2.8 THE BINOMIAL DISTRIBUTION

The *binomial distribution* is a pdf that applies under the following conditions:

1. The experiment has two possible outcomes, A and B.

2. The probability that any given observation results in an outcome of type A or B is constant, independent of the number of observations.

3. The occurrence of a type A event in any given observation does not affect the probability that the event A or B will occur again in subsequent observations.

Examples of such experiments are tossing a coin (heads or tails is the outcome), inspecting a number of similar items for defects (items are defective or not), and picking up objects from a box containing two types of objects.

The binomial distribution will be introduced with the help of the following experiment. Suppose that a box contains a large number of two types of objects, type A and type B. Let

p = probability that an object selected at random from this box is type A
$1 - p$ = probability that the randomly selected object is type B

An experimenter selects N objects at random.* The binomial distribution, giving the probability P_n that n out of the N objects are of type A, is

$$P_n = \frac{N!}{(N-n)!n!} p^n (1-p)^{N-n} \tag{2.45}$$

Example 2.8   A box contains 10,000 small metallic spheres, of which 2000 are painted white and the rest are painted black. A person removes 100 spheres from the box one at a time at random. What is the probability that 10 of these spheres are white?
Answer   The probability of picking one white sphere is

$$p = P(\text{white}) = \frac{2000}{10,000} = 0.2$$

The probability that 10 out of 100 selected spheres will be white is, according to Eq. 2.45,

$$P_{10} = \frac{100!}{(100-10)!\,10!}(0.2)^{10}(0.8)^{90} = 0.0034$$

Example 2.9   A coin is tossed three times. What is the probability that the result will be heads in all three tosses?
Answer   The probability of getting heads in one throw is 0.5. The probability of tossing the coin three times ($N = 3$) and getting heads in all three tosses ($n = 3$) is

$$P_3 = \frac{3!}{(3-3)!\,3!}(0.5)^3(1-0.5)^{3-3} = 0.125$$

---

* It is assumed that the box has an extremely large number of objects so that the removal of N of them does not change their number appreciably, or, after an object is selected and its type recorded, it is thrown back into the box. If the total number of objects is small, instead of Eq. 2.45, the hypergeometric density function should be used.

Of course, the same result could have been obtained in this simple case by using the multiplication law, Eq. 2.13:

$$P(\text{heads three times}) = (0.5)(0.5)(0.5) = 0.125$$

It is easy to show that the binomial distribution satisfies

$$\sum_{n=0}^{N} P_n = \sum_{n=0}^{N} \frac{N!}{(N-n)!n!} p^n (1-p)^{N-n} = 1 \tag{2.46}$$

The mean m is equal to

$$m = \overline{n} = \sum_{n=0}^{N} n P_n = PN \tag{2.47}$$

The variance $V(n)$ is

$$V(n) = \overline{(m-n)^2} = \sum_{n=0}^{N} (m-n)^2 P_n = m(1-p) = pN(1-p) \tag{2.48}$$

The standard deviation σ is

$$\sigma = \sqrt{V(N)} = \sqrt{m(1-p)} = \sqrt{p(1-p)N} \tag{2.49}$$

Figure 2.2 shows three binomial distributions for $N = 10$ and $p = 0.1, 0.4$, and 0.8. Notice that as $p \to 0.5$, the distribution tends to be symmetric around the mean.

2.9 THE POISSON DISTRIBUTION

The Poisson distribution applies to events whose probability of occurrence is small and constant. It can be derived from the binomial distribution by letting

$$N \to \infty$$
$$p \to 0$$

in such a way that the value of the average $m = Np$ stays constant. It is left as an exercise for the reader to show that under the conditions mentioned above, the binomial distribution takes the form known as the Poisson distribution,

$$P_n = \frac{m^n}{n!} e^{-m} \tag{2.50}$$

where P_n is the probability of observing the outcome n when the average for a large number of trials is m.

The Poisson distribution has wide applications in many diverse fields, such as decay of nuclei, persons killed by lightning, number of telephone calls received in a switchboard, emission of photons by excited nuclei, and appearance of cosmic rays.

Example 2.10 A radiation detector is used to count the particles emitted by a radioisotopic source. If it is known that the average counting rate is 20 counts/min, what is the probability that the next trial will give 18 counts/min?

Answer The probability of decay of radioactive atoms follows the Poisson distribution. Therefore, using Eq. 2.50,

$$P_{18} = \frac{20^{18}}{18!} e^{-20} = 0.0844 \approx 8\%$$

That is, if one performs 10,000 measurements, 844 of them are expected to give the result 18 counts/min.

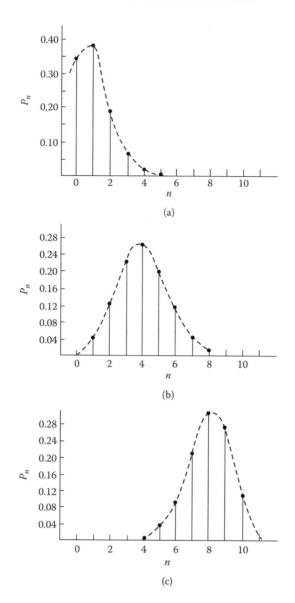

Figure 2.2 Three binomial distributions with $N = 10$ and (a) $p = 0.1$, (b) $p = 0.4$, and (c) $p = 0.8$.

Example 2.11 In a certain city with relatively constant population, the average number of people killed per year in automobile accidents is 75. What is the probability of having 80 auto accident fatalities during the coming year?

Answer The Poisson distribution applies. Therefore, using Eq. 2.50,

$$P_{80} = \frac{75^{80}}{80!} e^{-75} = 0.038 \approx 4\%$$

The Poisson distribution satisfies

$$\sum_{n=0}^{\infty} P_n = 1 \qquad (2.51)$$

The mean m is equal to

$$m = \bar{n} = \sum_{n=0}^{\infty} nP_n = m \qquad (2.52)$$

The variance is

$$V(n) = \overline{(m-n)^2} = \sum_{n=0}^{\infty} (m-n)^2 P_n = m \qquad (2.53)$$

The standard deviation σ is

$$\sigma = \sqrt{V(n)} = \sqrt{m} \qquad (2.54)$$

Figure 2.3 shows the Poisson distribution for three different means. It should be pointed out that as the mean increases, the Poisson distribution becomes symmetric around the mean. For $m = 20$, the distribution is already for all practical purposes symmetric around the mean, and it resembles the normal distribution, which is discussed next.

2.10 THE NORMAL (GAUSSIAN) DISTRIBUTION

Both the binomial and Poisson distributions apply to discrete variables, whereas most of the random variables involved in experiments are continuous. In addition, the use of discrete distributions necessitates the use of long or infinite series for the calculation of such parameters as the mean and the standard deviation (see Eqs. 2.47, 2.48, 2.52, 2.53). It would be desirable, therefore, to have a pdf that applies to continuous variables. Such a distribution is the normal or Gaussian distribution.

The normal distribution $G(x)$ is given by

$$G(x)\,dx = \frac{1}{(\sqrt{2\pi})\,\sigma} \exp\left[-\frac{(x-m)^2}{2\sigma^2} \right] dx \qquad (2.55)$$

where $G(x)\,dx$ = probability that the value of x lies between x and $x + dx$
$\quad\quad m$ = average of the distribution
$\quad\quad \sigma^2$ = variance of the distribution

Notice that this distribution, shown in Figure 2.4, has a maximum at $x = m$, is symmetric around m, is defined uniquely by the two parameters σ and m, and extends from $x = -\infty$ to $x = +\infty$. Equation 2.55 represents the shaded area under the curve of Figure 2.4. In general, the probability of finding the value of x between any two limits x_1 and x_2 is given by

$$G(x_1 \le x \le x_2) = \int_{x_1}^{x_2} G(x)\,dx \qquad (2.56)$$

The Gaussian given by Eq. 2.55 satisfies

$$\int_{-\infty}^{\infty} G(x)\,dx = 1 \qquad (2.57)$$

The average of the distribution is

$$\bar{x} = m = \int_{-\infty}^{\infty} xG(x)\,dx = m \qquad (2.58)$$

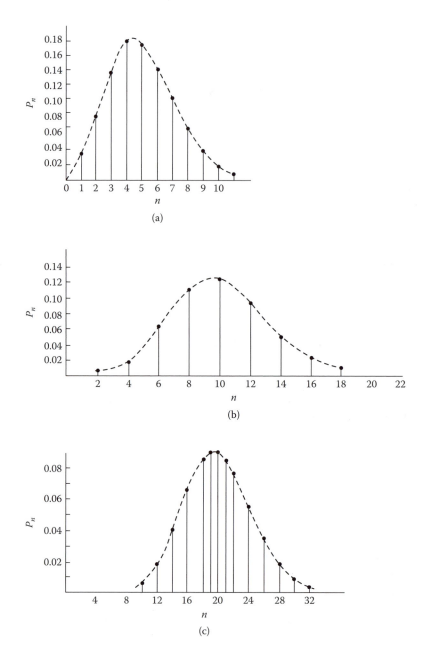

Figure 2.3 Three Poisson distributions: (a) $m = 5$, (b) $m = 10$, (c) $m = 20$.

The variance is

$$V(x) = \int_{-\infty}^{\infty} (x - m)^2 G(x) \, dx = \sigma^2 \tag{2.59}$$

The standard deviation is σ, already given as one of the two parameters defining the distribution; it is reated to the variance by

$$\sigma = \sqrt{V(x)} \tag{2.60}$$

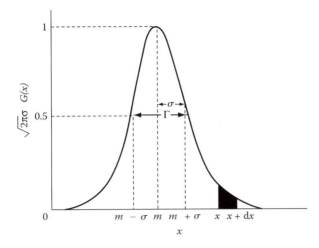

Figure 2.4 A normal (Gaussian) distribution.

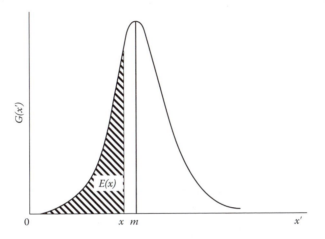

Figure 2.5 The cumulative normal distribution is equal to the shaded area under the Gaussian curve.

Three very important items associated with the Gaussian distribution are the following:

1. The cumulative normal distribution function, defined by

$$E(x) = \int_{-\infty}^{x} G(x') \, dx' = \int_{-\infty}^{x} \frac{1}{(\sqrt{2\pi})\sigma} \exp\left[-\frac{(x'-m)^2}{2\sigma^2}\right] dx' \tag{2.61}$$

The function $E(x)$ is very useful and is generally known as the error function (see Section 2.10.1). Graphically, the function $E(x)$ (Eq. 2.61) is equal to the shaded area of Figure 2.5. The function is sketched in Figure 2.6.

2. The *area under the curve of Figure 2.4 from* $x = m - \sigma$ to $x = m + \sigma$, given by

$$A_\sigma = \int_{m-\sigma}^{m+\sigma} G(x) \, dx = 0.683 \tag{2.62}$$

Equation 2.62 indicates that 68.3% of the total area under the Gaussian is included between $m - \sigma$ and $m + \sigma$. Another way of expressing this statement is to say that if a series of events follows the normal distribution, then it should be expected that 68.3% of the events will be located

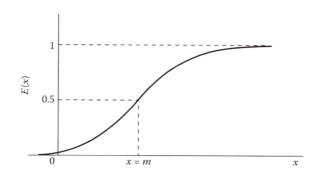

Figure 2.6 The error function.

between $m - \sigma$ and $m + \sigma$. As discussed later in Section 2.13, Eq. 2.62 is the basis for the definition of the "standard" error.

3. The *full width at half maximum (FWHM)*. The FWHM, usually denoted by the symbol Γ, is the width of the Gaussian distribution at the position of half of its maximum. The width Γ is slightly wider than 2σ (Figure 2.4). The correct relationship between the two is obtained from Eq. 2.55 by writing

$$G\left(m - \frac{\Gamma}{2}\right) = G\left(m + \frac{\Gamma}{2}\right) = \frac{1}{2}G(m)$$

Solving this equation for Γ gives

$$\Gamma = \left(2\sqrt{2\ln 2}\right)\sigma \approx 2.35\sigma \tag{2.63}$$

The width Γ is an extremely important parameter in measurements of the energy distribution of particles.

2.10.1 The Standard Normal Distribution

The evaluation of integrals involving the Gaussian distribution, such as those of Eqs. 2.56, 2.61, and 2.62, requires tedious numerical integration. The result of such integrations is a function of m and σ. Therefore, the calculation should be repeated every time m or σ changes. To avoid this repetition, the normal distribution is rewritten in such as way that

$$m = 0 \quad \text{and} \quad \sigma = 1$$

The resulting function is called the *standard normal distribution*. Integrals involving the Gaussian distribution, such as that of Eq. 2.61, have been tabulated based on the standard normal distribution for a wide range of x values. With the help of a simple transformation, it is very easy to obtain the integrals for any value of m and σ.

The standard normal distribution is obtained by defining the new variable.

$$t = \frac{x - m}{\sigma} \tag{2.64}$$

Substituting into Eq. 2.55, one obtains

$$G(t)dt = \frac{1}{\sqrt{2\pi}}e^{-t^2/2}\,dt \tag{2.65}$$

It is very easy to show that the Gaussian given by Eq. 2.65 has mean

$$\bar{t} = m = \int_{-\infty}^{\infty} tG(t)\,dt = 0$$

and variance

$$V(t) = \sigma^2 = \int_{-\infty}^{\infty} t^2 G(t)\, dt = 1$$

The cumulative standard normal distribution function, Eq. 2.61, is now written as

$$E(x) = \int_{-\infty}^{x} G(t)\, dt = \int_{-\infty}^{x} \frac{1}{\sqrt{2\pi}} e^{-t^2/2}\, dt \qquad (2.66)$$

or, in terms of the error function that is tabulated

$$E(x) = \frac{1}{2}\left(1 + \mathrm{erf}\frac{x}{\sqrt{2}}\right)$$

where

$$\mathrm{erf}\frac{x}{\sqrt{2}} = \sqrt{\frac{2}{\pi}} \int_{0}^{x} e^{-t^2/2}\, dt$$

Example 2.12 The uranium fuel of light water reactors is enclosed in metallic tubes with an average outside diameter (OD) equal to 20 mm. It is assumed that the OD is normally distributed around this average with a standard deviation $\sigma = 0.5$ mm. For safety reasons, no tube should be used with OD > 21.5 mm or OD < 18.5 mm. If 10,000 tubes are manufactured, how many of them are expected to be discarded because they do not satisfy the requirements given above?

Answer The probability that the OD of a tube is going to be less than 18.5 mm or greater than 21.5 mm is

$$G(x < 18.5) + G(x > 21.5) = \int_{-\infty}^{18.5} \frac{dx}{\sqrt{2\pi}(0.5)} \exp\left[-\frac{(x-20)^2}{2(0.5)^2}\right] + \int_{21.5}^{\infty} \frac{dx}{\sqrt{2\pi}(0.5)} \exp\left[-\frac{(x-20)^2}{2(0.5)^2}\right]$$

Graphically, the sum of these two probabilities is equal to the two shaded areas shown in Figure 2.7.

In terms of the standard normal distribution and also because the two integrals are equal, one obtains

$$G(x < 18.5) + G(x > 21.5) = 2\left[1 - \int_{-\infty}^{3} \frac{1}{\sqrt{2\pi}} e^{-t^2/2}\, dt\right]$$

where

$$t = \frac{x - 20}{0.5}$$

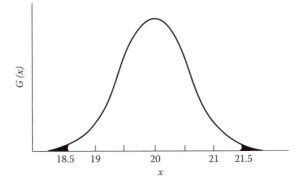

Figure 2.7 The shaded areas represent the fraction of defective rods, Example 2.12.

This last integral is tabulated in many books, handbooks, and mathematical tables (see bibliography of this chapter). From such tables, one obtains

$$\int_{-\infty}^{3} \frac{1}{\sqrt{2\pi}} e^{-t^2/2} \, dt = 0.99865$$

which gives

$$G(x < 18.5) + G(x < 21.5) = 0.0027$$

Therefore, it should be expected that under the manufacturing conditions of this example, 27 tubes out of 10,000 would be rejected.

2.10.2 Importance of the Gaussian Distribution for Radiation Measurements

The normal distribution is the most important distribution for applications in measurements. It is extremely useful because for almost any type of measurement that has been taken many times, the frequency with which individual results occur forms, to a very good approximation, a Gaussian distribution centered around the average value of the results. The greater the number of trials, the better their representation by a Gaussian. Furthermore, statistical theory shows that even if the original population of the results under study does not follow a normal distribution, their average does. That is, if a series of measurements of the variable $x_i|_{i-1,...,N}$ is repeated M times, the average values $\bar{x}_N|_{N-1,...,M}$ follow a normal distribution even though the x_i's may not. This result is known as the *central limit theorem* and holds for any random sample of variables with finite standard deviation.

In reality, no distribution of experimental data can be exactly Gaussian, since the Gaussian extends from $-\infty$ to $+\infty$. But for all practical purposes, the approximation is good and it is widely used because it leads to excellent results.

It is worth reminding the reader that both the binomial (Figure 2.2) and the Poisson (Figure 2.3) distributions resemble a Gaussian under certain conditions. This observation is particularly important in radiation measurements.

The results of radiation measurements are, in most cases, expressed as the number of counts recorded in a scaler. These counts indicate that particles have interacted with a detector and produced a pulse that has been recorded. The particles, in turn, have been produced either by the decay of a radioisotope or as a result of a nuclear reaction. In either case, the emission of the particle is statistical in nature and follows the Poisson distribution. However, as indicated in Section 2.9, if the average of the number of counts involved is more than about 20, the Poisson approaches the Gaussian distribution. For this reason, the individual results of such radiation measurements are treated as members of a normal distribution.

Consider now a Poisson and a Gaussian distribution having the same average, $m = 25$. Obviously, there is an infinite number of Gaussians with that average but with different standard deviations. The question one may ask is: "What is the standard deviation of the Gaussian that may represent the Poisson distribution with the same average?" The answer is that the Gaussian with $\sigma = \sqrt{m} = 5$ is almost identical with the Poisson. Table 2.1 presents values of the two distributions, and Figure 2.8 shows them plotted.

The following very important conclusion is drawn from this result. The outcomes of a series of radiation measurements are members of a Poisson distribution. They may be treated as members of a Gaussian distribution if the average result is more than $m = 20$. The standard deviation of that Gaussian distribution is $\sigma = \sqrt{m}$. Use of this conclusion is made in Section 2.17, which discusses statistics of radiation counting.

2.11 THE LORENTZIAN DISTRIBUTION

The Lorentzian distribution, which describes the resonances of nuclear reactions—in particular how the probability of interaction (cross section, see Chapter 4) changes as a function of particle energy—is given by

$$L(x) \, dx = \frac{1}{\pi} \frac{\Gamma/2}{(x-m)^2 + \Gamma^2/4} \tag{2.67}$$

Table 2.1
Comparison between a Poisson and a Gaussian Distribution Having the Same Mean ($m = 25$)

n	P_n (Poisson)	$G(n)$ (Gaussian) $\sigma = 5$
10	0.0004	0.0009
12	0.0017	0.0027
14	0.0059	0.0071
16	0.0154	0.0168
18	0.0316	0.0299
20	0.0519	0.0484
22	0.0702	0.0666
24	0.0795	0.0782
25	0.0795	0.0798
26	0.0765	0.0782
28	0.0632	0.0666
30	0.0454	0.0484
32	0.0286	0.0299
34	0.0159	0.0168
36	0.0079	0.0071
38	0.0035	0.0027

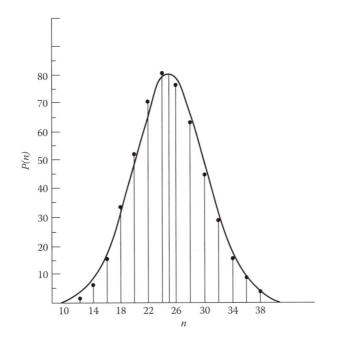

Figure 2.8   Comparison between a Poisson distribution with $m = 25$ and a Gaussian distribution with the same average and standard deviation $\sigma = \sqrt{m} = 5$.

where $L(x)\,dx$ is the probability that the value of x lies between x and $x + dx$. The Lorentzian is a symmetric function (Figure 2.9) centered around the value $x = m$. It can be easily shown that

$$\int_{-\infty}^{\infty} L(x)\,dx = 1$$

35

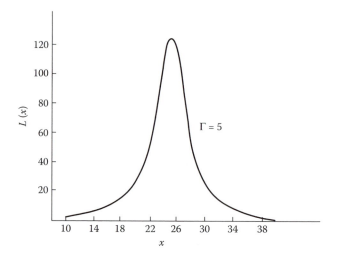

Figure 2.9 A Lorentzian distribution peaking at $x = 25$ and having a FWHM equal to 5.

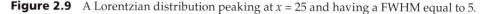

and that

$$\bar{x} = \int_{-\infty}^{\infty} x L(x)\, dx = m$$

Thus, the mean is given by the parameter m as expected from the symmetry of the function. One peculiar characteristic of the Lorentzian is the fact that its variance cannot be calculated. Indeed, the integral

$$\sigma^2 = V(x) = \int_{-\infty}^{\infty} (x - m)^2 L(x)\, dx$$

does not converge, which is the result of the slow decrease of the function away from the peak.

In the absence of a standard deviation, the parameter Γ is used for the description of the Lorentzian. The parameter Γ is *equal* to the FWHM of the function.

2.12 THE STANDARD, PROBABLE, AND OTHER ERRORS

Consider a measurement or series of measurements that gave the result R and its estimated error E. The experimenter reports the result as

$$R \pm E \tag{2.68}$$

in which case E is the absolute error (R and E have the same units), or as

$$R \pm \varepsilon\% \tag{2.69}$$

where $\varepsilon = (E/R)100$ = relative error (dimensionless). In most cases, the relative rather than the absolute error is reported.

Whether either Eq. 2.68 or 2.69 is used, the important thing to understand is that $R \pm E$ *does not mean* that the correct result has been bracketed between $R - E$ and $R + E$. It means only that *there is a probability* that the correct result has a value between $R - E$ and $R + E$. What is the value of this probability? There is no unanimous agreement on this matter, and different people use different values. However, over the years, two probability values have been used more frequently than others and have led to the definition to two corresponding errors, the standard and the probable error.

The standard error. If the result of a measurement is reported as $R \pm E_s$ and E_s is the standard error, then there is a 68.3% chance for the true result to have a value between $R - E_s$ and $R + E_s$.

The probable error. By definition, the probable error is equally likely to be exceeded or not. Therefore, if the result of a measurement is $R \pm E_p$ and E_p is the probable error, then there is a 50% chance for the true result to have a value between $R - E_p$ and $R + E_p$.

Both standard and probable errors are based on a Gaussian distribution. That is, it is assumed that the result R is the average of individual outcomes that belong to a normal distribution. This does not introduce any limitation in practice because, as stated in Section 2.10.2, the individual outcomes of a long series of any type of measurement are members of a Gaussian distribution.* With the Gaussian distribution in mind, it is obvious that the definition of the standard error is based on Eq. 2.62. If a result is R and the standard error is E_s, then $E_x = \sigma$.

$$\int_{R-E_s}^{R+E_s} \frac{1}{(\sqrt{2\pi})\sigma} \exp\left[-\frac{(x-R)^2}{2\sigma^2}\right] dx = 0.683$$

Correspondingly, the probable error E satisfies

$$\int_{R-E_p}^{R+E_p} \frac{1}{(\sqrt{2\pi})\sigma} \exp\left[-\frac{(x-R)^2}{2\sigma^2}\right] dx = 0.5$$

It can be shown that

$$E_p = 0.6745 E_s \tag{2.70}$$

The standard and probable errors are the most commonly used in reporting experimental results. Individual researchers may define other errors that represent a different percentage of the Gaussian. For example, the 95% error, E_{95}, is that which gives a 95% chance to have the true result bracketed between $R - E_{95}$ and $R + E_{95}$. It turns out that $E_{95} = 1.96\sigma$ (see Table 2.2).

2.13 THE ARITHMETIC MEAN AND ITS STANDARD ERROR

Although the true value of a quantity can never be determined, the error of the measurement can be reduced if the experiment is repeated many times.

Consider an experiment that has been repeated N times, where N is a large number, and produced the individual outcomes $n_i|_{i=1,...,N}$. Let the frequency of occurrence of n_i be P_{n_i}.† If one plots P_{n_i} versus n_i, the resulting curve resembles a Gaussian distribution as shown in Figure 2.10. The larger

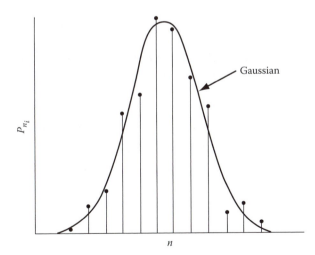

Figure 2.10 The distribution of the frequency of occurrence of individual results of a series of identical measurements tends to follow a Gaussian distribution.

---

* Exception: Radiation counting measurements with $m < 20$ obey the Poisson distribution.
† If $N = 1,000$ and n_i has occurred 15 times, $P_n = 15/1,000$.

the value of N, the more the histogram of Figure 2.10 coincides with a normal distribution. Assume that the dashed line of Figure 2.10 is an acceptable representation of the experimental results. Under these circumstances, how should the result of the measurement be reported and what is its standard error?

The result of the measurement is reported as the arithmetic average defined by

$$\bar{n} = \frac{n_1 + n_2 + \cdots + n_N}{N} = \sum_{i=1}^{N} \frac{n_i}{N} \tag{2.71}$$

This equation is the same as Eq. 2.31. As N increases, a better estimate of the true value of n is obtained—that is, the error of the measurement becomes smaller. The true value of n, which is also called the true mean, can only be obtained with an infinite number of measurements. Since it is impossible to perform an infinite number of trials, n is always calculated from Eq. 2.71.

The error of $\bar{n}$ depends on the way the individual measurements are distributed around $\bar{n}$—that is, it depends on the width of the Gaussian of Figure 2.10. As the width becomes smaller, the error gets smaller, and therefore the measurement is better. The standard error of $\bar{n}$ is defined in terms of the standard deviation of the distribution. Using Eq. 2.34 and setting $f(x_i) = 1/N$, the standard deviation of the distribution becomes

$$\sigma^2 = \sum_{i=1}^{N} \frac{(n_i - m)^2}{N} \tag{2.72}$$

With a finite number of measurements at our disposal, this equation for σ has to be modified in two ways. First, because the true mean m is never known, it is replaced by its best estimate, which is $\bar{n}$ (Eq. 2.71). Second, it can be generally shown that the best estimate of the standard deviation of N measurements is given by the following equation:

$$\sigma^2 = \frac{1}{N-1} \sum_{i=1}^{N} \left(n_i - \bar{n} \right)^2 \tag{2.73}$$

The differences between Eq. 2.72 and Eq. 2.73 are the use of $\bar{n}$ instead of m and the use of $N - 1$ in the denominator instead of N.* For a large number of measurements, it does not make any practical difference if one divides by N or $N - 1$. However, it makes a difference for small values of N. Using the extreme value of $N = 1$, one can show that division by N gives the wrong result. Indeed, dividing by N, one obtains

$$\sigma^2 = \frac{1}{N} \sum \left(n_1 - \bar{n} \right)^2 = \frac{1}{1} \sum \left(n_1 - n_1 \right)^2 = \frac{0}{1} = 0$$

Zero σ means zero error, which is obviously wrong. The error is never zero, certainly not in the case of one measurement. Division by $N - 1$, on the other hand, gives

$$\sigma^2 = \frac{1}{N-1} \sum \left(n_1 - n_1 \right)^2 = \frac{0}{0}$$

which, being indeterminate, is a more realistic value of the error based on a single measurement.

Since the N results are distributed as shown in Figure 2.10, 68.3% of the outcomes fall between $\bar{n} - \sigma$ and $\bar{n} + \sigma$ (see Eq. 2.62). Therefore, one additional measurement has a 68.3% chance of providing a result within $\bar{n} \pm \sigma$. For this reason, σ is called the standard deviation or *the standard error of a single measurement*. Is this equal to the standard error of $\bar{n}$? No, and here is why.

According to the definition of the standard error, if $\sigma_{\bar{n}}$ is the standard error of $\bar{n}$, it ought to have such a value that a new average $\bar{n}$ would have a 68.3% chance of falling between $\bar{n} - \sigma_{\bar{n}}$ and $\bar{n} + \sigma_{\bar{n}}$. To obtain the standard error of $\bar{n}$, consider Eq. 2.71 as a special case of Eq. 2.36a. The quantity $\bar{n}$ is a linear function of the uncorrelated random variables $n_1, n_2, ..., n_N$, each with standard deviation σ.

---

* The factor $N - 1$ is equal to the "degrees of freedom" or the number of independent data or equations provided by the results. The N independent outcomes constitute, originally, N independent data. However, after $\bar{n}$ is calculated, only $N - 1$ independent data are left for the calculation of σ.

Therefore

$$\bar{n} = \sum_{n=1}^{N} a_i n_i$$

where $a_i = 1/N$. Using Eq. 2.41, the standard deviation of $\bar{n}n$ is*

$$\sigma_{\bar{n}} = \sqrt{\sum_{i=1}^{N} a_i^2 \sigma_i^2} = \sqrt{\sum_{i=1}^{N} \frac{1}{N^2} \sigma^2} = \frac{\sigma}{\sqrt{N}} \tag{2.74}$$

If the series of N measurements is repeated, the new average will probably be different from $\bar{n}$, but it has a 68.3% chance of having a value between $n - \sigma_{\bar{n}}$ and $\bar{n} + \sigma_{\bar{n}}$. The result of the N measurements is

$$\bar{n} \pm \sigma_{\bar{n}} = \bar{n} \pm \frac{\sigma}{\sqrt{N}} \tag{2.75}$$

When a series of measurements is performed, it would be desirable to calculate the result in such a way that the error is a minimum. It can be shown that the average $\bar{n}$ as defined by Eq. 2.71 minimizes the quantity

$$\sum_{i=1}^{N} (\bar{n} - n_i)^2$$

which is proportional to the standard error. Finally, Eq. 2.75 shows that the error is reduced if the number of trials increases. However, that reduction is proportional to $1/\sqrt{N}$, which means that the number of measurements should be increased by a factor of 100 to be able to reduce the error by a factor of 10.

2.14 CONFIDENCE LIMITS

Consider a variable x_i that represents the value of the ith sample of a large population of specimens. The variable x_i may be the diameter of a sphere or the thickness of the cladding of a fuel rod or the length of the fuel rod. A designer may desire a certain diameter of the sphere or a certain thickness of the fuel cladding or a certain length of the fuel rod. What happens during actual fabrication is that the individual units are not exactly the same. The person who examines individual units as they are constructed, machined, or fabricated will find that there is a distribution of values for the quantity being examined. The average value is equal to that specified in the blueprints and is called the nominal value. Individual specimens, however, have values of x distributed around the nominal value x_n according to a Gaussian distribution,

$$G(x) = \frac{1}{(\sqrt{2\pi})\sigma} \exp\left[-\frac{(x - x_n)^2}{2\sigma^2} \right]$$

where x_n = nominal value of x = average value of x
σ = standard deviation of the distribution

The manufacturer of any product would like to know what the probability is that any one item will deviate from the nominal value by a certain amount. Or, setting some acceptable value of x, call it x_a, the manufacturer would like to know what is the probability that x will be bigger than x_a. Questions of this type come under the subject of "quality control."

The probability that x will exceed x_a is given by

$$P(x > x_a) = \int_{x_a}^{\infty} \frac{dx}{(\sqrt{2\pi})\sigma} \exp\left[-\frac{(x - x_n)^2}{2\sigma^2} \right] \tag{2.76}$$

---

* If the population of the events n_i is finite in size, then it can be shown that $\sigma_{\bar{n}}^2 = [(M-N)/(M-1)]\sigma^2/N$ where M is the total number of n_i's.

The acceptable value of x is usually expressed as

$$x_a = x_n + k\sigma \tag{2.77}$$

that is, the extreme acceptable value of x, x_a is allowed to be k standard deviations different from x_n. In terms of the standard normal distribution, Eq. 2.76 takes the form

$$P(t > k) = \int_k^\infty \frac{1}{\sqrt{2\pi}} e^{-t^2/2} dt \tag{2.78}$$

where

$$t = \frac{x - x_n}{\sigma} \tag{2.79}$$

and $P(t > k)$ = probability that x will exceed x_a by k standard deviations.

Table 2.2 gives values of $P(t > k)$ for several values of k. The values in Table 2.2 are interpreted as follows:

Consider $k = 1$. The probability that x will exceed x_a where $x_a = x_n + \sigma$ is 15.9%. If x is some property of a manufactured product, it is said that the confidence limit is, in this case, $1 - 0.159 = 0.841$ or 84.1%, that is, 84.1% of the specimens will have $x < x_a$ (Figure 2.11). If $k = 2$, the probability that x will exceed x_a is equal to 2.3%; therefore, the confidence limit is 97.7%.

In actual construction or fabrication of an item, the Gaussian distribution is determined by checking the variable x for a large number of specimens. An average value of x is calculated,

$$\bar{x} = \frac{1}{N}\sum_{i=1}^N x_i$$

and a standard deviation

$$\sigma = \sqrt{\frac{1}{N-1}}\sqrt{\sum_i^N (x_i - \bar{x})^2}$$

is obtained. The average $\bar{x}$ should be almost equal to the nominal value of x. A Gaussian distribution for this sample peaks at $\bar{x}$ and has a standard deviation σ. Knowing σ, the value of x_a is calculated from Eq. 2.77 after the confidence limit—the value of k—has been decided upon.

The use of the concept of confidence limits is widespread in industry. As a specific example, let us assume that x is the thickness of the cladding of a reactor fuel rod. The average (nominal) thickness is x_n. The reactor designer would like to be certain that a certain fraction of fuel rods will always have thickness within prescribed limits. Let us say that the designer desires a confidence limit of 99.87%. This means that no more than 13 rods out of 10,000 will be expected to have cladding thickness exceeding the nominal value by more than three standard deviations (Table 2.2).

Table 2.2
Probability Values and Confidence Limits

Number of Standard Deviations (k)	p(x > x_a)	Confidence Limit
0	0.500	50.0
1.0	0.159	84.1
1.285	0.100	90.0
1.5	0.067	93.3
1.645	0.050	95.0
1.96	0.025	97.5
2.0	0.023	97.7
2.5	0.006	99.4
3.0	0.0013	99.87

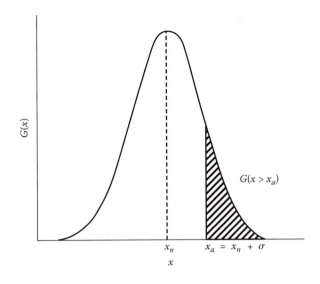

Figure 2.11 The probability that x will exceed x_a, where $x_a = x_n + \sigma$, is 15.9% (shaded area). The confidence limit is $1 - 0.159$, or 84.1%.

2.15 PROPAGATION OF ERRORS

2.15.1 Calculation of the Average and Its Standard Deviation

Sometimes an investigator has to determine a quantity that is a function of more than one random variable. In such cases, it is very important to know how to calculate the error of the complex quantity in terms of the errors of the individual random variables. This procedure is generally known as propagation of errors and is described in this section.

Consider the function $f(x_1, x_2, \ldots, x_M)$ which depends on the random variables $x_1, x_2, \ldots, x_M$. Generally, the values of $x_1, x_2, \ldots, x_M$ are determined experimentally and then the value of $f(x_1, x_2, \ldots, x_M)$ is calculated. For example,

1. $f(x_1, x_2) = x_1 \pm x_2$

2. $f(x_1, x_2) = x_1 x_2$

3. $f(x_1, x_2) = \dfrac{x_1}{x_2}$

4. $f(x_1, x_2) = \ln(x_1 + x_2)$

5. $f(x) = x^2$

6. $f(x_1, x_2, x_3) = \dfrac{(x_1 + x_2)}{x_3}$

It has already been mentioned that the x_i's are determined experimentally, which means that average values $\bar{x}_1, \bar{x}_2, \bar{x}_3, \ldots, \sigma_M$ are determined along with their standard errors $\sigma_1, \sigma_2, \ldots, \sigma_M$. Two questions arise:

1. What is the value of $f(x_1, \ldots, x_M)$ that should be reported?

2. What is the standard error of $f(x_1, \ldots, x_M)$?

It is assumed that the function $f(x_1, \ldots, x_M)$ can be expanded in a Taylor series around the averages $\bar{x}_i|_{i=1,\ldots,M}$:

$$f(x_1, x_2, \ldots, x_M) = f(\bar{x}_1, \bar{x}_2, \ldots, \bar{x}_M) + \sum_{i=1}^{M} (x_i - \bar{x}_i) \frac{\partial f}{\partial \bar{x}_i} + O(x_i - \bar{x}_i)^2$$

41

The notation used is that

$$\frac{\partial f}{\partial \overline{x}_i} = \frac{\partial f}{\partial x_i}\bigg|_{x_i = \overline{x}_i}$$

The term $O\left(x_i - \overline{x}_i\right)^2$ includes all the terms of order higher than first, and it will be ignored. Thus, the function is written

$$f(x_1, x_2, \ldots, x_M) = f(\overline{x}_1, \overline{x}_2, \ldots, \overline{x}_m) + \sum_{i=1}^{M} (x_i - \overline{x}_i) \frac{\partial f}{\partial \overline{x}_i} \tag{2.80}$$

Equation 2.80 is a special case of Eq. 2.36a. The average value of $f(x_1, \ldots, x_M)$, which is the value to be reported, is

$$\overline{f} = f(\overline{x}_1, \overline{x}_2, \ldots, \overline{x}_M) \tag{2.81}$$

The variance of $f(x_1, \ldots, x_M)$ is given by Eq. 2.40:

$$\sigma_f^2 = V(f) = \sum_{i=1}^{M} \left(\frac{\partial f}{\partial \overline{x}_i}\right)^2 \sigma_i^2 + 2 \sum_{i>i}^{M} \sum_{i}^{M} \left(\frac{\partial f}{\partial \overline{x}_i}\right)\left(\frac{\partial f}{\partial \overline{x}_i}\right) \rho_{ij} \sigma_i \sigma_j \tag{2.82}$$

where ρ_{ij} is the correlation coefficient given by Eq. 2.39.

The standard error $f(x_1, \ldots, x_M)$ is equal to the standard deviation

$$\sigma_{\overline{f}} = \sqrt{\sum_{i=1}^{M} \left(\frac{\partial f}{\partial \overline{x}_i}\right)^2 \sigma_i^2 + 2 \sum_{j>i}^{M} \sum_{i}^{M} \left(\frac{\partial f}{\partial \overline{x}_i}\right)\left(\frac{\partial f}{\partial \overline{x}_j}\right) \rho_{ij} \sigma_i \sigma_j} \tag{2.83}$$

Equations 2.81 and 2.83 are the answers to questions 1 and 2 stated previously. They indicate, first, that the average of the function is calculated using the average values of the random variables and, second, that its standard error is given by Eq. 2.83. Equation 2.83 looks complicated, but fortunately, in most practical cases, the random variables are uncorrelated—that is, $\rho_{ij} = 0$, and Eq. 2.83 reduces to

$$\sigma_{\overline{f}} = \sqrt{\sum_{i=1}^{M} \left(\frac{\partial f}{\partial \overline{x}_i}\right)^2 \sigma_i^2} \tag{2.84}$$

Unless otherwise specified, the discussion in the rest of this chapter will concern *only uncorrelated variables*. Therefore, Eqs. 2.81 and 2.84 will be used. The reader, however, should always keep in mind the assumption under which Eq. 2.84 is valid.

2.15.2 Examples of Error Propagation—Uncorrelated Variables

Examples of error propagation formulas for many common functions are given in this section. In all cases, uncorrelated variables are assumed.

Example 2.13 $f(x_1, x_2) = a_1 x_1 \pm a_2 x_2$, where a_1 and a_2 are constants

$$\overline{f} = a_1 \overline{x}_1 \pm a_2 \overline{x}_2$$

$$\sigma_{\overline{f}} = \sqrt{\left(\frac{\partial f}{\partial \overline{x}_1}\right)^2 \sigma_1^2 + \left(\frac{\partial f}{\partial \overline{x}_2}\right)^2 \sigma_2^2} = \sqrt{a_1^2 \sigma_1^2 + a_2^2 \sigma_2^2} \tag{2.85}$$

If $a_1 = a_2 = 1$, this example applies to the very common case of summation or difference of two variables.

Example 2.14 $f(x_1, x_2) = ax_1x_2$, where a is a constant

$$\bar{f} = a\bar{x}_1\bar{x}_2$$

$$\sigma_{\bar{f}} = a\sqrt{\bar{x}_2^2\sigma_1^2 + \bar{x}_1^2\sigma_2^2}$$

Example 2.15 $f(x_1, x_2) = ax_1/x_2$

$$\bar{f} = a\frac{\bar{x}_1}{\bar{x}_2}$$

$$\sigma_{\bar{f}} = a\sqrt{\frac{1}{\bar{x}_2^2}\sigma_1^2 + \frac{\bar{x}_1^2}{\bar{x}_2^4}\sigma_2^2}$$

The standard error for Examples 2.14 and 2.15 takes a simpler and easy-to-remember form for both the product and the quotient if it is expressed as the relative error. It is trivial to show that

$$\frac{\sigma_{\bar{f}}}{\bar{f}} = \sqrt{\left(\frac{\sigma_1}{\bar{x}_1}\right)^2 + \left(\frac{\sigma_2}{\bar{x}_2}\right)^2} \tag{2.86}$$

Thus, the relative error of the product ax_1x_2 or the quotient ax_1/x_2 is equal to the square root of the sum of the squares of the relative errors of the variables x_1 and x_2.

Example 2.16 $f(x) = x^m$, where m is some real number

$$\bar{f} = (\bar{x})^m$$

$$\sigma_{\bar{f}} = \left(\frac{\partial f}{\partial x}\right)\sigma_{\bar{x}} = m(\bar{x})^{m-1}\sigma_{\bar{x}}$$

or

$$\frac{\sigma_{\bar{f}}}{\bar{f}} = m\frac{\sigma_{\bar{x}}}{\bar{x}}$$

Example 2.17 $f(x) = e^{ax}$

$$\bar{f} = c^{a\bar{x}}$$

$$\sigma_{\bar{f}} = \left(\frac{\partial f}{\partial \bar{x}}\right)\sigma_{\bar{x}} = ae^{a\bar{x}}\sigma_{\bar{x}}$$

or

$$\frac{\sigma_{\bar{f}}}{\bar{f}} = a\sigma_{\bar{x}}$$

There is another very important use of Eq. 2.84, which has to do with the calculation of the variation of a function in terms of changes of the independent variables. Consider again the function $(x_1, x_2, \ldots, x_M)$ and assume that the variables $x_1, x_2, \ldots, x_M$ have changed by the amounts $\Delta x_1, \Delta x_2, \ldots, \Delta x_M$. The variation or change of $f(x_1, \ldots, x_M)\Delta$, is given by

$$\Delta f = \sqrt{\sum_{i=1}^{M}\left(\frac{\partial f}{\partial x_i}\right)^2 \Delta x_i^2} \tag{2.87}$$

Equation 2.87 should not be used if it is specified what the change of variable is, that is, if the change is a decrease or an increase. If the change is known, one should calculate the function $f(x_1, x_2, \ldots, x_M)$ using the new values of the x's and obtain Δf by subtracting the new from the old value.

Example 2.18   The speed of sound is obtained by measuring the time it takes for a certain sound signal to travel a certain distance. What is the speed of sound and its standard error if it takes the sound 2.5 ± 0.125 s to travel 850 ± 5 m?

Answer

$$f(x_1, x_2) = v = \frac{x}{t} = \frac{850}{2.5} = 340 \text{ m/s}$$

To calculate the error, use Eq. 2.86:

$$\frac{\sigma_v}{v} = \sqrt{\left(\frac{\sigma_x}{x}\right)^2 + \left(\frac{\sigma_t}{t}\right)^2} = \sqrt{\left(\frac{5}{850}\right)^2 + \left(\frac{0.125}{2.5}\right)^2} = 0.05 = 5\%$$

The result is 340 ± 17 m/s.

Example 2.19   A beam of photons going through a material of thickness x is attenuated in such a way that the fraction of photons traversing the material is $e^{-\mu x}$, where the constant μ is called the attenuation coefficient. If the thickness of the material changes by 10%, by how much will the emerging fraction of photons change? Take $x = 0.01$ m and $\mu = 15$ m^{-1}.

Answer   This is a case requiring the use of Eq. 2.87.

$$f(x) = e^{-\mu x}$$

$$\Delta f = \left(\frac{\partial f}{\partial x}\right)\Delta x = -\mu e^{-\mu x}\Delta x$$

$$\frac{\Delta f}{f} = -\mu \Delta x = -\mu x\left(\frac{\Delta x}{x}\right) = -(15)(0.01)(0.10) = -0.015$$

Therefore, if the thickness increases by 10%, the fraction of emerging photons decreases by 1.5%.

2.16 GOODNESS OF DATA—χ^2 CRITERION—REJECTION OF DATA

It is desirable when data are obtained during an experiment to be able to determine if the recording system works well or not. The experimenter should ask the question: Are all the obtained data true (due to the phenomenon studied), or are some or all due to extraneous disturbances that have nothing to do with the measurement? A number of tests have been devised for the purpose of checking how reliable the results are, that is, checking the "goodness of data." A good review for the basic applications of the chi-square statistic using counting data has been given by Tries.[1]

Before any tests are applied, an investigator should use common sense and try to avoid erroneous data. First, a good observer will never rely on a single measurement. He or she should repeat the experiment as many times as is feasible (but at least twice) and observe whether the results are reproducible or not. Second, the observer should check the results to see how they deviate from their average value. Too large or too small deviations are suspicious. The good investigator should be alert and should check such data very carefully. For example, if for identical, consecutive measurements one gets the following counts in a scaler:

10,000   10,000   10,000   10,002   9999   9998

the apparatus is not necessarily very accurate; it is probably faulty. In any event, a thorough check of the whole measuring setup should be performed.

The test that is used more frequently than any other to check the goodness of data is the χ^2 criterion (chi square), or Pearson's χ^2 test. The χ^2 test is based on the quantity

$$\chi^2 = \frac{\sum_{i=1}^{N} (\bar{n} - n_i)^2}{\bar{n}}$$

(2.88)

where $n_i|_{i=1,\dots,N}$ represents the results of N measurements with $\bar{n}$ being the average.

To apply the χ^2 test, one first calculates χ^2 using Eq. 2.88. Then, using Table 2.3, the corresponding probability is obtained. The meaning of the probability values are listed in Table 2.3. If the set of measurements is repeated, the value of χ^2 gives the probability to obtain a new χ^2 that is larger or smaller than the first value. For example, assume that $N = 15$ and $\chi^2 = 4.66$. From the table, the probability is 0.99, meaning that the probability for a new set of measurements to give a $\chi^2 < 4.66$ is less than $1 - 0.99$, that is, less than 1%. What this implies is that the data are clustered around the mean much closer than one would expect. Assume next that $N = 15$ and $\chi^2 = 29.14$. Again, from the table, the probability to get $\chi^2 > 29.14$ is only 1% or less. In this case, the data are scattered in

Table 2.3
Probability Table for χ^2 Criterion[a]

Degrees of Freedom[b] (N − 1)	Probability						
	0.99	0.95	0.90	0.50	0.10	0.05	0.01
2	0.020	0.103	0.211	1.386	4.605	5.991	9.210
3	0.115	0.352	0.584	2.366	6.251	7.815	11.345
4	0.297	0.711	1.064	3.357	7.779	9.488	13.277
5	0.554	1.145	1.610	4.351	9.236	11.070	15.086
6	0.872	1.635	2.204	5.348	10.645	12.592	16.812
7	1.239	2.167	2.833	6.346	12.017	14.067	18.475
8	1.646	2.733	3.490	7.344	13.362	15.507	20.090
9	2.088	3.325	4.168	8.343	14.684	16.919	21.666
10	2.558	3.940	4.865	9.342	15.987	18.307	23.209
11	3.053	4.575	5.578	10.341	17.275	19.675	24.725
12	2.571	5.226	6.304	11.340	18.549	21.026	26.217
13	4.107	5.892	7.042	12.340	19.812	22.363	27.688
14	4.660	6.571	7.790	13.339	21.064	23.685	29.141
15	5.229	7.261	8.547	14.339	22.307	24.996	30.578
16	5.812	7.962	9.312	15.338	23.542	26.296	32.000
17	6.408	8.672	10.085	16.338	24.769	27.587	33.409
18	7.015	9.390	10.865	17.338	25.989	28.869	34.805
19	7.633	10.117	11.651	18.338	27.204	30.144	36.191
20	8.260	10.851	12.443	19.337	28.412	31.410	37.566
21	8.897	11.591	13.240	20.337	29.615	32.671	38.932
22	9.542	12.338	14.041	21.337	30.813	33.924	40.289
23	10.196	13.091	14.848	22.337	32.007	35.172	41.638
24	10.856	13.848	15.659	23.337	33.196	36.415	42.980
25	11.534	14.611	16.473	24.337	34.382	37.382	44.314
26	12.198	15.379	17.292	25.336	35.563	38.885	45.642
27	12.879	16.151	18.114	26.336	36.741	40.113	46.963
28	13.565	16.928	18.939	27.336	37.916	41.337	48.278
29	14.256	17.708	19.768	28.336	39.087	42.557	49.588

[a] Calculated values of χ^2 will be equal to or greater than the values given in the table.
[b] See footnote on p. 38.

a pattern around the mean that is wider than one might expect. Finally, consider $N = 15$ and $\chi^2 = 13.34$. The probability is then 0.5, which means that, from a new set of measurements, it is equally probable to get a value of χ^2 that is smaller or larger than 13.34. Notice that the probability is close to 0.5 when $\chi^2 \sim N - 1$. In practice, a range of acceptable χ^2 values is selected in advance; then a set of data is accepted if χ^2 falls within this preselected range.

What should one do if the data fail the test? Should all, some, or none of the data be rejected? The answer to these questions is not unique, but rather depends on the criteria set by the observer and the type of measurement. If the data fail the test, the experimenter should be on the lookout for trouble. Some possible reasons for trouble are the following:

1. Unstable equipment may give inconsistent results, for example, spurious counts generated by a faulty component of an instrument.

2. External signals may be picked up by the apparatus and be "recorded." Sparks, radio signals, welding machines, etc., produce signals that may be recorded by a pulse type counting system.

3. If a number of samples are involved, widely scattered results may be caused by lack of sample uniformity.

4. A large χ^2 may result from one or two measurements that fall far away from the average. Such results are called the "outliers." Since the results are governed by the normal distribution, which extends from $-\infty$ to $+\infty$, in theory, at least, all results are possible. In practice, it is somewhat disturbing to have a few results that seem to be way out of line.

Should the outliers be rejected? And by what criterion? One should be conservative when rejecting data for three reasons:

1. The results are random variables following the Gaussian distribution. Therefore, outliers are possible.

2. As the number of measurements increases, the probability of an outlier increases.

3. In a large number of measurements, the rejection of an outlier has small effect on the average, although it makes the data look better by decreasing the dispersion.

One of the criteria used for data rejection is Chauvenet's criterion, stated as follows:

A reading or outcome may be rejected if it has a deviation from the mean greater than that corresponding to the $1 - 1/2N$ error, where N is the number of measurements.

Data used with Chauvenet's criterion are given in Table 2.4. For example, in a series of 10 measurements, $1 - 1/2N = 1 - 1/20 = 0.95$. If $n_i - \bar{n}$ exceeds the 95% error (1.96σ), then that reading could be rejected. In that case, a new mean should be calculated without this measurement and also a new standard deviation.

The use of Chauvenet's, or any other, criterion is not mandatory. It is up to the observer to decide if a result should be rejected or not.

Table 2.4
Data for Chauvenet's Criterion

Number of Measurements	Number of Standard Deviations Away from Average
2	1.15
3	1.38
4	1.54
5	1.65
10	1.96
15	2.13
25	2.33

2.17 THE STATISTICAL ERROR OF RADIATION MEASUREMENTS

Radioactive decay is a truly random process that obeys the Poisson distribution, according to which the standard deviation of the true mean m is $\sqrt{m}$. However, the true mean is never known and can never be found from a finite number of measurements. But is there a need for a large number of measurements?

Suppose one performs only one measurement and the result is n counts. The best estimate of the true mean, as a result of this single measurement, is this number n. If one takes this to be the mean, its standard deviation will be $\sqrt{n}$.

Indeed, this is what is done in practice. The result of a single count n is reported as $n \pm \sqrt{n}$, which implies that

1. The outcome n is considered the true mean.

2. The standard deviation is reported as the standard error of n.

The relative standard error of the count n is

$$\frac{\sigma_n}{n} = \frac{\sqrt{n}}{n} = \frac{1}{\sqrt{n}} \tag{2.89}$$

which shows that the relative error decreases if the number of counts obtained in the scaler increases. Table 2.5 gives several values of n and the corresponding percent standard error. To increase the number n, one either counts for a long time or repeats the measurement many times and combines the results. Repetition of the measurement is preferable to one single long count because by performing the experiment many times, the reproducibility of the results is checked.

Consider now a series of N counting measurements with the individual results $n_i|_{i=1,\ldots,N}$. It is assumed that the counts n_i were obtained under identical conditions and for the same counting time; thus, their differences are solely due to the statistical nature of radiation measurements. Each number n_i has a standard deviation $\sigma_i = \sqrt{n_i}$. The average of this series of measurements is, using Eq. 2.31,

$$\bar{n} = \frac{1}{N} \sum_{i=1}^{N} n_i \tag{2.31}$$

The standard error of $\bar{n}$ can be calculated in the following two ways:

1. The average $\bar{n}$ is the best estimate of a Poisson distribution of which the outcomes $n_i|_{i=1,\ldots,N}$ are members. The standard deviation of the Poisson distribution is (see Section 2.9) $\sigma = \sqrt{m} = \sqrt{\bar{n}}$. The standard error of the average is (see Eq. 2.75)

$$\sigma_{\bar{n}} = \frac{\sigma}{\sqrt{N}} = \sqrt{\frac{\bar{n}}{N}} \tag{2.90}$$

2. The average $\bar{n}$ may be considered a linear function of the independent variables n_i, each with standard error $\sqrt{n_i}$. Then, using Eq. 2.84, one obtains

$$\sigma_{\bar{n}} \sqrt{\sum_{i=1}^{N} \left(\frac{\partial \bar{n}}{\partial n_i} \right)^2 \sigma_{n_i}^2} = \sqrt{\sum_{i=1}^{N} \frac{1}{N^2} \left(\sqrt{n_i} \right)^2} = \sqrt{\sum_{i=1}^{N} \frac{1}{N^2} n_i} = \frac{1}{N} \sqrt{n_{\text{tot}}} \tag{2.91}$$

where $n_{\text{tot}} = n_1 + n_2 + \cdots + n_N =$ number of counts obtained from N measurements.

Table 2.5
Percent Standard Error of n
Counts

n	Standard Error of n (%)
100	10
1000	3.16
10,000	1
100,000	0.316
1,000,000	0.1

Table 2.6
Typical Results of a Counting Experiment[a]

Observation, i	Number of Counts Obtained in the Scaler, n_i	Square of Deviation $(n_i - \bar{n})^2$
1	197	81
2	210	16
3	200	36
4	198	64
5	205	1
6	195	121
7	190	256
8	220	196
9	215	81
10	230	576
Totals	2060	1428

[a] One could use Eqs. 2.73 and 2.74 for the calculation of σ and $\sigma_{\bar{n}}$. The result is

$$\sigma_{\bar{n}} = \sqrt{\frac{\sum_{i=1}^{N}(n_i - \bar{n})^2}{N(N-1)}} = \sqrt{\frac{1428}{10(9)}} = 3.98 \approx 4$$

For radiation measurements, use of Eqs. 2.90 and 2.91 is preferred.

It is not difficult to show that Eqs. 2.90 and 2.91 are identical.

In certain cases, the observer needs to combine results of counting experiments with quite different statistical uncertainties. For example, one may have to combine the results of a long and short counting measurement. Then the average should be calculated by weighting the individual results according to their standard deviations (see Bevington and Robinson, 2002; Eadie et al. 1971). The equation for the average is

$$\bar{n} = \frac{\sum_{i=1}^{N} n_i / \sigma_i^2}{\sum_{i=1}^{N} 1/\sigma_i^2} \tag{2.92}$$

Example 2.20 Table 2.6 presents typical results of 10 counting measurements. Using these data, the average count and its standard error will be calculated using Eqs. 2.31, 2.90, and 2.91. The average is

$$\bar{n} = \frac{1}{N}\sum_{i=1}^{N} n_i = \frac{2060}{10} = 206$$

Using Eq. 2.90 or Eq. 2.91, the standard error of $\bar{n}$ is

$$\sigma_{\bar{n}} = \sqrt{\frac{206}{10}} = \sqrt{\frac{2060}{10}} = 4.5$$

2.18 THE STANDARD ERROR OF COUNTING RATES

In practice, the number of counts is usually recorded in a scaler, but what is reported is the counting rate, that is, counts recorded per unit time. The following symbols and definitions will be used for counting rates:

G = number of counts recorded by the scaler in time t_G *with* the sample present = gross count

B = number of counts recorded by the scaler in time t_B *without* the sample = background count

$$g = \frac{G}{t_G} = \text{gross counting rate}$$

$$g = \frac{B}{t_B} = \text{background counting rate}$$

$$r = \text{net counting rate*} = \frac{G}{t_G} - \frac{B}{t_B} = g - b \qquad (2.93)$$

The standard error of the net counting rate can be calculated based on Eq. 2.84 and by realizing that r is a function of four independent variables G, t_G, and t_B:

$$\sigma_r = \sqrt{\left(\frac{\partial r}{\partial G}\right)^2 \sigma_G^2 + \left(\frac{\partial r}{\partial t_G}\right)^2 \sigma_{t_G}^2 + \left(\frac{\partial r}{\partial B}\right)^2 \sigma_B^2 + \left(\frac{\partial r}{\partial t_B}\right)^2 \sigma_{t_B}^2} \qquad (2.94)$$

The electronic equipment available today is such that the error in the measurement of time is, in almost all practical cases, much smaller than the error in the measurement of G and B.[†] Unless otherwise specified, σ_{t_G} and σ_{t_B} will be taken as zero. Then Eq. 2.94 takes the form

$$\sigma_r = \sqrt{\left(\frac{\partial r}{\partial G}\right)^2 \sigma_G^2 + \left(\frac{\partial r}{\partial B}\right)^2 \sigma_B^2} \qquad (2.95)$$

The standard errors of G and B are

$$\sigma_G = \sqrt{G}, \qquad \sigma_B = \sqrt{B}$$

Using Eqs. 2.93 and 2.95, one obtains for the standard error of the net counting rate,

$$\sigma_r = \sqrt{\frac{G}{t_G^2} + \frac{B}{t_B^2}} \qquad (2.96)$$

It is important to notice that in the equation for the net counting rate, the quantities G, B, t_G, and t_B are the independent variables, not g and b. The error of r will be calculated from the error in G, B, t_G, and t_B. It is very helpful to remember the following rule:

The statistical error of a certain count is determined from the number recorded by the scaler. That number is G and B, not the rates g and b.

Example 2.21   A radioactive sample gave the following counts:

$$G = 1000, \qquad t_G = 2 \text{ min}, \qquad B = 500, \qquad t_B = 10 \text{ min}$$

What is the net counting rate and its standard error?
Answer

$$r = \frac{G}{t_G} - \frac{B}{t_B} = \frac{1000}{2} - \frac{500}{10} = 500 - 50 = 450 \text{ counts/min}$$

$$\sigma_r = \sqrt{\frac{G}{t_G^2} + \frac{B}{t_B^2}} = \sqrt{\frac{1000}{2^2} + \frac{500}{10^2}} = \sqrt{250 + 5} = 16 \text{ counts/min}$$

$$r = 450 \pm 16 = 450 \pm 3.5\%$$

---

* When the counting rate is extremely high, the detector may be missing some counts. Then a "dead time" correction is necessary, in addition to background subtraction; see Section 2.21.
† The errors σ_{t_G} and σ_{t_B} may become important in experiments where very accurate counting time is paramount for the measurement.

A common error is that, since $r = g - b$, one is tempted to write

$$\sigma_r = \sqrt{\sigma_g^2 + \sigma_b^2} = \sqrt{\left(\sqrt{g}\right)^2 + \left(\sqrt{b}\right)^2} = \sqrt{g+b} = \sqrt{500+50} = 23 \text{ counts/min}$$

This result, $\sigma_r = 23$, is wrong because

$$\sigma_g \neq \sqrt{g} \quad \text{and} \quad \sigma_b \neq \sqrt{b}$$

The correct way to calculate the standard error based on g and b is to use

$$\sigma_g = \frac{\sqrt{G}}{t_G} = \frac{\sqrt{1000}}{2}, \quad \sigma_b = \frac{\sqrt{B}}{t_B} = \frac{\sqrt{500}}{10}$$

Then

$$\sigma_r = \sqrt{\sigma_g^2 + \sigma_b^2} = \sqrt{\frac{G}{t_G^2} + \frac{B}{t_B^2}} = 16 \text{ counts/min}$$

Usually, one determines G and B, in which case σ_r is calculated from Eq. 2.96. However, sometimes the background counting rate and its error have been determined earlier. In such a case, σ_r is calculated as shown in Ex. 2.22.

Example 2.22 A radioactive sample gave $G = 1000$ counts in 2 min. The background rate of the counting system is known to be $b = 100 \pm 6$ counts/min. What is the net counting rate and its standard error?
Answer

$$r = \frac{G}{t_G} - b = \frac{1000}{2} - 100 = 400 \text{ counts/min}$$

$$\sigma_r = \sqrt{\left(\frac{\partial r}{\partial G}\right)^2 \sigma_G^2 + \left(\frac{\partial r}{\partial b}\right)^2 \sigma_b^2} = \sqrt{\frac{G}{t_G^2} + \sigma_b^2} = \sqrt{\frac{1000}{4} + 6^2} = \sqrt{286},$$

$$= 17 \text{ counts/min}$$

In this problem, b and σ_b are given, not B and t_B. The standard error of the background rate has been determined by an earlier measurement. Obviously, b was not determined by counting for 1 min, because in that case, one would have

$$B = 100, \quad t_B = 1 \text{min}, \quad b = 100 \text{ counts/min}$$

$$\sigma_b = \frac{\sqrt{B}}{t_B} = \frac{\sqrt{100}}{1} = 10$$

2.18.1 Combining Counting Rates

If the experiment is performed N times with results

$$G_1, G_2, G_3, \ldots, G_N, \quad B_1, B_2, \ldots, B_N$$

for gross and background counts, the average net counting rate is

$$\bar{r} = \frac{1}{N}\sum_{i=1}^{N} r_i = \frac{1}{N}\sum_{i=1}^{N}\left(\frac{G_i}{t_{G_i}} - \frac{B_i}{t_{B_i}}\right)$$

In most cases, t_{G_i} and t_{B_i} are kept constant for all N measurements. That is, $t_{G_i} = t_G$ and $t_{B_i} = t_B$. Then

$$\bar{r} = \frac{1}{N}\sum_{i=1}^{N}\left(\frac{G_i}{t_{B_i}} - \frac{B_i}{t_{B_i}}\right) = \frac{1}{N}\left(\frac{G}{t_G} - \frac{B}{t_B}\right) \tag{2.97}$$

where

$$G = \sum_{i=1}^{N}G_i \quad \text{and} \quad B = \sum_{i=1}^{N}B_i$$

The standard error of the average counting rate is, using Eqs. 2.84 and 2.96,

$$\sigma_{\bar{r}} = \frac{1}{N}\sqrt{\sum_{i=1}^{N}\sigma_{r_i}^2} = \frac{1}{N}\sqrt{\sum_{i=1}^{N}\left(\frac{G_i}{t_G^2} + \frac{B_i}{t_B^2}\right)} = \frac{1}{N}\sqrt{\frac{G}{t_G^2} + \frac{B}{t_B^2}} \tag{2.98}$$

A special case. Sometimes the background rate is negligible compared to the gross counting rate. Then, Eq. 2.98 becomes

$$\sigma_{\bar{r}} = \frac{1}{N}\frac{\sqrt{G}}{t_G}$$

The relative standard error is

$$\frac{\sigma_r}{\bar{r}} = \frac{(1/N)\sqrt{G}/t_G}{G/Nt_G} = \frac{1}{\sqrt{G}}$$

This is the same as Eq. 2.89. Therefore, if the background is negligible, the relative standard error is the same for either the total count or the counting rate.

2.19 METHODS OF ERROR REDUCTION

In every radiation measurement, it is extremely important to perform it in such a way that the result is determined with the minimum possible error. In general, the first task of the investigator is to improve the counting apparatus by reducing the background as much as possible. Actually, the important quantity is the ratio b/g or b/r and not the absolute value of the background. Assuming that all possible improvements of background have been achieved, there is a procedure that, if followed, will result in a smaller error. Two such procedures will be discussed below. In addition, a method will be presented for the calculation of the counting time necessary to measure a counting rate with a desired degree of accuracy.

2.19.1 The Background Is Constant and There Is No Time Limit for Its Measurement

In this case, the background is measured for a long period to minimize the error introduced by it, that is, t_B is so long that

$$\sigma_r = \sqrt{\frac{G}{t_G^2} + \frac{B}{t_B^2}} \approx \sqrt{\frac{G}{t_G^2}}$$

Example 2.23 Suppose one obtains the following data:

$$G = 400, \quad t_G = 5\,\text{min}$$
$$B = 100, \quad t_B = 2.5\,\text{min}$$

Then

$$r = \frac{400}{5} - \frac{100}{2.5} = 40 \text{ counts/min}$$

$$\sigma_r = \sqrt{\frac{400}{5^2} + \frac{100}{2.5^2}} = 5.65 \text{ counts/min}$$

$$\frac{\sigma_r}{r} = \frac{5.65}{40} = 0.14 = 14\%$$

If the background is constant, this result can be improved by counting background for a long period of time, for example, 250 min. In that case, the result is

$$B = \frac{100}{2.4} \times 250 = 10,000 \text{ counts}, \quad t_B = 250 \text{ min}$$

$$r = \frac{400}{5} - \frac{10,000}{250} = 40 \text{ counts/min}$$

$$\sigma_r = \sqrt{\frac{400}{5^2} + \frac{10,000}{250^2}} = \sqrt{16 + 0.16} \approx 4$$

$$\frac{\sigma_r}{r} = \frac{4}{40} = 10\%$$

2.19.2 There Is a Fixed Time T Available for Counting Both Background and Gross Counts

In this case, the question is, what is the optimum time to be used for gross and background counting? Optimum time results in minimum statistical error for the net counting rate. The optimum time is determined as follows.

An estimate of the counting rates at the time of the measurement is obtained with a short count (not the final one). Assume that one obtained the approximate counting rates

$$g = \frac{G}{t_G}, \quad b = \frac{B}{t_B}$$

Then, from Eq. 2.96 and also using $G = gt_G$, $B = bt_B$,

$$\sigma_r^2 = \frac{g}{t_G} + \frac{b}{t_B}$$

The best times t_G and t_B are those that minimize σ_r or $(\sigma_r)^2$ subject to the constraint

$$t_B + t_G = T = \text{constant} \tag{2.99}$$

Considering σ_r^2 as a function of t_B and t_G, the minimum will be found by differentiating $(\sigma_r)^2$ and setting the differential equal to zero:

$$d(\sigma_r)^2 = -\frac{g}{t_G^2} dt_G - \frac{b}{t_B^2} dt_B = 0$$

Differentiating the constraint, Eq. 2.99, one finds

$$dt_G = -dt_B$$

Substituting this value of dt_G into $d(\sigma_r)^2$ gives

$$\frac{t_B}{t_G} = \sqrt{\frac{b}{g}}$$

Therefore, if there is a fixed time T for the measurement, the optimum counting times are determined from the two equations

$$t_G + t_B = T$$

$$\frac{t_B}{t_G} = \sqrt{\frac{b}{g}} \qquad (2.100)$$

2.19.3 Calculation of the Counting Time Necessary to Measure a Counting Rate with a Predetermined Statistical Error

Assume that the net counting rate of a radioactive sample should be measured with an accuracy of $a\%$, that is, $\sigma_r / r = a\%$. Also assume that a counting system is provided with a background counting rate b and standard error σ_b. Both b and σ_b have been reduced as much as possible for this system and have been determined earlier. The task is to determine the counting time t_G necessary to result in a percent standard error for the net counting rate. The time t_G is calculated as follows.

The net counting rate and its standard error are

$$r = \frac{G}{t_G} - b, \quad \sigma_r = \sqrt{\frac{G}{t_G^2} + \sigma_b^2} \qquad (2.101)$$

Therefore

$$\frac{\sigma_r}{r} = \frac{a}{100} = \frac{\sqrt{G/t_G^2 + \sigma_b^2}}{G/t_G - b} = \frac{\sqrt{g/t_G + \sigma_b^2}}{g - b}$$

Equation 2.101 solved for t_G gives

$$t_G = \frac{g}{(g-b)^2 (a/100)^2 - \sigma_b^2} \qquad (2.102)$$

It is assumed that an approximate gross counting rate is known.

Example 2.24   How long should a sample be counted to obtain the net counting rate with an accuracy of 1%? It is given that the background for the counting system is 100 ± 2 counts/min.
Answer   The first step is to obtain an approximate gross counting rate. Assume that the sample gave 800 counts in 2 min. Then $g = 800/2 = 400$ counts/min and, using Eq. 2.102,

$$t_G = \frac{400}{(400 - 100)^2 (0.01)^2 - 2^2} = 80 \text{ min}$$

Indeed, if one counts for 80 min, the error of r is going to be

$$\sigma_r = \sqrt{\frac{g}{t_G} + \sigma_b^2} = \sqrt{\frac{400}{80} + 2^2} = 3$$

$$\frac{\sigma_r}{r} = \frac{3}{300} = 0.01 = 1\%$$

2.19.4 Relative Importance of Error Components

In every measurement, the observer tries to reduce the experimental error as much as possible. If the quantity of interest depends on many variables, each with its own error, the effort to reduce the error should be directed toward the variable with the largest contribution to the final error.

Consider the quantity $Q = x + y - z$ and assume $x = 3$, $y = 2$, and $z = 1$. Also assume that the corresponding standard errors are

$$\sigma_x = 0.1, \quad \sigma_y = 0.23, \quad \sigma_z = 0.05$$

The standard error of Q is

$$\sigma_Q = \sqrt{\sigma_x^2 + \sigma_y^2 + \sigma_z^2} = \sqrt{0.1^2 + 0.23^2 + 0.05^2}$$
$$= \sqrt{0.01 + 0.0529 + 0.0025} = 0.26$$

From the relative magnitude of the errors, one can see that if it is necessary to reduce the error further, the effort should be directed toward reduction of σ_y first, σ_x second, and σ_z third. In fact, there is no need to reduce σ_z further before σ_x and σ_y reach the same magnitude as σ_z.

2.20 MINIMUM DETECTABLE ACTIVITY

The minimum detectable activity (MDA) is the smallest net count that can be reported with a certain degree of confidence that represents a true activity from a sample and is not a statistical variation of the background. The term MDA is not universally acceptable. In the general case, in measurements not necessarily involving radioactivity, other terms such as lowest detection limit have been used. Here, the notation and applications will be presented with the measurement of a radioactive sample in mind. A classical treatment of this work has been published elsewhere by Currie.[2] Other related articles include determination of the lowest limit of detection for personnel dosimetry systems,[3] a method for computing the decision level for samples containing radioactivity in the presence of background,[4] and evaluating a lower limit of detection.[5]

Obviously, MDA is related to low count rates. In such cases of low count rates, the person who performs the experiment faces two possible errors.

TYPE I error: To state that the true activity is greater than zero when, in fact, it is zero. If this is a suspected contaminated item, the person doing the measurement will report that the item is indeed contaminated when, in fact, it is not. This error is called *false positive*.

TYPE II error: To state that the true activity is zero when, in fact, it is not. Using the previous example, the person doing the measurement reports that the item is clean when, in fact, it is contaminated. This error is called *false negative*.

The outcomes of radiation measurements follow Poisson statistics, which become essentially Gaussian when the average is greater than about 20 (see Section 2.10.2). For this reason, the rest of this discussion will assume that the results of individual measurements follow a normal distribution and the confidence limits set will be interpreted with that distribution in mind. Following the notation used earlier,

B = background with standard deviation σ_B
G = gross signal with standard deviation σ_G
$n = G - B$ = net signal with standard deviation $\sigma_n = (\sigma_G^2 + \sigma_B^2)^{1/2}$

When the net signal is zero (and has a standard deviation $\sigma_n = \sigma_0$), a critical detection limit (CDL) is defined in terms of σ_0 with the following meaning:

1. A signal lower than CDL is not worth reporting.

2. The decision that there is nothing to report has a confidence limit of $1 - \alpha$, where α is a certain fraction of the normalized Gaussian distribution (Figure 2.12). Take as an example $\alpha = 0.05$. Then

$$\text{CDL} = k_\alpha \sigma_0 \tag{2.103}$$

with $k\alpha = 1.645$ (see Table 2.2). If $n < \text{CDL}$, one decides that the sample is not contaminated, and this decision has a 95% confidence limit.

The MDA should obviously be greater than the CDL. Keeping in mind that the possible MDA values also follow a normal distribution, a fraction β is established, meaning that a signal equal

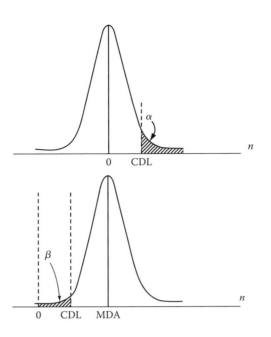

Figure 2.12 The meaning of the critical detection limit (CDL) and minimum detectable activity (MDA) in terms of the confidence limits defined by α and β.

to MDA is reported as a correct/true signal with a confidence limit $1 - \beta$. The value of MDA is given by

$$MDA = CDL + k_\beta \sigma_D \tag{2.104}$$

where σ_D is the standard deviation of MDA (Figure 2.12). Again, if $\beta = 0.05$, $k\beta = 1.645$.
   In most cases, in practice, $\alpha = \beta = 0.05$; then the CDL and MDA are defined with a 95% confidence limit. For radioactivity specifically, remember that

$$\sigma^2 = \sigma_G^2 + \sigma_B^2 = (\sigma_n^2 + \sigma_B^2 + \sigma_B^2) \tag{2.105}$$

and for $n = 0$, $\sigma_n = \sqrt{n} = 0$, and $\sigma = \sigma_0 = \sqrt{\sigma_B^2 + \sigma_B^2} = \sqrt{2}\,\sigma_B$. Then, if the 95% confidence limit is applied ($\alpha = \beta = 0.05$), CDL $= 1.645\,\sigma_0 = 2.326\,\sigma_B$. The value of MDA turns out to be (see Prob. 2.27)

$$MDA = k^2 + 2CDL = 2.71 + 4.653\sigma_B \tag{2.106}$$

Example 2.25   Consider the data of a single measurement to be $G = 465$ counts/min, $B = 400$/min. Assume that from previous measurements in that counting system it has been determined that $\sigma_B = 10$/min. The assumption is made that the background is constant. What does one report in this case?
Answer   The net count rate is $n = 465 - 400 = 65$ counts/min. The minimum detectable activity is, from Eq. 2.106, MDA $= 2.71 + 4.653 \times 10 = 49.2$. Since MDA < 65, one reports, with a 95% confidence limit, that this sample is radioactive.

In most cases, the second term of Eq. 2.106 is much larger than the first, and the MDA is taken as

$$MDA = 4.653\sigma_B \tag{2.107}$$

In using Eq. 2.106 or 2.107, the user should keep in mind the underlying assumption of the 95% confidence limit. The numerical factors will change if one chooses a different confidence limit.

2.21 DETECTOR DEAD-TIME CORRECTION AND MEASUREMENT OF DEAD TIME

Dead time or resolving time, of a counting system is defined as the minimum time that can elapse between the arrival of two successive particles at the detector and the recording of two distinct pulses. The components of dead time consist of the time it takes for the formation of the pulse in the detector itself and for the processing of the detector signal through the preamplifier-amplifier-discriminator-scaler (or preamplifier-amplifier-multichannel analyzer). With modern electronics, the longest component of dead time is that of the detector, and for this reason, the term "dead time" means the dead time of the detector. The dead-time component of the preamplifier-amplifier-discriminator-scaler can be ignored with most detectors.

Because of detector dead time, the possibility exists that some particles will not be recorded since the will not produce pulses for them. Pulses will not be produced because the detector will be "occupied" with the formation of the signal generated by particles arriving earlier. The counting loss of particles is particularly important in the case of high counting rates. Obviously, the observed counting rate should be corrected for the loss of counts due to detector dead time. The rest of this section presents the method for correction as well as a method for the measurement of the dead time.

Suppose τ is the dead time of the system and g the observed counting rate. The fraction of time during which the system is insensitive is $g\tau$. If n is the true counting rate, the number of counts lost is $n(g\tau)$. Therefore

$$n = g + ng\tau$$

and

$$n = \frac{g}{1 - g\tau} \tag{2.108}$$

Equation 2.108 corrects the observed gross counting rate g for the loss of counts due to the dead time of the detector. This is known as a nonparalyzable model where τ is fixed after each true event is recorded. In the paralyzable model while τ remains the same, for every true event that is not recorded, the dead time is extended by τ. Since the actual dead periods are not the same for each model a distribution function needs to be used for the paralyzable model to account for intervals between random events. Therefore, one gets

$$g = ne^{-n\tau}$$

An excellent treatise of nonparalyzable and paralyzable models are given by Knoll.[6]

Example 2.26   Suppose $\tau = 200$ μs and $g = 30,000$ counts/min. What fraction of counts is lost because of dead time? What is the true counting rate?
Answer   The true counting rate is

$$n = \frac{g}{1 - g\tau} = \frac{30,000/60}{1 - (30,000/60)(200 \times 10^{-6})}$$

or

$$n = 555.5 \text{ counts/s}$$

Therefore, dead time is responsible for loss of

$$\frac{555 - 500}{555} = \frac{55}{555} = 10\% \text{ of the counts}$$

Notice that the product $g\tau = 0.10$, that is, the product of the dead time and the gross counting rate, is a good indicator of the fraction of counts lost because of dead time.

The dead time is measured with the "two source" method as follows. Let n_1, n_2, n_{12} be the true gross counting rates from the first source only, from the second source only, and from both sources, respectively, and let n_b be the true background rate. Let the corresponding observed counting rates be g_1, g_2, g_{12}, b.

The following equation holds:

$$\left(\begin{array}{c}\text{True net}\\\text{counting rate}\end{array}\right)_1 + \left(\begin{array}{c}\text{True net}\\\text{counting rate}\end{array}\right)_2 = \left(\begin{array}{c}\text{True net}\\\text{counting rate}\end{array}\right)_{1+2}$$

$$(n_1 - n_b) \quad + \quad (n_2 - n_b) \quad = \quad (n_{12} - n_b)$$

or

$$n_1 + n_2 = n_{12} + n_b$$

Using Eq. 2.108,

$$\frac{g_1}{1 - g_1\tau} + \frac{g_2}{1 - g_2\tau} = \frac{g_{12}}{1 - g_{12}\tau} + \frac{b}{1 - b\tau} \tag{2.109}$$

It will be assumed now that, $b\tau \ll 1$ in which case,

$$\frac{b}{1 - b\tau} = b$$

(If $b\tau$ is not much less than 1, the instruments should be thoroughly checked for possible malfunction before proceeding with the measurement.)

The dead time τ can be determined from Eq. 2.109 after g_1, g_2, g_{12}, and b are measured. This is achieved by counting radioactive source 1, then sources 1 and 2 together, then only source 2, and finally the background after removing both sources. Equation 2.109 can be rearranged to give

$$(g_1 g_2 g_{12} + g_1 g_2 b - g_1 g_{12} b - g_2 g_{12} b)\tau^2 - 2(g_1 g_2 - g_{12}b)\tau + g_1 + g_2 - g_{12} - b = 0 \tag{2.110}$$

Equation 2.110 is a second degree algebraic equation that can be solved for τ. It was derived without any approximations.

If the background is negligible, Eq. 2.110 takes the form

$$g_1 g_2 g_{12}\tau^2 - 2g_1 g_2\tau + g_1 + g_2 - g_{12} = 0 \tag{2.111}$$

Solving for τ,

$$\tau = \frac{1}{g_{12}}\left[1 - \sqrt{1 - \frac{g_{12}}{g_1 g_2}(g_1 + g_2 - g_{12})}\right] \tag{2.112}$$

When dead-time correction is necessary, the net counting rate, called "true net counting rate," is given by

$$r = n - b - \frac{G/t_G}{1 - (G\tau/t_G)} - b \tag{2.113}$$

It is assumed that the true background rate has been determined earlier with the standard error σ_b. The standard error of r, σ_r, is calculated from Eq. 2.113 using Eq. 2.84. If the only sources of error are the gross count G and the background, the standard error of r is

$$\sigma_r = \sqrt{\left(\frac{1}{1-(G/t_G)\tau}\right)^4 \frac{G}{t_G^2} + \sigma_b^2} \tag{2.114}$$

If there is an error due to dead-time determination, a third term consisting of that error will appear under the radical of Eq. 2.114.

2.22 LOSS FREE COUNTING AND ZERO DEAD TIME

Radiation measurements can be classified into two major areas: those with low to medium dead times and those with high dead times. In environmental and most health physics applications radiation levels are low enough so that that typical counting systems can easily account for dead-time corrections. However, in high count rates, particularly in γ-ray spectroscopy, counting systems usually cannot correctly account for dead times above 50%. In recent years, loss free counting or zero dead-time correction modules have been integrated into γ-ray spectroscopy systems with very good results. A complete comprehensive review article of such systems has been completed by Westpahl.[7]

PROBLEMS

2.1 What is the probability when throwing a die three times of getting a four in any of the throws?

2.2 What is the probability when drawing one card from each of three decks of cards that all three cards will be diamonds?

2.3 A box contains 2000 computer cards. If five faulty cards are expected to be found in the box, what is the probability of finding two faulty cards in a sample of 250?

2.4 Calculate the average and the standard deviation of the probability density function $f(x) = 1/(b-a)$ when $a \leq x \leq b$. (This pdf is used for the calculation to round off errors.)

2.5 The energy distribution of thermal (slow) neutrons in a light-wave reactor follows very closely the Maxwell–Boltzmann distribution

$$N(E)dE = A\sqrt{E}e^{-E/kT}\,dE$$

where $N(E)\,dE$ = number of neutrons with kinetic energy between E and
$E + dE$
k = Boltzmann constant = 1.380662×10^{-23}J/°K
T = temperature, K
A = constant

Show that
(a) The mode of this distribution is $E = \frac{1}{2}kT$.
(b) The mean is $\bar{E} = \frac{3}{2}kT$.

2.6 If the average for a large number of counting measurements is 15, what is the probability that a single measurement will produce the result 20?

2.7 For the binomial distribution, prove

$$\text{(a)}\ \sum_{n=0}^{N} P_{(N)}^{(n)} = 1, \quad \text{(b)}\ \bar{n} = pN, \quad \text{(c)}\ \sigma^2 = m(1-p)$$

2.8 For the Poisson distribution, prove

$$\text{(a) } \sum_{n=0}^{\infty} P_n = 1, \quad \text{(b) } \bar{x} = m, \quad \text{(c) } \sigma^2 = m$$

2.9 For the normal distribution, show

$$\text{(a) } \int_{-\infty}^{\infty} P(x)\,dx = 1, \quad \text{(b) } \bar{x} = m, \quad \text{(c) the variance is } \sigma^2$$

2.10 If $n_1, n_2, \ldots, n_N$ are mutually uncorrelated random variables with a common variance σ^2, show that $\overline{(n_i - \bar{n})} = \dfrac{N-1}{N}\sigma^2$

2.11 Show that in a series of N measurements, the result R that minimizes the quantity

$$Q = \sum_{i=1}^{N} (R - n_i)^2$$

is $R = \bar{n}$, where $\bar{n}$ is given by Eq. 2.31.

2.12 Prove Eq. 2.62 using tables of the error function.

2.13 As part of a quality control experiment, the lengths of 10 nuclear fuel rods have been measured with the following results in meters:

2.60	2.62	2.65	2.58	2.61
2.62	2.59	2.59	2.60	2.63

What is the average length? What is the standard deviation of this series of measurements?

2.14 The average number of calls in a 911 switchboard is 4 calls per hour. What is the probability to receive 6 calls in the next hour?

2.15 At a uranium pellet fabrication plant the average pellet density is 17×10^3 kg/m³ with a standard deviation equal to 10^3 kg/m³. What is the probability that a given pellet has a density less than 14×10^3 kg/m³?

2.16 A radioactive sample was counted once and gave 500 counts in 1 min. The corresponding number for the background is 480 counts. Is the sample radio-active or not? What should one report based on this measurement alone?

2.17 A radioactive sample gave 750 counts in 5 min. When the sample was removed, the scaler recorded 1000 counts in 10 min. What is the net counting rate and its standard percent error?

2.18 Calculate the average net counting rate and its standard error from the data given below:

G	t_G (min)	B	t_B (min)
355	5	120	10
385	5	130	10
365	5	132	10

2.19 A counting experiment has to be performed in 5 min. The approximate gross and background counting rates are 200 counts/min and 50 counts/min, respectively.
   (a) Determine the optimum gross and background counting times.
   (b) Based on the times obtained in (a), what is the standard percent error of the net counting rate?

2.20 The strength of a radioactive source was measured with a 2% standard error by taking a gross count for time t min and a background for time $2t$ min.

Calculate the time t if it is given that the background is 300 counts/min and the gross count 45,000 counts/min.

2.21 The strength of radioactive source is to be measured with a detector that has a background of 120 ± 8 counts/min. The approximate gross counting rate is 360 counts/min. How long should one count if the net counting rate is to be measured with an error of 2%?

2.22 The buckling B^2 of a cylindrical reactor is given by

$$B^2 = \left(\frac{2.405}{R}\right)^2 = \left(\frac{\pi}{H}\right)^2$$

where R is the reactor radius and H is the reactor height.

If the radius changes by 2% and the height by 8%, by what% will B^2 change? Take $R = 1$ m, $H = 2$ m.

2.23 Using Chauvenet's criterion, should any of the scaler readings listed below be rejected?

115	121	103	151
121	105	75	103
105	107	100	108
113	110	101	97
110	109	103	101

2.24 As a quality control test in a nuclear fuel fabrication plant, the diameter of 10 fuel pellets has been measured with the following results (in mm): 9.50, 9.80, 9.75, 9.82, 9.93, 9.79, 9.81, 9.65, 9.99, 9.57. Calculate (a) the average diameter, (b) the standard deviation of this set of measurements, (c) the standard error of the average diameter, (d) should any of the results be rejected based on the Chauvenet criterion?

2.25 Using the data of Prob. 2.13, what is the value of accepted length x_a if the confidence limit is 99.4%?

2.26 An environmental sample has been collected for determination of ^{210}Po content. The sample is chemically separated and counted in an instrument with the following results 60 days after sampling:

Chemical Yield	80%
Counting Efficiency	20%
Sample Counts (gross)	20 counts
Sample Count Time	30 min
Background Counts	10 counts
Background Count Time	30 min
Half Life of Po-210	138 days

(a) What was the sample ^{210}Po net counting rate at the time of the sampling?
(b) What is the standard error of value determined in part (a)?
(c) The lower limit of detection (LLD) at the 95% confidence level has been defined as: LLD $= 1.645(2\sqrt{2})S_b$ where S_b is the standard deviation. Calculate the LLD for this determination.
(d) Does the activity level of this sample exceed the LLD for this determination?

2.27 Prove that for radioactivity measurements the value of MDA is given by the equation MDA $= k^2 + 2CDL$, if $k_\alpha = k_\beta = k$. Hint: when $n =$ MDA, the variance $\sigma^2 = $ MDA $+ \sigma_0^2$.

2.28 A sample was counted for 5 min and gave 2250 counts; the background, also recorded for 5 min, gave 2050 counts. Is this sample radioactive? Assume confidence limits of both 95% and 90%.

2.29 Determine the dead time of a detector based on the following data obtained with the two-source method:

$$g_2 = 14{,}000 \text{ counts/min} \qquad g_{12} = 26{,}000 \text{ counts/min}$$
$$g_2 = 15{,}000 \text{ counts/min} \qquad b = 50 \text{ counts/min}$$

2.30 If the dead time of a detector is 100 μs, what is the observed counting rate if the loss of counts due to dead time is equal to 5%?

2.31 Calculate the true net activity and its standard percent error for a sample that gave 70,000 counts in 2 min. The dead time of the detector is 200 μs. The background is known to be 100 ± 1 counts/min.

2.32 Calculate the true net activity and its standard error based on the following data:

$$G = 100{,}000 \text{ counts} \qquad \text{obtained in 10 min}$$
$$B = 10{,}000 \text{ counts} \qquad \text{obtained in 100 min}$$

The dead time of the detector is 150 μs.

BIBLIOGRAPHY

Bevington, P. R., and Robinson, D. K., *Data Reduction and Error Analysis for the Physical Sciences*, 3rd ed., McGraw-Hill, Boston, 2002.

Devore, J. L., *Probability and Statistics for Engineering and the Sciences*, Brooks/Cole, Belmont, CA, 2009.

Eadie, W. T., Dryard, D., James, F. E., Roos, M., and Sadoulet, B., *Statistical Methods in Experimental Physics*, North-Holland, Amsterdam, 1971.

Fornasini, P., *The Uncertainty in Physical Measurements: An Introduction to Data Analysis in the Physics Laboratory*, Springer Science, New York, 2008.

REFERENCES

1. Tries, M.A., *Health Phys.* **77**:441 (1999).

2. Currie, L. A., *Anal. Chem.* **40**:586 (1968).

3. Robertson, P. L., and Carlson, R. D., *Health Phys.* **62**:2 (1992).

4. Borak, T. B., and Kirchner, T. B., *Health Phys.* **69**:892 (1995).

5. Fong, S. H., and Alvarez, J. L., *Health Phys.* **72**:282 (1997).

6. Knoll, G. F., *Radiation Detection and Measurement*, 3rd ed., pp 119–127; 636–642, Wiley, New York, 2000.

7. Westphal, G. P., *J. Radioanal. Nucl. Chem.* **275**:677 (2008).

3

Review of Atomic and Nuclear Physics

3.1 INTRODUCTION

This chapter reviews the concepts of atomic and nuclear physics relevant to radiation measurements. It should not be considered a comprehensive discussion of any of the subjects presented. For in-depth study, the reader should consult the references listed at the end of the chapter. If a person has studied and understood this material, this chapter could be skipped without loss of continuity.

This review is not presented from the historical point of view. Atomic and nuclear behavior and the theory and experiments backing it are discussed as we understand them today. Emphasis is given to the fact that the current "picture" of atoms, nuclei, and subatomic particles is only a model that represents our best current theoretical and experimental evidence. This model may change in the future if new evidence is obtained pointing to discrepancies between theory and experiment.

3.2 ELEMENTS OF RELATIVISTIC KINEMATICS

The special theory of relativity developed by Einstein in 1905 is based on two simple postulates.

First Postulate The laws of nature and the results of all experiments performed in a given frame of reference (system of coordinates) are independent of the translational motion of the system as a whole.

Second Postulate The speed of light in vacuum is independent of the motion of its source.

These two postulates, simple as they are, predict consequences that were unthinkable at that time. The most famous predictions of the special theory of relativity are as follows:

1. The mass of a body changes when its speed changes.

2. Mass and energy are equivalent ($E = mc^2$).

Einstein's predictions were verified by experiment a few years later, and they are still believed to be correct today.

The main results of the special theory of relativity will be presented here without proof, using the following notations:

M = rest mass of a particle (or body)
M^* = mass of a particle in motion
v = speed of the particle
c = speed of light in vacuum = 3×10^8 m/s
T = kinetic energy of the particle
E = total energy of the particle

According to the theory of relativity, the mass of a moving particle (or body) changes with its speed according to the equation

$$M^* = \frac{M}{\sqrt{1-(v/c)^2}} = \frac{M}{\sqrt{1-\beta^2}} \tag{3.1}$$

or

$$M^* = \gamma M \tag{3.2}$$

where

$$\beta = \frac{v}{c} \tag{3.3}$$

and

$$\gamma = \frac{1}{\sqrt{1-\beta^2}} \tag{3.4}$$

Equation 3.1 shows that

1. As the speed of a moving particle increases, its mass also increases, thus making additional increase of its speed more and more difficult.

2. It is impossible for any mass to reach a speed equal to or greater than the speed of light in vacuum.*

The total energy of a particle of mass M^* is

$$E = M^* c^2 \tag{3.5}$$

Equation 3.5 expresses the very important concept of equivalence of mass and energy. Since the total energy E consists of the rest mass energy plus the kinetic energy, Eq. 3.5 may be rewritten as

$$E = M^* c^2 = T + Mc^2 \tag{3.6}$$

Combining Eqs. 3.2 and 3.6, one obtains the relativistic equation for the kinetic energy

$$T = (\gamma - 1)Mc^2 \tag{3.7}$$

The quantity γ, which is defined by Eq. 3.4 ($\gamma = M^* c^2 / Mc^2$), indicates how many times the mass of the particle has increased, relative to its rest mass, because of its motion. For large moving masses, the relativistic mass increase is too small to measure. Thus, without the availability of subatomic particles such as electrons and protons, it would be extremely difficult to verify this part of Einstein's theory.

The equation that relates the linear momentum and the total energy of a particle is

$$E^2 = (Mc^2) + (pc)^2 \tag{3.8}$$

where

$$\mathbf{p} = M^* v = \gamma Mv \tag{3.9}$$

is the linear momentum. Combining Eqs. 3.6 and 3.8, one obtains

$$T = \sqrt{(Mc^2)^2 + (pc)^2} - Mc^2 \tag{3.10}$$

or

$$p = \frac{1}{c}\sqrt{T^2 + 2TMc^2} \tag{3.11}$$

Equation 3.10 is used for the determination of the kinetic energy if the momentum is known, while Eq. 3.11 gives the momentum if the kinetic energy is known.

For small values of β (Eq. 3.3)—that is, for small speeds—the equations of relativity reduce to the equations of Newtonian (classical) mechanics. In classical mechanics, the mass is constant, and T and p are given by

$$T = \frac{1}{2}Mv^2 \tag{3.12}$$

$$\mathbf{p} = Mv \tag{3.13}$$

---

* The speed of light in a medium with index of refraction n is c/n; thus, it is possible for particles to move faster than with c/n in certain media (see Cerencov Radiation, Evans).

If the kinetic energy of a particle is a considerable fraction of its rest mass energy, Eqs. 3.7 and 3.9 should be used for the determination of T and p. Then the particle is *relativistic*. If, on the other hand, $\beta \ll 1$, the particle is *nonrelativistic*, and Eqs. 3.12 and 3.13 may be used.

Example 3.1 What is the mass increase of a bullet weighing 0.010 kg and traveling at twice the speed of sound?

Answer The speed of the bullet is $\upsilon \approx 700$ m/s. Using Eqs. 3.2 and 3.4,

$$\frac{M^*}{M} = \gamma = \frac{1}{\sqrt{1-\beta^2}} \approx 1 + \frac{1}{2}\beta^2 = 1 + \frac{1}{2}\left(\frac{700}{3\times10^8}\right)^2 = 1 + 2.72\times10^{-12}$$

The mass increase is

$$M^* - M = 2.72\times10^{-12}\, M = 2.72\times10^{-14}\text{ kg}$$

which is almost impossible to detect.

Example 3.2 An electron has a kinetic energy of 200 keV. (a) What is its speed? (b) What is its new mass relative to its rest mass?

Answer The rest mass energy of the electron is 511 keV. Since $T/mc^2 = 200/511 = 0.391$, relativistic equations should be used, (a) The speed of the electron is obtained with the help of Eqs. 3.7 and 3.4. Equation 3.7 gives

$$\gamma = 1 + \frac{T}{mc^2} = 1 + 0.391$$

and from Eq. 3.4 one obtains

$$\beta = \sqrt{1 + \frac{1}{\gamma^2}} = 0.695$$

Therefore

$$\upsilon = \beta c = (0.695)(3\times10^8 \text{ m/s}) = 2.085\times10^8 \text{ m/s}$$

(b) The new mass relative to the rest mass has already been determined because

$$\gamma = \frac{m^*c^2}{mc^2} = 1.391$$

that is, the mass of this electron increased 39.1%.

It is instructive to calculate the speed of this electron using the classical method of Eq. 3.12 to see the difference:

$$\upsilon = \sqrt{\frac{2T}{m}} = \sqrt{\frac{2Tc^2}{mc^2}} = \left(\sqrt{\frac{2T}{mc^2}}\right)c = \left(\sqrt{\frac{2(200)}{511}}\right)c = 0.885c$$

Thus, the classical equation determines the speed with an error

$$\frac{\upsilon_{cl} - \upsilon_{rel}}{\upsilon_{rel}} = \frac{0.885c - 0.695c}{0.695c} = 27\%$$

Example 3.3   What is the kinetic energy of a neutron with speed 6×10^7 m/s? What is its mass increase?
Answer   For this particle,

$$\beta = \frac{v}{c} = \frac{6 \times 10^7}{3 \times 10^8} = 0.2$$

Using Eqs. 3.4 and 3.7,

$$T = (\gamma - 1)Mc^2 = \left(\frac{1}{\sqrt{1-\beta^2}} - 1\right)Mc^2 = \left(\frac{1}{\sqrt{1-0.2^2}} - 1\right)Mc^2$$

$$= (1.021-1)Mc^2 = 0.021Mc^2 + (0.021)939.55\,\text{MeV} = 19.73\,\text{MeV}$$

$$= 3.16 \times 10^{-12}\,\text{J}$$

The mass increase is $M*/M = \gamma = 1.021$, that is, a 2.1% mass increase.

3.3 ATOMS

To the best of our knowledge today, every atom consists of a central positively charged nucleus around which negative electrons revolve in stable orbits. Considered as a sphere, the atom has a radius of the order of 10^{-10} m and the nucleus has a radius of the order of 10^{-14} m. The number of electrons is equal to the number of positive charges of the nucleus; thus, the atom is electrically neutral (in its normal state).

The number of positive elementary charges in the nucleus is called the *atomic number* and is indicated by Z. The atomic number identifies the chemical element. All atoms of an element have the same chemical properties.

The atomic electrons move around the nucleus as a result of the attractive electrostatic Coulomb force between the positive nucleus and the negative charge of the electron. According to classical electrodynamics, the revolving electrons ought to continuously radiate part of their energy, follow a spiral orbit, and eventually be captured by the nucleus. Obviously, this does not happen: atoms exist and are stable.

The available experimental evidence points toward the following facts regarding the motion of atomic electrons:

1. Bound atomic electrons revolve around the nucleus in stable orbits without radiating energy. Every orbit corresponds to certain electron energy and is called an *energy state*.

2. Only certain orbits (only certain energies) are allowed. That is, the energy states of the bound electrons form a discrete spectrum, as shown in Figure 3.1. This phenomenon is called *quantization*.

3. If an electron moves from an orbit (state) of energy E_i to another of energy E_f, then (and only then) electromagnetic radiation, an X-ray, is emitted with frequency v such that

$$v = \frac{E_i - E_f}{h} \tag{3.14}$$

where h is Planck's constant.

The energy of the X-ray depends on the atomic number:

$$E_x = hv = E_i - E_f = k(Z-a)^2 \tag{3.15}$$

where k and a are constants. X-ray energies range from a few eV for the light elements to a few hundreds of keV for the heaviest elements.

Every atom emits characteristic X-rays with discrete energies that identify the atom like fingerprints. For every atom, the, the X-rays are identified according to the final state of the electron transition that produced them. Historically, the energy states of atomic electrons are

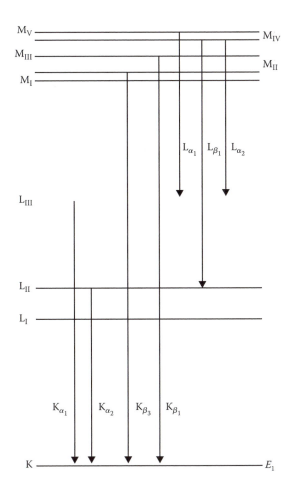

Figure 3.1 An atomic energy level diagram showing X-ray nomenclature (not drawn to scale). E_1 = lowest energy state = ground state.

characterized by the letters K, L, M, N, etc. The K state or K orbit or K shell is the lowest energy state, also called the ground state. The X-rays that are emitted as a result of electronic transitions to the K state, from any other initial state, are called K X-rays (Figure 3.1). Transitions to the L state give rise to L X-rays and so on. K_α and K_β X-rays indicate transitions from L to K and M to K states, respectively.

A bound atomic electron may receive energy and move from a state of energy E_1 to another of higher energy E_2. This phenomenon is called excitation of the atom (Figure 3.2). An excited atom moves preferentially to the lowest possible energy state. In times of the order of 10^{-8} s, the electron that jumped to E_2 or another from another state will fall to E_1 and an X-ray will be emitted.

An atomic electron may receive enough energy to leave the atom and become a free particle. This phenomenon is called *ionization*, and the positive entity left behind is called an ion. The energy necessary to cause ionization is the *ionization potential*. The ionization potential is not the same for all the electrons of the same atom because the electrons move at different distances from the nucleus. The closer the electron is to the nucleus, the more tightly bound it is and the greater its ionization potential becomes. Table 3.1 lists ionization potentials of the least bound electron for certain elements.

When two or more atoms join together and form a molecule, their common electrons are bound to the molecule. The energy spectrum of the molecule is also discrete, but more complicated than that shown in Figure 3.1.

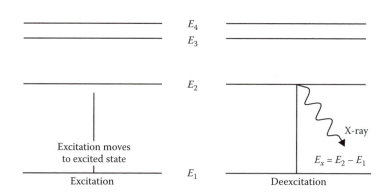

Figure 3.2 Excitation and deexcitation of the atom.

Table 3.1
Ionization Potential for the Least Bound Electron of Certain Elements

Element	Ionization Potential (eV)	Element	Ionization Potential (eV)
H	13.6	Ne	21.56
He	24.56	Na	5.14
Li	5.4	A	15.76
Be	9.32	Fe	7.63
B	8.28	Pb	7.42
C	11.27	U	4.0

3.4 NUCLEI

At the present time, all experimental evidence indicates that nuclei consist of neutrons and protons, which are particles known as *nucleons*. Nuclei then consist of nucleons. Some of the properties of a neutron, a proton, and an electron, for comparison, are listed in Table 3.2. A free proton—outside the nucleus—will eventually pick up an electron and become a hydrogen atom, or it may be absorbed by a nucleus. A free neutron either will be absorbed by a nucleus or will decay according to the equation

$$n \rightarrow p^+ + e^- + \bar{v}$$

that is, it will be transformed into a proton by emitting an electron and another particle called an *antineutrino*.

A nucleus consists of A particles,

$$A = N + Z$$

where A = mass number
$\qquad N$ = number of neutrons
$\qquad Z$ = number of protons = atomic number of the element

A nuclear species X is indicated as

$$^A_Z X$$

where X = chemical symbol of the element. For example,

$$^{16}_{8} O$$

Isobars are nuclides that have the same A.

Table 3.2
Neutron–Proton Properties

Rest Mass	Electron	Neutron	Proton
kg	9.109558×10^{-31}	1.674928×10^{-27}	1.672622×10^{-27}
MeV	0.511	939.552	938.258
U		1.008665	1.007276
Charge	$-e$	0	$+e$

Isotopes are nuclides that have the same Z. They are nuclei of the same chemical element. They have the same chemical but slightly different physical properties, due to their difference in mass. The nuclear properties change drastically from isotope to isotope.

Isotones are nuclides that have the same N, that is, the same number of neutrons.

Isomers are two different energy states of the same nucleus.

The different atomic species are the result of different combinations of one type of particle—the electron. There are 92 natural elements found in nature. Since 1940, 20 more have been artificially produced for a total of 115 elements. The different nuclides, on the other hand, are the result of different combinations of two kinds of particles, neutrons and protons, and so there are many more possibilities. There are more than 700 known nuclides.

Experiments have determined that nuclei are almost spherical, with a volume proportional to the mass number A and a radius approximately equal to*

$$R = 1.3 \times 10^{-15} A^{1/3} \text{ in meters} \tag{3.16}$$

The mass of the nucleus with mass number A and atomic number Z, indicated as $M_N(A, Z)$, is equal to

$$M_N(A,Z) = ZM_p + NM_n - B(A,Z)c^2 \tag{3.17}$$

where M_p = mass of the proton
M_n = mass of the neutron
$B(A, Z)$ = binding energy of the nucleus.

The binding energy is equal to the energy that was released when the N neutrons and Z protons formed the nucleus. More details about the binding energy are given in the next section.

The unit used for the measurement of nuclear mass is equal to 1/12 of the mass of the isotope $^{13}_{6}C$. Its symbol is u (formerly amu for atomic mass unit):

$$1\,u = \frac{1}{12}(\text{mass of } ^{12}_{6}C) = 1.660540 \times 10^{-27}\,kg = 931.481\,MeV$$

In many experiments, what is normally measured is the atomic, not the nuclear, mass. To obtain the atomic mass, one adds the mass of all the atomic electrons (see next section). A table of atomic masses of many isotopes is given in Appendix B. The mass may be given in any of the following three ways:

1. Units of u

2. Kilograms

3. Energy units (MeV or J), in view of the equivalence of mass and energy.

3.5 NUCLEAR BINDING ENERGY

The mass of a nucleus is given by Eq. 3.18 in terms of the masses of its constituents. That same equation also defines the binding energy of the nucleus:

$$B(A,Z) = \left[ZM_p + NM_n - M_N(A,Z) \right]c^2 \tag{3.18}$$

---

* For nonspherical nuclei, the radius given by Eq. 3.16 is an average.

The factor c^2, which multiplies the mass to transform it into energy, will be omitted from now on. It will always be implied that multiplication or division by c^2 is necessary to obtain energy from mass or vice versa. Thus, Eq. 3.18 is rewritten as

$$B(A,Z) = ZM_p + NM_n - M_N(A,Z) \tag{3.19}$$

The meaning of $B(A, Z)$ may be expressed in two equivalent ways:

1. The binding energy $B(A, Z)$ of a nucleus is equal to the mass transformed into energy when the Z protons and the $N = A - Z$ neutrons got together and formed the nucleus. An amount of energy equal to the binding energy was released when the nucleus was formed.

2. The binding energy $B(A, Z)$ is equal to the energy necessary to break the nucleus apart into its constituents, Z free protons and N free neutrons.

As mentioned in Section 3.4, atomic masses rather than nuclear masses are measured in most cases. For this reason, Eq. 3.19 will be expressed in terms of atomic masses by adding the appropriate masses of atomic electrons. If one adds and subtracts Zm in Eq. 3.19,

$$B(A,Z) = ZM_p + Zm + NM_n - M_N(A,Z) - Zm$$
$$= Z(M_p + m) + NM_n - [M_N(A,Z) + Zm] \tag{3.20}$$

Let
$\quad M_H$ = mass of the hydrogen atom
$\quad\quad B_e$ = binding energy of the electron in the hydrogen atom
$B_e(A,Z)$ = binding energy of all the electrons of the atom whose nucleus has mass $M_N(A,Z)$
$M(A,Z)$ = mass of the atom with nuclear mass equal to $M_N(A,Z)$

Then

$$M_H = M_p + m - B_e \tag{3.21}$$

$$M(A,Z) = M_N(A,Z) + Zm - B_e(A,Z) \tag{3.22}$$

Combining Eqs. 3.20, 3.21, and 3.22, one obtains

$$B(A,Z) = ZM_H + NM_n - M(A,Z) - B_e(A,Z) + ZB_e \tag{3.23}$$

Unless extremely accurate calculations are involved, the last two terms of Eq. 3.23 are neglected. The error introduced by doing so is insignificant because ZB_e and $B_e(A,Z)$ are less than a few keV and they tend to cancel each other, while $B(A,Z)$ is of the order of MeV. Equation 3.23 is, therefore, usually written as

$$B(A,Z) = ZM_H + NM_n - M(A,Z) \tag{3.24}$$

Example 3.4 What is the total binding energy of $^4_2$He?
Answer Using Eq. 3.24 and data from Appendix B,

$$B(4,2) = 2M_H + 2M_n - M(4,2)$$
$$= [2(1.00782522) + 2(1.00866544) - 4.00260361]u$$
$$= 0.03037771u = (0.0303771u)931.478\,\text{MeV/u}$$
$$= 28.296\,\text{MeV} = 4.53 \times 10^{-12}\,\text{J}$$

Example 3.5   What is the binding energy of the nucleus $^{238}_{92}U$?
Answer

$$B(238,92) = \left[92(1.00782522) + 146(1.00866544) - 238.05076\right]u$$
$$= 1.93431448\,u = (1.93431448\,u)931.478\,MeV/u$$
$$= 1801.771\,MeV = 2.886 \times 10^{-10}\,J$$

The energy necessary to remove one particle from the nucleus is the *separation* or *binding energy* of that particle for that particular nuclide. A "particle" may be a neutron, a proton, an alpha particle, a deuteron, and so on. The separation or binding energy of a nuclear particle is analogous to the ionization potential of an electron. If a particle enters the nucleus, an amount of energy equal to its separation energy is released.

The separation or binding energy of a neutron (B_n) is defined by the equation

$$B_n = M\left[(A-1),Z\right] + M_n - M(A,Z) \tag{3.25}$$

Using Eq. 3.24, Eq. 3.25 is written as

$$B_n = B(A,Z) - B\left[(A-1),Z\right] \tag{3.26}$$

which shows that the energy of the last neutron is equal to the difference between the binding energies of the two nuclei involved. Typical values of B_n are a few MeV (<10 MeV).

The separation or binding energy of a proton is

$$B_p = M(A-1,Z-1) + M_H - M(A,Z) \tag{3.27}$$

or, using Eq. 3.24,

$$B_p = B(A,Z) - B(A-1,Z-1) \tag{3.28}$$

The separation energy for an alpha particle is

$$B_\alpha = M(A-4,Z-2) + M_{He} - M(A,Z) \tag{3.29}$$

or, using Eq. 3.24,

$$B_\alpha = B(A,Z) - B(A-4,Z-2) - B(4,2) \tag{3.30}$$

Example 3.6   What is the separation energy of the last neutron of the ^4_2He nucleus?
Answer   Using data from Appendix *B* and Eq. 3.25, one obtains

$$B_n = M(3,2) + M_n - M(4,2)$$
$$= \left[(3.016030 + 1.008665 - 4.002604)u\right]931.478\,MeV/u$$
$$= 0.022091(931.478\,MeV) = 20.58\,MeV = 3.3 \times 10^{-12}\,J$$

If the average binding energy per nucleon,

$$b(A,Z) = \frac{B(A,Z)}{A} \tag{3.31}$$

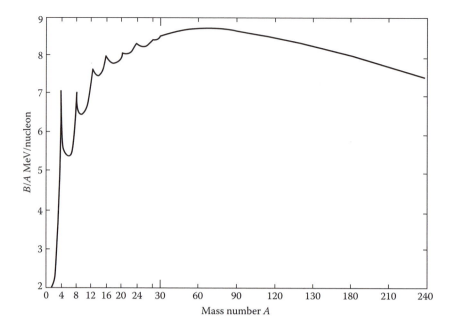

Figure 3.3 The change of the average binding energy per nucleon with mass number A. Notice the change in scale after $A = 30$. (From *The Atomic Nucleus* by R. D. Evans. Copyright © 1955 by McGraw-Hill. Used with the permission of McGraw-Hill Book Company.)

is plotted as a function of A, one obtains the result shown in Figure 3.3. The average binding energy changes relatively little, especially for $A > 30$. Notice that Figure 3.3 has a different scale for $A < 30$.

Figure 3.3 is very important because it reveals the processes by which energy may be released in nuclear reactions. If one starts with a very heavy nucleus ($A = 240$) and breaks it into two medium-size nuclei (fission), energy will be released because the average binding energy per nucleon is larger for nuclides in the middle of the periodic table than it is for heavy nuclides. On the other hand, if one takes two very small nuclei ($A = 2, 3$) and fuses them into a larger one, energy is again released due to similar increase in the average binding energy per nucleon.

3.6 NUCLEAR ENERGY LEVELS

Neutrons and protons are held together in the nucleus by nuclear forces. Although the exact nature of nuclear forces is not known, scientists have successfully predicted many characteristics of nuclear behavior by assuming a certain form for the force and constructing nuclear models based on that form. The success of these models is measured by how well their predicted results agree with the experiment. Many nuclear models have been proposed, each of them explaining certain features of the nucleus; but as of today, no model exists that explains all the facts about all the known nuclides.

All the nuclear models assume that the nucleus, like the atom, can exist only in certain discrete energy states. Depending on the model, the energy states may be assigned to the nucleons—neutrons and protons—or the nucleus as a whole. The present discussion of nuclear energy levels will be based on the second approach.

The lowest possible energy state of a nucleus is called the ground state (Figure 3.4). In Figure 3.4, the ground state is shown as having negative energy to indicate a bound state. The ground state and all the excited states below the zero energy level are called bound states. If the nucleus finds itself in any of the bound states, it deexcites after a time of the order of 10^{-12}–10^{-10} s by dropping to a lower state. Deexcitation is accompanied by the emission of a photon with energy equal to the difference between the energies of the initial and final states. Energy states located above the zero energy level are called virtual energy levels. If the nucleus obtains enough energy to be

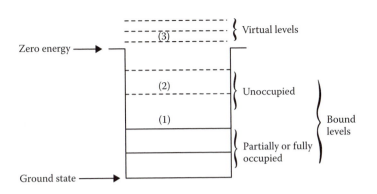

Figure 3.4 Bound and virtual nuclear energy levels.

raised to a virtual level, it may deexcite either by falling to one of the bound levels or by emitting a nucleon.

Studies of the energy levels of all the known nuclides reveal the following:

1. The distance between nuclear energy levels is of the order of keV to MeV. By contrast, the distance between atomic levels is of the order of eV.

2. The distance between levels decreases as the excitation energy increases (Figure 3.5). For very high excitation energies, the density of levels becomes so high that it is difficult to distinguish individual energy levels.

3. As the mass number A increases, the number of levels increases; that is, heavier nuclei have more energy levels than lighter nuclei (in general—there may be exceptions).

4. As A increases, the energy of the first excited state decreases (again, in general—exceptions exist). For example, ^{9}Be: first excited state is at 1.68 MeV, ^{56}Fe: first excited state is at 0.847 MeV, and ^{238}U: first excited state is at 0.044 MeV.

3.7 ENERGETICS OF NUCLEAR DECAYS

This section discusses the energetics of α, β, and γ decay, demonstrating how the kinetic energies of the products of the decay can be calculated from the masses of the particles involved. In all cases, it will be assumed that the original unstable nucleus is at rest—that is, it has zero kinetic energy and linear momentum. This assumption is very realistic because the actual kinetic energies of nuclei due to thermal motion are of the order of kT (of the order of eV), where k is the Boltzmann constant and T the temperature (Kelvin), while the energy released in most decays is of the order of MeV.

In writing the equation representing the decay, the following notation will be used:

$$M = \text{atomic mass (or } Mc^2 = \text{rest mass energy)}$$
$$E_\gamma = \text{energy of a photon}$$
$$T_i = \text{kinetic energy of a particle type } i$$
$$P_i = \text{linear momentum of a particle type } i.$$

3.7.1 Gamma Decay

In γ decay, a nucleus goes from an excited state to a state of lower energy and the energy difference between the two states is released in the form of a photon. Gamma decay is represented by

$$^A_Z X^* \rightarrow {}^A_Z X + \gamma$$

where $^A_Z X^*$ indicates the excited nucleus.

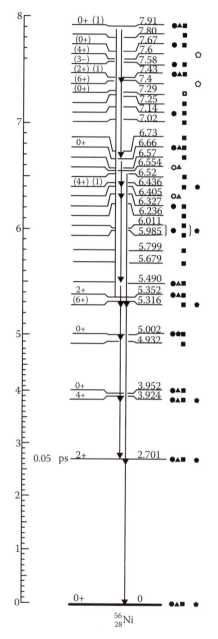

Figure 3.5 The energy levels of $^{56}_{28}$Ni. In this diagram, the zero energy has been switched to the ground state of the nucleus. The numbers on the right-hand column show the energy of each level in MeV. (From Lederer, C. M., and Shirley, V. S. (eds.), *Table of Isotopes*, 7th ed., Wiley-Interscience, New York, 1978, p. 163.)

Applying conservation of energy and momentum for the states before and after the decay, we have*

---

* Equations in this chapter are written in terms of atomic, not nuclear, masses. This notation introduces a slight error because the binding energy of the atomic electrons is not taken into account (see Section 3.5).

Conservation of energy:

$$M^*(A,Z) = M(A,Z) + T_M + E_\gamma \tag{3.32}$$

Conservation of momentum:

$$0 = \mathbf{P}_M + \mathbf{P}_\gamma \tag{3.33}$$

Using these two equations and the nonrelativistic form of the kinetic energy of the nucleus,

$$T_N = \frac{1}{2}MV^2 = \frac{P_M^2}{2M} = \frac{P_\gamma^2}{2M} = \frac{E_\gamma^2}{2Mc^2} \tag{3.34}$$

Use has been made of the relationship $E_\gamma = P_\gamma c$ (the photon rest mass is zero).

Equation 3.34 gives the kinetic energy of the nucleus after the emission of a photon of energy E_γ. This energy is called the *recoil energy*.

The recoil energy is small. Consider a typical photon of 1 MeV emitted by a nucleus with $A = 50$. Then, from Eq. 3.34,

$$T_N \approx \frac{1^2 \, (\text{MeV})^2}{2(50)(932)(\text{MeV})} \approx 11 \, \text{eV}$$

Most of the time, this energy is neglected and the gamma energy is written as

$$E_\gamma = M^*(A,Z) - M(A,Z)$$

However, there are cases where the recoil energy may be important, for example, in radiation damage studies.

Sometimes the excitation energy of the nucleus is given to an atomic electron instead of being released in the form of a photon. This type of nuclear transition is called internal conversion (IC), and the ejected atomic electron is called an internal conversion electron.

Let T_i be the kinetic energy of an electron ejected from shell i and B_i be the binding energy of an electron in shell i. Equation 3.32 now takes the form

$$M^*(A,Z) = M(A,Z) + T_i + B_i + T_N \tag{3.35}$$

Even though the electron has some nonzero rest mass energy, it is so much lighter than the nucleus that $T_N \ll T_i$. Consequently, T_N is neglected and Eq. 3.35 is written as

$$T_i = M^*(A,Z) - M(A,Z) - B_i \tag{3.36}$$

If $Q = M^*(A,Z) - M(A,Z)$ = energy released during the transition, then

$$T_i = Q - B_i$$

When internal conversion occurs, there is a probability than an electron from the K shell, L shell, or another shell, may be emitted. The corresponding equations for the electron kinetic energies are

$$\begin{aligned} T_K &= Q - B_K, \\ T_L &= Q - B_L, \\ T_M &= Q - B_M, \text{ etc.} \end{aligned} \tag{3.37}$$

Therefore, a nucleus that undergoes internal conversion is a source of groups of monoenergetic electrons with energies given by Eqs. 3.37. A typical internal conversion electron spectrum is shown in Figure 3.6. The two peaks correspond to K and L electrons. The diagram on the right

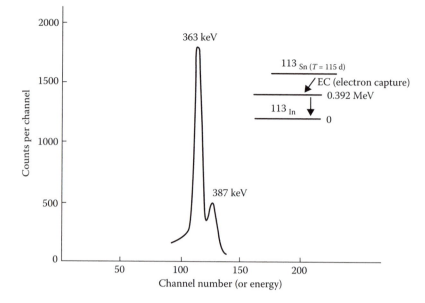

Figure 3.6 The internal conversion spectrum of ^{113}Sn. The two peaks correspond to K electrons (363 keV) and L electrons (387 keV).

(Figure 3.6) shows the transition energy to be 392 keV. The K-shell energy is then $B_K = 392 - 363 = 29$ keV and the L-shell binding energy is $B_L = 392 - 387 = 5$ keV. Let:

λ_e = probability that internal conversion will occur
λ_γ = probability that a photon will be emitted
λ_i = probability that an electron from shell i will be emitted
λ = total probability for γ decay

Then*

$$\lambda_e = \lambda_K + \lambda_L + \lambda_M + \cdots \tag{3.38}$$

and

$$\lambda = \lambda_\gamma + \lambda_e \tag{3.39}$$

For most nuclei, $\lambda_e = 0$, but there is no γ-decaying nucleus for which $\lambda_\gamma = 0$. This means radioisotopes that internally convert, emit gammas as well as electrons. After an atomic electron is emitted, the empty state that was created will quickly be filled by another electron that "falls in" from the outer shells. Because of such a transition, an X-ray is emitted. Therefore, internally converting nuclei *emit γ-rays, electrons,* and *X-rays.*

Radioisotopes that undergo internal conversion are the only sources of monoenergetic electrons, except for accelerators. They are very useful as instrument calibration sources. Three isotopes frequently used are ^{113}Sn, ^{137}Cs, and ^{207}Bi.

3.7.2 Alpha Decay

Alpha decay is represented by the equation

$$^A_Z X \rightarrow \,^{A-4}_{Z-2}X + \,^4_2He$$

Applying conservation of energy and momentum,

$$M(A,Z) = M(A-4, Z-2) + M(4, 2) + T_N + T_\alpha \tag{3.40}$$

---

* Tables of isotopes usually give, not the values of the different λ's, but the so-called IC coefficients, which are the ratios λ_K/λ_γ, λ_L/λ_γ, etc. (see Ref. 2).

and

$$0 = P_\alpha + P_M \tag{3.41}$$

The energy that becomes available as a result of the emission of the alpha particle is called the *decay energy* Q_α, defined by

$$Q_\alpha = (\text{mass of parent}) - (\text{mass of decay products})$$
$$Q_\alpha = M(A,Z) - M(A-4, Z-2) - M(4,2) \tag{3.42}$$

Obviously, for α decay to occur, Q_α should be greater than zero. Therefore, α decay is possible only when

$$M(A,Z) > M(A-4, Z-2) + M(4,2) \tag{3.43}$$

If the daughter nucleus is left in its ground state, after the emission of the alpha, the kinetic energy of the two products is (from Eq. 3.40),

$$T_N + T_\alpha = Q_\alpha \tag{3.44}$$

In many cases, the daughter nucleus is left in an excited state of energy E_i, where i indicates the energy level. Then, Eq. 3.44 becomes

$$T_N + T_\alpha = Q_\alpha - E_i \tag{3.45}$$

which shows that the available energy (Q_α) is decreased by the amount E_i.

The kinetic energies T_α and T_N can be calculated from Eqs. 3.41 and 3.44. The result is

$$T_\alpha = \frac{M(A-4, Z-2)}{M(A-4, Z-2) + M(4,2)}(Q_\alpha - E_i) \approx \frac{A-4}{A}(Q_\alpha - E_i) \tag{3.46}$$

$$T_N = \frac{M_\alpha}{M(A-4, Z-2) + M(4,2)}(Q_\alpha - E_i) \approx \frac{4}{A}(Q_\alpha - E_i) \tag{3.47}$$

Example 3.7   What are the kinetic energies of the alphas emitted by ^{238}U?

Answer   The decay scheme of $^{238}_{92}U$ is shown in Figure 3.7. After the alpha is emitted, the daughter nucleus, $^{234}_{90}Th$, may be left in one of the two excited states at 0.16 MeV and 0.048 MeV or go to the ground state.

The decay energy Q_α is (Eq. 3.42)

$$Q_\alpha = M(238, 92) - M(234, 90) - M(4,2)$$
$$= 238.050786 - 234.043594 - 4.002603$$
$$= 0.004589\,u = 0.004589 \times 931.481\,\text{MeV} = 4.27\,\text{MeV}$$

Depending on the final state of $^{234}_{90}Th$, the energy of the alpha particle is

$$T_\alpha = \frac{234}{238}Q_\alpha = 4.20\,\text{MeV}$$

$$T_\alpha = \frac{234}{238}(Q_\alpha - 0.048) = 4.15\,\text{MeV}$$

$$T_\alpha = \frac{234}{238}(Q_\alpha - 0.16) = 4.04\,\text{MeV}$$

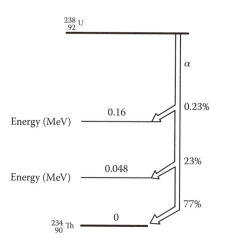

Figure 3.7 The alpha decay scheme of ^{238}U. The percentage values give the probability of decay through the corresponding level. (From Lederer, C. M., and Shirley, V. S. (eds.), *Table of Isotopes*, 7th ed., Wiley-Interscience, New York, 1978, p. 163.)

3.7.3 Beta Decay

In β decay, a nucleus emits an electron or a positron and is transformed into a new element. In addition to the electron or the positron, a neutral particle with rest mass zero is also emitted. There are two types of β decay, β^- and β^+.

β^- *Decay.* This type of decay is represented by

$$_Z^A X \rightarrow {}_{Z+1}^A X + \beta^- + \bar{v}$$

where β^- = negative beta particle = electron and $\bar{v}$ = antineutrino.

Historically, the name β particle has been given to electrons that are emitted by nuclei undergoing beta decay. The antineutrino ($\bar{v}$) is a neutral particle with rest mass so small that it is taken equal to zero in our discussion.

The energy equation of β^- decay is

$$M_N(A,Z) = M_N(A,Z+1) + m + T_{\beta^-} + T_{\bar{v}} + T_N \tag{3.48}$$

where $M_N(A, Z)$ is the *nuclear* mass and m is the electron rest mass. Using atomic masses, Eq. 3.48 becomes (see Section 3.5)

$$M(A,Z) = M(A,Z+1) + T_{\beta^-} + T_{\bar{v}} + T_N \tag{3.49}$$

The momentum equation is

$$0 = \mathbf{P}_N + \mathbf{P}_{\beta^-} + \mathbf{P}_{\bar{v}} \tag{3.50}$$

The β^- decay energy, Q_{β^-}, is defined as

$$Q_{\beta^-} = M(A,Z) - M(A,Z+1) \tag{3.51}$$

The condition for β^- decay to be possible is

$$M(A,Z) - M(A,Z+1) > 0 \tag{3.52}$$

In terms of Q_{β^-}, Eq. 3.49 is rewritten in the form

$$T_{\beta^-} + T_{\bar{v}} + T_N = Q_{\beta^-} \tag{3.53}$$

Equations 3.50 and 3.53 show that three particles, the nucleus, the electron, and the antineutrino, share the energy Q_{β^-}, and their total momentum is zero.

There is an infinite number of combinations of kinetic energies and momenta that satisfy these two equations and as a result, the energy spectrum of the betas is continuous.

In Eq. 3.53, the energy of the nucleus, T_M, is much smaller than either T_{β^-} or $T_{\bar{v}}$ because the nuclear mass is huge compared to that of the electron or the antineutrino. For all practical purposes, T_M can be neglected and Eq. 3.53 takes the form

$$T_{\beta^-} + T_{\bar{v}} = Q_{\beta^-} \tag{3.54}$$

As in the case of α decay, the daughter nucleus may be left in an excited state after the emission of the β^- particle. Then, the energy available to become kinetic energy of the emitted particles is less. If the nucleus is left in the ith excited state E_i, Eq. 3.54 takes the form

$$T_{\beta^-} + T_{\bar{v}} = Q_{\beta^-} - E_i = E_{max} \tag{3.55}$$

According to Eq. 3.54, the electron and the antineutrino share the energy Q_{β^-} (or E_{max}) and there is a certain probability that either particle may have an energy within the limits

$$0 \leq T_{\beta^-} \leq E_{max} \tag{3.56}$$

$$E_{max} \geq T_{\bar{v}} \geq 0 \tag{3.57}$$

which means that the beta particles have a continuous energy spectrum. Let $\beta(T)\,dT$ be the number of beta particles with kinetic energy between T and $T + dT$. The function $\beta(T)$ has the general shape shown in Figure 3.8. The energy spectrum of the antineutrinos is the complement of that shown in Figure 3.8, consistent with Eq. 3.57. The continuous energy spectrum of β^- particles should be contrasted with the energy spectrum of internal conversion electrons shown by Figure 3.6.

As stated earlier, beta particles are electrons. The practical difference between the terms *electrons* and *betas* is this: A beam of electrons of energy T consists of electrons each of which has the kinetic energy T. A beam of beta particles with energy E_{max} consists of electrons that have a continuous energy spectrum (Figure 3.8) ranging from zero up to a maximum kinetic energy E_{max}.

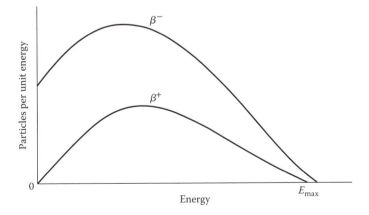

Figure 3.8 A typical beta energy spectrum (shows shape only; does not mean that β^- intense than β^+).

Figure 3.9 shows the β^- decay scheme of the isotope $^{137}_{55}$Cs. For an example of a Q_{β^-} calculation, consider the decay of $^{137}_{55}$Cs:

$$Q_{\beta^-} = M(55,137) - M(56,137) = (136.90682 - 136.90556u = 0.00126u)$$

$$= 0.0012625(931.478 \text{ MeV}) = 1.1760 \text{ MeV} = 1.36 \times 10^{-13} \text{ J}$$

If the $^{137}_{56}$Ba is left in the 0.6616-MeV state (which happens 93.5% of the time), the available energy is

$$E_{max} = 1.1760 - 0.6616 = 0.5144 \text{ MeV}$$

For many calculations, it is necessary to use the average energy of the beta particles, $\bar{E}_{\beta^-}$. An accurate equation for $\bar{E}_{\beta^-}$ has been developed,[1] but in practice the average energy is taken to be

$$\bar{E}_{\beta^-} = \frac{E_{max}}{3}$$

β^+ *Decay.* The expression representing β^+ decay is

$$^A_Z X \rightarrow {}^A_{Z-1}X + \beta^+ + v$$

where β^+ = positron and N = neutrino.
The energy equation of β^+ decay is

$$M_N(A,Z) = M_N(A,Z-1) + m + T_\beta + T_v + T_N \tag{3.58}$$

Using atomic masses, Eq. 3.58 becomes

$$M(A,Z) = M(A,Z-1) + 2m + T_\beta + T_v + T_N \tag{3.59}$$

The momentum equation is

$$0 = P_N + P_{\beta^+} + P_v \tag{3.60}$$

The β^+ decay energy is

$$Q_{\beta^+} = M(A,Z) - M(A,Z-1) - 2m \tag{3.61}$$

The condition for β^+ decay to be possible is

$$M(A,Z) - M(A,Z-1) - 2m > 0 \tag{3.62}$$

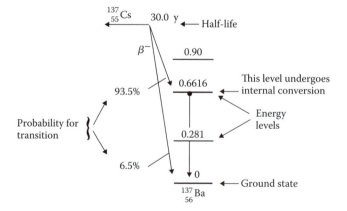

Figure 3.9 The decay scheme of ^{137}Cs. The Q value of the β^- decay is –1.176 MeV. Probability for each transition is given in %. (From Lederer, C. M., and Shirley, V. S. (eds.), *Table of Isotopes*, 7th ed., Wiley-Interscience, New York, 1978, p. 163.)

A comparison of Eqs. 3.52 and 3.62 shows that β^- decay is possible if the mass of the parent is just bigger than the mass of the daughter nucleus, while β^+ decay is possible only if the parent and daughter nuclear masses differ by at least $2mc^2 = 1.022$ MeV.

The energy spectrum of β^+ particles is continuous, for the same reasons the β^- spectrum is, and similar to that of β^- decay (Figure 3.8). The average energy of the positrons from β^+ decay, $\bar{E}_{\beta^+}$, is also taken to be equal to $E_{max}/3$ unless extremely accurate values are needed, in which case the equation given in Reference 1 should be used.

A typical β^+ decay scheme is shown in Figure 3.10.

Electron Capture. In some cases, an atomic electron is captured by the nucleus and a neutrino is emitted according to the equation

$$M_N(A,Z)+m = M_N(A,Z-1)+T_v+B_e \tag{3.63}$$

In Eq. 3.63, all the symbols have been defined before except B_e, binding energy of the electron captured by the nucleus. This transformation is called *electron capture* (EC). In terms of atomic masses, Eq. 3.63 takes the form

$$M(A,Z)= M(A,Z-1)+T_v \tag{3.64}$$

The energy Q_{EC} released during EC is

$$Q_{EC} = M(A,Z)-M(A,Z-1) \tag{3.65}$$

The condition for EC to be possible is

$$M(A,Z)-M(A,Z-1)>0 \tag{3.66}$$

Electron capture is an alternative to β^+ decay. Comparison of Eqs. 3.61 and 3.66 shows that nuclei that cannot experience β^+ decay can undergo EC, since a smaller mass difference is required for the latter process. Of course, EC is always possible if β^+ decay is. For example, ^{22}Na (Figure 3.10) decays both by β^+ and EC.

After EC, there is a vacancy left behind that is filled by an electron falling in from a higher orbit. Assuming that a K electron was captured, an L electron may fill the empty state left behind. When this happens, an energy approximately equal to $B_K - B_L$ becomes available (where B_K and B_L are the binding energy of a K or L electron, respectively). The energy $B_K - B_L$ may be emitted as a K X-ray called *fluorescent radiation,* or it may be given to another atomic electron. If this energy is given to an L electron, that particle will be emitted with kinetic energy equal to $(B_K - B_L) - B_L = B_K - 2B_L$. Atomic electrons emitted in this way are called *Auger electrons.*

Whenever an atomic electron is removed and the vacancy left behind is filled by an electron from a higher orbit, there is a competition between the emission of Auger electrons and fluorescent

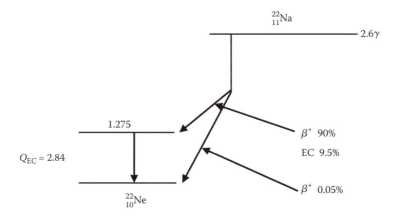

Figure 3.10 The decay scheme of ^{22}Na. Notice that it is Q_{EC} that is plotted, not Q_{β^+}. (From Lederer, C. M., and Shirley, V. S. (eds.), *Table of Isotopes,* 7th ed., Wiley-Interscience, New York, 1978, p. 163.)

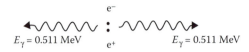

$$E_\gamma = 0.511 \text{ MeV} \qquad E_\gamma = 0.511 \text{ MeV}$$

Figure 3.11 Electron–positron annihilation.

radiation. The number of X-rays emitted per vacancy in a given shell is the *fluorescent yield*. The fluorescent yield increases with atomic number.

3.7.4 Particles, Antiparticles, and Electron–Positron Annihilation

Every known subatomic particle has a counterpart called the *antiparticle*. A charged particle and an antiparticle have the same mass, and opposite charge. If a particle is neutral—for example, the neutron—its antiparticle is still neutral. Then their difference is due to some other property, such as magnetic moment. Some particles, like the photon, are identical with their own antiparticles. An antiparticle cannot exist together with the corresponding particle: when an antiparticle meets a particle, the two react and new particles appear.

Consider the example of the electron and the "antielectron," which is the positron. The electron and the positron are identical particles except for their charge, which is equal to e but negative and positive, respectively. The rest mass of either particle is equal to 0.511 MeV. A positron moving in a medium loses energy continuously, as a result of collisions with atomic electrons (see Chapter 4). Close to the end of its track, the positron combines with an atomic electron, the two annihilate, and photons appear with a total energy equal to $2mc^2$. At least two photons should be emitted for conservation of energy and momentum to be satisfied (Figure 3.11). Most of the time, two photons, each with energy 0.511 MeV, are emitted. As a result, every positron emitter is also a source of 0.511-MeV annihilation gammas.

3.7.5 Complex Decay Schemes

For many nuclei, more than one mode of decay is positive. Users of radio isotopic sources need information about particles emitted, energies, and probabilities of emission. Many books on atomic and nuclear physics contain such information, and the most comprehensive collection of data on this subject can be found in the *Table of Isotopes* by Lederer and Shirley.[2] Figure 3.12 shows an example of a complex decay scheme taken from that book.

3.8 THE RADIOACTIVE DECAY LAW

Radioactive decay is spontaneous change of a nucleus. The change may result in a new nuclide or simply change the energy of the nucleus. If there is a certain amount of a radioisotope at hand, there is no certainty that in the next second "so many nuclei will decay" or "none will decay." One can talk of the probability that a nucleus will decay in a certain period of time.

The probability that a given nucleus will decay per unit time is called the *decay constant* and is indicated by the letter λ. For a certain species, λ is

1. The same for all the nuclei

2. Constant, independent of the number of nuclei present

3. Independent of the age of the nucleus

Consider a certain mass m of a certain radioisotope with decay constant λ. The number of atoms (or nuclei) in the mass m is equal to

$$N = m\frac{N_A}{A} \tag{3.67}$$

where $N_A = 6.022 \times 10^{23} =$ Avogadro's number and $A =$ atomic weight of the isotope.

This number of atoms decreases with time, due to the decay according to

Decrease per unit time = decay per unit time

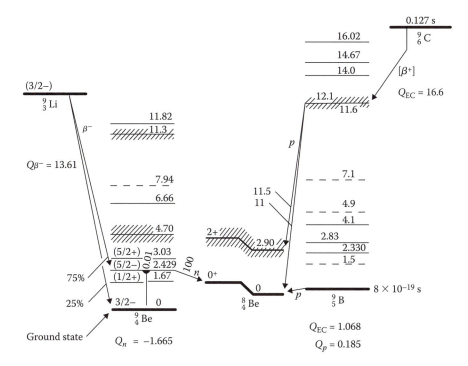

Figure 3.12 A complex decay scheme. For complete explanation of all the symbols and numbers see Ref. 2. Half-life is given for each element's ground state, and energy of each level is given at intermediate states. Q_n is the neutron separation energy. Transition probabilities are indicated as percentages. (From Lederer, C. M., and Shirley, V. S. (eds.), *Table of Isotopes*, 7th ed., Wiley-Interscience, New York, 1978, p. 163.)

or mathematically,

$$-\frac{\mathrm{d}N(t)}{\mathrm{d}t} = \lambda N(t) \tag{3.68}$$

The solution of this equation is

$$N(t) = N(0)\mathrm{e}^{-\lambda t} \tag{3.69}$$

where $N(0)$ is the number of atoms at $t = 0$.

The probability that a nucleus *will not decay* in time t—that is, it will survive time t—is given by the ratio of

$$\frac{\text{atoms not decaying in time } t}{\text{atoms at } t = 0} = \frac{N(0)\mathrm{e}^{-\lambda t}}{N(0)} = \mathrm{e}^{-\lambda t} \tag{3.70}$$

The probability that the nucleus will decay between t and $t + \mathrm{d}t$ is $p(t)\,\mathrm{d}t =$ (probability to survive to time 0) (probability to decay in $\mathrm{d}t$) $= \mathrm{e}^{-\lambda t}\,\mathrm{d}t$. The average lifetime $\bar{t}$ of the nucleus is given by

$$\bar{t} = \frac{\int_0^\infty t p(t)\,\mathrm{d}t}{\int_0^\infty p(t)\,\mathrm{d}t} = \frac{\int_0^\infty t\mathrm{e}^{-\lambda t}\,\mathrm{d}t}{\int_0^\infty \mathrm{e}^{-\lambda t}\,\mathrm{d}t} = \frac{1}{\lambda} \tag{3.71}$$

One concept used extensively with radioisotopes in the *half-life T*, defined as the time it takes for half of a certain number of nuclei to decay. Thus, using Eq. 3.69,

$$\frac{N(T)}{N(0)} = \frac{1}{2} = e^{-\lambda t}$$

which then gives the relationship between λ and t:

$$t = \frac{\ln 2}{\lambda} \tag{3.72}$$

For a sample of $N(t)$ nuclei at time t, each having decay constant λ, the expected number of nuclei decaying per unit time is

$$A(t) = \lambda N(t) \tag{3.73}$$

where $A(t)$ = activity of the sample at time t.

The units of activity are the Becquerel (Bq), equal to 1 decay/s, or the Curie (Ci) equal to 3.7×10^{10} Bq. The Becquerel is the SI unit defined in 1977.

The term *specific activity* (SA) is used frequently. It may have one of the two following meanings:

1. For solids,

$$SA = \frac{\text{activity}}{\text{mass}} \; (\text{Bq/kg or Ci/g})$$

2. For gases or liquids,

$$SA = \frac{\text{activity}}{\text{volume}} \; (\text{Bq/m}^3 \text{ or Ci/cm}^3)$$

Example 3.8 What is the SA of ^{60}Co?
Answer The SA is

$$SA = \frac{A}{m} = \frac{\lambda N}{m} = \frac{\ln 2}{Tm} m \frac{N_A}{A} = \frac{(\ln 2)(6.022 \times 10^{23})}{(5.2 \text{ y})(3.16 \times 10^7 \text{ s/y})(0.060 \text{ kg})}$$
$$= 4.23 \times 10^{16} \text{ Bq/kg} = 1.14 \times 10^3 \text{ Ci/g}$$

Example 3.9 What is the SA of a liquid sample of 10^{-3} m^3 containing 10^{-6} kg of ^{32}P?
Answer The SA is

$$SA = \frac{A}{V} = \frac{\lambda N}{V} = \frac{\ln 2}{VT} m \frac{N_A}{A} = \frac{(\ln 2)(10^{-6} \text{ kg})(6.022 \times 10^{23})}{(10^{-3} \text{ m}^3)(14.3 \text{ d})(86400 \text{ s/d})(0.032 \text{ kg})}$$
$$= 1.05 \times 10^{16} \text{ Bq/m}^3 = 0.285 \text{ Ci/cm}^3$$

There are isotopes that decay by more than one mode. Consider such an isotope decaying by the modes 1, 2, 3, ..., i (e.g., alpha, beta, gamma, etc., decay), and let

λ_i = probability per unit time that the nucleus will decay by the ith mode

The total probability of decay (total decay constant) is

$$\lambda = \lambda_1 + \lambda_2 + \cdots + \lambda_i + \cdots \tag{3.74}$$

If the sample contains $N(t)$ atoms at time t, the number of decays per unit time by the ith mode is

$$A_i(t) = \lambda_i N(t) = \lambda_i N(0)e^{-\lambda t} \tag{3.73a}$$

The term *partial half-life* is sometimes used to indicate a different decay mode. If T_i is the partial half-life for the ith decay mode, using Eqs. 3.72 and 3.74, one obtains

$$\frac{1}{T} = \frac{1}{T_1} + \frac{1}{T_2} + \frac{1}{T_3} + \cdots + \frac{1}{T_i} + \cdots \tag{3.75}$$

It should be pointed out that it is the total decay constant that is used by Eqs. 3.69 and 3.73a, and not the partial decay constants.

Example 3.10   The isotope ^{252}Cf decays by alpha decay and by spontaneous fission. The total half-life is 2.646 years and the half-life for alpha decay is 2.731 years. What is the number of spontaneous fissions per second per 10^{-3} kg (1 g) of ^{252}Cf?
Answer   The spontaneous fission activity is

$$A_{sf} = \lambda_{sf} N = \frac{\ln 2}{T_{sf}} m \frac{N_A}{A}$$

The spontaneous fission half-life is, using Eq. 3.75,

$$T_{sf} = \frac{T_t T_\alpha}{T_\alpha - T_t} = \frac{(2.646 \times 2.731)}{2.731 - 2.646} = 85 \text{ y} = 2.68 \times 10^9 \text{ s}$$

Therefore,

$$A_{sf} = \frac{\ln 2}{2.68 \times 10^9 \text{s}} (10^{-3} \text{ kg}) \frac{6.022 \times 10^{22}}{252 \times 10^3} \text{ atoms/kg} = 6.17 \times 10^{11} \text{ sf/s}$$

Sometimes the daughter of a radioactive nucleus may also be radioactive and decay to a third radioactive nucleus. Thus, a radioactive chain

$$N_1 \rightarrow N_2 \rightarrow N_3 \rightarrow \cdots$$

is generated. An example of a well-known series is that of ^{235}U, which through combined α and β^- decays ends up as an isotope of lead. The general equation giving the number of atoms of the ith isotope, at time t in terms of the decay constants of all the other isotopes in the chain was developed by Bateman.[3] If $N_i(0)$ is the number of atoms of the ith isotope of the series at time $t = 0$ and

$$N_i(0) = 0, \quad i > 1$$

then the Bateman[3] equation takes the form

$$N_i(t) = \lambda_1 \lambda_2 \cdots \lambda_{i-1} N_1(0) \sum_{j=1}^{i} \frac{e^{-\lambda_j t}}{\prod_{k \neq j} (\lambda_k - \lambda_j)} \tag{3.76}$$

Example 3.11 Apply the Bateman equation for the second and third isotope in a series.
Answer

(a) $N_2(t) = \lambda_1 N_1(0) \left(\dfrac{e^{-\lambda_1 t}}{\lambda_2 - \lambda_1} + \dfrac{e^{-\lambda_2 t}}{\lambda_1 - \lambda_2} \right) = \dfrac{\lambda_1 N_1(0)}{(\lambda_2 - \lambda_1)} \left(e^{-\lambda_1 t} - e^{-\lambda_2 t} \right)$

(b) $N_3(t) = \lambda_1 \lambda_2 N_1(0) \left[\dfrac{e^{-\lambda_1 t}}{(\lambda_2 - \lambda_1)(\lambda_3 - \lambda_1)} + \dfrac{e^{-\lambda_2 t}}{(\lambda_1 - \lambda_2)(\lambda_3 - \lambda_2)} + \dfrac{e^{-\lambda_3 t}}{(\lambda_1 - \lambda_3)(\lambda_2 - \lambda_3)} \right]$

3.9 NUCLEAR REACTIONS

3.9.1 General Remarks

A *nuclear reaction* is an interaction between two particles, a fast bombarding particle, called the *projectile*, and a slower or stationary *target*. The products of the reaction may be two or more particles. For the energies considered here (<20 MeV), the products are also two particles (with the exception of fission, which is discussed in the next section).

If x_1, X_2 are the colliding particles and x_3, X_4 are the products, the reaction is indicated as

$$\prescript{A_1}{Z_1}{x_1} + \prescript{A_2}{Z_2}{X_2} \rightarrow \prescript{A_3}{Z_3}{x_3} + \prescript{A_4}{Z_4}{X_4}$$

or

$$X_2(x_1, x_3)X_4$$

The particles in parentheses are the light particles, x_1 being the projectile. Another representation for the reason is based on the light particles only, in which case the reaction shown above is indicated as an (x_1, x_3) reaction. For example, the reaction

$$\prescript{1}{0}{n} + \prescript{10}{5}{B} \rightarrow \prescript{4}{2}{He} + \prescript{7}{3}{Li}$$

may be indicated as $\prescript{10}{5}{B}(n,\alpha)\prescript{7}{3}{Li}$ or simply as an (n, α) reaction.

Certain quantities are conserved when a nuclear reaction takes place. Four are considered here. For the reaction shown above, the following quantities are conserved:

Charge:

$$Z_1 + Z_2 = Z_3 + Z_4$$

Mass number:

$$A_1 + A_2 = A_3 + A_4$$

Total energy:

$$E_1 + E_2 = E_3 + E_4 \ \left(\text{rest mass plus kinetic energy}\right)$$

Linear momentum:

$$P_1 + P_2 = P_3 + P_4$$

Many nuclear reactions proceed through the formation of a *compound nucleus*. The compound nucleus, formed after particle x_1 collides with X_2, is highly excited and lives for a time of the order of 10^{-12}–10^{-14} s before it decays to x_3 and X_4. A compound nucleus may be formed in more than one way and may decay by more than one mode that does not depend on the mode of formation. Consider the example of the compound nucleus $\prescript{14}{7}{N}$:

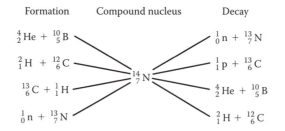

The modes of formation and decay of ^{14}N are shown in the form of an energy-level diagram in Figure 3.13. No matter how the compound nucleus is formed, it has an excitation energy equal to the separation energy of the projectile (α, n, p, etc.) plus a fraction of the kinetic energy of the two particles. Since the separation energy is of the order of MeV, it is obvious that the compound nucleus has considerable excitation energy even if the projectile and the target have zero kinetic energy.

Exactly what happens inside the compound nucleus is not known. It is believed—and experiment does not contradict this idea—that the excitation energy of the compound nucleus is shared

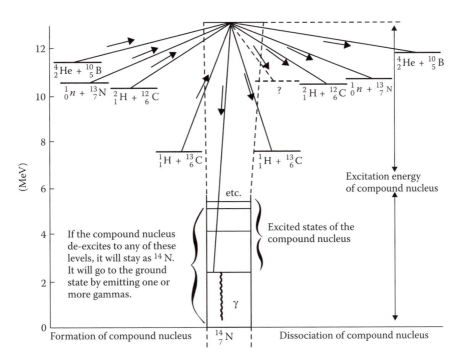

Figure 3.13 Different modes of formation and decay of the component nucleus. For clarity, the diagram shows that the compound nucleus has the same excitation energy regardless of the way it is formed. This is not necessarily the case.

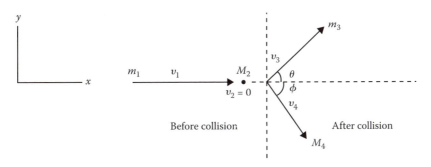

Figure 3.14 The kinematics of the reaction $M_2 (m_1, m_3)M_4$.

quickly by all the nucleons $(A_1 + A_2)$. There is continuous exchange of energy among all the nucleons until one of them (or a cluster of them) obtains energy greater than its separation energy and is able to leave the compound nucleus, becoming a free particle.

3.9.2 Kinematics of Nuclear Reactions

In this section, the following two questions will be answered:

1. Given the masses m_1, M_2, m_3, M_4, and the kinetic energies of the projectile (m_1) and the target (M_2), how can one calculate the kinetic energies of the products with masses m_3 and M_4?

2. What is the minimum kinetic energy the particles with masses m_1, M_2 ought to have to be able to initiate the reaction?

The discussion will be limited to the case of a stationary target, the most commonly encountered in practice.

Consider a particle of mass m_1 having speed v_1 (kinetic energy T_1) hitting a stationary particle of Mass M_2. The particles m_3, M_4 are produced as a result of this reaction with speeds v_3, v_4 (kinetic energies T_3, T_4), as shown in Figure 3.14. Applying conservation of energy and linear momentum, one has

Energy:

$$m_1 + T_1 + M_2 = m_3 + T_3 + M_4 + T_4 \tag{3.77}$$

Momentum, x axis:

$$m_1 v_1 = m_3 v_3 \cos\theta + M_4 v_4 \cos\phi \tag{3.78}$$

Momentum, y axis:

$$m_3 v_3 \sin\theta = M_4 v_4 \sin\phi \tag{3.79}$$

The quantity

$$Q = m_1 + M_2 - m_3 - M_4 \tag{3.80}$$

is called the *Q* value of the reaction. If $Q > 0$, the reaction is called *exothermic* or *exoergic*. If $Q < 0$, then the reaction is called *endothermic* or *endoergic*.

Assuming nonrelativistic kinematics, in which case $T = (1/2)mv^2$, Eqs. 3.77 to 3.79 take the form

$$T_1 + Q = T_3 + T_4 \tag{3.81}$$

$$\sqrt{2m_1 T_1} = \sqrt{2m_3 T_3} \cos\theta + \sqrt{2M_4 T_4} \cos\phi \tag{3.82}$$

$$\sqrt{2m_3 T_3} \sin\theta = \sqrt{2M_4 T_4} \sin\phi \tag{3.83}$$

Equations 3.81–3.83 have four unknowns T_3, T_4, ϕ, and θ, so they cannot be solved to give a unique answer for any single unknown. In practice, one expresses a single unknown in terms of a second one—for example, T_3 as a function of θ, after eliminating T_4 and ϕ. Such an expression, although straightforward, is complicated. Two cases of special interest are the following.

Case 1: $\theta = 0$, $\phi = 180°$. In this case, the particles m_3 and M_4 are emitted along the direction of motion of the bombarding particle (Figure 3.15). Equations 3.81 and 3.82 take the form

$$T_1 + Q = T_3 + T_4 \tag{3.84}$$

$$\sqrt{m_1 T_1} = \sqrt{m_3 T_3} - \sqrt{M_4 T_4} \tag{3.85}$$

and they can be solved for T_3 and T_4. These values of T_3 and T_4 give the maximum and minimum kinetic energies of particles m_3 and M_4.

Example 3.12 Consider the reaction

$$^1_0 n + {}^{14}_7 N \rightarrow {}^4_2 He + {}^{11}_5 B$$

with the nitrogen being at rest and the neutron having energy 2 MeV. What is the maximum kinetic energy of the alpha particle?

Answer The *Q* value of the reaction is

$$Q = (14.003074 + 1.008665 - 4.002603 - 11.009306) \times 931.481 \text{ MeV}$$
$$= -0.158 \text{ MeV}$$

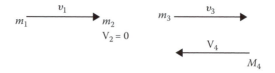

Figure 3.15 A case where the reaction products are emitted 180° apart.

Solving Eqs. 3.84 and 3.85 for T_3, one obtains a quadratic equation for T_3 (T_3 in MeV),

$$T_3^2 - 2.577T_3 + 1.482 = 0$$

which gives two values of T_3:

$$T_{3,1} = 1.710 \text{ MeV}, \qquad T_{3,2} = 0.866 \text{ MeV}$$

The corresponding values of T_4 are

$$T_{4,1} = 0.132 \text{ MeV}, \qquad T_{4,2} = 0.976 \text{ MeV}$$

The two pairs of values correspond to the alpha being emitted at $\theta = 0$ ($T_3 = 1.709$ MeV = max. kinetic energy) or $\theta = 180°$ ($T_2 = 0.865$ MeV = min. kinetic energy). Correspondingly, the boron nucleus is emitted at $\phi = 180°$ or $\phi = 0°$. One can use the momentum balance equation (Eq. 3.85) to verify this conclusion.

Case 2: $\theta = 90°$. In this case, the reaction looks as shown in Figure 3.16. The momentum vectors form a right triangle as shown on the right of Figure 3.16.

Therefore, Eqs. 3.81 and 3.82 take the form

$$T_1 + Q = T_3 + T_4 \tag{3.86}$$

$$m_1 T_1 + m_3 T_3 = M_4 T_4 \tag{3.87}$$

Again, one can solve for T_3 and T_4. The value of T_3 is

$$T_3 = \frac{(M_4 - m_1)T_1 + M_4 Q}{m_3 + M_4} \tag{3.88}$$

Example 3.13 What is the energy of the alpha particle in the reaction

$$_0^1 n + {}_7^{14}N \rightarrow {}_2^4 He + {}_5^{11}B$$

if it is emitted at 90°? Use $T_1 = 2$ MeV, the same as in Example 3.12.
Answer Using Eq. 3.88,

$$T_3 = \frac{(11-1)2 + 11(-0.163)}{15} \text{ MeV} = 1.214 \text{ MeV}$$

The value of the minimum (threshold) energy necessary to initiate a reaction can be understood with the help of Figure 3.17. When the particle m_1 enters the target nucleus M_2, a compound

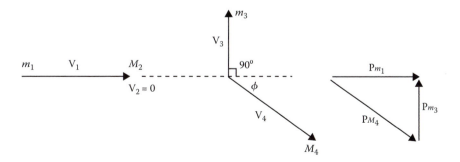

Figure 3.16 A case where the reaction products are emitted 90° apart.

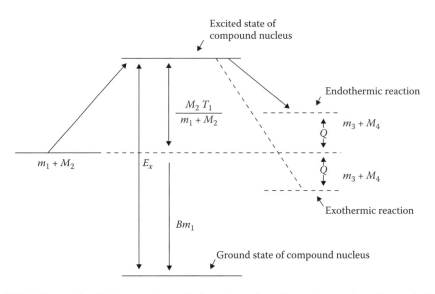

Figure 3.17 Energy-level diagram for endothermic and exothermic reactions. For endothermic reactions, the threshold energy is equal to $[(m_1 + M_2)/M_2]|Q|$.

nucleus is formed with excitation energy equal to

$$Bm_1 + \frac{M_2}{m_1 + M_2}T_1 \tag{3.89}$$

where Bm_1 = binding energy of particle m_1 and $M_2T_1/(m_1 + M_2)$ = part of the incident particle kinetic energy available as excitation energy of the compound nucleus.

Only a fraction of the kinetic energy T_1 is available as excitation energy, because the part

$$\left(\frac{m_1}{m_1 + M_2}T_1\right)$$

becomes kinetic energy of the compound nucleus (see Evans or any other book on nuclear physics), and as such is not available for excitation.

If the reaction is exothermic ($Q > 0$), it is energetically possible for the compound nucleus to deexcite by going to the state ($m_3 + M_4$) (Figure 3.17), even if $T_1 \approx 0$. For an endothermic reaction, however, energy at least equal to $|Q|$ should become available (from the kinetic energy of the projectile). Therefore, the kinetic energy T_1 should be such that

$$\frac{M_2}{m_1 + M_2}T_1 \geq |Q| \tag{3.90}$$

or the threshold kinetic energy for the reaction is

$$T_{1,\text{th}} = \frac{m_1 + M_2}{M_2}|Q| \tag{3.91}$$

3.10 FISSION

Fission is the reaction in which a heavy nucleus splits into two heavy fragments. In the fission process, net energy is released, because the heavy nucleus has less binding energy per nucleon that the fission fragments, which belong to the middle of the periodic table. In fact, for $A > 85$ the binding energy per nucleon decreases (Figure 3.3); therefore any nucleus with $A > 85$

would go to a more stable configuration by fissioning. Such "spontaneous" fission is possible but very improbable. Only very heavy nuclei ($Z > 92$) undergo spontaneous fission at a considerable rate.

For many heavy nuclei ($Z \geq 90$), fission takes place if an amount of energy at least equal to a critical energy E_c is provided in some way, as by neutron or gamma absorption. Consider, as an example, the nucleus ^{235}U (Figure 3.18). If a neutron with kinetic energy T_n is absorbed, the compound-nucleus ^{236}U has excitation energy equal to (Eq. 3.89)

$$B_n + \frac{A}{A+1}T_n$$

If $B_n + AT_n/(A+1) \geq E_c$, fission may occur and the final state is the one shown as fission products in Figure 3.18. For ^{236}U, $E_c = 5.3$ MeV and $B_n = 6.4$ MeV. Therefore, even a neutron with zero kinetic energy may induce fission, if it is absorbed. For ^{239}U, which is formed when a neutron is absorbed by ^{238}U, $B_n = 4.9$ MeV and $E_c = 5.5$ MeV. Therefore, fission cannot take place unless the neutron kinetic energy satisfies

$$T_n > \frac{A+1}{A}(E_c - B_n) = \frac{239}{238}(5.5 - 4.9) \approx 0.6 \text{ MeV}$$

The fission fragments are nuclei in extremely excited states with mass numbers in the middle of the periodic system. They have a positive charge of about $20e$ and they are neutron-rich. This happens because the heavy nuclei have a much higher neutron-proton ratio than nuclei in the middle of the periodic table.

Consider as an example ^{236}U (Figure 3.19). Assume that it splits into two fragments as follows:

$$Z_1 = 48, \quad N_1 = 80, \quad A_1 = 128$$
$$Z_2 = 44, \quad N_2 = 64, \quad A_2 = 108$$

The two fission fragments have a neutron-proton ratio higher than what stability requires for their atomic mass. They get rid of the extra neutron either by directly emitting neutrons or by β^- decay.

A nucleus does not always split in the same fashion. There is a probability that each fission fragment (A, Z) will be emitted, a process called *fission yield*. Figure 3.20 shows the fission yield for ^{235}U fission. For thermal neutrons, the "asymmetric" fission is favored. It can be shown that asymmetric fission yields more energy. As the neutron energy increases, the excitation energy of the compound nucleus increases. The possibilities for fission are such that it does not make much difference, from an energy point of view, whether the fission is symmetric or asymmetric. Therefore, the probability of symmetric fission increases.

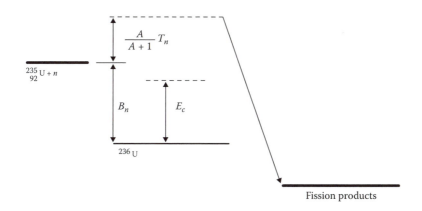

Figure 3.18 The fission of ^{235}U induced by neutron absorption.

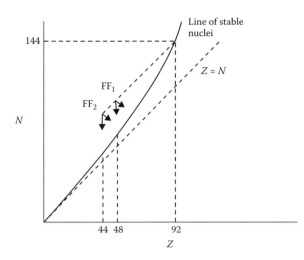

Figure 3.19 The fission fragments FF_1 and FF_2 from ^{236}U fission are neutron-rich. They reduce their neutron number either by beta decay or by neutron emission.

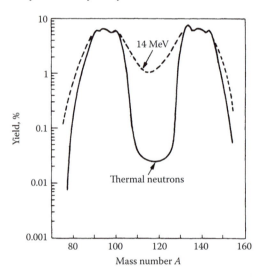

Figure 3.20 ^{235}U fission yield for thermal neutron induced fission. The dashed line indicates the yield when the fission is induced by 14-MeV neutrons. (Data from Dilorio, G. J., Direct physical measurement of mass yields in thermal fission of Uranium-235, PhD Thesis, University of Illinois at Urbana–Champaign, 1976.).

The fission fragments deexcite by emitting neutrons, betas, and gammas, and most of the fragments stay radioactive long after the fission takes place. The important characteristics of the particles emitted by fission fragments are as follows:

1. *Betas.* About six β^- particles are emitted per fission, carrying a total average energy of 7 MeV.

2. *Gammas.* About seven gammas are emitted at the time of fission. These are called *prompt gammas*. At later times, about seven to eight more gammas are released, called *delayed gammas*. Photons carry a total of about 15 MeV per fission.

3. *Neutrons.* The number of neutrons per fission caused by thermal neutrons is between two and three. This number increases linearly with the kinetic energy of the neutron inducing the fission. The average energy of a neutron emitted in fission is about 2 MeV. More than 99% of the neutrons are emitted at the time of fission and are called *prompt neutrons*. A very small fraction is emitted as *delayed neutrons*. Delayed neutrons are very important for the control of nuclear reactors.

Table 3.3
Fission Products

Particle	Number/Fission	MeV/Fission
Fission fragments	2	160–170
Neutrons	2–3	5
Gammas (prompt)	7	8
Gammas (delayed)	7	7
Betas	6	7
Neutrinos	6	11
Total		198–208

4. Neutrinos. About 11 MeV are taken away by neutrinos, which are also emitted during fission. This energy is the only part of the fission energy yield that completely escapes. It represents about 5% of the total fission energy.
   Table 3.3 summarizes the particles and energies involved in fission.*

PROBLEMS

3.1 What is the speed of a 10-MeV electron? What is its total mass, relative to its rest mass?

3.2 What is the speed of a proton with a total mass equal to $2Mc^2$? (M is the proton rest mass).

3.3 What is the kinetic energy of a neutron that will result in 1% error difference between relativistic and classical calculation of its speed?

3.4 What is the mass of an astronaut traveling with speed $v = 0.8c$? Mass at rest is 70 kg.

3.5 What is the kinetic energy of an alpha particle with a total mass 10% greater than its rest mass?

3.6 What would the density of graphite be if the atomic radius were 10^{-13} m? [Atomic radius (now) 10^{-10} m; density of graphite (now) 1600 kg/m³.]

3.7 Calculate the binding energy of the deuteron. [$M(^1\text{H}) = 1.007825$ u; $M(^2\text{H}) = 2.01410$ u.]

3.8 One of the most stable nuclei is ^{55}Mn. Its nuclidic mass is 54.938 amu. Determine its total binding energy and average binding energy per nucleon.

3.9 Calculate the separation energy of the last neutron of ^{241}Pu. [$M(^{240}\text{Pu}) = 240.053809$ u; $M(^{241}\text{Pu}) = 241.056847$ u.]

3.10 Assume that the average binding energy per nucleon (in some new galaxy) changes with A as shown in the following figure:

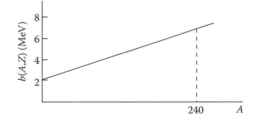

(a) Would fission or fusion or both release energy in such a world?

(b) How much energy would be released if a tritium (^3H) nucleus and a helium (^4He) nucleus combined to form a lithium nucleus? [$M(^3\text{H}) = 3.016050$ u; $M(^4\text{He}) = 4.002603$ u; $M(^7\text{Li}) = 7.016004$ u.]

---

* Tritium is sometimes produced in fission. In reactors fueled with ^{235}U, it is produced at the rate of 8.7×10^{-6} tritons per fission. The most probable kinetic energy of the tritons is about 7.5 MeV.

3.11 A simplified diagram of the ^{137}Cs decay is shown in the figure below. What is the recoil energy of the nucleus when the 0.6616-MeV gamma is emitted?

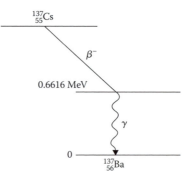

3.12 The isotope ^{239}Pu decays by alpha emissions to ^{235}U as shown in the following figure.

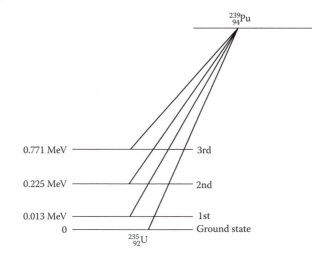

(a) What is Q_α?

(b) What is the kinetic energy of the alphas if the ^{235}U nucleus is left in the third excited state? What is the kinetic energy of the alphas if the ^{235}U nucleus is left in the ground state? [$M(^{235}$U$) = 235.043926$ u; $M(^{239}$Pu$) = 239.052159$ u]

3.13 Consider the isotopes ^{63}Zn and ^{63}Cu. Is β^+ decay possible? Is EC possible? What is Q_{β^+}? What is Q_{EC}? [$M(^{63}$Cu$) = 62.929597$ u: $M(^{63}$Zn$) = 62.933212$ u]

3.14 The Q value for the ^{3}He(n,p) reaction is 0.76 MeV. What is the nuclidic mass of ^{3}He?

3.15 ^{13}N decays by b^+ emission. The maximum kinetic energy of the b^+ is 1.20 MeV. What is the nuclidic mass of ^{13}N?

3.16 ^{14}C is believed to be made in the upper atmosphere by an (n,p) process on ^{14}N. What is Q for this reaction?

3.17 The isotope $^{11}_{4}$Be decays to $^{11}_{5}$B. What are the maximum and average kinetic energy of the betas? [$M(^{11}$Be$) = 11.021658$ u; $M(^{11}$B$) = 11.009306$ u]

3.18 Natural uranium contains the isotopes ^{234}U, ^{235}U, and ^{238}U, with abundances and half-lives as shown below:

	Half-Life (Years)	Abundance (%)
^{234}U	2.47×10^5	0.0057
^{235}U	7.10×10^8	0.71
^{238}U	4.51×10^9	99.284

(a) What is the alpha specific activity of natural uranium?

(b) What fraction of the activity is contributed by each isotope?

3.19 The isotope ^{210}Po generates 140,000 W/kg thermal power due to alpha decay. What is the energy of the alpha particle? ($T_\alpha = 138.4$ d)

3.20 How many years ago did the isotope ^{235}U make up 3% of natural uranium?

3.21 What is the specific alpha activity of ^{239}Pu? (For ^{239}Pu: $T_{sf} = 5.5 \times 10^{15}$ y, $T_{tot} = 2.44 \times 10^4$ y)

3.22 Consider the reaction $^{7}_{3}\text{Li}(p,n)^{7}_{4}\text{Be}$. What is the Q value for this reaction? If a neutron is emitted at 90° (in LS) with kinetic energy 2 MeV, what is the energy of the incident proton? [$M(^7\text{Li}) = 7.016004$ u; $M(^7\text{Be}) = 7.016929$ u]

3.23 What is the necessary minimum kinetic energy of a proton to make the reaction $^{4}_{2}\text{He}(p,d)^{3}_{3}\text{He}$ possible? (^{4}He at rest).

3.24 A 1-MeV neutron collides with a stationary $^{13}_{7}\text{N}$ nucleus. What is the maximum kinetic energy of the emerging proton?

3.25 What is the threshold gamma energy for the reaction

$$\gamma + {}^{12}_{6}\text{C} \rightarrow 3\left({}^{4}_{2}\text{He}\right)$$

3.26 What is the energy expected to be released as a result of a thermal neutron induced fission in ^{239}Pu if the two fission fragments have masses $M_1 = 142$ u and $M_2 = 95$ u?

BIBLIOGRAPHY

Evans, R. D., *The Atomic Nucleus*, McGraw-Hill, New York, 1955.

Heyde, K., *Basic Ideas and Concepts in Nucelar Physics: An Introductory Approach*, 3rd ed., Institute of Physics Publishing, Bristol, UK, 2004.

Jevremovic, T., *Nuclear Principles in Engineering*, Springer, New York, NY, 2005.

Loveland, W., Morrisey, D. J., and Seaborg, G. T., *Modern Nuclear Chemistry*, Wiley, Hoboken, NJ, 2005.

Mayo, R. M., *Nuclear Concepts for Engineers*, American Nuclear Society, La Grange Park, IL, 1998.

Shultis, J. K., and Faw, R. E., *Fundamentals of Nuclear Science and Engineering*, Marcel Dekker, New York, NY, 2002.

REFERENCES

1. Stamatelatos, M. G., and England, T. R., *Nucl. Sci. Eng.* **63**:304 (1977).

2. Lederer, C. M., and Shirley, V. S. (eds.), *Table of Isotopes*, 7th ed., Wiley-Interscience, New York, 1978.

   (NB: Table of Isotopes has been updated Firestone, R. B., Coral M. Baglin, C. M. Chu, S. Y. F, *Table of Isotopes,* 8th ed., John Wiley & Sons, 1999.)

3. Bateman, H., *Proc. Cambridge Philos. Soc.* **15**:423 (1910).

4. Dilorio, G. J., "Direct Physical Measurement of Mass Yields in Thermal Fission of Uranium-235", Ph.D., University of Illinois at Urbana-Champaign, 1976.

4

Energy Loss and Penetration of Radiation through Matter

4.1 INTRODUCTION

This chapter discusses the mechanisms by which ionizing radiation interacts and loses energy as it moves through matter. The study of this subject is extremely important for radiation measurements because the detection of radiation is based on its interactions and the energy deposited in the material of which the detector is made. Therefore, to be able to build detectors and interpret the results of the measurement, we need to know how radiation interacts and what the consequences are of the various interactions.

The topics presented here should be considered only an introduction to this extensive subject. Emphasis is given to that material considered important for radiation measurements. The range of energies considered is shown in Table 1.1.

For the discussion that follows, ionizing radiation is divided into three groups:

1. Charged particles: electrons (e^-), positrons (e^+), protons (p), deuterons (d), alphas (α), heavy ions ($A > 4$)

2. Photons: gammas (γ) or X-rays

3. Neutrons (n)

The division into three groups is convenient because each group has its own characteristic properties and can be studied separately.

A charged particle moving through a material interacts, primarily, through Coulomb forces, with the negative electrons and the positive nuclei that constitute the atoms of that material. As a result of these interactions, the charged particle loses energy continuously and finally stops after traversing a finite distance, called the *range*. The range depends on the type and energy of the particle and on the material through which the particle moves. The probability of a charged particle going through a piece of material without an interaction is practically zero. This fact is very important for the operation of charged-particle detectors.

Neutrons and gammas have no charge. They interact with matter in ways that will be discussed below, but there is a finite nonzero probability that a neutron or a γ-ray may go through any thickness of any material without having an interaction. As a result, no finite range can be defined for neutrons or gammas.

4.2 MECHANISMS OF CHARGED PARTICLE ENERGY LOSS

Charged particles traveling through matter lose energy in the following ways:

1. In Coulomb interactions with electrons and nuclei

2. By emission of electromagnetic radiation (bremsstrahlung)

3. In nuclear interactions

4. By emission of Cerenkov radiation.

For charged particles with kinetic energies considered here, nuclear interactions may be neglected, except for heavy ions ($A > 4$) (see Section 4.7).

Cerenkov radiation constitutes a very small fraction of the energy loss. It is important only because it has a particle application in the operation of Cerenkov detectors (see Evans, 1972). Cerenkov radiation is visible electromagnetic radiation emitted by particles traveling in a medium, with speed greater than the speed of light in that medium.

4.2.1 Coulomb Interactions

Consider a charged particle traveling through a certain material, and consider an atom of that material. As shown in Figure 4.1, the fast charged particle may interact with the atomic electrons or the nucleus of the atom. Since the radius of the nucleus is approximately 10^{-14} m and the radius

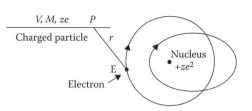

Figure 4.1 A fast charged particle of mass M and charge ze interacts with the electrons of an atom.

of the atom is 10^{-10} m, one might expect that

$$\frac{\text{Number of interactions with electrons}}{\text{number of interactions with nuclei}} = \frac{(R^2)\ \text{atom}}{(R^2)\ \text{nucleus}} = \frac{(10^{-10})^2}{(10^{-14})^2} \approx 10^8$$

This simplified argument indicates that collisions with atomic electrons are more important than those with nuclei. Nuclear collisions will not be considered here.

Looking at Figure 4.1, at a certain point in time the particle is at point P and the electron at E. If the distance between them is r, the coulomb force is $F = k(ze^2/r^2)$, where ze is the charge of the particle and k is a constant that depends on the units. The action of this force on the electron, over a period, may result in the transfer of energy from the moving charged particle to the bound electron. Since a bound atomic electron is in a quantized state, the result of the passage of the charged particle may be ionization or excitation.

Ionization occurs when the electron obtains enough energy to leave the atom and become a free particle with kinetic energy equal to

$$(KE)_e = (\text{energy given by particle}) - (\text{ionization potential})$$

The electron freed from the atom acts like any other moving charged particle. It may cause ionization of another atom if its energy is high enough. It will interact with matter, lose its kinetic energy, and finally stop. Fast electrons produced by ionizing collisions are called δ rays.

The ionization leaves behind a positive ion, which is a massive particle compared to an electron. If an ion and an electron move in a gas, the ion will move much slower than the electron. Eventually, the ion will pick up an electron from somewhere and will become a neutral atom again.

Excitation takes place when the electron acquires enough energy to move to an empty state in another orbit of higher energy. The electron is still bound, but it has moved from a state with energy E_1 to one with E_2, thus producing an excited atom. In a short period of time, of the order of 10^{-8}–10^{-10} s, the electron will move to a lower energy state, provided there is one empty. If the electron falls from E_2 to E_1, the energy $E_2 - E_1$ is emitted in the form of an X-ray with frequency $v = (E_2 - E_1)/h$.

Collisions that result in ionization or excitation are called *inelastic collisions*. A charged particle moving through matter may also have elastic collisions with nuclei or atomic electrons. In such a case, the incident particle loses the energy required for conservation of kinetic energy and linear momentum. Elastic collisions are not important for charged-particle energy loss and detection.

4.2.2 Emission of Electromagnetic Radiation (Bremsstrahlung)

Every free charged particle that accelerates or decelerates loses part of its kinetic energy by emitting electromagnetic radiation. This radiation is called *bremsstrahlung*, which in German means braking radiation. Bremsstrahlung is not a monoenergetic radiation. It consists of photons with energies from zero up to a maximum equal to the kinetic energy of the particle.

Emission of bremsstrahlung is predicted not only by quantum mechanics but also by classical physics. Theory predicts that a charge that is accelerated radiates energy with intensity proportional to the square of its acceleration. Consider a charged particle with charge ze and mass M moving in a certain material of atomic number Z. The Coulomb force between the particle and a nucleus of the material is $F \sim zeZe/r^2$, where $r = $ distance between the two charges. The acceleration of the incident charged particle is $a = F/M \sim zZe^2/M$. Therefore the intensity of the emitted radiation I is

$$I \propto a^2 \sim \left(\frac{zZe^2}{M}\right)^2 \sim \frac{z^2Z^2}{M^2} \tag{4.1}$$

This expression indicates the following:

1. For two particles traveling in the same medium, the lighter particle will emit a much greater amount of bremsstrahlung than the heavier particle (other things being equal).

2. More bremsstrahlung is emitted if a particle travels in a medium with high atomic number Z than in one with low atomic number.

For charged particles with energies considered here, the kinetic energy lost as bremsstrahlung might be important for electrons only. Even for electrons, it is important for high-Z materials like lead ($Z = 82$). For detailed treatment of the emission of bremsstrahlung, the reader should consult the references listed at the end of the chapter.

4.3 STOPPING POWER DUE TO IONIZATION AND EXCITATION

A charged particle moving through a material exerts Coulomb forces on many atoms simultaneously. Every atom has many electrons with different ionization and excitation potentials. As a result of this, the moving charged particle interacts with a tremendous number of electrons—millions. Each interaction has its own probability for occurrence and for a certain energy loss. It is impossible to calculate the energy loss by studying individual collisions. Instead, an average energy loss is calculated per unit distance traveled. The calculation is slightly different for electrons or positrons than for heavier charged particles like p, d, and α, for the following reason.

It was mentioned earlier that most of the interactions of a charged particle involve the particle and atomic electrons. If the mass of the electron is taken as 1, then the masses of the other common heavy* charged particles are the following:

Electron mass = 1
Proton mass ≈ 1840
Deuteron mass ≈ 2(1840)
Alpha mass ~ 4(1840).

If the incoming charged particle is an electron or a positron, it may collide with an atomic electron and lose all its energy in a single collision because the collision involves two particles of the same mass. Hence, incident electrons or positrons may lose a large fraction of their kinetic energy in one collision. They may also be easily scattered to large angles, as a result of which their trajectory is zigzag (Figure 4.2). Heavy charged particles, on the other hand, behave differently. On the average, they lose smaller amounts of energy per collision. They are hardly deflected by atomic electrons, and their trajectory is almost a straight line.

Assuming that all the atoms and their atomic electrons act independently, and considering only energy lost to excitation and ionization, the average energy loss[†] per unit distance traveled by the particle is given by Eqs. 4.2, 4.3, and 4.4. (For their derivation, see Evans, 1972; Segré, 1968; Roy and Reed, 1968.)

Electron or positron trajectory

Heavy particle trajectory

Figure 4.2 Possible electron and heavy particle trajectories.

---

* In this discussion, "heavy" particles are all charged particles except electrons and positrons.
† Since $E = T + Mc^2$ and Mc^2 = constant, $dE/dx = dT/dx$; thus, Eqs. 4.2–4.4 express the kinetic as well as the total energy loss per unit distance.

Stopping power due to ionization–excitation for p, d, t, α:

$$\frac{dE}{dx}(\text{MeV/m})^{\ddagger} = 4\pi r_0^2 z^2 \frac{mc^2}{\beta^2} NZ \left[\ln\left(\frac{2mc^2}{I}\beta^2\gamma^2 \right) - \beta^2 \right] \tag{4.2}$$

Stopping power due to ionization–excitation for electrons:

$$\frac{dE}{dx}(\text{MeV/m}) = 4\pi\gamma r_0^2 \frac{mc^2}{\beta^2} NZ \left\{ \ln\left(\frac{\beta\gamma\sqrt{\gamma-1}}{I}mc^2 \right) + \frac{1}{2\gamma^2}\left[\frac{(\gamma-1)^2}{8} + 1 - (\gamma^2 + 2\gamma - 1)\ln 2 \right] \right\} \tag{4.3}$$

Stopping power due to ionization–excitation for positrons:

$$\frac{dE}{dx}(\text{MeV/m}) = 4\pi r_0^2 \frac{mc^2}{\beta^2} NZ \left\{ \ln\left(\frac{\beta\gamma\sqrt{\gamma-1}}{I}mc^2 \right) \right.$$
$$\left. - \frac{\beta^2}{24}\left[23 + \frac{14}{\gamma+1} + \frac{10}{(\gamma+1)^2} + \frac{4}{(\gamma+1)^3} \right] + \frac{\ln 2}{2} \right\} \tag{4.4}$$

where $r_0 = e^2/mc^2 = 2.818 \times 10^{-15}$ m = classical electron radius

$$4\pi r_0^2 = 9.98 \times 10^{-29} \text{ m}^2 \approx 10^{-28} \text{ m}^2 = 10^{-24} \text{ cm}^2$$

mc^2 = rest mass energy of the electron = 0.511 MeV

$$\gamma = (T + Mc^2)/Mc^2 = 1/\sqrt{1-\beta^2}$$

T = kinetic energy = $(\gamma - 1)Mc^2$
M = rest mass of the particle
$\beta = v/c$ c = speed of light in vacuum = 2.997930×10^8 m/s $\approx 3 \times 10^8$ m/s
N = number of atoms/m^3 in the material through which the particle moves
$N = \rho(N_A/A)$ N_A = Avogadro's number = 6.022×10^{23} atoms/mol A = atomic weight
Z = atomic number of the material
z = charge of the incident particle ($z = 1$ for e^-, e^+, p, d; $z = 2$ for α)
I = mean excitation potential of the material

An approximate equation for I, which gives good results for $Z > 12$,[1] is

$$I(\text{eV}) = (9.76 + 58.8Z^{-1.19})\, Z \tag{4.5}$$

Table 4.1 gives values of I for many common elements.

Many different names have been used for the quantity dE/dx: names like energy loss, specific energy loss, differential energy loss, or stopping power. In this text, the term *stopping power* will be used for dE/dx given by Eq. 4.2 to 4.4, as well as for a similar equation for heavier charged particles presented in Section 4.7.2.

It should be noted that the stopping power

1. Is independent of the mass of the particle

2. Is proportional to z^2 [(charge)^2] of particle

3. Depends on the speed v of particle

4. Is proportional to the density of the material (N).

For low kinetic energies, dE/dx is almost proportional to $1/v^2$. For relativistic energies, the term in brackets predominates and dE/dx increases with kinetic energy. Figure 4.3 shows the general behavior of dE/dx as a function of kinetic energy. For all particles, dE/dx exhibits a minimum

---

‡ In SI units, the result would be J/m; 1 MeV = 1.602×10^{-13} J.

Table 4.1
Values of Mean Excitation Potentials for Common Elements and Compounds[a]

Element	I (eV)	Element	I (eV)
H	20.4	Fe	281*
He	38.5	Ni	303*
Li	57.2	Cu	321*
Be	65.2	Ge	280.6
B	70.3	Zr	380.9
C	73.8	I	491
N	97.8	Cs	488
O	115.7	Ag	469*
Na	149	Au	771*
Al	160*	Pb	818.8
Si	174.5	U	839*

[a] Values of I with * are from experimental results of Refs. 2 and 3. Others are from Refs. 4 and 5.

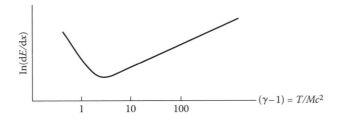

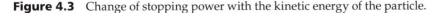

Figure 4.3 Change of stopping power with the kinetic energy of the particle.

that occurs approximately at $\gamma \approx 3$. For electrons, $\gamma = 3$ corresponds to $T = 1$ MeV; for alphas, $\gamma = 3$ corresponds to $T \simeq 7452$ MeV; for protons, $\gamma = 3$ corresponds to $T \cong 1876$ MeV. Therefore, for the energies considered here (see Table 1.1), the dE/dx for protons and alphas will always increase, as the kinetic energy of the particle decreases (Figure 4.3, always on the left of the curve minimum); for electrons, depending on the initial kinetic energy, dE/dx may increase or decrease as the electron slows down.

Equations 4.3 and 4.4, giving the stopping power for electrons and positrons, respectively, are essentially the same. Their difference is due to the second term in the bracket, which is always much smaller than the logarithmic term. For an electron and positron with the same kinetic energy, Eqs. 4.3 and 4.4 provide results that are different by about 10% or less. For low kinetic energies, dE/dx for positrons is larger than that for electrons; at about 2000 keV, the energy loss is the same; for higher kinetic energies, dE/dx for positron is less than that for electrons.

As stated earlier, Eqs. 4.2–4.4 disregard the effect of forces between atoms and atomic electrons of the attenuating medium. A correction for this *density effect*[6,7] has been made, but it is small and it will be neglected here. The density effect reduces the stopping power slightly.

Equations 4.2–4.4 are not valid for very low energies. In the case of Eq. 4.2, a nuclear shell correction is applied (see Ziegler et al., 2008), which appears in the brackets as a negative term and becomes important at low energies ($T \le 100$ keV). Even without this correction, the value in brackets takes a negative value when $(2mc^2\beta^2\gamma^2)/I \le 1$. The value of this term depends on the medium because of the presence of the ionization potential I. As an example, for oxygen ($I = 89$ eV) this term becomes less than 1 for $T < 40$ keV.

For electrons of very low kinetic energy, Eq. 4.3, takes the form (see Roy and Reed, 1968)

$$\frac{dE}{dX} \approx 4\pi r_0^2 \frac{mc^2}{\beta^2} NZ \ln\left(\frac{mc^2\beta^2}{I} \sqrt{\frac{2.7182}{8}} \right), \quad |\beta \ll 1| \tag{4.5a}$$

Again for oxygen, the argument of the logarithm becomes less than 1 for electron kinetic energy $T < 76$ eV. For positrons, the low-energy limit of the validity of Eq. 4.4 is equal to the positron energy for which the whole value within brackets is less than zero.

Example 4.1   What is the stopping power for a 5-MeV alpha particle moving in silicon?
Answer   For silicon, $A = 28$, $Z = 14$, $\rho = 2.33$ kg/m³,

$$\gamma = \frac{5 + 4(931.5)}{4(931.5)} = 1.00134, \quad \beta^2 = 1 - \frac{1}{\gamma^2} = 0.00268$$

$$\frac{dE}{dx} = 10^{-28} \times 2^2 \frac{0.511}{0.00268} 2.33 \times 10^3 \frac{0.6022 \times 10^{24}}{28 \times 10^{-3}}$$

$$\times 14 \left(\ln \frac{2(0.511)(0.00268)(1.00134)^2}{172 \times 10^{-6}} - 0.00268 \right)$$

$$= 1.48 \times 10^5 \text{ MeV/m} = 2.37 \times 10^{-8} \text{ J/m} = 0.148 \text{ MeV/}\mu\text{m}$$

Or, in terms of MeV/(g/cm²),

$$\frac{dE}{dx} = 1.48 \times 10^5 \text{ MeV/m} = 1480 \text{ MeV/cm/}(2.33\text{g/cm}^3)$$

$$= 635.2 \text{ MeV/}(g/cm^2)$$

Example 4.2   What is the stopping power for a 5-MeV electron moving in silicon?
Answer   For an electron,

$$\gamma = \frac{5 + 0.511}{0.511} = 10.785, \quad \beta = \sqrt{1 - \frac{1}{\gamma^2}} = 0.9957, \quad \beta^2 = 0.9914$$

$$\frac{dE}{dx} = 10^{-28} \frac{0.511}{0.9914} 2.33 \times 10^3 \frac{0.6022 \times 10^{24}}{28 \times 10^{-3}}$$

$$\times 14 \left[\ln \frac{0.9957(10.785)\sqrt{9.785}(0.511)}{172 \times 10^{-6}} + \frac{1}{2 \times 10.785^2} \right.$$

$$\left. \times \left(\frac{9.785^2}{8} + 1 - (10.785^2 + 2 \times 10.785 - 1) \ln 2 \right) \right]$$

$$= 403.5 \text{ MeV/m} = 4.035 \text{ MeV/cm} = 6.46 \times 10^{-11} \text{ J/m}$$

In Example 4.2, the stopping power for the 5-MeV electron is, in terms of MeV/(g/cm²),

$$\frac{4.035 \text{ MeV/cm}}{2.33 \text{ g/cm}^3} = 1.73 \text{ MeV/}(g/cm^2)$$

Notice the huge difference in the value of stopping power for an alpha versus an electron of the same kinetic energy traversing the same material.

Tables of dE/dx values are usually given in units of MeV/(g/cm²) [or in SI units of J/(kg/m²)]. The advantage of giving the stopping power in these units is the elimination of the need to define the density of the stopping medium that is necessary, particularly for gases. The following simple equation gives the relationship between the two types of units:

$$\frac{1}{\rho(g/cm^3)} \frac{dE}{dx} (\text{MeV/cm}) = \frac{dE}{dx} \left[\text{MeV/}(g/cm^2) \right] \tag{4.6}$$

4.4 ENERGY LOSS DUE TO BREMSSTRAHLUNG EMISSION

The calculation of energy loss due to emission of bremsstrahlung is more involved than the calculation of energy loss due to ionization and excitation. Here, an approximate equation will be given for electrons or positrons only, because it is for these particles that energy loss due to emission of radiation may be important.

For electrons or positrons with kinetic energy T(MeV) moving in a material with atomic number Z, the energy loss due to bremsstrahlung emission, $(dE/dx)_{rad}$, is given in terms of the ionization and excitation energy loss by Eq. 4.7 (see Evans, 1972).

$$\left(\frac{dE}{dx}\right)_{rad} = \frac{ZT\,(\text{MeV})}{750}\left(\frac{dE}{dx}\right)_{ion}$$
(4.7)

where $(dE/dx)_{ion}$ is the stopping power due to ionization–excitation (Eq. 4.3 or 4.4).

Example 4.3 Consider an electron with $T = 5$ MeV. What fraction of its energy is lost as bremsstrahlung as it starts moving (a) in aluminum and (b) in lead?
Answer (a) If it travels in aluminum ($Z = 13$),

$$\left(\frac{dE}{dx}\right)_{rad} = \frac{13(5)}{750}\left(\frac{dE}{dx}\right)_{ion} = 0.09\left(\frac{dE}{dx}\right)_{ion}$$

That is, the rate of energy loss due to radiation is about 9% of $(dE/dx)_{ion}$.
(b) For the same electron moving in lead ($Z = 82$),

$$\left(\frac{dE}{dx}\right)_{rad} = \frac{82(5)}{750}\left(\frac{dE}{dx}\right)_{ion} = 0.55\left(\frac{dE}{dx}\right)_{ion}$$

In this case, the rate of radiation energy loss is 55% of $(dE/dx)_{ion}$.

Equation 4.7, relating radiation to ionization energy loss, is a function of the kinetic energy of the particle. As the particle slows down, T decreases and $(dE/dx)_{rad}$ also decreases. The total energy radiated as bremsstrahlung is approximately equal in MeV to*

$$T_{rad} = 4.0 \times 10^{-4} ZT^2$$
(4.8)

Example 4.4 What is the total energy radiated by the electron of Ex. 4.3?
Answer Using Eq. 4.8,

(a) In aluminum: $T_{rad} = (4.0 \times 10^{-4})(13)5^2 = 0.130$ MeV
(b) In lead: $T_{rad} = (4.0 \times 10^{-4})(82)5^2 = 0.820$ MeV

The total stopping power for electrons or positrons is given by the sum of Eqs. 4.3 or 4.4 and 4.7:

$$\left(\frac{dE}{dx}\right)_{tot} = \left(\frac{dE}{dx}\right)_{ion} + \left(\frac{dE}{dx}\right)_{rad} = \left(1 + \frac{ZT}{750}\right)\left(\frac{dE}{dx}\right)_{ion}$$
(4.9)

If the particle moves in a compound or a mixture, instead of a pure element, an effective atomic number Z_{ef} should be used in Eqs. 4.7 and 4.8. The value of Z_{ef} is given by

$$Z_{ef} = \frac{\sum_{i=1}^{L}(w_i/A_i)Z_i^2}{\sum_{i=1}^{L}(w_i/A_i)Z_i}$$
(4.10)

---

* The coefficient 4.0×10^{-4} used in Eq. 4.8 is not universally accepted (see Evans, 1972).

where L = number of elements in the compound or mixture
$\quad w_i$ = weight fraction of ith element
$\quad A_i$ = atomic weight of ith element
$\quad Z_i$ = atomic number of ith element

For a compound with molecular weight M, the weight fraction is given by

$$w_t = \frac{N_i A_i}{M} \tag{4.11}$$

where N_i is the number of atoms of the ith element in the compound.

4.5 CALCULATION OF dE/dx FOR A COMPOUND OR MIXTURE

Equations 4.2 to 4.4 give the result of the stopping power calculation if the particle moves in a pure element. If the particle travels in a compound or a mixture of several elements, the stopping power is given by

$$\left(\frac{1}{\rho}\frac{dE}{dx}\right)_{compound} = \sum_i w_i \frac{1}{\rho_i}\left(\frac{dE}{dx}\right)_i \tag{4.12}$$

where ρ = density of compound or mixture
$\quad \rho_i$ = density of the ith element
$1/\rho_i (dE/dx)_i$ = stopping power in MeV/(kg/m^2) for the ith element, as calculated using Eqs. 4.2 to 4.4 and 4.6.

Example 4.5 What is the stopping power for a 10-MeV electron moving in air? Assume that air consists of 21% oxygen and 79% nitrogen.
Answer Equation 4.12 will be used, but first dE/dx will have to be calculated for the two pure gases. Using Eq. 4.3,

$$\gamma = \frac{T+mc}{mc^2} = \frac{10+0.511}{0.511} = 20.569$$

$$\beta = \sqrt{\frac{\gamma^2-1}{\gamma^2}} = 0.9988, \quad \beta = 0.9976$$

For oxygen,

$$\frac{1}{\rho}\left(\frac{dE}{dx}\right)_o = 10^{-28}\left(\frac{0.511}{0.9976}\right)\frac{0.6022\times10^{24}}{16\times10^{-3}}(8)$$

$$\times\left[\ln\frac{0.9988(20.569)\sqrt{19.569}(0.511)}{115.7\times10^{-6}}\right.$$

$$\left.+\frac{1}{2(20.569)^2}\left(\frac{19.569^2}{8}+1-\left[(20.569)^2+2(20.569)-1\right]\ln 2\right)\right]$$

$$= 0.194 \text{ MeV/}(kg/m)^2$$

$$= 3.10\times10^{-14} \text{ J/}(kg/m)^2 = 1.94 \text{ MeV/}(g/cm)^2$$

For nitrogen,

$$\frac{1}{\rho}\left(\frac{dE}{dx}\right)_N = 10^{-28}\left(\frac{0.511}{0.9976}\right)\frac{0.6022\times10^{24}}{14\times10^{-3}}(7)$$

$$\times\left[\ln\frac{0.9988(20.569)\sqrt{19.569}(0.511)}{97.8\times10^{-6}}+\frac{1}{2(20.569)^2}\right.$$

$$\left.\times\left(\frac{19.569^2}{8}+1-\left[\frac{19.569^2}{8}(20.569)^2+2(20.569)-1\right]\ln 2\right)\right]$$

$$= 0.194 \text{ MeV/}(kg/m^2) = 3.14\times10^{-14} \text{ J/}(kg/m^2) = 1.94 \text{ MeV/}(g/cm^2)$$

For air,

$$\left(\frac{1}{\rho}\frac{dE}{dx}\right)_{air} = 0.21\left(\frac{1}{\rho}\frac{dE}{dx}\right)_{O} + 0.79\left(\frac{1}{\rho}\frac{dE}{dx}\right)_{N}$$

$$= \left[0.21(3.10\times10^{-14}) + 0.79(3.10\times10^{-14})\right] J/(kg/m^2)$$

$$= 3.14\times10^{-14} \ J/(kg/m^2) = 1.96 \ MeV/(g/cm^2)$$

$$\left(\frac{dE}{dx}\right)_{air} = 3.14\times10^{-14} \ J/(kg/m^2) \ (1.29 \ kg/m^3)$$

$$= 4.05\times10^{-14} \ J/m = 0.253 \ MeV/m$$

4.6 RANGE OF CHARGED PARTICLES

A charged particle moving through a certain material loses its kinetic energy through interactions with the electrons and nuclei of the material. Eventually, the particle will stop, pick up the necessary number of electrons from the surrounding matter, and become neutral. For example,

$$p^+ + e^- \rightarrow \text{hydrogen atom}$$

$$\alpha^{2+} + 2e^- \rightarrow \text{He atom}$$

The total distance traveled by the particle is called the *path length*. The path length S, shown in Figure 4.4, is equal to the sum of all the partial path lengths S_i. The thickness of material that just stops a particle of kinetic energy T, mass M, and charge z is called the *range R* of the particle in that material. It is obvious that $R \leq S$. For electrons, which have a zigzag path, $R < S$. For heavy charged particles, which are very slightly deflected, $R \approx S$.

Range is distance, and its basic dimension is length (m). In addition to meters, another common unit used for range is kg/m² (or g/cm²). The relationship between the two is

$$R \ (kg/m^2) = [R(m)]\left[\rho \ (kg/m^3)\right] \tag{4.13}$$

where ρ is the density of the material in which the particle travels. The range measured in kg/m² is independent of the state of matter. That is, a particle will have the same range in kg/m² whether it moves in ice, water, or steam. Of course, the range measured in meters will be different.

The range is an average quantity. Particles of the same type with the same kinetic energy moving in the same medium will not stop after traveling exactly the same thickness R. Their path length will not be the same either. What actually happens is that the end points of the path lengths will be distributed around an average thickness called the range. To make this point more clear, two experiments will be discussed dealing with transmission of charged particles. Heavy particles and electrons–positrons will be treated separately. There are some minor differences for range equations between electrons and positrons. However, no different equations are

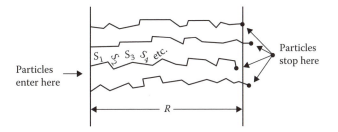

Figure 4.4 Pathlength (S) and range (R). The end points of the pathlengths are distributed around an average thickness that is the range.

presented here because we deal mostly with electrons and not positrons and the differences are small enough to be unimportant for the applications considered here

4.6.1 Range of Heavy Charged Particles (p, d, t, α; $1 \le A \le 4$)

Consider a parallel beam of heavy charged particles all having the same energy and impinging upon a certain material (Figure 4.5). The thickness of the material may be changed at will. On the other side of the material, a detector records the particles that traverse it. It is assumed that the particle direction does not change and that the detector will record all particles that go through the material, no matter how low their energy is. The number of particles $N(t)$ traversing the thickness t changes, as shown in Figure 4.6.

In the beginning, $N(t)$ stays constant, even though t changes. Beyond a certain thickness, $N(t)$ starts decreasing and eventually goes to zero. The thickness for which $N(t)$ drops to half its initial value is called the *mean range R*. The thickness for which $N(t)$ is practically zero is called the *extrapolated range R_e*. The difference between R and R_e is about 5% or less. Unless otherwise specified, when range is used, it is the mean range R.

Semiempirical formulas have been developed that give the range as a function of particle kinetic energy. For alpha particles, the range in air at normal temperature and pressure is given by

$$R\,(\text{mm}) = \exp\left[1.61\sqrt{T\,(\text{MeV})}\right], \qquad 1 < T \le 4\ \text{MeV}$$
$$R\,(\text{mm}) = (0.05T + 2.85)T^{3/2}\ \text{MeV}, \quad 4 < T \le 15\ \text{MeV} \tag{4.14}$$

where T is the kinetic energy of the particle in MeV. Figure 4.7 gives the range of alphas in silicon.

If the range is known for one material, it can be determined for any other by applying the Bragg–Kleeman rule*:

$$\frac{R_1}{R_2} = \frac{\rho_2}{\rho_1}\sqrt{\frac{A_1}{A_2}} \tag{4.15}$$

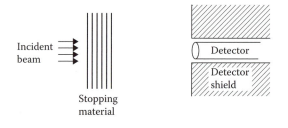

Figure 4.5   Particle transmission experiment.

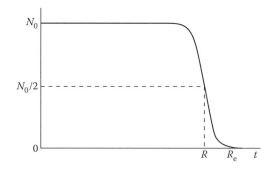

Figure 4.6   The number of heavy charged particles (α, p, d, t) transmitted through thickness t.

---

* The Bragg–Kleeman rule does not hold for electron or positron ranges.

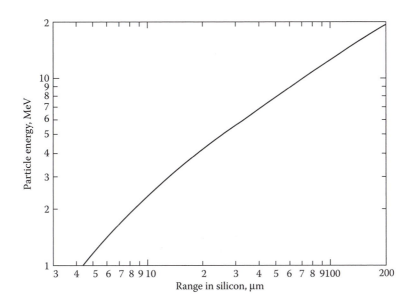

Figure 4.7 Range–energy curve for alpha particles in silicon (Reference 8).

where ρ_i and A_i are the density and atomic weight, respectively, of material i. For a compound or mixture, an effective molecular weight is used, obtained from the equation

$$\sqrt{A_{\text{ef}}} = \left(\sum_{i-1}^{L} \frac{w_i}{\sqrt{A_i}} \right)^{-1} \tag{4.16}$$

where the quantities w_i A_i and L have the same meaning as in Eq. 4.10.

Example 4.6 What is the effective molecular weight for water? What is it for air?
Answer For H_2O (11% H, 89% O),

$$\sqrt{A_{\text{ef}}} = \left(\frac{0.11}{\sqrt{1}} + \frac{0.89}{\sqrt{16}} \right)^{-1} = 3, \quad A_{\text{ef}} = 9$$

For air (22.9% O, 74.5% N, 2.6% Ar),

$$\sqrt{A_{\text{ef}}} = \left(\frac{0.229}{\sqrt{16}} + \frac{0.745}{\sqrt{14}} + \frac{0.026}{\sqrt{40}} \right)^{-1} = 3.84, \quad A_{\text{ef}} = 14.74$$

Using the Bragg–Kleeman rule (Eq. 4.15), with air at normal temperature and pressure as one of the materials ($\rho = 1.29$ kg/m³, $\sqrt{A_{\text{ef}}} = 3.84$), one obtains

$$R \text{ (mm)} = (3.36 \times 10^{-1}) \frac{\sqrt{A_{\text{ef}}}}{r \text{ (kg/m}^3)} R_{\text{air}} \text{ (mm)} \tag{4.17}$$

There are two ways to obtain the range of alphas in a material other than air and silicon:

1. The range in air should be obtained first, using Eq. 4.14, and then the range in the material of interest should be calculated using Eq. 4.17.
2. The range in silicon could be read from Figure 4.7, and then the range in the material of interest should be calculated using Eq. 4.15.

Example 4.7   What is the range of a 3-MeV alpha particle in gold?
Answer   The range of this alpha in silicon is (Figure 4.7) $R = 12.5\ Mm = 12.5 \times 10^{-6}$ m. Using Eq. 4.15, the range in gold is

$$R_{Au} = (12.5 \times 10^{-6}) \frac{2.33 \times 10^3}{19.32 \times 10^3} \sqrt{\frac{197}{28}} = 10^{-6}\ \text{m} = 4\ \mu\text{m}$$

Or, using Eqs. 4.14 and 4.17,

$$R_{Au} = (3.2 \times 10^{-1}) \frac{\sqrt{197}}{19.32 \times 10^3} \exp(1.61\sqrt{3}) = 3.8 \times 10^{-3}\ \text{mm} = 3.8\ \mu\text{m}$$

Example 4.8   What is the range of a 10-MeV alpha particle in aluminum?
Answer   From Figure 4.7, the range in silicon is $R = 72\ \mu\text{m} = 7.2 \times 10^{-5}\ \mu\text{m}$. Using Eq. 4.15, the range in aluminum is

$$R_{Ai} = (72\ \mu\text{m}) \frac{2.33 \times 10^3}{19.32 \times 10^3} \sqrt{\frac{27}{28}} = 60.7\ \mu\text{m}$$

Or, using Eqs. 4.14 and 4.17,

$$R_{Ai} = (3.36 \times 10^{-1}) \frac{\sqrt{27}}{2.7 \times 10^3} [0.05(10) + 2.85]10^{3/2}\ \text{mm}$$

$$= 6.85 \times 10^{-5}\ \text{m} = 68.5\ \mu\text{m}$$

The difference of 8 µm is within the range of accuracy of the Bragg–Kleeman rule and the ability to read a log–log graph.

The range of protons in *aluminum* has been measured by Bichsel.[9] His results are represented very well by the following two equations:

$$R\ (\mu\text{m}) = 14.217^{-1.5874}, \quad 1\ \text{MeV} < T \leq 2.7\ \text{MeV} \tag{4.18}$$

$$R\ (\mu\text{m}) = 10.5 \frac{T^2}{0.68 + 0.434 \ln T}, \quad 2.7\ \text{MeV} \leq T \leq 20\ \text{MeV} \tag{4.19}$$

For other materials, Eq. 4.15 should be used after the range in aluminum is determined from Eqs. 4.18 and 4.19. A very comprehensive paper dealing with proton stopping power, as well as range, for many materials is that of Janni.[4]

The range of protons and deuterons can be calculated from the range of an alpha particle *of the same speed* using the formula

$$R(p,d) = 4 \frac{M(p,d)}{M_\alpha} R_\alpha - 2\ (\text{mm, air}) \tag{4.20}$$

where R_α = range in air of an alpha particle having the same speed as the deuteron or the proton
M = mass of the particle (1 for proton, 2 for deuteron)
M_α = mass of alpha particle = 4

For materials other than air, the Bragg–Kleeman rule (Eq. 4.17) should be used.

The alpha and proton or deuteron ranges are related by the same speed rather than the same kinetic energy due to the dependence of dE/dx on the speed of the particle.

Example 4.9   What is the range of the 5-MeV deuteron in air?

Answer   Equation 4.20 will be used, but first the range of an alpha particle with speed equal to that of a 5-MeV deuteron will have to be calculated. The kinetic energy of an alpha particle with the *same speed* as that of the deuteron will be found using the corresponding equations for the kinetic energy. Since $T = (1/2)MV^2$ for these nonrelativistic particles,

$$T_\alpha = \frac{M_\alpha}{M_d}T_d = 2T_d = 10\,\text{MeV}$$

The range of a 10-MeV alpha particle (in air) is (Eq. 4.14)

$$R_\alpha = (0.05T + 2.85)\,T^{3/2} = 106\,\text{mm} = 0.106\,\text{m}$$

The range of the 5-MeV deuteron (in air) is then (Eq. 4.20)

$$R_d = 2R_\alpha - 2 = 210\,\text{mm} = 0.21\,\text{m}$$

Example 4.10   What is the range of a 5-MeV deuteron in aluminum?

Answer   Using the Bragg–Kleeman rule (Eq. 4.17) the result of the previous example,

$$R_d\,(\text{Al}) = (3.36 \times 10^{-1})\frac{\sqrt{27}}{2.7 \times 10^3}210\,\text{mm} = 0.136\,\text{mm} = 136\,\mu\text{m}$$

4.6.2 Range of Electrons and Positrons

Electrons and positrons behave in essentially the same way with regard to energy loss, slowing down, and penetration through matter. Small differences exist; one was indicated when dE/dx was discussed in Section 4.3. Small differences in the values of the range between electrons and positrons should also be expected, and indeed this is the case. Most of the range measurements have been performed with electrons because electrons are used much more frequently than positrons in radiation measurements. For this reason, from this point on, only electrons will be discussed. The reader should be aware that the results are equally applicable for positrons, to a first approximation, but for very accurate results the references listed at the end of this chapter should be consulted.

If the experiment shown in Figure 4.5 and discussed in Section 4.6.1 is repeated with the incident beam consisting of monoenergetic electrons, the result will look as shown in Figure 4.8. For electrons, the transmission curve does not have a flat part. It decreases gradually to a level which is the background. The range* is equal to the thickness of the material, which

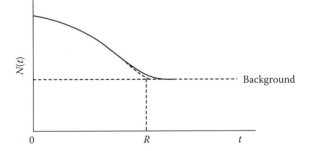

Figure 4.8   The number of electrons transmitted through thickness t. Experiment setup shown in Figure 4.5.

---

* In many texts, this is called the "extrapolated" range. Since only one type of range is used, there is no need to carry along the world "extrapolated."

is defined by the point where the linear extrapolation of the transmission curve meets the background.

The semiempirical equation giving the range of electrons for the energy range 0.3–30 MeV has been developed by Tabata, Ito, and Okabe,[10] based on the experimental results available until 1972. This equation, indicated from now on as the TIO equation, has the following form:

$$R(kg/m^2) = a_1 \left(\frac{\ln[1 + a_2(\gamma - 1)]}{a_2} - \frac{a_3(\gamma - 1)}{1 + a_4(\gamma - 1)^{a_5}} \right)$$ (4.21)

where

$$a_1 = \frac{2.335\ A}{Z^{1.209}}, \quad a_3 = 0.9891 - (3.01 \times 10^{-4}\ Z)$$ (4.22)

$$a_2 = 1.78 \times 10^{-4}\ Z, \quad a_4 = 1.468 - (1.180 \times 10^{-2}\ Z), \quad a_5 = \frac{1.232}{Z^{0.109}}$$

A, Z, and γ have been defined in Section 4.3.

Figures 4.9 and 4.10 show results based on Eq. 4.21, as well as experimental points.

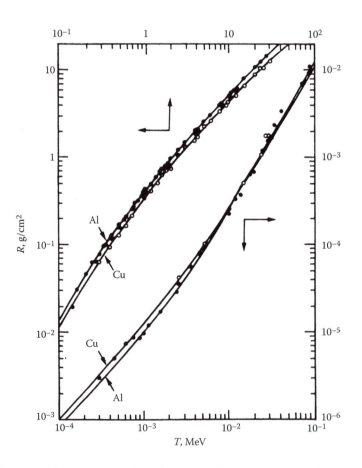

Figure 4.9 The range of electrons as a function of their kinetic energy as obtained by using Eq. 4.21. The solid circles are experimental data for Al; the open circles are for Cu. (From Tabata, T., et al., *Nucl. Instrum. Meth.*, 103, 85, 1972.)

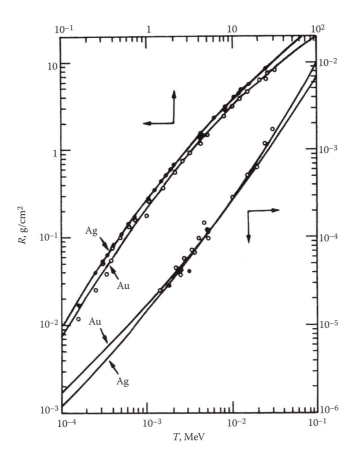

Figure 4.10 The range of electrons as a function of their kinetic energy as obtained by using Eq. 4.21. The solid circles are experimental data for Ag; the open circles are for Au. (From Tabata, T., et al., *Nucl. Instrum. Meth.*, 103, 85, 1972.)

In the case of absorbers that are mixtures or compounds, the atomic number Z and atomic weight A to be used in Eq. 4.22 are given by

$$Z_{ef} = \sum_{i}^{L} w_i Z_i \tag{4.23}$$

$$A_{ef} = Z_{ef} \left(\sum_{i}^{L} w_i \frac{Z_i}{A_i} \right)^{-1} \tag{4.24}$$

where w_i is the weight fraction of element with atomic number Z_i and atomic weight A_i.

Example 4.11 What is the range of 1-MeV electrons in gold? ($Z = 79$, $A = 197$.)
Answer Using Eqs. 4.21 and 4.22,

$$a_1 = \frac{2.335(197)}{79^{1.209}} = 2.336$$

$$a_2 = (1.78 \times 10^{-4})(79) = 0.01406$$

$$a_3 = 0.9891 - (3.01 \times 10^{-4})(79) = 0.965$$

$$a_4 = 1.468 - (1.180 \times 10^{-2})(79) = 0.5358$$

111

$$a_5 = \frac{1.232}{79^{0.109}} = 0.765$$

$$\gamma = \frac{1.511}{0.511} = 2.957$$

$$R = 2.336\left(\frac{\ln(1+0.0275)}{0.01406} - \frac{1.8885}{1.895}\right) = 2.18\,\text{kg/m}^2 = 0.218\,\text{g/cm}^2$$

Since the density of gold is 19.3×10^3 kg/m³, the range in μm is

$$R = \frac{2.18\,\text{kg/m}^2}{19.3\times10^3\,\text{kg/m}^3} = 1.13\times10^{-4}\,\text{m} = 113\,\mu\text{m}$$

Example 4.12 What is the range of 1-MeV electrons in aluminum? ($Z = 13$, $A = 27$.)
Answer Again, using Eqs. 4.21 and 4.22,

$$a_1 = \frac{2.335(27)}{13^{1.209}} = 2.837$$
$$a_2 = (1.78\times10^{-4})(13) = 2.314\times10^{-3}$$
$$a_3 = 0.9891 - (3.01\times10^{-4})(13) = 0.985$$
$$a_4 = 1.468 - (1.180\times10^{-2})(13) = 1.3146$$
$$a_5 = \frac{1.232}{13^{0.109}} = 0.9315$$

$$R = 2.837\left(\frac{\ln(1.0045)}{0.00231} - \frac{1.928}{3.457}\right) = 3.93\,\text{kg/m}^2 = 0.393\,\text{g/cm}^2$$

Since the density of aluminum is 2.7×10^3 kg/m³, the range in μm is

$$R = \frac{3.93\,\text{kg/m}^2}{2.7\times10^3\,\text{kg/m}^3} = 1.46\times10^{-3}\,\text{m} = 1460\,\mu\text{m}$$

Another formula for the range of protons and electrons is given by Shultis and Faw (Radiation Shielding, p. 71).

Table 4.1a from Shultis and Faw give constants for several months.

Table 4.1a
Constants for the Empirical Formula $y = a + bx + cx^2$ Relating Charged Particle Energy and Range in Which $y = \log_{10} R$ (g/cm²) and $x = \log_{10} T$ (MeV), $T =$ Initial Energy of the Particle (MeV)

Material	Protons			Electrons		
	a	b	c	a	b	c
Aluminum	−2.3829	1.3494	0.19670	−0.27957	1.2492	−0.18247
Iron	−2.2262	1.2467	0.22281	−0.23199	1.2165	−0.19504
Gold	−1.8769	1.1663	0.20568	−0.13552	1.1292	−0.20889
Air	−2.5207	1.3729	0.21045	−0.33545	1.2615	−0.18124
Water	−2.5814	1.3767	0.20954	−0.38240	1.2799	−0.17378
Tissue	−2.5839	1.3851	0.20710	−0.37829	1.2803	−0.17374
Bone	−2.5154	1.3775	0.20466	−0.33563	1.2661	−0.17924

Example 4.12a   What is the range of 2 MeV electrons in Al?
Answer $x = \log_{10} 2 = 0.301$
$y = 0.280 + 1.25 \times 0.301 - 0.182 \times (0.301)^2$

$$R(g/cm^2) = 10^{0.08} = 1.20 \ g/cm^2, \quad R(cm) = \frac{1.20 \ g/cm^2}{2.7 \ g/cm^3} = 0.44 \ cm$$

4.6.3 Transmission of Beta Particles

Beta particles have a continuous energy spectrum extending from zero energy up to maximum kinetic energy E_{max} (see Section 3.7.3). If the transmission experiment shown in Figure 4.5 is repeated with an incident beam of β particles, the result will look as shown in Figure 4.11. The number of betas $N(t)$ transmitted through a thickness t is very closely represented by

$$N(t) = N(0)e^{-\mu t} \tag{4.25}$$

where μ is called the mass absorption coefficient.

The value of μ has been determined experimentally as a function of the maximum beta energy and is given by

$$\mu \ (m^2/kg) = 1.7 E_{max}^{-1.14} \tag{4.26}$$

where E_{max} is in MeV. Notice that μ is given in units of m²/kg; therefore the thickness t in the exponent of Eq. 4.25 should be in kg/m². The exponential transmission law represented by Eq. 4.25 is the result of experimental observation. There is no theory predicting it. The range of β particles is calculated using Eq. 4.21 for kinetic energy equal to E_{max}.

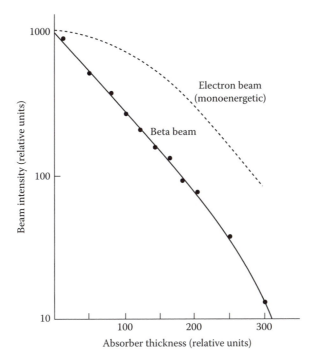

Figure 4.11   Transmission of betas. A corresponding curve for monoenergetic electrons is also shown with $E_{mono} = E_{max}$.

Example 4.13 What fraction of 2-MeV betas will go through a single Al foil of thickness 0.1 mm?
Answer The mass absorption coefficient is, using Eq. 4.26.

$$\mu = 1.7 \times 2^{-1.14} = 0.7714 \, m^2/kg$$

The fraction transmitted is, using Eq. 4.25,

$$\frac{N(t)}{N(0)} = e^{-\mu t} = \exp[-0.7714 \, m^2/kg(0.1 \times 10^{-3} \, m) \, (2.7 \times 10^{-3} \, kg/m^3)]$$

$$= \exp(-0.208) = 0.81$$

Therefore, 81% of the betas will go through this foil.

4.6.4 Energy Loss after Traversing a Material of Thickness $t < R$

One is often required to calculate the energy loss of a charged particle after it traverses a material of thickness t. The first step in solving such a problem is to calculate the range of the particle in that medium. If the range is $R < t$, the particle stopped in the medium and the total energy lost is equal to the initial energy of the particle. If $R > t$, the energy loss ΔE is given by

$$\Delta E = \int_0^t \left(\frac{dE}{dx}\right) dx \tag{4.27}$$

where dE/dx is the total stopping power (ionization-excitation plus radiation loss). If $t = R$, one may take dE/dx as constant and obtain

$$\Delta E = \left(\frac{dE}{dx}\right)_0 t, \quad t \ll R \tag{4.28}$$

where $(dE/dx)_0$ is the stopping power calculated for the initial energy of the particle.

If the thickness t is a considerable fraction of the range, dE/dx cannot be considered constant. Then, Eq. 4.27 should be integrated using the appropriate form of dE/dx. Since the stopping power is a complicated expression, the integration cannot be carried out by hand. A numerical integration can be performed by a computer. In most cases, however, the following approach gives adequate results.

The thickness t is divided into N segments of length Δx_i, where

$$\sum_{i=1}^{N} \Delta x_i = t$$

Equation 4.27 takes the form

$$\Delta E = \sum_{i=1}^{N} \left(\frac{dE}{dx}\right)_i \Delta x_i \tag{4.29}$$

where $(dE/dx)_i$ is the stopping power calculated for the kinetic energy of the particle at the beginning of the segment Δx_i.

There is no general rule as to the best value of the number of segments N. Obviously, N should be such that $(dE/dx)_i$ changes by a small but acceptable amount as the particle travels the segment Δx_i.

Example 4.14 What is the energy loss of a 10-MeV electron going through 15 mm of aluminum?
Answer Using Eq. 4.21 or Figure 4.9, the range of a 10-MeV electron in aluminum is $R = 20.4$ mm. The particle will emerge, but the thickness of the absorber is a considerable fraction of the range. Therefore, one should use Eq. 4.29.

If one chooses $N = 5$ and equal segments, Eq. 4.29 takes the form

$$\Delta E = \sum_{i=1}^{5} \left(\frac{dE}{dx}\right)_i \Delta x_i, \quad \Delta x_i = 3 \, mm$$

The table below shows how the calculation proceeds:

i	T_i (MeV)	dE/dx (MeV/mm)	$(\Delta E)_i$ (MeV)	$T_{i+1} = T_i - (\Delta E)_i$
1	10	0.605	1.815	8.185
2	8.185	0.568	1.704	6.481
3	6.481	0.530	1.590	4.891
4	4.891	0.492	1.476	3.415
5	3.415	0.457	1.373	2.042

Total energy loss is 7.958 MeV. Using $(dE/dx)_i$, the energy loss would have been equal to 0.605 MeV/mm × 15 mm = 9.075 MeV, which is overestimated by about 14%.

4.7 STOPPING POWER AND RANGE OF HEAVY IONS ($Z > 2$, $A > 4$)

4.7.1 Introduction

The equations presented in Sections 4.3–4.6 for energy loss and range of charged particles were derived with the assumption that the charge of the particle does not change as the particle traverses the medium. This assumption is certainly valid for electrons, positrons, protons, and deuterons ($Z = 1$). It holds well for alphas too ($Z = 2$). However, for $Z > 2$, the charge of the particle cannot be assumed constant, and for this reason, the energy loss and range calculations require special treatment.

Consider an atom or an ion with speed greater than the orbital velocity of its own electrons. If this particle enters a certain medium, the atomic electrons will be quickly removed from the atom or ion, leaving behind a bare nucleus. The nucleus will keep moving through the medium, continuously losing energy in collisions with the electrons of the medium.* It is probable that the ion will capture an electron in one of these collisions. It is also probable that the electron will be lost in another collision. As the ion slows down and its speed becomes of the same order of magnitude as the orbital speeds of the atomic electrons, the probability for electron capture increases, while the probability for electron loss decreases. When the ion slows down even farther and is slower than the orbiting electrons, the probability of losing an electron becomes essentially zero, while the probability of capturing one becomes significant. As the speed of the ion continues to decrease, a third electron is captured, then a fourth, and so on. At the end, the ion is slower than the least bound electron. By that time, it is a neutral atom. What is left of its kinetic energy is exchanged through nuclear and not electronic collisions. The neutral atom is considered as stopped when it either combines chemically with one of the atoms of the material or is in thermal equilibrium with the medium.

4.7.2 The dE/dx Calculation

The qualitative discussion of Section 4.7.1 showed how the charge of a heavy ion changes as the ion slows down in the medium. It is this variation of the charge that makes the energy loss calculation very difficult. There is no single equation given dE/dx for all heavy ions and for all stopping materials. Instead, dE/dx is calculated differently, depending on the speed of the ion relative to the speed of the orbital electrons.

The stopping power is written, in general, as the sum of two terms:

$$\frac{dE}{dx} = \left(\frac{dE}{dx}\right)_e + \left(\frac{dE}{dx}\right)_n \tag{4.30}$$

where $(dE/dx)_e$ = electronic energy loss
$(dE/dx)_n$ = nuclear energy loss

An excellent review of the subject is presented by Northcliffe[11] and Lindhard, Scharff, and Schiott.[12] The results are usually presented as universal curves in terms of two dimensionless quantities, the distance s and the energy ε, first introduced by Lindhard et al.[12] and defined as follows:

$$s = 4\pi a^2 N \frac{M_1 M_2 x}{(M_1 + M_2)^2} \tag{4.31}$$

---

* Collisions with nuclei are not important if the particle moves much faster than the atomic electrons.

$$\varepsilon = \frac{a}{r_0}\left(\frac{M_2}{Z_1 Z_2 (M_1 + M_2)}\right)\frac{T}{mc^2} \tag{4.32}$$

where* $a = 0.8853 a_0 (Z_1^{2/3} + Z_2^{2/3})^{-1/2}$
$\quad x$ = actual distance traveled
$\quad a_0 = h^2/me^2$ = Bohr radius = 5.29×10^{-11} m
$\quad Z_1, M_1$ = charge and mass of incident particle
$\quad Z_2, M_2$ = charge and mass of stopping material

The parameters N, r_0, and mc^2 have been defined in Section 4.3.

At high ion velocities, $v \gg v_0 Z_1^{2/3}$, were $v_0 = e^2/\hbar$ orbital velocity of the electron in the hydrogen atom, the nuclear energy loss is negligible. The particle has an effective charge equal to Z_1, and the energy loss is given by an equation of the form

$$\frac{dE}{dx} \sim \frac{Z_1^2 Z_2}{v^2 A_2}\ln\frac{2mv^2}{I} \tag{4.33}$$

which is similar to Eq. 4.2.

At velocities of the order of $v \approx v_0 Z_1^{2/3}$, the ion starts picking up electrons and its charge keeps decreasing. The energy loss through nuclear collisions is still negligible.

In the velocity region $v \leq v_0 Z_1^{2/3}$, the electronic energy loss equation takes the form[10]

$$\left(\frac{d\varepsilon}{d\rho}\right)_e = k\varepsilon^n \tag{4.34}$$

where

$$k = \xi_e \frac{0.0793 Z_1 Z_2 (A_1 + A_2)^{3/2}}{(Z_1^{2/3} + Z_2^{2/3})^{3/4} A_1^{3/2} + A_2^{1/2}}$$

and n has a value very close to $(1/2)$.[12,13] The constant k depends on Z and A only, not on energy, and its value is less than 1. Some typical values are given in Table 4.2.

Table 4.3 shows the kinetic energy per unit atomic mass, as well as the kinetic energy, of several ions for $v \approx v_0 Z_1^{2/3}$.

The electronic stopping power for different ions and stopping materials is obtained by using the following semiempirical approach.

The ratio of stopping power for two ions having the same velocity and traveling in the same medium is given by (using Eq. 4.33)

$$\frac{(dE/dx)_{T_1 Z_1 A_1}}{(dE/dx)_{T_2 Z_2 A_2}} = \frac{Z_1^2}{Z_2^2} \tag{4.35}$$

The application of Eq. 4.35 to heavy ions should take into account the change of the charge Z_1 as the ion slows down. This is accomplished by replacing Z_1 with an effective charge,

$$Z_{eff} = \eta Z_1$$

Table 4.2
Values of k Used in Eq. 4.34

Z_1	A_1	Z_2	A_1	k
10	50	13	27	0.085
20	60	79	197	0.022
92	238	79	197	0.162

---

* The number $0.8853 = (9\pi^2)^{1/3}/2^{7/3}$ is called the Thomas–Fermi constant.

Table 4.3
The Kinetic Energy of Heavy Ions for Several Values of
$v = v_0 Z^{2/3}v$

Ion	Z_1	$v_0 Z_1^{2/3}(\times 10^{-7})$(m/s)	$B (\times 10^2)$	T/A_1	T (MeV)
C	6	0.72	2.4	0.27	3.3
Al	13	1.2	4.04	076	20.7
Ni	28	2.0	6.7	2.13	126
Br	35	2.3	7.8	2.87	230
Ag	47	2.8	9.5	4.27	461
I	53	3.1	10.3	5.02	638
Au	79	4.0	13.4	8.6	1694
U	92	4.5	14.9	10.57	2515

where η is a parameter that depends on energy. The second particle in Eq. 4.35 is taken to be the proton ($Z_2 = A_2 = 1$), thus leading to the form[14–16]

$$\frac{(dE/dx)_{Z_1 A_1 T_1}}{(dE/dx)_{pT_p}} = \frac{\eta^2 Z_1^2}{\eta_p^2} \tag{4.36}$$

where the effective proton charge η_p is given by Eq. 4.37, reported by Booth and Grant,[16] and T_p is the proton kinetic energy in MeV:

$$\eta_p^2 = \left[1 - \exp(-150T_p)\right]\exp(-0.835e^{-14.5T_p}) \tag{4.37}$$

Equations giving the value of η have been reported by many investigators.[14–17] The most recent equation reported by Forster et al.[17] valid for $8 \leq Z_1 \leq 20$ and for $v/v_0 > 2$ is

$$\eta = 1 - A(Z_1) \exp\left(-0.879\frac{v}{v_0}Z_1^{-0.65}\right) \tag{4.38}$$

with

$$A(Z_1) = 1.035 - 0.4\exp(-0.16Z_1)$$

The proton stopping power is known.[18] Brown[19] has developed an equation of the form

$$\ln\left(\frac{dE}{dx}\right)_p = a + b\ln T_p + C(\ln T_p)^2 \tag{4.39}$$

by least squares fittting the data of Northcliffe and Schilling.[18] The most recent data are those of Janni.[4]

The experimental determination of dE/dx is achieved by passing ions of known initial energy through a thin layer of a stopping material and measuring the energy loss. The thickness Δx of the material should be small enough that $dE/dx \approx \Delta E/\Delta x$. Unfortunately, such a value of Δx is so small, especially for very heavy ions, that the precision of measuring Δx is questionable and the uniformity of the layer has an effect on the measurement. Typical experimental results of stopping power are presented in Figure 4.12. The data of Figure 4.12 come from Ref. 13. The solid line is based on the following empirical equation proposed by Bridwell and Bucy[20] and Bridwell and Moak[21]:

$$\frac{dE}{dx}[\text{MeV}/(\text{kg m}^2)] = \frac{2064.5}{A_2}\sqrt{\frac{TA_1 Z_2}{Z_1}} \tag{4.40}$$

where T is the kinetic energy of the ion in MeV.

For a compound or mixture, dE/dx can be obtained by using Eq. 4.12 with $(dE/dx)_i$ obtained from Eq. 4.36 or Eq. 4.40.

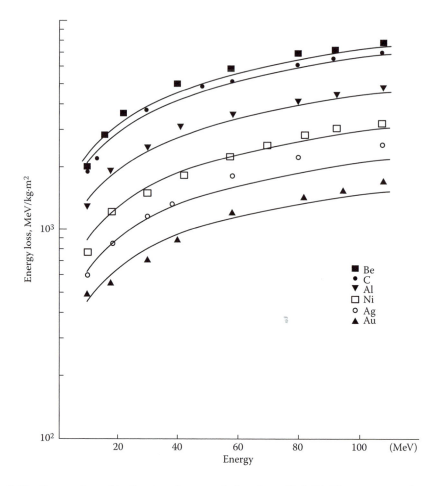

Figure 4.12 Energy loss of iodine ions in several absorbers (Ref. 13). The curves are based on Eq. 4.40.

At velocities $v < v_0 Z_1^{2/3}$, the energy loss through nuclear elastic collisions becomes important. The so-called *nuclear stopping power* is given by the following approximate expression[10]:

$$\left(\frac{d\varepsilon}{d\rho}\right)_n = \frac{1}{2\varepsilon}\ln(1.294\varepsilon) \qquad (4.41)$$

While the electronic stopping power $(d\varepsilon/d\rho)_e$ continuously decreases as the ion speed v decreases, the nuclear stopping power increases as v decreases, goes through a maximum, and then decreases again (Figure 4.13).

4.7.3 Range of Heavy Ions

The range of heavy ions has been measured and calculated for many ions and for different absorbers. But there is no single equation—either theoretical or empirical—giving the range in all cases. Heavy ions are hardly deflected along their path, except very close to the end of their track, where nuclear collisions become important. Thus the range R, which is defined as the depth of penetration along the direction of incidence, will be almost equal to the path length, the actual distance traveled by the ion. With this observation in mind, the range is given by

$$R = \int_0^E \frac{dE}{(dE/dx)_e + (dE/dx)_n} \qquad (4.42)$$

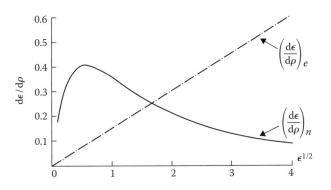

Figure 4.13 The electronic and nuclear energy loss as a function of the dimensionless energy ε (Ref. 12).

Results of calculations based on Eq. 4.42 are given by many authors. Based on calculations described in Ref. 12, Siffert and Coche[22] present universal graphs for several heavy ions in silicon (Figures 4.14 and 4.15).

The range of a heavy ion in a compound or mixture is calculated from the range in pure elements by using the equation[23,24]

$$R \ (\text{kg/m}^2) = \left(\sum_i \frac{w_i}{R_i} \right)^{-1} \tag{4.43}$$

where R_i = range, in kg/m², in element I
$\quad w_i$ = weight fraction of ith element

4.8 INTERACTIONS OF PHOTONS WITH MATTER

Photons, also called X-rays or γ-rays, are electromagnetic radiation. Considered as particles, they travel with the speed of light c and they have zero rest mass and charge. The relationship between the energy of a photon, its wavelength λ, and frequency is

$$E = h\nu = h\frac{c}{\lambda} \tag{4.44}$$

There is no clear distinction between X-rays and γ-rays. The term X-rays is applied generally to photons with $E < 1$ MeV. Gammas are the photons with $E > 1$ MeV. In what follows, the terms photon, γ, and X-ray will be used interchangeably.

X-rays are generally produced by atomic transitions such as excitation and ionization. Gamma rays are emitted in nuclear transitions. Photons are also produced as bremsstrahlung, by accelerating or decelerating charged particles. X-rays and γ-rays emitted by atoms and nuclei are monoenergetic. Bremsstrahlung has a continuous energy spectrum.

There is a long list of possible interactions of photons, but only the three most important ones will be discussed here: the photoelectric effect, Compton scattering, and pair production.

4.8.1 The Photoelectric Effect

The photoelectric effect is an interaction between a photon and a bound atomic electron. As a result of the interaction, the photon disappears and one of the atomic electrons is ejected as a free electron, called the *photoelectron* (Figure 4.16). The kinetic energy of the electron is

$$T = E_\gamma - B_e \tag{4.45}$$

where E_γ = energy of the photon
$\quad B_e$ = binding energy of the electron

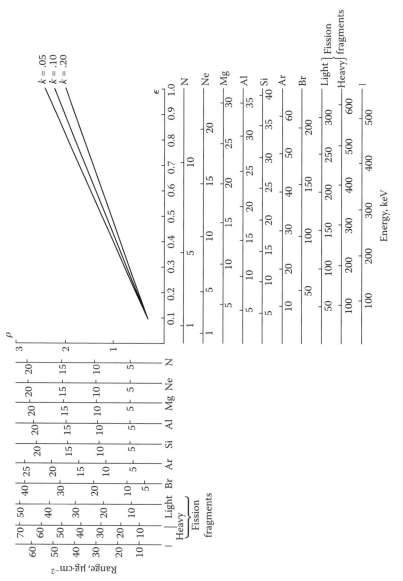

Figure 4.14 Universal range–energy plot for $\epsilon < 1$. It allows determination of range in silicon for many heavy ions (Reference 22).

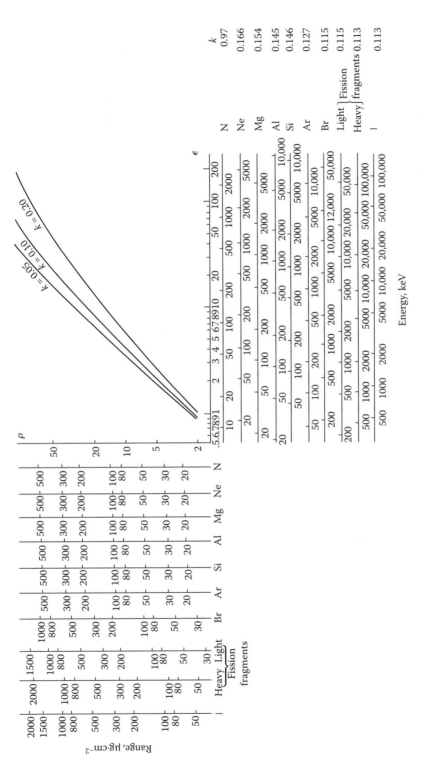

Figure 4.15 Universal range–energy plot for $\varepsilon > 1$. It allows determination of range in silicon for many heavy ions (Reference 22).

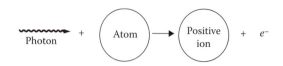

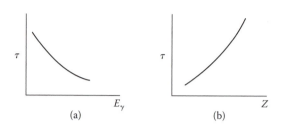

Figure 4.16 The photoelectric effect.

Figure 4.17 Dependence of the photoelectric cross section on (a) photon energy and (b) atomic number of the material.

The probability of this interaction occurring is called the *photoelectric cross section* or *photoelectric coefficient*. Its calculation is beyond the scope of this book, but it is important to discuss the dependence of this coefficient on parameters such as $E\gamma$, Z, and A. The equation giving the photoelectric coefficient may be written as

$$\tau \; (\mathrm{m}^{-1}) = aN \frac{Z^n}{E_\gamma^m}[1 - O(Z)] \tag{4.46}$$

where τ = probability for photoelectric effect to occur per unit distance traveled by the photon
    a = constant, independent of Z and E_γ
  m, n = constants with a value of 3 to 5 (their value depends on E_γ; see Evans, 1972)
      N, Z have been defined in Section 4.3.
    O = the order of magnitude

The second term in brackets indicates correction terms of the first order in Z. Figure 4.17 shows how the photoelectric coefficient changes as a function of E_γ and Z. Figure 4.17 and Eq. 4.46 show that the photoelectric effect is more important for high-Z material, that is, more probable in lead ($Z = 82$) than in Al ($Z = 13$). It is also more important for $E_\gamma = 10$ keV than $E_\gamma = 500$ keV (for the same material). Using Eq. 4.46, one can obtain an estimate of the photoelectric coefficient of one element in terms of that of another. If one takes the ratio of T for two elements, the result for photons of the *same energy* is

$$\tau_2(\mathrm{m}^{-1}) = \tau_1 \frac{\rho_2}{\rho_1}\left(\frac{A_1}{A_2}\right)\left(\frac{Z_2}{Z_1}\right)^n \tag{4.47}$$

where ρ_i and A_i are density and atomic weight, respectively, of the two elements, and τ_1 and τ_2 are given in m^{-1}. If τ_1 and τ_2 are given in m^2/kg, Eq. 4.47 takes the form

$$\tau_2(\mathrm{m}^2/\mathrm{kg}) = \tau_1 \frac{A_1}{A_2}\left(\frac{Z_2}{Z_1}\right)^n \tag{4.47a}$$

4.8.2 Compton Scattering or Compton Effect

The *Compton effect* is a collision between a photon and a free electron. Of course, under normal circumstances, all the electrons in a medium are not free but bound. If the energy of the photon, however, is of the order of keV or more, while the binding energy of the electron is of the order of eV, the electron may be considered free.

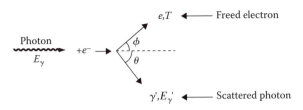

Figure 4.18 The Compton effect.

The photon does not disappear after a Compton scattering. Only its direction of motion and energy change (Figure 4.18). The photon energy is reduced by a certain amount that is given to the electron. Therefore, conservation of energy gives (assuming the electron is stationary before the collision):

$$T = E_\gamma - E_{\gamma'} \tag{4.48}$$

If Eq. 4.48 is used along with the conservation of momentum equations, the energy of the scattered photon as a function of the scattering angle θ can be calculated. The result is (see Evans, 1972).

$$E_{\gamma'} = \frac{E_\gamma}{1 + (1 - \cos\theta)E_\gamma/mc^2} \tag{4.49}$$

Using Eqs. 4.48 and 4.49, one obtains the kinetic energy of the electron:

$$T = \frac{(1 - \cos\theta)E_\gamma/mc^2}{1 + (1 - \cos\theta)E_\gamma/mc^2} E_\gamma \tag{4.50}$$

A matter of great importance for radiation measurement is the maximum and minimum energy of the photon and the electron after the collision. The minimum energy of the scattered photon is obtained when $\theta = \pi$. This, of course, corresponds to the maximum energy of the electron. From Eq. 4.49,

$$E_{\gamma', \min} = \frac{E_\gamma}{1 + 2E_\gamma/mc^2} \tag{4.51}$$

and

$$T_{\max} = \frac{2E_\gamma/mc^2}{1 + 2E_\gamma/mc^2} E_\gamma \tag{4.52}$$

The maximum energy of the scattered photon is obtained for $\theta = 0$, which essentially means that the collision did not take place. From Eqs. 4.49 and 4.50,

$$E_{\gamma', \max} = E_\gamma$$
$$T_{\min} = 0$$

The conclusion to be drawn from Eq. 4.51 is that the minimum energy of the scattered photon is greater than zero. Therefore, *in Compton scattering, it is impossible for all the energy of the incident photon to be given to the electron*. The energy given to the electron will be dissipated in the material within a distance equal to the range of the electron. The scattered photon may escape.

Example 4.15 A 3-MeV photon interacts by Compton scattering, (a) What is the energy of the photon and the electron if the scattering angle of the photon is 90°? (b) What if the angle of scattering is 180°?

Answer (a) Using Eq. 4.49,

$$E_\gamma = \frac{3}{1+(1-0)3/0.511} = 0.437 \text{ MeV}$$

$$T = 3 - 0.437 = 2.563 \text{ MeV}$$

(b) Using Eq. 4.51,

$$E_{\gamma,\text{min}} = \frac{3}{1+(2)3/0.511} = 0.235 \text{ MeV}$$

$$T = 3 - 0.235 = 2.765 \text{ MeV}$$

Example 4.16 What is the minimum energy of the γ-ray after Compton scattering if the original photon energy is 0.511 MeV, 5 MeV, 10 MeV, or 100 MeV?

Answer The results are shown in the table below (Eq. 4.51 has been used):

E_γ	$(E_\gamma)_{\text{min}}$	T
0.511	0.170	0.341
5	0.243	4.757
10	0.25	9.75
100	0.25	99.75

The probability that Compton scattering will occur is called the *Compton coefficient* or the *Compton cross section*. It is a complicated function of the photon energy, but it may be written in the form

$$\sigma(\text{m}^{-1}) = NZf(E_\gamma) \tag{4.53}$$

where σ = probability for Compton interaction to occur per unit distance
$f(E_\gamma)$ = a function of E_γ

If one writes the atom density N explicitly, Eq. 4.53 takes the form

$$\sigma \sim \rho \frac{N_A}{A} Zf(E_\gamma) \sim \rho \left(\frac{N_A}{A}\right)\frac{A}{2} f(E_\gamma) \sim \rho \frac{N_A}{2} f(E_\gamma) \tag{4.54}$$

In deriving Eq. 4.54, use has been made of the fact that for most materials, except hydrogen, $A \approx 2Z$ to $A \approx 2.6Z$. According to Eq. 4.54, the probability for Compton scattering to occur is almost independent of the atomic number of the material. Figure 4.19 shows how σ changes as a function of Ey and Z. If the Compton cross section is known for one element, it can be calculated for any other by using Eq. 4.53 (for photons of the same energy):

$$\sigma_2(\text{m}^{-1}) = \sigma_1 \left(\frac{\rho_2}{\rho_1}\right)\left(\frac{A_1}{A_2}\right)\left(\frac{Z_2}{Z_1}\right) \tag{4.55}$$

where σ_1 and σ_2 are given in m⁻¹. If σ_1 and σ_2 are given in m²/kg, Eq. 4.55 takes the form

$$\sigma_2(\text{m}^2/\text{kg}) = \sigma_1 \left(\frac{A_1}{A_2}\right)\left(\frac{Z_2}{Z_1}\right) \tag{4.55a}$$

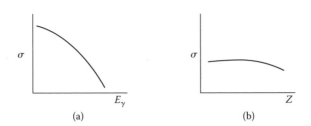

Figure 4.19 Dependence of the Compton cross section on (a) photon energy and (b) atomic number of the material.

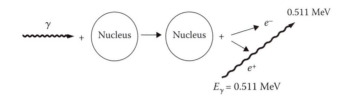

Figure 4.20 Pair production. The gamma disappears and a positron–electron pair is created. Two 0.511-MeV photons are produced when the positron annihilates.

4.8.3 Pair Production

Pair production is an interaction between a photon and a nucleus. As a result of the interaction, the photon disappears and an electron–positron pair appears (Figure 4.20). Although the nucleus does not undergo any change as a result of this interaction, its presence is necessary for pair production to occur. A γ-ray will not disappear in empty space by producing an electron–positron pair.*

Conservation of energy gives the following equation for the kinetic energy of the electron and the positron:

$$T_{e^-} + T_{e^+} = E_\gamma - (mc^2)_{e^-} - (mc^2)_{e^+} = E_\gamma - 1.022 \text{ MeV} \tag{4.56}$$

The available kinetic energy is equal to the energy of the photon minus 1.022 MeV, which is necessary for the production of the two rest masses. Electron and positron share, for all practical purposes, the available kinetic energy, that is,

$$T_{e^-} = T_{e^+} = \frac{1}{2}(E_\gamma - 1.022 \text{ MeV}) \tag{4.57}$$

Pair production eliminates the original photon, but two photons are created when the positron annihilates (see Section 3.7.4). These annihilation gammas are important in constructing a shield for a positron source as well as for the detection of gammas (see Chapter 12).

The probability for pair production to occur, called the *pair production coefficient or cross section* is a complicated function of E_γ and Z (see Evans, 1972; Roy and Reed, 1968). It may be written in the form

$$\kappa \text{ (m}^{-1}) = NZ^2 f(E_\gamma, Z) \tag{4.58}$$

where κ is the probability for pair production to occur per unit distance traveled and $f(E_\gamma, Z)$ is a function that changes slightly with Z and increases with E_γ.

Figure 4.21 shows how κ changes with E_γ and Z. It is important to note that κ has a threshold at 1.022 MeV and increases with E_γ and Z. Of the three coefficients (τ and σ being the other two), κ is the only one increasing with the energy of the photon.

---

* Pair production may take place in the field of an electron. The probability for that to happen is much smaller and the threshold for the gamma energy is $4mc^2 = 2.04$ MeV.

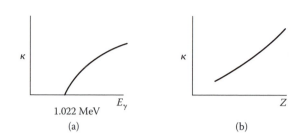

Figure 4.21 Dependence of the pair production cross section on (a) photon energy and (b) atomic number of the material.

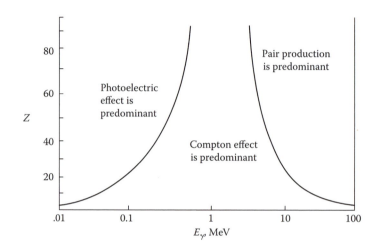

Figure 4.22 The relative importance of the three major gamma interactions. (From *The Atomic Nucleus* by R. D. Evans. Copyright © 1972 by McGraw-Hill. Used with the permission of McGraw-Hill Book Company.)

If the pair production cross section is known for one element, an estimate of its value can be obtained for any other element by using Eq. 4.58 (for photons of the same energy).

$$\kappa_2 \ (\mathrm{m}^{-1}) = \kappa_1 \left(\frac{\rho_2}{\rho_1} \right) \left(\frac{A_1}{A_2} \right) \left(\frac{Z_2}{Z_1} \right)^2 \tag{4.59}$$

where κ_1 and κ_{12} are given in m^{-1}. If κ_1 and κ_2 are given in m^2/kg, Eq. 4.59 takes the form

$$\kappa_2 \ (\mathrm{m}^2/\mathrm{kg}) = \kappa_1 \left(\frac{A_1}{A_2} \right) \left(\frac{Z_2}{Z_1} \right)^2 \tag{4.59a}$$

4.8.4 Total Photon Attenuation Coefficient

When a photon travels through matter, it may interact through any of the three major ways discussed earlier. (For pair production, $E_\gamma > 1.022$ MeV.) There are other interactions, but they are not mentioned here because they are not important in the detection of gammas.

Figure 4.22 shows the relative importance of the three interactions as E_γ and Z change. Consider a photon with $E = 0.1$ MeV. If this particle travels in carbon ($Z = 6$), the Compton effect is the predominant mechanism by which this photon interacts. If the same photon travels in iodine ($Z = 53$), the photoelectric interaction prevails. For a photon of 1 MeV, the Compton effect predominates regardless of Z. If a photon of 10 MeV travels in carbon, it will interact mostly through Compton scattering. The same photon moving in iodine will interact mainly through pair production.

The total probability for interaction M, called the total *linear* attenuation coefficient, is equal to the sum of the three probabilities:

$$\mu \ (m^{-1}) = \tau + \sigma + \kappa \tag{4.60}$$

Physically, μ is the probability of interaction per unit distance.

There are tables that give μ for all the elements, for many photon energies.*

Most of the tables provide μ in units of m²/kg (or cm²/g), because in these units the density of the material does not have to be specified. If μ is given in m²/kg (or cm²/g), it is called the *total mass attenuation coefficient*. The relationship between linear and mass coefficients is

$$\mu \ (m^2/kg) = \frac{\mu \ (m^{-1})}{\rho \ (kg/m^3)} \tag{4.61}$$

Figure 4.23 shows the individual coefficients as well as the total mass attenuation coefficient for lead, as a function of photon energy. The total mass attenuation coefficient shows a minimum because as E increases, τ decreases, κ increases, and σ does not change appreciably. However, the minimum of μ does not fall at the same energy for all elements. For lead, μ shows a minimum at $E_\gamma \sim 3.5$ MeV; for aluminum, the minimum is at 20 MeV; and for NaI, the minimum is at 5 MeV.

If a parallel beam of monoenergetic photons with intensity $I(0)$ strikes a target of thickness t (Figure 4.24), the number of photons, $I(t)$, emerging without having interacted in the target is given by

$$I(t) = I(0)e^{-\mu t} \tag{4.62}$$

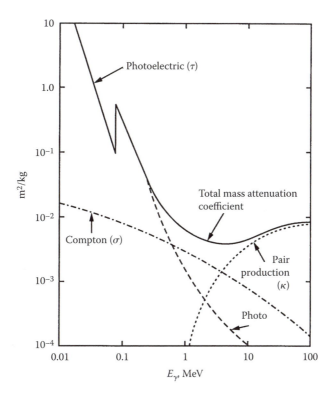

Figure 4.23 Mass attenuation coefficients for lead ($Z = 82$, $\rho = 11.35 \times 10^3$ kg/m³).

* Tables of mass attenuation coefficients are given in Appendix D.

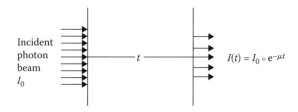

Figure 4.24 The intensity of the transmitted beam (only particles that did not interact) decreases exponentially with material thickness.

The probability that a photon will traverse thickness t without an interaction is

$$\frac{\text{number transmitted}}{\text{number incident}} = \frac{I(0)\,e^{-\mu t}}{I(0)} = e^{-\mu t}$$

Based on this probability, the average distance between two successive interactions, called the *mean free path* (mfp) (λ), is given by

$$\lambda(\text{m}) = \frac{\int_0^\infty x e^{-\mu x}\,dx}{\int_0^\infty e^{-\mu x}\,dx} = \frac{1}{\mu} \tag{4.63}$$

Thus, the mean free path is simply the inverse of the total linear attenuation coefficient. If $\mu = 10\ \text{m}^{-1}$ for a certain γ-ray traveling in a certain medium, then the distance between two successive interactions of this gamma in that medium is $\lambda = 1/\mu = 1/10\ \text{m} = 0.10\ \text{m}$.

The total mass attenuation coefficient for a compound or a mixture is calculated by the same method used for $(dE/dx)_c$ in Section 4.5. It is easy to show (see Prob. 4.15) that

$$\mu_c(\text{m}^2/\text{kg}) = \sum_i w_i \mu_i(\text{m}^2/\text{kg}) \tag{4.64}$$

where μ_c = total *mass* attenuation coefficient for a compound or a mixture
$\qquad w_i$ = weight fraction of the ith element in the compound
$\qquad \mu_i$ = total *mass* attenuation coefficient of the ith element

Example 4.17 What is the total mass attenuation coefficient for 1.25-MeV gammas in NaI?
Answer For this compound, the following data apply:

$$\text{Na: } \mu = 0.00546\ \text{m}^2/\text{kg}, \quad w = \frac{23}{150} = 0.153$$

$$\text{I: } \mu = 0.00502\ \text{m}^2/\text{kg}, \quad w = \frac{127}{150} = 0.847$$

Using Eq. 4.64,

$$\mu\ (\text{NaI}) = 0.00546(0.153) + 0.00502(0.847) = 0.00509\ \text{m}^2/\text{kg} = 0.0509\ \text{cm}^2/\text{g}$$

The density of NaI is $3.67 \times 10^3\ \text{kg/m}^3$; hence,

$$\mu\ (\text{m}^{-1}) = 0.00509\ \text{m}^3/\text{kg}(3.67 \times 10^3\ \text{kg/m}^3) = 18.567\ \text{m}^{-1} = 0.187\ \text{cm}^{-1}$$

4.8.5 Photon Energy Absorption Coefficient

When a photon has an interaction, only part of its energy is absorbed by the medium at the point where the interaction took place. Energy given by the photon to electrons and positrons is considered absorbed at the point of interaction because the range of these charged particles is short. However, X-rays, Compton-scattered photons, or annihilation gammas may escape. The fraction of photon energy that escapes is important when one wants to calculate heat generated due to gamma absorption in shielding materials or gamma radiation dose to humans (see Chapter 16). The gamma energy deposited in any material is calculated with the help of an energy absorption coefficient defined in the following way.

The *gamma energy absorption coefficient* is, in general, that part of the total attenuation coefficient that, when multiplied by the gamma energy, will give the energy deposited at the point of interaction. Equation 4.60 gives the total attenuation coefficient. The *energy absorption coefficient* μ_a is*

$$\mu_a = \tau + \frac{T_{av}}{E_\gamma}\sigma + \kappa \tag{4.65}$$

where T_{av} is the average energy of the Compton electron and μ_a may be a linear or mass energy absorption coefficient, depending on the units (see Section 4.8.4).

In writing Eq. 4.65, it is assumed that

1. If photoelectric effect or pair production takes place, all the energy of the gamma is deposited there.

2. If Compton scattering occurs, only the energy of the electron is absorbed. The Compton-scattered gamma escapes.

In the case of photoelectric effect, assumption (1) is valid. For pair production, however, it is questionable because only the energy $E_\gamma - 1.022$ MeV is given to the electron-positron pair. The rest of the energy, equal to 1.022 MeV, is taken by the two annihilation gammas, and it may not be deposited in the medium. There are cases when Eq. 4.65 is modified to account for this effect.[25] Gamma absorption coefficients, as defined by Eq. 4.65, are given in Appendix D.

Example 4.18 A 1Ci ^{137}Cs source is kept in a large water vessel. What is the energy deposited by the gammas in H_2O at a distance 0.05 m from the source?

Answer ^{137}Cs emits a 0.662-MeV gamma. The mass absorption coefficient for this photon in water is (Appendix D) 0.00327 m^2/kg. The total mass attenuation coefficient is 0.00862 m^2/kg. The energy deposited at a distance of 0.05 m from the source is ($Ed = \phi\mu_a E_\gamma$)

$$E_d\left(\frac{MeV}{(kg\,s)}\right) = \frac{S}{4\pi r^2}e^{-\mu r}\left(\frac{\gamma}{(m^2\,s)}\right)[\mu_a\,(m^2/kg)][E\,(MeV/\gamma)]$$

$$= \frac{3.7 \times 10^{10}}{4\pi(0.05)^2}(e^{-0.00862(10^3)0.05})\,0.00327(0.662)$$

$$= 1.66 \times 10^9\,MeV/(kg\,s) = 2.65 \times 10^{-4}\,J/(kg\,s)$$

This result needs to be multiplied by 0.85 which is the intensity of the ^{137}Cs gamma ray to give a final answer of 2.2.5 × 10^{-4} J/(kg s).

4.8.6 Buildup Factors

Consider a point isotropic monoenergetic gamma source at a distance r from a detector, as shown in Figure 4.25, with a shield of thickness t between source and detector. The total gamma beam hitting the detector consists of two components.

---

* A more detailed definition of the energy absorption coefficient is given by Shultis and Faw.

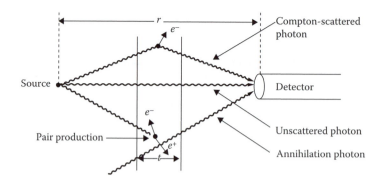

Figure 4.25 If a point isotropic source is placed behind a shield of thickness t, both scattered and unscattered photons will hit the detector.

1. The *unscattered beam* (ϕ_u) consists of those photons that go through the shield without any interaction. If the source strength is $S(\gamma/s)$, the intensity of the unscattered beam or the unscattered photon flux is given by the simple and exact expression

$$\phi_u = \frac{S}{4\pi r^2} e^{-\mu t} \tag{4.66}$$

2. The *scattered beam* (ϕ_s) consists of scattered incident photons and others generated through interactions in the shield (e.g., X-rays and annihilation gammas). The calculation of the scattered beam is not trivial, and there is no simple expression like Eq. 4.62 representing it.

The total flux hitting the detector is

$$\phi_{\text{tot}} = \phi_u + \phi_s \tag{4.67}$$

Obviously, for the calculation of the correct energy deposition by gammas, either for the determination of heating rate in a certain material or the dose rate to individuals, the total flux should be used. Experience has shown that rather than calculating the total flux using Eq. 4.67, there are advantages to writing the total flux in the form

$$\phi_{\text{tot}} = B\phi_u \tag{4.68}$$

where B is a buildup factor, definted and computed in such a way that Eq. 4.68 gives the correct total flux. Combining Eqs. 4.67 and 4.68, one obtains

$$B = \frac{\phi_{\text{tot}}}{\phi_u} = 1 + \frac{\phi_s}{\phi_u} \geq 1 \tag{4.69}$$

How will B be determined? Equation 4.69 will be used, of course, but that means one has to determine the scattered flux. Then where is the advantage of using B? The advantage comes from the fact that B values for a relatively small number of cases can be computed and tabulated and then, by interpolation, one can obtain the total flux using Eq. 4.68 for several other problems. In other words, the use of the buildup factor proceeds in two steps:

1. Buildup factor values are tabulated for many cases.

2. The appropriate value of B that applies to a case under study is chosen and used in Eq. 4.68 to obtain the total flux.

In general, the buildup factor depends on the energy of the photon, on the mean free paths traveled by the photon in the shield, on the geometry of the source (parallel beam or point isotropic), and on the geometry of the attenuating medium (finite, infinite, slab, etc.).

The formal definition of B upon which its calculation is based is

$$B(E, \mu r) = \frac{\text{quantity of interest due to total flux}}{\text{quantity of interest due to unscattered flux}}$$

Quantities of interest and corresponding buildup factors are shown in Table 4.4.

The mathematical formulas for the buildup factors are (assuming a monoenergetic, E_0, point isotropic source) as follows:

Number buildup factor:

$$B_N(E_0, \mu r) = \frac{\int_0^{E_0} \phi_{\text{tot}}(r, E)\, dE}{(S/4\pi r^2)e^{-\mu r}} \tag{4.70}$$

Energy deposition buildup factor:

$$B_E(E_0, \mu r) = \frac{\int_0^{E_0} \phi_{\text{tot}}(r, E)E\mu_a^{\text{med}}(E)\, dE}{(S/4\pi r^2)e^{-\mu r}\mu_a^{\text{med}}(E_0)E_0} \tag{4.71}$$

Dose buildup factor:

$$B_D(E_0, \mu r) = \frac{\int_0^{E_0} \phi_{\text{tot}}(r, E)E\mu_a^{\text{tis}}(E)dE}{(S/4\pi r^2)e^{-\mu r}\mu_a^{\text{tis}}(E_0)E_0} \tag{4.72}$$

In Eqs. 4.70–4.72, the photon flux $\phi(r, E)$ is a function of space r and energy E, even though all photons start from the same point with the same energy E_0. Since $B(E, \mu r)$ expresses the effect of scattering as the photons travel the distance r, it should not be surprising to expect $B(E, \mu r) \rightarrow 1$ as $\mu r \rightarrow 0$.

Note that the only difference between energy and dose buildup factors is the type of gamma absorption coefficient used. For energy deposition, one uses the absorption coefficient for the medium in which energy deposition is calculated; for dose calculations, one uses the absorption coefficient in tissue.

Extensive calculations of buildup factors have been performed,[26–31] and the results have been tabulated for several gamma energies, media, and distances. In addition, attempts have been made to derive empirical analytic equations. Two of the most useful formulas are as follows:

Berger formula:

$$B(E, \mu r) = 1 + a(E)\mu r e^{b(E)\mu r} \tag{4.73}$$

Taylor formula:

$$B(E, \mu r) = A(E)e^{-a_1(E)\mu r} + [1 - A(E)]e^{-a_2(E)\mu r} \tag{4.74}$$

The constants $a(E)$, $b(E)$, $A(E)$, $a_1(E)$, $a_2(E)$ have been determined by fitting the results of calculations to these analytic expressions. Appendix E provides some values for the Berger formula constants. The best equations for the gamma buildup factor representation are based on the so-called

Table 4.4
Types of Buildup Factors

Quantity of Interest	Corresponding Buildup Factor
Flux ϕ (γ/cm²·s)	Number buildup factor
Energy deposited in medium	Energy deposition buildup factor
Dose (absorbed)	Dose buildup factor

"geometric progression" (G–P)[32] form. The G–P function has the form

$$B(E, x) = 1 + (b-1)(K^x - 1)/(K-1), \quad K \neq 1$$
$$= 1 + (b-1)x, \quad K = 1 \tag{4.75}$$

$$K(x) = cx^a + d\frac{\tanh\left[(x/X_k) - 2\right] - \tanh(-2)}{1 - \tanh(-2)} \tag{4.76}$$

where $x = \mu r$ = distance traveled in mean free paths
b = value of B for $x = 1$
K = multiplication factor per mean free paths
a, b, c, d, X_k = parameters that depend on E

Extensive tables of these constants are given in Ref. 31. The use of the buildup factor is shown in Example 4.19. More examples are provided in Chapter 16 in connection with dose–rate calculations.

Example 4.19 A 1-Ci ^{137}Cs source is kept in a large water tank. What is the energy deposition by the Cs gammas at a distance of 0.5 m from the source?
Answer Using the data of Example 4.18, the distance traveled by the 0.662-MeV photons in water is $\mu r = (0.00862 \text{ m}^2/\text{kg}) (0.5 \text{ m}) (10^3 \text{ kg/m}^3) = 4.31$ mean free path. From Ref. 32, the energy deposition buildup factor is $B(0.662, 4.31) = 13.5$.
The energy deposition is

$$E_d\left(\frac{\text{MeV}}{\text{kg s}}\right) = \frac{3.7 \times 10^{10}}{4\pi(0.5)^2} e^{-4.31}(0.00327)(0.662)13.5 = 4.62 \times 10^6 \text{ MeV/(kg s)}$$
$$= 7.4 \times 10^{-7} \text{ J/(kg s)}$$

4.9 INTERACTIONS OF NEUTRONS WITH MATTER

Neutrons, with protons, are the constituents of nuclei (see Section 3.4). Since a neutron has no charge, it interacts with nuclei only through nuclear forces. When it approaches a nucleus, it does not have to go through a Coulomb barrier, as a charged particle does. As a result, the probability (cross section) for nuclear interactions is higher for neutrons than for charged particles. This section discusses the important characteristics of neutron interactions, with emphasis given to neutron cross sections and calculation of interaction rates.

4.9.1 Types of Neutron Interactions

The interactions of neutrons with nuclei are divided into two categories: scattering and absorption.
Scattering. In this type of interaction, the neutron interacts with a nucleus, but both particles reappear after the reaction. A scattering collision is indicated as an (n, n) reaction or as

$$n + {}_Z^A X \rightarrow {}_Z^A X + n$$

Scattering may be elastic or inelastic. In elastic scattering, the total *kinetic* energy of the two colliding particles is conserved. The kinetic energy is simply redistributed between the two particles. In inelastic scattering, part of the kinetic energy is given to the nucleus as an excitation energy. After the collision, the excited nucleus will return to the ground state by emitting one or more γ-rays.
Scattering reactions are responsible for neutron's slowing down in reactors. Neutrons emitted in fission have an average energy of about 2 MeV. The probability that neutrons will induce fission is much higher if the neutrons are very slow—"thermal"—with kinetic energies of the order of eV. The fast neutrons lose their kinetic energy as a result of scattering collisions with nuclei of a "moderating" material, which is usually water or graphite.

Absorption. If the interaction is an absorption, the neutron disappears, but one or more other particles appear after the reaction takes place. Table 4.5 illustrates some examples of absorptive reactions.

4.9.2 Neutron Reaction Cross Sections

Consider a monoenergetic parallel beam of neutrons hitting a thin target* of thickness t (Figure 4.26). The number of reactions per second, R, taking place in this target may be written as

$$R(\text{reactions/s}) = \left(\begin{array}{c}\text{neutrons per m}^2\text{s}\\\text{hitting the target}\end{array}\right)\left(\begin{array}{c}\text{targets exposed}\\\text{to the beam}\end{array}\right)$$
$$\times \left(\begin{array}{c}\text{probability of interaction}\\\text{per n/m}^2\text{ per nucleus}\end{array}\right)$$

or

$$R = I[n/(\text{m}^2\text{s})][N(\text{nuclei/m}^3)][a(\text{m}^2)][t(\text{m})][\sigma(\text{m}^2)] \tag{4.77}$$

where I, a, and t are shown in Figure 4.26. The parameter σ, called the *cross section*, has the following physical meaning:

$$\sigma\ (\text{m}^2) = \begin{array}{l}\text{probability that an interaction will occur per target nucleus}\\\text{per neutron per m}^2\text{ hitting the target}\end{array}$$

Table 4.5
Absorptive Reactions

Reaction	Name
$n + {}_Z^A X \rightarrow {}_{Z-1}^A Y + p$	(n, p) reaction
$n + {}_Z^A X \rightarrow {}_{Z-2}^{A-3} Y + {}_2^4 He$	(n, α) reaction
$n + {}_Z^A X \rightarrow {}_Z^{A-1} X + 2n$	$(n, 2n)$ reaction
$n + {}_Z^A X \rightarrow {}_Z^{A+1} X + \gamma$	(n, γ) reaction
$n + {}_Z^A X \rightarrow {}_{Z_1}^{A_1} Y_1 + {}_{Z_2}^{A_2} Y_2 + n + n + \cdots$	fission

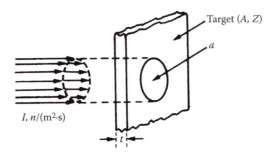

Figure 4.26 A parallel neutron beam hitting a thin target: a = area of target struck by the beam.

---

* A thin target is one that does not appreciably attenuate the neutron beam (see Eq. 4.80).

The unit of σ is the barn (b):

$$1\,\text{b} = 10^{-24}\,\text{cm}^2 = 10^{-28}\,\text{m}^2$$

Since the nuclear radius is approximately 10^{-15}–10^{-14} m, 1 b is approximately equal to the cross-sectional area of a nucleus.

Neutron cross sections are defined separately for each type of reaction and isotope. For the reactions discussed in Section 4.9.1, one defines, for example,

σ_s = elastic scattering cross section
σ_i = inelastic scattering cross section
σ_a = absorption cross section
σ_γ = capture cross section
σ_f = fission cross section

The total cross section—that is, the total probability that a reaction of any type will take place—is equal to the sum of all the σ's:

$$\sigma_{\text{tot}} = \sigma_s + \sigma_i + \sigma_\gamma + \sigma_f + \cdots \tag{4.78}$$

In the notation used here, $\sigma_a = \sigma_\gamma + \sigma_f$.

Neutron cross sections depend strongly on the energy of the neutron as well as on the atomic weight and atomic number of the target nucleus.

Figures 4.27 and 4.28 show the total cross section for two isotopes over the same neutron energy range. Notice the vast difference between the two σ's, both in terms of their variation with energy and their value in barns. (All available information about cross sections as a function of energy for all isotopes is contained in the evaluated nuclear data files [known as ENDF] stored at the Brookhaven National Laboratory, Upton, NY.)

The cross section σ (b) is called the *microscopic* cross section. Another form of the cross section, also frequently used, is the *macroscopic* cross section Σ (m^{-1}), defined by the equation

$$\Sigma_i\,(\text{m}^{-1}) = N(\text{nuclei/m}^3)[\sigma_i(\text{m}^2)] \tag{4.79}$$

and having the following physical meaning:

Σ_i = probability that an interaction of type i will take place per unit distance of travel of a neutron moving in a medium that has N nuclei/m^3.

The macroscopic cross section is analogous to the linear attenuation coefficient of γ-rays (Section 4.8.4). If a parallel beam of monoenergetic neutrons with intensity $I(0)$ impinges upon a material of thickness t, the number of neutrons that emerges without having interacted in the material is (see Figure 4.24)

$$I(t) = I(0)e^{-\Sigma_t t} \tag{4.80}$$

where $\Sigma_t = \Sigma_s + \Sigma_i + \Sigma_a + \cdots$ = total macroscopic neutron cross section.

As with γ-rays, $e^{-\Sigma_t t}$ = probability that the neutron will travel distance t without an interaction. The average distance between two successive interactions, the mean free path λ, is

$$\lambda = \frac{\int_0^\infty x e^{-\Sigma x}\,dx}{\int_0^\infty e^{-\Sigma x}\,dx} = \frac{1}{\Sigma} \tag{4.81}$$

Example 4.20 What are the macroscopic cross sections Σ_s, Σ_a, and Σ_t for thermal neutrons in graphite? The scattering cross section is $\sigma_s = 4.8$ b and the absorption cross section is $\sigma_\alpha = 0.0034$ b. What is the mean free path?

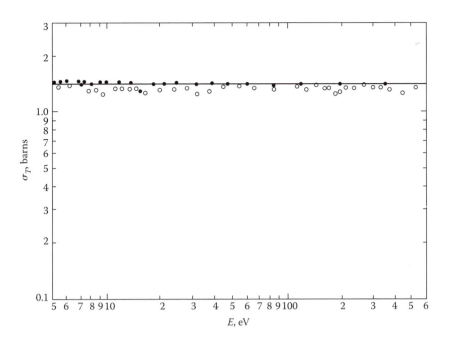

Figure 4.27 The total neutron cross section of ^{27}Al from 5 to 600 eV. (From BNL-325; http://www.nndc.bnl.gov)

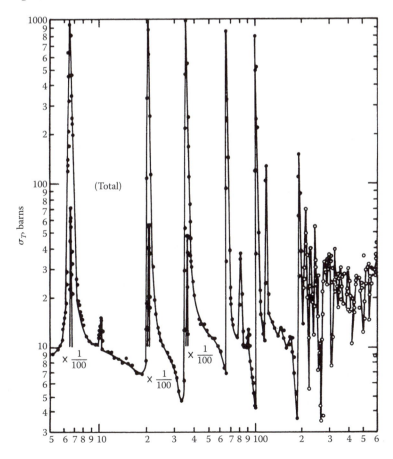

Figure 4.28 The total cross section of ^{238}U from 5 to 600 eV. (From BNL-325; http://www.nndc.bnl.gov)

Answer   For graphite, $\rho = 1.6 \times 10^3$ kg/m³ and $A = 12$. Therefore,

$$N = \rho \frac{N_A}{A} = (1.6 \times 10^3) \frac{0.6023 \times 10^{24}}{12 \times 10^{-3}} = 0.0803 \times 10^{30} \text{ atoms/m}^3$$

Using Eq. 4.79,

$$\Sigma_s = (0.0803 \times 10^{30})(4.8 \times 10^{-28}) = 38.5 \text{ m}^{-1} = 0.385 \text{ cm}^{-1}$$
$$\Sigma_a = (0.0803 \times 10^{30})(0.0034 \times 10^{-28}) = 0.027 \text{ m}^{-1} = 0.00027 \text{ cm}^{-1}$$
$$\Sigma_t = \Sigma_s + \Sigma_a = 38.53 \text{ m}^{-1} = 0.3853 \text{ cm}^{-1}$$

The mean free path is

$$\lambda = \frac{1}{\Sigma_t} = 0.0259 \text{ m} = 2.59 \text{ cm}$$

For a mixture of several isotopes, the macroscopic cross section Σ_i is calculated by

$$\Sigma_i = \sum_j N_j \sigma_{ij} \tag{4.82}$$

where σ_{ij} = microscopic cross section of isotope j for reaction type I

$$N_j = \frac{w_j \rho N_A}{A_j}$$

w_j = weight fraction of jth isotope in the mixture
ρ = density of mixture

Equation 4.82 assumes that all the isotopes act independently, that is, that the chemical-crystal binding forces are negligible. In certain cases, especially for thermal neutrons, these binding forces play an important role and cannot be neglected. In those cases, Eq. 4.82 does not apply.

Example 4.21   What is the total macroscopic absorption cross section of natural uranium? Natural uranium consists of 0.711% ^{235}U, and the rest is, essentially, ^{238}U. For thermal neutrons, the absorption cross sections are $\sigma_\alpha(^{235}\text{U}) = 678$ b and $\sigma_a(^{238}\text{U}) = 2.73$ b.
Answer   The density of uranium is 19.1×10^3 kg/m³. Therefore, using Eq. 4.82,

$$\Sigma_a(U) = 0.00711(19.1 \times 10^3) \frac{0.6023 \times 10^{24}}{235 \times 10^{-3}}(678 \times 10^{-28})$$
$$+ 0.99289(19.1 \times 10^3) \frac{0.6023 \times 10^{24}}{238 \times 10^{-3}}(273 \times 10^{-28})$$
$$+ (23.6 + 13.1) \text{ m}^{-1} = 36.7 \text{ m}^{-1} = 0.367 \text{ cm}^{-1}$$

4.9.3 The Neutron Flux

The neutron flux is a scalar quantity that is used for the calculation of neutron reaction rates. In most practical cases, the neutron source does not consist of a parallel beam of neutrons hitting a target. Instead, neutrons travel in all directions and have an energy (or speed) distribution. A case in point is the neutron environment inside the core of a nuclear reactor. Neutron reaction rates are calculated as follows in such cases.

Consider a medium that contains neutrons of the same speed v, but moving in all directions. Assume that at some point in space the neutron density is n (neutrons/m³). If a target is placed at that point, the interaction rate R [reactions/(m^3 s)] will be equal to

$$R = \left(\begin{array}{c}\text{distance travelled by all}\\\text{neutrons in 1 m}^3\end{array}\right)\left(\begin{array}{c}\text{probability of interaction per unit}\\\text{distance traveled by one neutron}\end{array}\right)$$

or

$$R = [n(\text{neutrons/m}^3)v(\text{m/s})][\Sigma(\text{m}^{-1})] = nv\ \Sigma[\text{reaction/(m}^3\text{s})]$$

The product nv, which has the units of neutrons/(m^2 s) and represents the total pathlength traveled per second by all the neutrons in 1 m^3, is called the *neutron flux ϕ*:

$$\phi = nv[n/(\text{m}^2\text{ s})] \tag{4.83}$$

Although the units of neutron flux are $n/(\text{m}^2\text{ s})$, the value of the flux $\phi(r)$ at a particular point r does not represent the number of neutrons that would cross 1 m^2 placed at point r. The neutron flux is equal to the number of neutrons crossing 1 m^2 in 1 s, *only* in the case of a parallel beam of neutrons. Using Eq. 4.83, the expression for the reaction rate becomes

$$R_i = \phi\ \Sigma_i[(\text{reactions of type } i)/(\text{m}^3\text{ s})] \tag{4.84}$$

Example 4.22   What is the fission rate at a certain point inside a nuclear reactor where the neutron flux is known to be $\phi = 2.5 \times 10^{14}$ neutrons/(m^2 s), if a thin foil of ^{235}U is placed there? The fission cross section for ^{235}U is $\sigma_f = 577$ b.
Answer   The macroscopic fission cross section is

$$\Sigma_f = N\sigma_f = (19.1 \times 10^3)\frac{0.6023 \times 10^{24}}{235 \times 10^{-3}}(577 \times 10^{-28})$$

$$= 2824 \text{ m}^{-1} = 28.24 \text{ cm}^{-1}$$

and

$$R_f = \phi\Sigma_f = (2.5 \times 10^{14})2824 = 7.06 \times 10^{17} \text{ fissions/(m}^3\text{ s})$$

$$= 7.06 \times 10^{11} \text{ fissions/(cm}^3\text{ s})$$

Another quantity related to the flux and used in radiation exposure calculations is the *neutron fluence F*, defined by

$$F(n/\text{m}^2) = \int_t \phi(t)\ dt \tag{4.85}$$

with the limits of integration taken over the time of exposure to the flux $\phi(t)$.

4.9.4 Interaction Rates of Polyenergetic Neutrons

Equation 4.84 gives the reaction rate for the case of monoenergetic neutrons. In practice, and especially for neutrons produced in a reactor, the flux consists of neutrons that have an energy spectrum extending from $E = 0$ up to some maximum energy E_{max}. In such a case, the reaction rate is written in terms of an average cross section. Let:

$\phi(E)dE$ = neutron flux consisting of neutrons with kinetic energy between
$\quad\quad\quad E$ and $E + dE$

$\sigma_i(E)$ = cross section for reaction type i for neutrons with kinetic energy E

$\quad N$ = number of targets per m^3 (stationary targets)

The reaction rate is

$$R[\text{reactions}/(\text{m}^3\text{s})] = \int dE\, \phi(E)N\sigma_i(E) \tag{4.86}$$

where the integration extends over the neutron energies of interest. The total flux is

$$\phi = \int \phi(E)\, dE \tag{4.87}$$

In practice, an average cross section is defined in such a way that, when is multiplied by the total flux, it gives the reaction rate of Eq. 4.86, that is,

$$R = \int_E\int_E dE\, \phi(E)N\sigma_i(E) = \phi\overline{\Sigma}_i = \phi N\overline{\sigma}_i \tag{4.88}$$

from which the definition of the average cross section is

$$\overline{\sigma}_i = \frac{\int dE\, \phi(E)\sigma_i(E)}{\int \phi(E)\, dE} \tag{4.89}$$

The calculation of average cross sections is beyond the scope of this text. The reader should consult the proper books on reactor physics. The main purpose of this short discussion is to alert the reader to the fact that when polyenergetic neutrons are involved, an appropriate *average* cross section should be used for the calculation of reaction rates.

PROBLEMS

4.1 Calculate the stopping power due to ionization and excitation of a 2-MeV electron moving in water. What is the radiation energy loss rate of this particle? What is the total energy radiated?

4.2 Calculate the stopping power in aluminum for a 6-MeV alpha particle.

4.3 What is the radiation (bremsstrahlung) energy loss rate of a 2 MeV electron moving in Fe? What fraction of the kinetic energy of this electron will be lost as radiation?

4.4 The window of a Geiger–Muller detector is made of mica and has a thickness of 0.02 kg/m^2 ($\rho = 2.6 \times 10^3$ kg/m^3). For mica composition, use $NaAl_3Si_3O_{10}(OH)_2$.

(a) What is the minimum electron energy that will just penetrate this window?

(b) What is the energy loss, in MeV/mm, of an electron with the kinetic energy determined in (a) moving in mica?

(c) What is the energy loss, in MeV/mm, of a 6-MeV alpha particle moving in mica?

(d) Will a 6-MeV alpha particle penetrate this mica window?

4.5 Beta particles emitted by ^{32}P($E_{max} = 1.7$ MeV) are counted by a gas detector. Assuming that the window of the detector causes negligible energy loss, what gas pressure is necessary to stop all the betas inside the detector if the length of the detector is 100 mm? Assume that the gas is argon.

4.6 What is the kinetic energy of an alpha particle that will just penetrate the human skin? For the skin, assume $t = 1$ mm; $\rho = 10^3$ kg/m^3; 65% O, 18% C, 10% H, 7% N.

4.7 Repeat Prob. 4.6 with an electron.

4.8 What is the range of a 3 MeV electron in tissue? For tissue composition use: 11% H, 65% O, 24% C.

4.9 Assuming that a charged particle loses energy linearly with distance, derive the function $T = T(x)$, where $T(x) = $ kinetic energy of the particle after going through thickness x. The initial kinetic energy is T_0, and the range is R.

4.10 A beam of 6-MeV alpha particles strikes a gold foil with thickness equal to one-third of the alpha range. What is the total energy loss of the alpha as it goes through this foil?

4.11 What is the energy deposited in a piece of paper by a beam of 1.5-MeV electrons? Assume that the paper has the composition CH_2, thickness 0.1 mm, and density 800 kg/m^3. The incident parallel electron beam consists of 10^8 electrons/(m^2 s). Give your result in MeV/(cm^2 s) and J/(m^2 s).

4.12 What is the range of 10-MeV proton in air at 1 atm? What is the range at 10 atm?

4.13 What is the range of a 4-MeV deuteron in gold?

4.14 A 1.5-MeV gamma undergoes Compton scattering. What is the maximum energy the Compton electron can have? What is the minimum energy of the scattered photon?

4.15 The energy of a Compton photon scattered to an angle of 180° is 0.8 MeV. What is the energy of the incident photon?

4.16 Prove that a gamma scattered by 180°, as a result of a Compton collision, cannot have energy greater than $mc^2/2$, where $mc^2 = 0.511$ MeV is the rest mass energy of the electron.

4.17 Prove that the attenuation coefficient of gammas for a compound or a mixture can be written as

$$\mu(\mathrm{m}^2/\mathrm{kg}) = \sum_{i=1}^{H} w_i \mu_i (\mathrm{m}^2/\mathrm{kg})$$

where w_i = weight fraction of ith element
M_i = total mass attenuation coefficient of ith element

4.18 A gamma-emitting isotope is detected by a Ge detector that is 5 mm thick. The isotope emits two gammas: 1.5 MeV and 3.5 MeV. Determine for each photon, assuming they move parallel to the detector axis,

(a)   The fraction of gammas that traverses the detector without any interaction

(b)   The average distance traveled in the detector before the first interaction takes place

(c)   Fraction of interactions in the detector that are photoelectric

(d)   Fraction of interactions that are pair production.

Attenuation coefficients:

E	μ (cm^2/g)	κ (cm^2/g)	T (cm^2/g)
1.5	0.0465	3.29×10^{-4}	4.41×10^{-4}
3.5	0.0340	9.1×10^{-5}	5.78×10^{-3}

4.19 A parallel beam of gammas impinges upon a multiple shield consisting of successive layers of concrete, Fe, and Pb, each layer having thickness 100 mm. Calculate the fraction of gammas traversing this shield. The total attenuation coefficients are μ(concrete) = 0.002 m^2/kg, μ(Fe) = 0.004 m^2/kg, and μ(Pb) = 0.006 m^2/kg; $\rho_{concrete}$ = 2.3 × 10^3 kg/m^3.

4.20 A researcher is using a 100 mCi ^{22}Na source for an experiment. Health Physics requirements dictate that he use a Pb shield of such thickness that the total gamma flux outside the shield is no more than 2000 g/cm$^{2\cdot}$s. Determine the thickness of the shield.

4.21 Assume that a parallel beam of 3-MeV gammas and a parallel beam of 2-MeV neutrons impinge upon a piece of lead 50 mm thick. What fraction of γ's and what fraction of neutrons will emerge on the other side of this shield without any interaction? Based on your result, what can you say about the effectiveness of lead as a shield for γ's or neutrons? [σ(2 MeV) = 3.5 b]

4.22 What are the capture, fission, and total macroscopic cross sections of uranium enriched to 90% in ^{235}U for thermal neutrons? (ρ = 19.1 × 10^3 kg/m^3)

$$^{235}\text{U: } \sigma_\gamma = 101 \text{ b} \qquad \sigma_f = 577 \text{ b} \qquad \sigma_s = 8.3 \text{ b}$$
$$^{238}\text{U: } \sigma_\gamma = 2.7 \text{ b} \qquad \sigma_f = 0 \qquad \sigma_s = 8 \text{ b}$$

4.23 What is the average distance a thermal neutron will travel in 90% enriched uranium (see Prob. 4.22) before it has an interaction?

4.24 The water in a pressurized-water reactor contains dissolved boron. If the boron concentration is 800 parts per million, what is the mean free path of thermal neutrons? The microscopic cross sections are

$$H_2O: \quad \sigma_s = 103 \text{ b} \qquad \sigma_a = 0.65 \text{ b}$$
$$\text{Boron:} \quad \sigma_s = 4 \text{ b} \qquad \sigma_a = 759 \text{ b}$$

BIBLIOGRAPHY

Ahmed, S. N., *Physics and Engineering of Radiation Detection*, Academic Press, Elsevier, 2007.

Brookhaven National Lab, National Nuclear Data Center, http://www.nndc.bnl.gov/

Evans, R. D., *The Atomic Nucleus*, McGraw-Hill, New York, 1972.

Hussein, E. M. A., *Radiation Mechanics: Principles and Practice*, Elsevier, Amsterdam, 2007.

Leroy, C., and Rancoita, P-G., *Principles of Radiation Interaction in Matter and Radiation*, World Scientific Publishing, Singapore, 2004.

Mughabbhab, S., *Atlas of Neutron Resonances: Resonances Parameters and Thermal Cross Sections, Z=1–100*, Elsevier, 2006.

Roy, R. R., and Reed, R. D., *Interactions of Photons and Leptons with Matter*, Academic Press, New York, 1968.

Segré, E., *Nuclei and Particles*, W. A. Benjamin, New York, 1968.

Shultis, J. K., and Faw, R. E., *Radiation Shielding*, American Nuclear Society, La Grange Park, IL, 2000.

Shultis, J. K., and Faw, R. E., *Fundamentals of Nuclear Science and Engineering*, Marcel Dekker, New York, 2002.

Ziegler, J. F., Biersack, J. P., and Ziegler, M. D., *SRIM: The Stopping and Range of Ions in Matter*, Self-Publishing; http://www.lulu.com/, 2008.

REFERENCES

1. Berger, M. J., and Seltzer, S. M., NASA SP-3012, 1964.

2. Anderson, H. H., Sorensen, H., and Vadja, P., *Phys. Rev.* **180**:383 (1969).

3. Sorensen, H., and Anderson, H. H., *Phys. Rev.* **8B**:1854 (1973).

4. Janni, J. F., *Atomic Data Nucl. Data Tables* **27**:147–339 (1982).

5. Berger, M. J., and Seltzer, S. M., NBSIR 82–2550A, 1983.

6. Sternheimer, R. M., *Phys. Rev.* **88**:851 (1952).

7. Sternheimer, R. M., *Phys. Rev.* **103**:511 (1956).

8. Williamson, C. F., Baujot, J. P., and Picard, J., CEA-R-3042 (1966).

9. Bichsel, H., *Phys. Rev.* **112**:1089 (1958).

10. Tabata, T., Ito, R., and Okabe, S., *Nucl. Instrum. Meth.* **103**:85 (1972).

11. Northcliffe, L. C., *Ann. Rev. Nucl. Sci.* **13**:67 (1963).

12. Lindhard, J., Scharff, M., and Schiott, H. E., *Fys. Med.* **33**:14 (1963).

13. Moak, C. D., and Brown, M. D., *Phys. Rev.* **149**:244 (1966).

14. Brown, M. D., and Moak, C. D., *Phys. Rev.* **6B**:90 (1972).

15. Betz, G., Isele, H. J., Rossle, E., and Hortig, G., *Nucl. Instrum. Meth.* **123**:83 (1975).

16. Booth, W., and Grant, I. S., *Nucl. Phys.* **63**:481 (1965).

17. Forster, J. S., Ward, D., Andrews, H. R., Ball, G. G., Costa, G. J., Davies, W. G., and Mitchell, J. V., *Nucl. Instrum. Meth.*, **136**:349 (1976).

18. Northcliffe, L. C., and Schilling, R. F., *Nucl. Data* **A7**:223 (1970).

19. Brown, M. D., "Interaction of Uranium Ions in Solids," Ph.D. thesis, University of Tennessee, 1972 (unpublished).

20. Bridwell, L., and Bucy, S., *Nucl. Sci. Eng.* **37**:224 (1969).

21. Bridwell, L., and Moak, C. D., *Phys. Rev.* **156**:242 (1967).

22. Siffert, P., and Coche, A., "General Characteristics of the Interactions of Nuclear Radiations with Matter and their Consequences," in G. Bertolini and A. Coche (eds.), *Semiconductor Detectors*, Wiley, New York, 1968, pp. 279–300.

23. Hakim, M., and Shafrir, N. H., *Nucl. Sci. Eng.* **48**:72 (1972).

24. Gesini, G., Lucarini, G., and Rustichelli, R., *Nucl. Instrum. Meth.* **127**:579 (1975).

25. Hubbell, J. H., NSRDS-NBS 29, August 1969.

26. Goldstein, H., and Wilkins, J. E., Jr., NYO-3075, Nuclear Development Associates, Inc., 1954.

27. Jaeger, R. G. (ed.), *Engineering Compendium on Radiation Shielding* V1, IAEA, New York, 1968.

28. Trubey, D. K., ORNL-RSIC-10, 1966.

29. Eisenhauer, C. M., and Simmons, G. L., *Nucl. Sci. Eng.* **56**:263 (1975).

30. Kuspa, J. P., and Tsoulfanidis, N., *Nucl. Sci. Eng.* **52**:117 (1973).

31. Chilton, A. B., *Nucl. Sci. Eng.* **59**:436 (1979).

32. Trubey, D. K., "New Gamma Buildup Factor Data for Point Kernel Calculations: ANS-6.4.3 Standard Reference Data," ORNL/RSIC-49, 1988.

5

Gas-Filled Detectors

5.1 INTRODUCTION

Gas-filled detectors operate by utilizing the ionization produced by radiation as it passes through a gas. Typically, such a detector consists of two electrodes to which a certain electrical potential is applied. The space between the electrodes is filled with a gas (Figure 5.1). Ionizing radiation, passing through the space between the electrodes, dissipates part or all of its energy by generating electron-ion pairs. Both electrons and ions are charge carriers that move under the influence of the electrical field. Their motion induces a current on the electrodes, which may be measured (Figure 5.1a). Or, through appropriate electronics, the charge produced by the radiation may be transformed into a pulse, in which case particles are counted individually (Figure 5.1b). The first type of detector (Figure 5.1a) is called *current* or *integrating chamber*; the second type (Figure 5.1b) is called *pulse chamber*. To get an idea of what charges and currents one might expect to measure, consider this representative example.

For most gases, the average energy required to produce an electron–ion pair is about 30 eV. This number takes into account all collisions, including those that lead to excitation. If a 3-MeV alpha and beta particle deposits all its energy in the detector, it will produce, on the average,

$$\frac{3 \times 10^6}{30} = 10^5 \text{ electron–ion pairs}$$

A typical gas-filled detector* has a capacitance of about 50 pF, and the charge will be collected in a time of the order of 1 μs. If all the charge created by the 3-MeV particle is collected, the voltage and current expected are of the order of

$$V = \frac{Q}{C} \approx \frac{10^5 \times 1.6 \times 10^{-19} \text{C/el}}{50 \times 10^{-12} \text{F}} \approx 0.5 \times 10^{-3} \text{ V} \approx 0.5 \text{ mV}$$

$$i = \frac{Q}{t} \approx \frac{10^5 \times 1.6 \times 10^{-19}}{10^{-6}} \text{A} \sim 1.6 \times 10^{-8} \text{ A}$$

In an ionized gas without an electric field, electrons and positive ions will move at random with an average kinetic energy equal to $(3/2)kT$, where k = Boltzmann's constant and T = temperature of the gas (Kelvin). When an electric field is present, both electrons and positive ions acquire a net velocity component along the lines of the electric field. Electrons move toward the positive electrode, positive ions toward the negative one. The force on either charge carrier is the same and equal to $F = Ee$, where E = electric field intensity, but the acceleration is quite different. The acceleration a is equal to F/M, where M is the mass of the ion or electron. Therefore, the acceleration of an electron will be thousands of times larger than the acceleration of an ion. The time it takes the electrons to reach the positive electrode of a typical detector is about 1 μs. The corresponding time for the positive ions is about 1 ms, a thousand times longer.

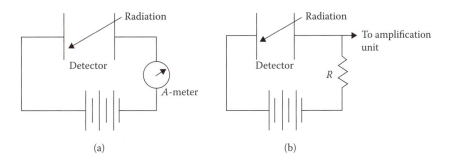

Figure 5.1 A typical gas-filled detector: (a) the direct current produced in the circuit is measured; (b) individual pulses are detected.

---

* Although the correct term is gas-filled detector, the short term *gas counter* is also used.

The discussion up to this point has been limited to the effects of the ionization produced directly by the incident particle. This is called *primary* ionization. There are types of gas-filled detectors in which the electric field is so strong that the electrons of the primary ionization acquire enough kinetic energy between collisions to produce new electron–ion pairs. These new charges constitute the *secondary* ionization. Primary and secondary ionization are generated within such a short period of time that they contribute to one and the same pulse.

5.2 RELATIONSHIP BETWEEN HIGH VOLTAGE AND CHARGE COLLECTED

Assume that the following experiment is performed (Figure 5.2). A radioactive source of constant intensity is placed at a fixed distance from a gas-filled detector. The high voltage (HV) applied to the detector may be varied with the help of a potentiometer. An appropriate meter measures the charge collected per unit time. If the HV applied to the detector is steadily increased, the charge collected per unit time changes as shown in Figure 5.3. The curve of Figure 5.3 is divided into five regions, which are explained as follows:

Region I. When the voltage is very low, the electric field in the detector is not strong, electrons and ions move with relatively slow speeds, and their recombination rate is considerable. As V increases, the field becomes stronger, the carriers move faster, and their recombination rate decreases up to the point where it becomes zero. Then, all the charge created by the ionizing radiation is being collected ($V = V_I$). Region I is called the *recombination* region.

Region II. In region II, the charge collected stays constant despite a change in the voltage because the recombination rate is zero and no new charge is produced. This is called the *ionization* region.

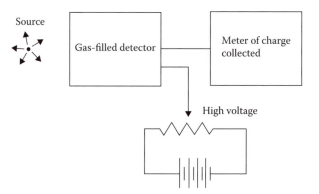

Figure 5.2 Experimental setup for the study of the relationship between high voltage applied and charge collected.

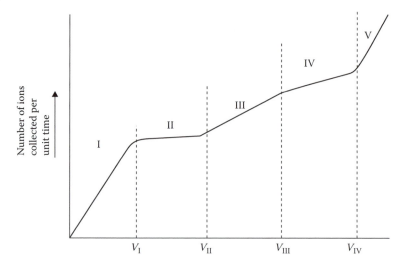

Figure 5.3 The relationship between voltage applied to the detector and charge collected.

144

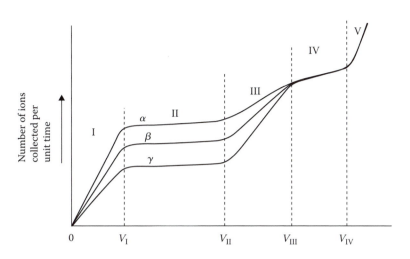

Figure 5.4 The relationship between charge collected and applied voltage for three different types of particles. In region IV, the curve increases slightly but is the same for all particles.

Region III. In this region, the collected charge starts increasing because the electrons produce secondary ionization that results in charge multiplication. The electric field is so strong, in a certain fraction of the detector volume, that electrons from the primary ionization acquire enough energy between collisions to produce additional ionization. The gas *multiplication factor*—that is, the ratio of the total ionization produced divided by the primary ionization—is, for a given voltage, independent of the primary ionization. Thus the output of the detector is proportional to the primary ionization. The pulse height at the output is proportional to the energy dissipated inside the detector; therefore particle identification and energy measurement are possible. This region is, appropriately enough, called the *proportional* region.

Region IV. In this region, the electric field inside the detector is so strong that a single electron-ion pair generated in the chamber is enough to initiate an avalanche of electron-ion pairs. This avalanche will produce a strong signal with shape and height independent of the primary ionization and the type of particle, a signal that depends only on the electronics of the detector. Region IV is called the *Geiger–Müller* (GM) region.

Region V. If the applied voltage is raised beyond the value V_{IV}, a single ionizing event initiates a continuous discharge in the gas, and the device is not a particle detector anymore. No gas-filled detector should operate with voltage $V > V_{IV}$.

If the graph discussed above is obtained using an α, β, or γ source, the results will be as shown in Figure 5.4.

5.3 VARIOUS TYPES OF GAS-FILLED DETECTORS

Gas-filled detectors take their name from the voltage region ion which they operate. No detector operates in region I of Figure 5.3, because a slight change in voltage will change the signal.

Ionization chambers operate in region II. No charge multiplication takes place. The output signal is proportional to the particle energy dissipated in the detector; therefore measurement of particle energy is possible. Since the signal from an ionization chamber is not large, only strongly ionizing particles such as alphas, protons, fission fragments, and other heavy ions are detected by such detectors. The voltage applied is less than 1000 V.

Proportional counters operate in region III. Charge multiplication takes place, but the output signal is still proportional to the energy deposited in the counter. Measurement of particle energy is possible. Proportional counters may be used for the detection of any charged particle.

Identification of the type of particle is possible with both ionization and proportional counters. An alpha particle and an electron having the same energy and entering either of the detectors, will give a different signal. The alpha particle signal will be bigger than the electron signal. The voltage applied to proportional counters ranges between 800 and 2000 V.

GM counters operate in region IV. GM counters are very useful because their operation is simple and they provide a very strong signal, so strong that a preamplifier is not necessary. They can be used with any kind of ionizing radiation (with different levels of efficiency). The disadvantage of

145

GM counters is that their signal is independent of the particle type and its energy. *Therefore, a GM counter provides information only about the number of particles.* Another minor disadvantage is their relatively long dead time (200–300 ms). (For more details about dead time, see Section 5.6.2.) The voltage applied to GM counters ranges from 500 to 2000 V.

Gas-filled detectors may be constructed in any of three basic geometries: parallel plate, cylindrical, or spherical (Figure 5.5). In a parallel-plate chamber, the electric field (neglecting edge effects) is uniform, with strength equal to

$$E = \frac{V_0}{d} \tag{5.1}$$

In the cylindrical chamber, the voltage is applied to a very thin wire, a few mills of an inch in diameter, stretched axially at the center of the cylinder. The cylinder wall is usually grounded. The electric field is, in this case,

$$E(r) = \frac{V_0}{\ln(b/a)} \frac{1}{r} \tag{5.2}$$

where a = radius of the central wire
b = radius of the counter
r = distance from the center of the detector

It is obvious from Eq. 5.2 that very strong electric fields can be maintained inside a cylindrical detector close to the central wire. Charge multiplication is achieved more easily in a cylindrical than in a plate-type gas detector. For this reason, proportional and GM counters are manufactured with cylindrical geometry.

In a spherical detector, the voltage is applied to a small sphere located at the center of the detector. The wall of the detector is usually grounded. The electric field is

$$E(r) = V_0 \frac{ab}{b-a} \frac{1}{r^2} \tag{5.3}$$

where a, b, and r have the same meaning as in cylindrical geometry. Strong fields may be produced in a spherical counter, but this type of geometry is not popular because of construction difficulties.

A detector filled with a gas at a certain pressure may operate in any of the regions II–IV discussed earlier, depending on a combination of the following parameters:

1. Size of the detector

2. Size of wire (in cylindrical detectors)

3. Gas type

4. Gas pressure

5. Level of high voltage.

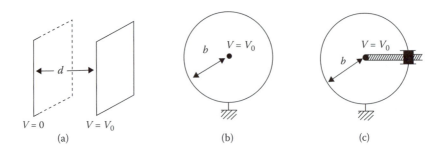

Figure 5.5 The different geometries of gas-filled detectors: (a) parallel plate; (b) cylindrical; (c) spherical.

Normally, gas detectors are manufactured to operate in one region only. The user buys an ionization counter, a proportional counter, or a GM counter. The manufacturer has selected the combination of variables 1–4 listed above that results in the desired type of gas-filled detector. The last variable, the high voltage applied, is not a fixed number, but a range of values. The range is specified by the manufacturer, but the user decides on the best possible value of HV.

The rest of this chapter discusses the special characteristics of the three types of gas-filled detectors.

5.4 IONIZATION CHAMBERS

5.4.1 Pulse Formation in an Ionization Chamber

The formation and shape of the signal in an ionization chamber will be analyzed for a parallel-plate detector as shown in Figure 5.1b. The analysis is similar for a cylindrical or a spherical chamber.

Consider the ionization chamber shown in Figure 5.6. The two parallel plates make a capacitor with capacitance C, and with the resistor R an RC circuit is formed. A constant voltage V_0 is applied on the plates. The time-dependent voltage $V(t)$ across the resistor R represents the signal. The objective of this section is to obtain the function $V(t)$.

Assume that one electron–ion pair has been formed at a distance x_0 from the collecting plate (collector). The electron and the ion start moving in the electric field, and they acquire kinetic energy at the expense of the electrostatic energy stored in the capacitance of the chamber. If the charge moves a distance dx, conservation of energy requires that

$$\text{Work on charges} = \text{Change in electrostatic energy}$$

$$eE(dx^+ + dx^-) = d\left(\frac{Q^2}{2C}\right) = \frac{Q}{C}\,dQ \approx V_0(dQ^- + dQ^+) \tag{5.4}$$

where E = electric field intensity
$\quad\quad\quad Q$ = charge on chamber plates
dQ^+, dQ^- = changes in positive, negative charge, respectively

It is assumed that the change in the charge (dQ) is so small that the voltage V_0 stays essentially constant. The voltage $V(t)$ across the resistor R is the result of this change in the charge and is given by

$$V(t) = \frac{1}{C}\int_0^t dQ(t) = \frac{1}{C}\int_0^t (dQ^+ + dQ^-) \tag{5.5}$$

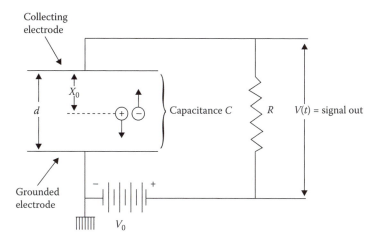

Figure 5.6 The electronic circuit of a parallel-plate ionization chamber.

Substituting in Eq. 5.5 the value of dQ from Eq. 5.4, one obtains

$$V(t) = \frac{1}{C}\int_0^t \frac{e}{V_0} E\,(dx^+ - dx^-)$$

(5.6)

Let

w^+ = drift velocity of positive ions
w^- = drift velocity of electrons

In general, the drift velocity is a function of the *reduced field strength* E/p, where p is the gas pressure in the chamber.

The derivation up to this point is independent of the chamber geometry. To proceed further requires substitution of the value of the electric field from either Eq. 5.1, 5.2, or 5.3. For a plate-type ionization chamber the field is constant (Eq. 5.1), independent of x, and so is the drift velocity. Therefore, Eq. 5.6 becomes

$$V(t) = \frac{e}{Cd}\int_0^t (w^+ + w^-)\,dt = -\frac{e}{Cd}(w^-t + w^+t)$$

(5.7)

The drift velocity of the electron is a few thousand times more than the velocity of the ion,[*] which means the electron will reach the collector plate before the ion has hardly moved. Let

$T^{(+)}$ = time it takes for an ion to reach the cathode
$T^{(-)}$ = time it takes for an electron to reach the collector (anode)

Typical values of these times are

$$T^{(+)} \approx \text{ms} \quad T^{(-)} \approx \mu\text{s}$$

Equation 5.7 shows that for $t < T^{(-)}$, the voltage $V(t)$ changes linearly with time (Figure 5.7):

$$V(t) = -\frac{e}{Cd}(w^- + w^+)t \quad 0 < t \le T^{(-)}$$

(5.8)

For $T > T^{(-)}$ the signal is

$$V(t) = -\frac{e}{Cd}(x_0 + w^+t) \quad t > T^{(-)}$$

(5.9)

Finally, after $t = T^{(+)}$, the ion reaches the grounded cathode and the signal reaches its maximum (negative) value, which is

$$V(T^+) = -\frac{e}{Cd}x_0 \quad t > T^{(+)}$$

(5.10)

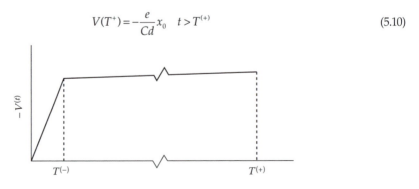

Figure 5.7 The voltage pulse generated by an ionization chamber.

---

[*] Typical values of drift velocities are $w^+ \approx 10$ m/s, $w^- = 10^4 - 10^5$ m/s.

If N electron–ion pairs are produced, the final voltage will be

$$V(T^+) = -\frac{Ne}{Cd}x_0, \quad t = T^{(+)} \tag{5.11}$$

For $t > T^{(+)}$ the pulse decays with decay constant RC (see Section 10.3).

The pulse profile of Figure 5.7 was derived under the assumption that all ion pairs were produced at $x = x_0$. Actually, the ionization is produced along the track traveled by the incident particle. The final pulse will be the result of the superposition of many pulses with different $T^{(-)}$ values. Because of this effect, the sharp change in slope at $t = T^{(-)}$ will disappear and the pulse will be smoother.

The pulse of Figure 5.7 is not suitable for counting individual particles because it does not decay quickly enough. A pulse-type detector should produce a signal that decays faster than the average time between the arrival of two successive particles. For example, if the counting rate is 1000 counts/min, a particle arrives at the detector, on the average, every 1/1000 min (60 ms).

In Figure 5.7, the pulse could be stopped at time $t = T^{(+)}$ by electronic means. Such a technique would produce pulses with height proportional to the total charge generated in the detector, but with a duration of a few hundreds of microseconds, which is unacceptably long. The method used in practice is to "chop off" the pulse at time $t = T^{(-)}$, which amounts to stopping the pulse after only the electrons are collected. The signal is then fed into a RC circuit that, as described in Chapter 10, changes the pulse as shown in Figure 5.8.

Let $V_i(t)$ be the signal at the output of the detector that is used as an input to an RC circuit. From Eq. 5.8,

$$V_i(t) = -\frac{e}{Cd}(w^- + w^+)t = kt \tag{5.8a}$$

Using this signal as an input, the output voltage across the resistor R_0 is (see Sections 10.3 and 10.4), for $0 \le t \le T^{(-)}$ (Figure 5.7),

$$V_0(t) = kC_0R_0(1 - e^{-t/C_0R_0}) \tag{5.12}$$

For $t > T^{(-)}$, $V_i(t)$ is essentially constant, and

$$V_0(t) = \frac{kC_0R_0(1 - (e^{-T^{(-)}}/C_0R_0))\, e^{-t}}{R_0C_0} \tag{5.13}$$

The signal $V_0(t)$ is shown in Figure 5.8b. Usually, the RC circuit is the first stage of the preamplifier, which accepts the signal of the ionization chamber.

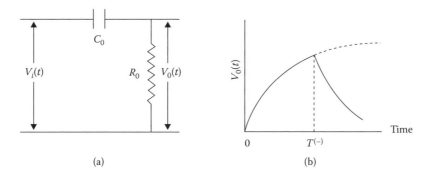

(a)   (b)

Figure 5.8 (a) The signal $V_i(t)$ is fed into the RC circuit, (b) the output of the RC circuit decays quickly with a decay constant R_0C_0.

149

The disadvantage of the signal in Figure 5.8b is that its maximum value depends on the position where the ionization was produced. Indeed, from Eq. 5.12, one obtains for $t = T^{(-)}$ noting that $k = -e(w^- + w^+)/Cd \approx -ew^-/Cd$,

$$\text{since } w^- \gg w^+, \quad T^{(-)} \ll C_0 R_0, \quad \text{and} \quad T^{(-)} = \frac{x_0}{w^-}$$

$$V(T^{(-)}) = -(ew^-/Cd)\,C_0 R_0 (1 - e^{-T^{(-)}/C_0 R_0}) \approx -(ex_0/Cd) \tag{5.14}$$

Thus the peak value of the pulse in Figure 5.8b depends on x_0. This disadvantage can be corrected in several ways. One is by placing a grid between the two plates and keeping it at an intermediate voltage V_g ($0 < V_g < V_0$). For more details about the "gridded" ionization chamber, the reader should consult the references at the end of this chapter.

The analysis of the pulse formation in a cylindrical or a spherical detector follows the same approach. The results are slightly different because the electric field is not constant (see Eqs. 5.2 and 5.3), but the general shape of the signal is that shown in Figure 5.7. (See Franzen and Cochran, 1962; and Kowalski for detailed calculations of the pulse shapes for the three geometries of gas-filled chambers.)

5.4.2 Current Ionization Chambers

An ionization chamber of the *current* type measures the average ionization produced by many incoming particles. This is achieved by measuring directly the electrical current generated in the chamber, using either a sensitive galvanometer for currents of 10^{-8} A or higher (Figure 5.9), or an electrometer (sometimes with an amplifier) for currents less than 10^{-8} A. In the case of the electrometer, as shown in Figure 5.10, the current is determined by measuring the voltage drop across the known resistance R. The voltage drop may be measured by the electrometer directly or after some amplification.

For current ionization chambers, it is very important to know the relationship between applied voltage and output current (for a constant radiation source). This relationship, which is shown in Figure 5.11, consists of regions I and II of the graph of Figure 5.3. The proper operating voltage of the ionization chamber is that for which all the ionization produced by the incident radiation is

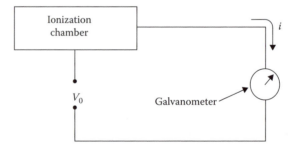

Figure 5.9 Measurement of the current produced by an ionization chamber by using a galvanometer.

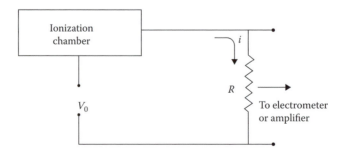

Figure 5.10 Measurements of the current produced by an ionization chamber by using an electrometer.

measured. If this is the case, a slight increase of the applied voltage will result in negligible change of the measured current. The voltage is then called the saturation voltage (V_s), and the corresponding current is called saturation current. The value of the saturation current depends on the intensity and type of the radiation source (Figure 5.11). It also depends, for the same radiation source, on the size and geometry of the chamber as well as on the type and pressure of the gas used. If one considers different gases, other things being equal, the highest current will be produced by the gas with the lowest average energy needed for the production of one electron–ion pair. Typical energies for common gases are given in Table 5.1.

During measurements of the ionization current with an electrometer, one would like to know the response of the measuring instrument if the signal from the ionization chamber changes. Assume that the current of the chamber changes suddenly from a value of i_1 to i_2. The response of the electrometer is obtained by considering the equivalent electronic circuit of Figure 5.10, shown in Figure 5.12. The capacitor C represents the combined capacitance of the chamber and everything else. The resistor R represents a corresponding total resistance for the circuit. The signal to be measured is the voltage $V(t)$, where *for $t \leq 0$*,

$$V_1 = i_R R = i_1 R \tag{5.15}$$

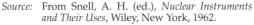

Figure 5.11 The ionization chamber current as a function of applied voltage.

Table 5.1
Average Energy Needed for Production of One Electron–Ion Pair

Gas	Energy per Pair (eV)
H	36.3
He	42.3
A	26.4
Air	34
CO_2	32.9
C_2H_6 (Ethane)	24.8
CH_4	27.3

Source: From Snell, A. H. (ed.), *Nuclear Instruments and Their Uses*, Wiley, New York, 1962.

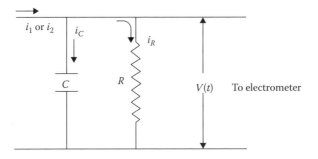

Figure 5.12 The equivalent electronic circuit of Figure 5.10.

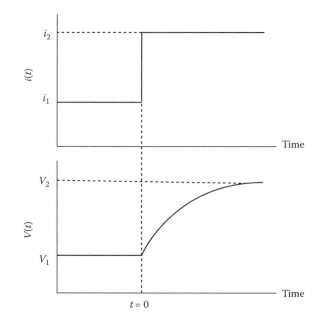

Figure 5.13 Response of an electrometer to a step change of the ionization current.

At $t = 0$, the current changes instantaneously from i_1 to i_2, and the voltage will eventually become

$$V_2 = i_2 R \tag{5.16}$$

During the transition period, Kirchhoff's first law gives

$$i_2 = i_C + i_R = \frac{dQ(t)}{dt} + \frac{V(t)}{R} = C\frac{dV(t)}{dt} + \frac{V(t)}{R}$$

or

$$\frac{dV(t)}{dt} + \frac{1}{RC}V(t) = \frac{i_2}{C} \tag{5.17}$$

The solution of this differential equation, with the initial condition given by Eq. 5.15, is

$$V(t) = i_2 R + R(i_1 - i_2)e^{-t/RC} \tag{5.18}$$

The function given by Eq. 5.18 is shown in Figure 5.13. The response of the electrometer is exponential with a rate of change determined by the time constant RC. For fast response, the time constant should be as short as practically possible.

5.5 PROPORTIONAL COUNTERS

5.5.1 Gas Multiplication in Proportional Counters

When the electric field strength inside a gas-filled detector exceeds a certain value, the electrons that move in such a field acquire, between collisions, sufficient energy to produce new ions. Thus, more electrons will be liberated, which in turn will produce more ions. The net effect of this process is multiplication of the primary ionization. The phenomenon is called *gas multiplication*.* To achieve the high field intensity needed for gas multiplication without excessive applied voltage,

---

* Also called gas gain or charge amplification.

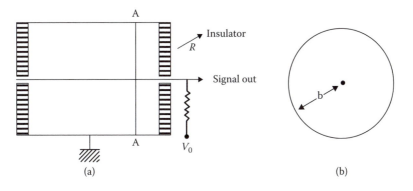

Figure 5.14 (a) A cylindrical gas-filled detector. (b) Cross section of the detector at AA.

chambers operating in this mode are usually cylindrical with a very thin wire stretched axially at the center of the detector (Figure 5.14). The wall of the detector is normally grounded and a positive voltage is applied to the central wire. In such a geometry, the electrostatic field inside the chamber is radial and its intensity is

$$E(r) = \frac{V_0}{\ln(b/a)} \frac{1}{r} \tag{5.2}$$

The field intensity increases rapidly as the wire is approached. Since the radius a of the wire is a few mills of an inch and thousands of times smaller than the radius b of the detector, an extremely strong electric field is produced in a fraction of the chamber's volume. This volume is so small that the probability that the incident radiation will produce an electron ion pair in it is negligible.

In addition to the secondary electrons produced by collisions, electrons are also produced by two other processes as given below:

1. Photoelectric interactions

2. Bombardment of the cathode surface by positive ions.

The photoelectric interactions are caused by photons that are produced in the detector as a result of the ionization and excitation of the atoms and molecules of the gas. If the chamber is filled with a monatomic gas, these photons produce photoelectrons only when they strike the cathode (wall of cylinder) because they do not have enough energy to ionize the atoms of the gas. If the detector is filled with a gas mixture, however, photons emitted by molecules of one gas may ionize molecules of another.

Electrons are also emitted when the positive ions, which are produced in the chamber, reach the end of their journey and strike the cathode. The significance of this effect depends on the type of material covering the surface of the cathode and, more important, on the type of the gas filling the chamber.

The production of electrons by these processes results in the generation of successive avalanches of ionization because all the electrons, no matter how they are produced, migrate in the direction of the intense electric field and initiate additional ionization. The gas multiplication factor M, which is equal to the *total* number of free electrons produced in the detector when *one* pair is produced by the incident radiation, is calculated as follows. Let:

N = total number of electrons set free per primary electron–ion pair

δ = average number of photoelectrons produced per ion pair generated in the detector ($\delta \ll 1$)

The initial avalanche of N electrons will produce δN photoelectrons. Each photoelectron produces a new avalanche of N new electrons; therefore the second avalanche consists of δN^2 electrons. The third avalanche will have δN^3 electrons, and so on. The total number of electrons per initial ion pair produced is then

$$M = N + \delta N^2 + \delta N^3 + \cdots$$

The magnitude of δN depends on the applied voltage. If $\delta N < 1$, the gas multiplication factor is

$$M = \frac{N}{1 - \delta N} \qquad (5.19)$$

It should be noted that

1. If $\delta N \ll 1$, the photoelectric effect is negligible and $M = N$ = initial gas multiplication (first avalanche).
2. If $\delta N < 1$, M can become much larger than N.
3. If $\delta N \geq 1$, $M \to \infty$, which means that a self-supporting discharge occurs in the detector.

The gas multiplication factor M is a function of the ratio $V_0/\ln(b/a)$ and the product Pa, where P is the pressure of the gas in the detector (Rossi and Staub, 1949). Experimental results of M values for two gases are shown in Figures 5.15 and 5.16. Diethorn[1] has obtained the equation

$$\ln M = \frac{V \ln 2}{\Delta V \ln(b/a)} \ln \frac{V}{\overline{K} Pa \ln(b/a)} \qquad (5.20)$$

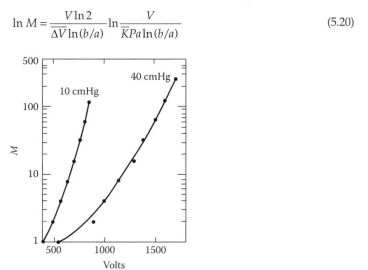

Figure 5.15 Gas multiplication M versus voltage. Gas is 93.6% pure argon ($a = 0.005$ in, $b = 0.435$ in, at two different pressures). (From Rossi, B. B., and Staub, H. H., *Ionization Chambers and Counters*, McGraw-Hill, New York, 1949.)

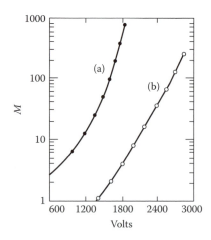

Figure 5.16 Gas multiplication M versus voltage. Gas is BF_3. (a) $a = 0.005$ in, $b = 0.75$ in, $P = 10$ cmHg. (b) $a = 0.005$ in, $b = 0.78$ in, $P = 80.4$ cmHg. (From Rossi, B. B., and Staub, H. H., *Ionization Chambers and Counters*, McGraw-Hill, New York, 1949.)

where $\overline{\Delta V}$ and $\overline{K}$ are constants of the gas. Equation 5.20 has been tested and found to be valid.[2–4] As Figures 5.15 and 5.16 show, M increases almost exponentially with applied voltage.

One method by which the strong dependence of M on applied voltage is reduced is by adding a small amount of a polyatomic organic gas in the gas of the detector. One popular mixture is 10% CH_4 and 90% argon. The organic gases, called "quenching" gases, stabilize the operation of the detector by reducing the effect of the secondary processes. They achieve this because organic polyatomic molecules

1. Dissociate rather than produce electrons when they hit the cathode

2. Dissociate when they absorb a photon

3. Have lower ionization potential than the molecules of the main gas; as a result, they are ionized in collisions with ions of the main gas and thus prevent the ions from reaching the cathode

The total charge produced in a proportional detector is

$$Q = MNe = M\frac{\Delta E}{w}e \tag{5.21}$$

where ΔE = energy of the incident particle dissipated in the detector and
w = average energy required for production of one electron–ion pair

Equation 5.21 indicates that Q (output) is proportional to the energy deposited in the detector (ΔE). This is the reason why such detectors are called proportional. The proportionality holds, however, only if the gas multiplication factor M is constant, independent of the primary ionization. The question then arises, under what conditions is this true?

A proportional counter is strictly proportional as long as the space charge due to the positive ions does not modify too much the electric field around the wire. The magnitude of the space charge is a function of the primary ionization and the gas multiplication. If the primary ionization is very small, the value of M may be 10^5 to 10^6 before the space charge affects the proportionality. On the other hand, if the primary ionization is too strong, the critical value of M is smaller. It has been reported[5] that there is a critical maximum value of the charge produced by the multiplication process beyond which proportionality does not hold. That number, obviously, depends on the counter (size, types of gas, etc.).

The events that produce the avalanches of electrons in a proportional counter are statistical in nature. The final multiplication factor M will not be constant but will show statistical fluctuations. The probability that the multiplication will have the value M is, according to Snyder,[6] equal to

$$P(M) = \frac{1}{\overline{M}}\exp\left(-\frac{M}{\overline{M}}\right) \tag{5.22}$$

where $\overline{M}$ is the mean multiplication factor. The variance of M is, from Eq. 5.22,

$$\sigma_M^2 = \overline{M}^2 \tag{5.23}$$

5.5.2 The Pulse Shape of a Proportional Counter

The shape of the pulse of a proportional counter is understood as one follows the events that lead to the formation of the pulse. A cylindrical counter will be considered, such as that shown in Figure 5.14.

Assume that the incident particle generated N electron–ion pairs at a certain point inside the counter. The electrons start moving toward the wire (anode). As soon as they reach the region of the strong field close to the wire, they produce secondary ionization. Since all the secondary ionization is produced in the small volume surrounding the wire, the amplitude of the output pulse is independent of the position of the primary ionization. The electrons of the secondary ionization are collected quickly by the wire, before the ions have moved appreciably. The ion contribution to the pulse is negligible because the ions cross only a very small fraction of the potential difference on their way to the anode. The pulse developed in the central wire is almost entirely due to the motion of the ions. As the ions move toward the cathode, the voltage pulse on the wire begins to rise: quickly at first, when the ions are crossing the region of the intense electric field and slower later, when the ions move into the region of low-intensity field. The voltage pulse as a function of time is given by (Kowalski, 1970)

$$V(t) = \frac{Q}{2C\ln(b/a)}\ln\left(1 + \frac{b^2}{a^2}\frac{t}{t_{ion}}\right)$$

(5.24)

where Q is given by Eq. 5.21
   C = capacitance of the counter
   t_{ion} = time it takes the ions to reach the cathode

The equation for t_{ion} is (Kowalski)

$$t_{ion} = \frac{P\ln(b/a)}{2V_0\mu_{ion}}(b^2 - r^2)$$

(5.25)

where P = gas pressure
   μ_{ion} = ion mobility in the field of the counter*
   r = point where the ion was produced

The pulse $V(t)$ is shown by the solid line of Figure 5.17. The pulse rises quickly and reaches half of its maximum in time of the order of microseconds. Then it bends and rises at a much slower rate, until about a millisecond later it reaches its final value, Q/C.

The pulse of Figure 5.17 was derived under the assumption that all the ions were produced at the same point. In reality, the ions are produced along the track of the incident particle. This modifies the shape of the pulse during its initial rise but it leaves it virtually unaffected during the later period.

The pulse of Figure 5.17 is unacceptably long, even for a modest counting rate. As in the case of the ionization chamber, the pulse is "chopped off" at some convenient time with the help of a differentiating circuit (Chapter 10). The result will be a pulse shown by the dashed line in Figure 5.17.

5.5.3 The Change of Counting Rate with High Voltage—The High-Voltage Plateau

When a detector is used for the study of a phenomenon involving counting of particles, the investigator would like to be certain that changes in the counting rate are due to changes in the phenomenon under study and not due to changes of the environment such as atmospheric pressure, temperature, humidity, or voltage. For most radiation measurements, all these factors may be neglected except voltage changes.

Consider a gas-filled detector. For its operation, it is necessary to apply HV, usually positive, which may range from +300 to +3000 V, depending on the detector. For the specific detector used in an experiment, the observer would like to know by what fraction the counting rate will change if the HV changes by a certain amount. It is highly desirable to have a system for which the change

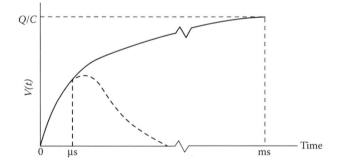

Figure 5.17 The voltage pulse of a proportional counter.

---

* The ion mobility is the proportionality constant between the drift velocity and the reduced field; thus
   $w^+ = \mu^+ (E/P)$.

in the counting rate is negligible, when the HV changes for a reason beyond the control of the investigator (e.g., change in the 110 V provided by the outlet on the wall, which may, in turn, cause a fluctuation in the output of the HV power supply). For this reason, the response of a counting system to such variations ought to be known. This information is provided by the HV plateau of the detector. The determination of the HV plateau will be discussed below for a proportional counter. However, the experiment and the results are equally applicable for a GM counter.

The HV plateau is obtained by performing the experiment sketched in Figure 5.18. A radioactive source, emitting a certain type of particles, is placed at a fixed distance from the detector. The signal from the detector is amplified with the help of a preamplifier and an amplifier. It is then fed through a discriminator, and pulses above the discriminator level are counted by the scaler. The counting rate of the scaler is recorded as a function of the HV, the only variable changed. The result of the experiment is shown in Figure 5.19 (lower curve). Also shown in Figure 5.19 (upper curve) is a part of the graph of Figure 5.3 from regions II (ionization) and III (proportional) with the ordinate now shown as pulse height, which is, of course, proportional to the number of ions collected per unit time. The dashed line represents the discriminator level. The shape of the HV plateau is explained as follows.

For very low voltage ($V < V_A$) the counting rate is zero. The source is there, ionization is produced in the detector, pulses are fed into the amplifier and the discriminator, but the scaler does not receive any signal because all the pulses are below the discriminator level. Hence, the counting rate is zero. As the HV increases beyond V_A, more ionization is produced in the detector, some pulse heights generated in it are above the discriminator level, and the counting rate starts increasing. The counting rate keeps increasing with HV, since more and more pulses are produced with a height above the discriminator level. This continues up to the point when $V \approx V_B$. For $V > V_B$, the ionization is still increasing, the pulse height is also increasing, but all the pulses are now above the discriminator level. Since all the pulses are counted, each *pulse being recorded as one regardless of its height*, the counting rate does not change. This continues up to $V \approx V_C$. Beyond that point, the counting rate will start increasing again because the HV is so high that spurious and double pulses may be generated. The detector should not be operated beyond $V = V_C$.

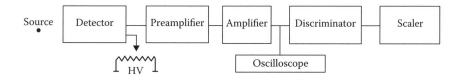

Figure 5.18 Experimental arrangement for the determination of the HV plateau.

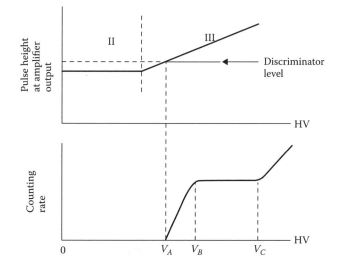

Figure 5.19 The HV plateau (lower curve).

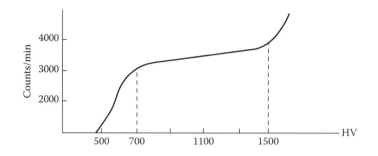

Figure 5.20 The HV plateau used in Example 5.1.

The region of the graph between V_B and V_c is called the *HV plateau*. It represents the operational range of the detector. Although the manufacturer of the detector provides this information to the investigator, it is standard (and safe) practice to determine the plateau of a newly purchased detector before it is used in an actual measurement for the first time.

The plateau of Figure 5.19 is shown as completely flat. For most detectors, the plateau has a positive slope that may be due to spurious counts or to increasing efficiency of the detector, or to both of these effects. Investigation of proportional counters[7] showed that the positive slope is the result of an increase in detector efficiency. For GM counters, on the other hand, the slope of the plateau is due to the production of more spurious counts.

The performance of a detector is expressed in terms of the slope of the plateau given in the form

$$\text{Plateau slope} = \frac{\Delta r/r}{\Delta V} \tag{5.26}$$

where $\Delta r/r$ is the relative change of the counting rate r for the corresponding change in voltage ΔV. Frequently, Eq. 5.26 is expressed in percent change of the counting rate per 100 V change of the high voltage, that is,

$$\text{Plateau slope} = \frac{100(\Delta r/r)}{\Delta V}(100) = 10^4 \frac{\Delta r/r}{\Delta V} \tag{5.27}$$

Example 5.1 What is the change of counting rate per 100 V of the plateau for a detector having the plateau shown in Figure 5.20?
Answer The plateau extends from about 700 to 1500 V. The slope over that region is (using Eq. 5.27),

$$\frac{10^4(r_2 - r_1)/r_1}{V_2 - V_1} = \frac{10^4(3800 - 3000)/3000}{1500 - 700} = 3.3\% \text{ per } 100\,\text{V}$$

The location of the plateau of a proportional counter depends on the type of particles being detected. If a source emits two types of particles with significantly different primary ionization, two separate plateaus will be obtained, with the plateau corresponding to the more ionizing particles appearing first. Figure 5.21 shows such a plateau for a proportional counter detecting alpha and beta particles. The existence of two plateaus is a consequence of the fact that in the proportional region, differentiation of the ionization produced by different types of particles is still possible (see region III in Figure 5.4). In the GM region, this distinction is lost and for this reason GM counters have only one HV plateau regardless of the type of incident radiation (region IV of Figure 5.4).

5.6 GEIGER–MÜLLER COUNTERS

5.6.1 Operation of a GM Counter and Quenching of the Discharge

A GM counter is a gas-filled detector that operates in region IV of Figure 5.3. Its construction and operation are in many ways similar to those of a proportional counter. The GM counter is usually cylindrical in shape, like most of the proportional counters. The electric field close to the central

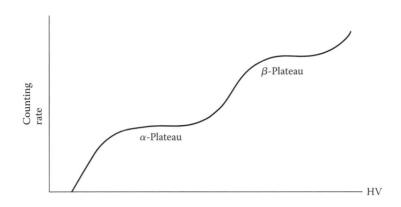

Figure 5.21   Alpha and beta plateaus of a proportional counter.

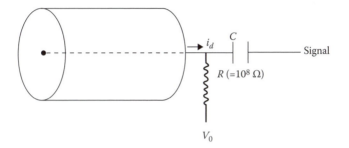

Figure 5.22   The circuit used for external quenching of a GM counter.

wire is so strong that $N\delta \approx 1$ (see Section 5.5.1) and the gas multiplication factor M is extremely high. In a GM counter, a single primary electron–ion pair triggers a great number of successive avalanches. Therefore, the output signal is independent of the primary ionization.

The operation of the GM counter is much more complicated than that of the proportional counter. When the electrons are accelerated in the strong field surrounding the wire, they produce, in addition to a new avalanche of electrons, considerable excitation of the atoms and molecules of the gas. These excited atoms and molecules produce photons when they deexcite. The photons, in turn, produce photoelectrons in other parts of the counter. Thus the avalanche, which was originally located close to the wire, spreads quickly in most of the counter volume. During all this time, the electrons are continuously collected by the anode wire, while the much slower moving positive ions are still in the counter and form a positive sheath around the anode. When the electrons have been collected, this positive sheath, acting as an electrostatic screen, reduces the field to such an extent that the discharge should stop. However, this is not the case because the positive ions eject electrons when they finally strike the cathode, and since by that time the field has been restored to its original high value, a new avalanche starts and the process just described is repeated. Clearly, some means are needed by which the discharge is permanently stopped or "quenched." Without quenching, a GM tube would undergo repetitive discharging. There are two general methods of quenching the discharge.

In *external quenching*, the operating voltage of the counter is decreased, after the start of the discharge until the ions reach the cathode, to a value for which the gas multiplication factor is negligible. The decrease is achieved by a properly chosen RC circuit as shown in Figure 5.22. The resistance R is so high that the voltage drop across it due to the current generated by the discharge (i_d) reduces the voltage of the counter below the threshold needed for the discharge to start (the net voltage is $V_0 - i_d R$). The time constant RC, where C represents the capacitance between anode and ground, is much longer than the time needed for the collection of the ions. As a result, the counter is inoperative for an unacceptably long period. Or, in other words, its dead time is too long.

The *self-quenching* method is accomplished by adding to the main gas of the counter a small amount of a polyatomic organic gas or a halogen gas.

The organic gas molecules, when ionized, lose their energy by dissociation rather than by photoelectric processes. Thus, the number of photoelectrons, which would spread and continue the avalanche, is greatly reduced. In addition, when the organic ions strike the surface of the cathode, they dissociate instead of causing the ejection of new electrons. Therefore, new avalanches do not start.

GM counters using an organic gas as a quenching agent have a finite lifetime because of the dissociation of the organic molecules. Usually, the GM counters last for 10^8–10^9 counts. The lifetime of a GM detector increases considerably if a halogen gas is used as the quenching agent. The halogen molecules also dissociate during the quenching process, but there is a certain degree of regeneration of the molecules, which greatly extends the useful lifetime of the counter.

5.6.2 The Pulse Shape and the Dead Time of a GM Counter

The signal of a GM counter is formed in essentially the same way as the signal of a proportional counter and is given by the same equation, Eq. 5.24. For GM counters the signal is the result of the sum of the contributions from all the positive ion avalanches produced throughout the volume of the counter. The final pulse is similar in shape to that shown in Figure 5.17, except that the pulse rises much slower. The shape and height of GM counter pulses are not very important because the pulse is only used to signal the presence of the particle and nothing else. However, how one pulse affects the formation of the next one is important.

As discussed in Section 5.6.1, during the formation of a pulse, the electric field in the counter is greatly reduced because of the presence of the positive ions around the anode. If a particle arrives during that period, no pulse will be formed because the counter is insensitive. The insensitivity lasts for a certain time, called the dead time of the counter. Then, the detector slowly recovers, with the pulse height growing exponentially during the *recovery* period. This is illustrated in Figure 5.23, which shows the change of the voltage and pulse for a typical GM counter. Typical values of dead time are from 100 to 300 μs. If the dead time is 100 μs and the counting rate is

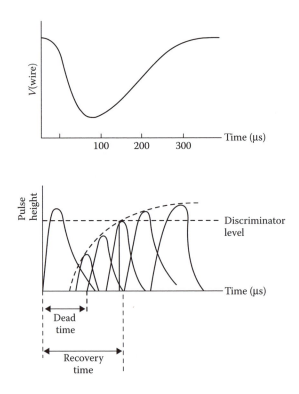

Figure 5.23 Dead time and recovery time for a GM counter.

500 counts/s, there is going to be a 5% loss of counts due to dead time. Correction for dead time is described in Section 2.2.1.

5.7 GAS-FLOW DETECTORS

The gas-filled detectors described so far are all sealed. That is, the counter is a closed volume filled with a gas at a certain pressure. The radiation source is placed outside the detector; therefore, the particles have to penetrate the wall of the detector to be counted. In doing so, some particles may be absorbed by the wall and some may be backscattered; in the case of charged particles, they will all lose a certain fraction of their energy. To minimize these effects, most commercial gas-filled detectors have a thin window through which the radiation enters the detector. The window may still be too thick for some alpha and low-energy beta particles. For this reason, detectors have been developed with the capability of having the source placed inside the chamber.

Gas-filled detectors of this type are called *gas-flow detectors*. Their name comes from the fact that the gas flows continuously through the detector during operation. This is necessary because the detector cannot be sealed if the source is placed inside the chamber.

Gas-flow detectors come in different geometries. Probably the most common one is that of the hemispherical detector as shown in Figure 5.24. The high voltage is applied to a wire attached to the top of the hemisphere. The gas flows slowly through the detector, the flow rate being controlled by a regulator. At the exit, the gas goes through a liquid (e.g., some oil) and forms bubbles as it comes out. The formation of the bubbles indicates that the gas is flowing, and the rate of bubble formation gives an idea of the gas-flow rate.

Counting with gas-flow detectors involves the following steps:

1. The chamber is opened and the sample is placed in its designated location inside the chamber.

2. The chamber is closed.

3. Gas from the gas tank is allowed to flow rapidly through the volume of the detector and purge it (for a few minutes).

4. After the detector is purged, the gas-flow rate is considerably reduced, to a couple of bubbles per second, and counting begins.

There are two advantages in placing the sample inside the detector:

1. The particles do not have to penetrate the window of the detector, where they might be absorbed, scattered out of the detector, or lose energy.

2. Close to 50% of the particles emitted by the source have a chance to be recorded in a hemispherical detector, or close to 100% in a spherical detector. If the source is placed outside the detector, there are always less than 50% of the particles entering the detector.

A hemispherical detector is also called a 2π detector, while a spherical detector with the source located at its center is called a 4π detector. Figure 5.24 shows a 2π counter.

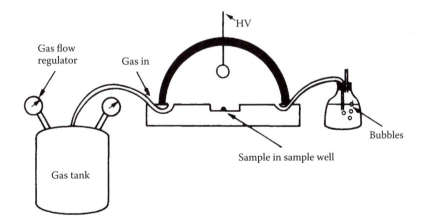

Figure 5.24   A hemispherical (2π) gas-flow detector.

Gas-flow detectors may operate as proportional or GM counters. In fact, there are commercial models that may operate in one or the other region depending on the voltage applied and the gas used. In a proportional gas-flow detector, the gas is usually methane or a mixture of argon and methane. In the GM region, the gas is a mixture of argon and isobutane.

In some gas-flow detector models, there is provision for placing a very thin window between the sample and the sensitive volume of the detector to reduce the effects of slight contamination of the sample well or of static charges that interfere with the measurement. In the detector of Figure 5.24, the thin window will be placed on top of the sample well. A different arrangement is shown in Figure 5.25.

Gas-flow detectors are used as low-background alpha–beta detection systems. Requirements for low-background measurements arise in cases where the level of activity from the sample is very low, compared to background. Examples of such cases are samples that monitor contamination of water supplies or of air or ground.

There are commercially available systems that have a background counting rate of less than 1 count/min for betas and a considerably lower rate for alphas. Such a low background is achieved by shielding the detector properly (surrounding it with lead) and using electronic means to reject most of the background radiation. A system offered by one of the manufacturers uses two detectors. The first is the gas-flow detector and the second is a cosmic ray detector (Figure 5.26). The two detectors are operated in *anticoincidence* (see Section 10.8), which means that events due to particles going through both detectors (e.g., cosmic rays or other radiation from the environment) will not be counted. Only pulses produced by the activity of the sample in the gas-flow detector will be recorded.

Discrimination between alphas and betas can be achieved in many ways. The two methods most frequently used with gas-flow detectors are based on range and energy differences. Before

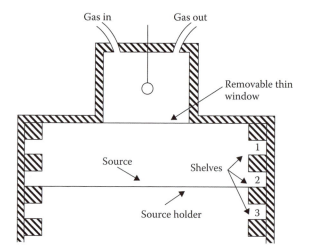

Figure 5.25 A gas-flow detector with removable thin window and movable source holder.

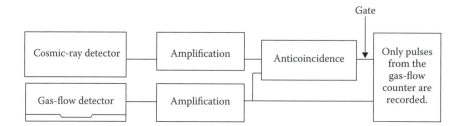

Figure 5.26 A low-background alpha–beta counting system utilizing two detectors and anticoincidence. The anticoincidence output gates the scaler to count only pulses from the gas-flow detector.

these methods are discussed, the reader should recall that the maximum energy of most beta emitters is less than 2 MeV while the energy of alphas from most alpha emitters is 5–6 MeV.

Because the range of alphas is much shorter than that of betas, a sample can be analyzed for alpha and beta activity by counting it twice: once with a thin foil covering it to stop the alphas, and a second time without the foil to record alphas and betas.

Energy discrimination is based on the difference in pulse height produced by the two types of particles: the alphas, being more energetic, produce higher pulses; thus, a simple discriminator at an appropriate level can reject the beta pulses.

5.7.1 The Long-Range Alpha Detector (LRAD)

A variation of the gas-flow detector has been developed[8,9] for the detection of alpha contamination. Common alpha particle detectors are limited by the short range of alphas in air. For example, the range of a 6-MeV alpha in air at normal temperature and pressure is about 46 mm. To circumvent this limitation, the LRAD does not measure the alphas directly. Instead, as shown schematically in Figure 5.27, the ions created by the alphas in air are transported, with the help of airflow, and directed into an ion chamber. There, the current created by the ions is measured by an electrometer. Since the number of ions produced is proportional to the strength of the alpha source, the signal of the electrometer is also proportional to the alpha source strength.

In principle, a similar detector could be developed for any particle that produces ions. However, particles like electrons, gammas, and neutrons generate a much smaller number of ions than alpha particles do, traveling over the same distance. For this reason, an LRAD-type detector would have a smaller sensitivity for these other particles than for alphas. Of course, an LRAD-type detector would operate satisfactorily for the detection of protons, deuterons, and other heavy ions.

5.7.2 Internal Gas Counting

An alternative to the gas-flow detector is *internal gas counting*, which is used with low-energy β-emitters. In internal gas counting, a gaseous form of the radioisotope is introduced into the detector (usually a proportional counter) along with the counting gas. As with gas-flow detectors, by having the source inside the detector, losses in the window are avoided and an increase in efficiency is achieved by utilizing a 4π geometry.

Internal gas counting requires that corrections be made for wall and end effects and for the decrease in electric field intensity at the ends.[10–12] One way to reduce the end effect is to use a spherical proportional counter,[13] in which the anode wire is stretched along a diameter and the cathode is, of course, spherical. The electric field inside the sphere is

$$E = \frac{V}{\ln(b/a)}\frac{1}{r} \tag{5.2}$$

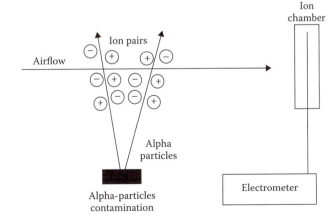

Figure 5.27 The LRAD detects ions generated by alphas in air with the help of an ion chamber and an electrometer.

At a certain distance r from the anode, the electric field becomes stronger at the ends of the anode because b, the radius of the cathode, gets smaller. However, the supports of the wire tend to reduce the field. By property adjusting the supports, one may make the field uniform. In cylindrical detectors, corrections for end effects are applied by a length-compensation method.[10]

Internal gas counting is used for the production of standards. Using this technique, the National Institute of Standards and Technology produced standards of ^{3}H, ^{14}C, ^{37}A, ^{85}K, ^{131m}Xe, and ^{133}Xe.

5.8 RATE METERS

A rate meter is a device that measures the average rate of incoming pulses. Rate meters are used for continuous monitoring of an event, where the average counting rate versus time rather than the instantaneous counting rate is needed.

The basic operation of a rate meter is to feed a known charge per pulse into a capacitor that is shunted by a resistor (Figure 5.28). Let

$$r = \text{counting rate (pulses/s)}$$
$$q = \text{charge per pulse}$$
$$V = \text{voltage across capacitor}$$
$$R = \text{resistance}$$
$$Q = \text{capacitor charge}$$

The net rate of change of Q with respect to time is given by

$$\frac{dQ}{dt} = (\text{charge fed by pulses/s}) - (\text{charge flowing through resistor})$$

or

$$\frac{dQ}{dt} = rq - \frac{Q}{RC} \tag{5.28}$$

The solution of this differential equation with the initial condition $Q(0) = 0$ is

$$Q(t) = rqRC(1 - e^{-t/RC}) \tag{5.29}$$

or, if one writes the result in terms of the output voltage,

$$V(t) = \frac{Q(t)}{C} = rqR(1 - e^{-t/RC}) \tag{5.30}$$

For time $t \gg RC$, equilibrium is reached and the value of the voltage is

$$V_{\infty} = rqR \tag{5.31}$$

The signal of a rate meter is the voltage V_{∞} given by Eq. 5.31. Notice that V_{∞} is independent of the capacitance C and proportional to the counting rate r. The voltage V_{∞} is measured with an appropriate voltmeter.

If a pulse-type detector is used, the counts accumulated in the scaler have a statistical uncertainty that is calculated as shown in Chapter 2. If a rate meter is used, what is the uncertainty of the measurement? To obtain the uncertainty, one starts with Eq. 5.29, which gives the charge of the capacitor C. It is important to note that the charge changes exponentially with time. Thus, the

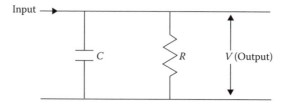

Figure 5.28   The circuit of a rate meter.

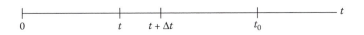

Figure 5.29 Pulses arriving during Δt, and at t, contribute to σ_Q at $t = t_0$.

contribution of the charge from a pulse arriving at $t = 0$ is not instantaneous but continues for a period.

Consider an observation point t_0 (Figure 5.29). The standard deviation σ_Q of the charge collected at $t = t_0$ is the result of contributions from pulses having arrived earlier. If the counting rate is r, the number of pulses in a time interval Δt is, on the average, $r\,\Delta t$. The statistical uncertainty of this number is $\pm \sqrt{r\Delta t}$, or the uncertainty of the charge is $\pm q\sqrt{r\Delta t}$. One can show that a single pulse arriving at time t contributes to the signal at time $t = t_0$, an amount of charge equal to $q\exp[-(t_0-t)\,RC]$. Therefore, the variance of the charge at time $t = t_0$ is

$$\sigma_Q^2 = \int_0^{t_0} (q\sqrt{r\,dt}\,e^{-(t_0-t)/RC})^2 \tag{5.32}$$

Integration of Eq. 5.32 gives the result

$$\sigma_Q^2 = 0.5q^2 rRC(1 - e^{-2t_0/RC}) \tag{5.33}$$

For $t_0 \gg RC$, Eq. 5.33 takes the form

$$\sigma_Q = q\sqrt{\frac{r(RC)}{2}} \tag{5.34}$$

At equilibrium, $Q = rqRC$ (from Eq. 5.29); therefore,

$$\sigma_r = \frac{\sigma_Q}{qRC} = \sqrt{\frac{r}{2RC}} \tag{5.35}$$

and

$$\frac{\sigma_r}{r} = \sqrt{\frac{1}{2RCr}} \tag{5.36}$$

The quantity RC is the time constant of the circuit shown in Figure 5.28. Equation 5.36 states that any instantaneous reading on a rate meter has a relative standard error equal to that of a total number of counts obtained by counting for a time equal to $2RC$ (assuming the background is negligible).

5.9 GENERAL COMMENTS ABOUT CONSTRUCTION OF GAS-FILLED DETECTORS

This section summarizes the important characteristics of gas-filled detectors.

Geometry. Parallel-plate detectors are almost exclusively ionization chambers. The intense fields needed for gas multiplication can be produced only in cylindrical or spherical geometry.

In the cylindrical geometry, which is the most frequently used, the strong electric field exists close to the central wire. The wire is usually made of tungsten or platinum. It has a diameter of 25–100 µm (few mills of an inch); it must be uniform in radius, without any bends or kinks, and be placed concentrically with the outer cylinder. Of particular importance is the smoothness of the central wire. Any kinks or tiny specks of material attached to its surface amount to pointed tips where very high electric fields are generated. Such a high field is a source of spurious discharges that interfere with counting.

Gases and pressures used. For ionization chambers, almost any gas or pressure may be used. Even atmospheric air has been used.

For proportional or GM counters, the noble gases—argon in particular—are normally used. A small percentage of additional gases is also used for quenching purposes. In proportional counters, methane is frequently added to the main gas. The so-called P-10 mixture, consisting of 90% argon and 10% methane, is extensively used. Another mixture is 4% isobutane and 96% helium. Several gas pressures have been used. As Figures 5.15 and 5.16 show, the gas multiplication depends on the pressure. Usually the pressure is less than 1 atmosphere. Of course, gas-flow counters operate at ambient pressure.

As discussed in Section 5.6.1, the quenching gas in a GM counter is either an organic polyatomic molecule such as ethyl alcohol, or a halogen such as bromine or chlorine. A typical mixture is 0.1% chlorine in neon. The gas pressure in a GM counter is, in most cases, less than 1 atmosphere. The pressure affects the operating voltage.

Counter window. When the source is placed outside the detector, it is very important for the radiation to enter the detector after traversing as thin a wall material as possible. Any material in the path of radiation may scatter, absorb, or cause energy loss. This is particularly critical in the measurement of alphas and low-energy betas, which have a very short range. It is not important for neutron and gamma detectors.

All detectors have walls as thin as possible (or practical), but in addition, many commercial designs have an area on the surface of the detector designated as the "window," consisting of a very thin material. In cylindrical detectors, the window is usually the front end of the cylinder (the other end houses electrical connectors). There are some cylindrical detectors with windows located on the cylindrical surface.

Materials and thicknesses of windows are

1. Glass, down to 0.30–0.40 kg/m^2 (100 μm)

2. Aluminum, 0.25–0.30 kg/m^2 (100 μm)

3. Steel, 0.60–0.80 kg/m^2 (80 μm)

4. Mica, 0.01 kg/m^2 (3 μm)

5. Mylar (plain or aluminized), 0.01 kg/m^2

6. Special ultra thin membranes or foils, ~10^{-3} kg/m^2

5.10 APPLICATIONS OF GAS-FILLED DETECTORS

While there has been an emphasis in this chapter on the fundamentals of gas-filled detectors, a wide variety of applications of such systems has been developed in nuclear instrumentation. Following is not an all-inclusive list of such applications:

- A new active method for continuous radon measurements based on a multiple cell proportional counter[14]

- Application of nuclear reaction analysis for the fluorine content measurements under the aging investigations of gas-filled particle detectors[15]

- Readout electronics for X-ray imaging using gas-filled detectors[16]

- A novel gas-filled detector for synchrotron radiation applications[17]

- Studies of ^{3}He and isobutane mixture as neutron proportional counter gas[18]

- Modeling of ionization produced by fast charged particles in gases[19]

- A first mass production of gas electron multipliers[20]

- Neutron detector development at Brookhaven[21]

- A low-pressure gas detector for heavy-ion tracking and particle identification[22]

- The status of gas-filled detector developments at a third generation synchrotron source (ESRF)[23]

- High-performance, imaging, thermal neutron detectors[24]

PROBLEMS

5.1 Sketch the HV plateau of a detector, if all the pulses out of the amplifier have exactly the same height.

5.2 How would the sketch of Problem 5.1 change if there are two groups of pulses out of the amplifier (two groups, two different pulse heights)?

5.3 Sketch counting rate versus discriminator threshold, assuming that the electronic noise consists of pulses in the range $0 < V < 0.1$ V and all the pulses due to the source have height equal to 1.5 V.

5.4 In a cylindrical gas counter with a central wire radius equal to 25 μm (0.001 in.), outer radius 25 mm (~1 in.), and 1000 V applied between anode and cathode, what is the distance from the center of the counter at which an electron gains enough energy in 1 mm of travel to ionize helium gas? (Take 23 eV as the ionization potential of helium.)

5.5 A GM detector with a mica window is to be used for measurement of ^{14}C activity. What should the thickness of the window be if it is required that at least 90% of the ^{14}C betas enter the detector?

5.6 What is the minimum pressure required to stop 6-MeV alphas inside the argon atmosphere of a spherical gas counter with a 25-mm radius? Assume the alpha source is located at the center of the detector.

5.7 You are asked to construct a cylindrical Ar-filled detector of such length that the beta particles emitted by a ^{32}P source and traveling parallel to the axis of the counter are just stopped in it (assume negligible detector window thickness). What should be the length of the detector if the Ar pressure is 20 Atm?

5.8 What is the ratio of the saturation ionization currents for a chamber filled with CH_4 (other things being equal)?

5.9 Show that the variance of M is equal to $\overline{M}^2$ if the probability distribution is given by Eq. 5.22.

5.10 Calculate the maximum value of the positive ion time given by Eq. 5.25 for a cylindrical detector with a cathode radius equal to 19 mm (~0.75 in.) and a central anode wire with a radius of 25 μm (~0.001 in.). The high voltage applied is 1000 V; the pressure of the gas is 13.3 kPa (10 cmHg), and the mobility of the ions is 13.34 Pa m^2/(V s).

5.11 The observed counting rate of a detector is 22,000 counts/min. What is the error in the true counting rate if the dead time is 300 μs and no dead-time correction is applied?

BIBLIOGRAPHY

Ahmed, S. N., *Physics and Engineering of Radiation Detection*, Academic Press, Elsevier, Amsterdam, 2007.

Eichholz, G. G., and Poston, J. W., *Nuclear Radiation Detection*, Lewis Publishers, Chelsea, Michigan, 1985.

Fenyves, E., and Haiman, O., *The Physical Principles of Nuclear Radiation Measurements*, Academic Press, New York, 1969.

Franzen, W., and Cochran, L. W., "Pulse Ionization Chambers and Proportional Counters," in A. H. Snell (ed.), *Nuclear Instruments and Their Uses*, Wiley, New York, 1962.

Grupen, K., and Schwartz, B., *Particle Detectors*, 2nd ed., Cambridge University Press, 2008.

Hussein, E. M. A., *Radiation Mechanics: Principles and Practice*, Elsevier, Amsterdam, 2007.

Kleinknecht, K., *Detectors for Particle Radiation*, 2nd ed., Cambridge University Press, 1998.

Knoll, G. F., *Radiation Detection and Measurement*, 3rd ed., Wiley, New York, 2000.

Kowalski, E., *Nuclear Electronics*, Springer-Verlag, New York, and Heidelberg, Berlin, 1970.

Leroy, C., and Rancoita, P-G., *Principles of Radiation Interaction in Matter and Radiation*, World Scientific Publishing, Singapore, 2004.

Rossi, B. B., and Staub, H. H., *Ionization Chambers and Counters*, McGraw-Hill, New York, 1949.

REFERENCES

1. Diethorn, W., NYO-0628, 1956.

2. Kiser, R. W., *Appl. Sci. Res.* **8B**:183 (1960).

3. Williams, W., and Sara, R. I., *Int. J. Appl. Rad. Isotopes* **13**:229 (1962).

4. Bennett, E. F., and Yule, T. J., ANL-7763, 1971.

5. Hanna, G. C., Kirkwood, H. W., and Pontecorvo, B., *Phys. Rev.* **75**:985 (1949).

6. Snyder, H. S., *Phys. Rev.* **72**:181 (1947).

7. Champion, P. J., *Nucl. Instrum. Meth.* **112**:75 (1973).

8. MacArthur, D. W., Allander, K. S., Bounds, J. A., Butterfield, K. B., and McAtee, J. L., *Health Phys.* **63**:324 (1992).

9. MacArthur, D. W., Allander, K. S., Bounds, J. A., and McAtee, J. L., *Nucl. Technol.* **102**:270 (1993).

10. Mann, W. B., Seliger, H. H., Marlow, W. F., and Medlock, R. W., *Rev. Sci. Instrum.* **31**:690 (1960).

11. Garfunkel, S. B., Mann, W. B., Schima, F. J., and Unterweger, M. P., *Nucl. Instrum. Meth.* **112**:59 (1973).

12. Bambynek, W., *Nucl. Instrum. Meth.* **112**:103 (1973).

13. Benjamin, P. W., Kemsholl, C. D., and Redfearn, J., *Nucl. Instrum. Meth.* **59**:77 (1968).

14. Mazed, D., Ciolini, R., Curzio, G., and Del Gratta, A., *Nucl. Instrum. Meth. Phys. Res., Section A*, **582**:535 (2007).

15. Krivchitch, A. G., and Lebedev, V. M., *Nucl. Instrum. Meth. Phys. Res., Section A*, **581**:167 (2007).

16. Hervé, C., Le Caër, T., Cerrai, J., and Scaringella, B., *Nucl. Instrum. Meth. Phys. Res., Section A*, **580**:1428 (2007).

17. Kocsis, M., Boesecke, P., Carbone, D., Herve, Becker, B., Diawara, Y., Durst, R., *et al.*, *Nucl Instrum. Meth. Phys. Res., Section A*, **563**:172 (2006).

18. Desai, S. S., and Shaikh, A. M., *Nucl. Instrum. Meth. Phys. Res., Section A*, **554**:474 (2005).

19. Smirnov, I. B., *Nucl. Instrum. Meth Phys. Res., Section A*, **515**:439 (2003).

20. Barbeau, P. S., Collar, J. I., Geissinger, J. D., Miyamoto, J., Shipsey, I., and Yang, R., *Nucl. Instrum. Meth. Phys. Res., Section A*, **513**:362 (2003).

21. Yu, B., Harder, J. A., Mead, J. A., Radeka, V., Schaknowski, N. A., and Smith, G.C. *Nucl. Instrum. Meth. Phys. Res., Section A*, **495**:216 (2002).

22. Cunsolo, A., Cappuzzello, F., Foti, A., Gangnant, P., Lazzaro, A., Libin, J. F., Melita, A. L., *et al.*, *Nucl. Instrum. Meth. Phys. Res., Section A*, **495**:416 (2002).

23. Menhard Kocsis, M., *Nucl. Instrum. Meth. Phys. Res., Section A*, **471**:103 (2001).

24. Radeka, V., Schaknowski, N. A., Smith, G. C., and Yu, B., *Nucl. Instrum. Meth. Phys. Res., Section A*, **419**:642 (1998).

6

Scintillation Detectors

6.1 INTRODUCTION

Scintillators are materials—solids, liquids, gases—that produce sparks or scintillations of light when ionizing radiation passes through them. The first solid material to be used as a particle detector was a scintillator. It was William Crookes who in 1903 was the first person to observe scintillations with alpha particles impinging on a ZnS screen.[1] In 1910, in Rutherford's experimental setup, alpha particles hit a zinc sulfide screen and produced scintillations, which were counted with or without the help of a microscope—a very inefficient process, inaccurate and time consuming. The method was abandoned for about 30 years and was remembered again when advanced electronics made possible amplification of the light produced in the scintillator.

The amount of light produced in the scintillator is very small. It must be amplified before it can be recorded as a pulse or in any other way. The amplification or multiplication of the scintillator's light is achieved with a device known as the *photomultiplier tube* (or *phototube*). Its name denotes its function: it accepts a small amount of light, amplifies it many times, and delivers a strong pulse at its output. Amplifications of the order of 10^6 are common for many commercial photomultiplier tubes. Apart from the phototube, a detection system that uses a scintillator is no different from any other (Figure 6.1).

The operation of a scintillation detector may be divided into two broad steps:

1. Absorption of incident radiation energy by the scintillator and production of photons in the visible part of the electromagnetic spectrum.

2. Amplification of the light by the photomultiplier tube and production of the output pulse.

The sections that follow analyze these two steps in detail. The different types of scintillators are divided, for the present discussion, into three groups:

1. Inorganic scintillators

2. Organic scintillators

3. Gaseous scintillators

6.2 INORGANIC (CRYSTAL) SCINTILLATORS

Most of the inorganic scintillators are crystals of the alkali metals, in particular alkali iodides, that contain a small concentration of an impurity. Some examples are NaI(Tl), CsI(Tl), CaI(Na), LiI(Eu), and CaF_2(Eu). The element in parentheses is the impurity or activator. Although the activator has a relatively small concentration—for example, thallium in NaI(Tl) is 10^{-3} on a per mole basis—it is the agent that is responsible for the luminescence of the crystal. In 1948 Hofstatder[2] discovered the crystal detector and in 1950 published his findings on the properties of scintillation detectors.[3]

6.2.1 The Mechanism of the Scintillation Process

The luminescence of inorganic scintillators can be understood in terms of the allowed and forbidden energy bands of a crystal. The electronic energy states of an atom are discrete energy levels, which in an energy-level diagram are represented as discrete lines. In a crystal, the allowed energy

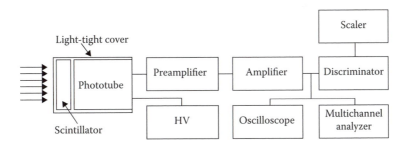

Figure 6.1 A detection system using a scintillator.

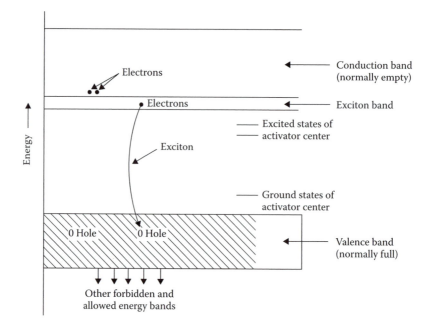

Figure 6.2 Allowed and forbidden energy bands of a crystal.

states widen into bands (Figure 6.2). In the ground state of the crystal, the uppermost allowed band that contains electrons is completely filled. This is called the *valence band*. The next allowed band is empty (in the ground state) and is called the *conduction band*. An electron may obtain enough energy from incident radiation to move from the valence to the conduction band. Once there, the electron is free to move anywhere in the lattice. The removed electron leaves behind a hole in the valence band, which can also move. Sometimes, the energy given to the electron is not sufficient to raise it to the conduction band. Instead, the electron remains electrostatically bound to the hole in the valence band. The electron–hole pair thus formed is called an *exciton*. In terms of energy states, the exciton corresponds to elevation of the electron to a state higher than the valence but lower than the conduction band. Thus, the exciton states form a thin band, with the upper level coinciding with the lower level of the conduction band (Figure 6.2). The width of the exciton band is of the order of 1 eV, whereas the gap between valence and conduction bands is of the order of 8 eV.

In addition to the exciton band, energy states may be created between valence and conduction bands because of crystal imperfections or impurities. Particularly important are the states created by the activator atoms such as thallium. The activator atom may exist in the ground state or in one of its excited states. Elevation to an excited state may be the result of a photon absorption, or of the capture of an exciton, or of the successive capture of an electron and a hole. The transition of the impurity atom from the excited to the ground state, if allowed, results in the emission of a photon in times of the order of 10^{-8} s. If this photon has a wavelength in the visible part of the electromagnetic spectrum, it contributes to a scintillation. Thus, production of a scintillation is the result of the occurrence of these events:

1. Ionizing radiation passes through the crystal.

2. Electrons are raised to the conduction band.

3. Holes are created in the valence band.

4. Excitons are formed.

5. Activation centers are raised to the excited states by absorbing electrons, holes, and excitons.

6. Deexcitation is followed by the emission of a photon.

The light emitted by a scintillator is primarily the result of transitions of the activator atoms, and not of the crystal. Since most of the incident energy goes to the lattice of the crystal—eventually becoming heat—the appearance of luminescence produced by the activator atoms means that energy is transferred from the host crystal to the impurity. For NaI(Tl) scintillators, about 12% of the incident energy appears as thallium luminescence.[4] An excellent explanation of the principle of photoemission mechanism has been presented by Lempicki in 1995 in his review article on the physics of inorganic scintillators.[5]

The magnitude of light output and the wavelength of the emitted light are two of the most important properties of any scintillator. The light output affects the number of photoelectrons generated at the input of the photomultiplier tube (see Section 6.5), which in turn affects the pulse height produced at the output of the counting system. Information about the wavelength is necessary in order to match the scintillator with the proper photomultiplier tube. Emission spectra of NaI(Tl), CsI(Na), and CsI(Tl) are shown in Figure 6.3. Also shown in Figure 6.3 are the responses of two phototube cathode materials. Table 6.1 gives the most important properties of some inorganic scintillators.

The light output of the scintillators depends on temperature. Figure 6.4 shows the temperature response of NaI(Tl), Cs(Tl), and CsI(Na).

6.2.2 Time Dependence of Photon Emission

Since the photons are emitted as a result of decays of excited states, the time of their emission depends on the decay constants of the different states involved. Experiments show that the emission of light follows an exponential decay law of the form

$$N(t) = N_0 e^{-t/T} \tag{6.1}$$

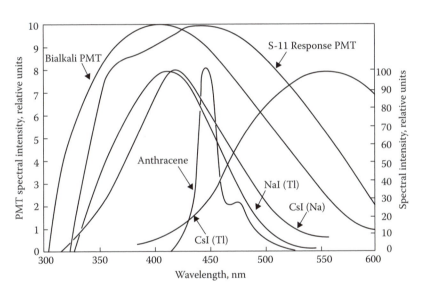

Figure 6.3 Emission spectra of NaI(Tl), CsI(Tl), CsI(Na), and anthracene, compared to the spectral response of two photocathode materials. PMT, photomultiplier tube. (From Harshaw Research Laboratory Report, Harshaw Chemical Company, 1978.)

Table 6.1
Properties of Certain Inorganic Scintillators

Material	Wavelength of Maximum Emission (nm)	Scintillation Efficiency (Relative, %)	Decay Time (μs)	Density (10^3 kg/m³)
NaI(Tl)	410	100	0.23	3.67
CaF$_2$(Eu)	435	50	0.94	3.18
CsI(Na)	420	80	0.63	4.51
CsI(Tl)	565	45	1.00	4.51
Bi$_4$Ge$_3$O$_{12}$	480	8	0.30	7.13
CdWO$_4$	530	20	0.90	7.90
^{6}LiI(Eu)	470	30	0.94	3.49

173

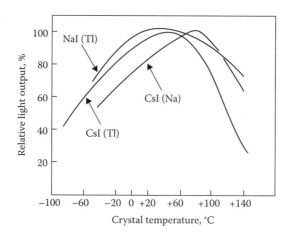

Figure 6.4 Temperature dependence of light output of NaI(Tl), CsI(Tl), and CsI(Na). (From Harshaw Research Laboratory Report.)

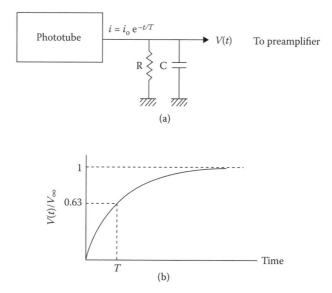

Figure 6.5 (a) A voltage pulse results from the exponential current, (b) the shape of the pulse for $RC \gg T$.

where $N(t)$ is the number of photons emitted at time t and T the decay time of the scintillator (see Table 6.1).

Most of the excited states in a scintillator have essentially the same lifetime T. There are, however, some states with longer lifetimes contributing a slow component in the decay of the scintillator known as *afterglow*. It is present to some extent in all inorganic scintillators and may be important in certain measurements where the integrated output of the phototube is used. Two scintillators with negligible afterglow are $CaF_2(Eu)$ and $Bi_4Ge_3O_{12}$ (bismuth orthogermanate).

In a counting system using a scintillator, the light produced by the crystal is amplified by a photomultiplier tube and is transformed into an electric current having the exponential behavior given by Eq. 6.1. This current is fed into an RC circuit as shown in Figure 6.5, and a voltage pulse is produced of the form

$$V(t) = V_\infty (e^{-t/RC} - e^{-t/T}) \tag{6.2}$$

In practice, the value of RC is selected to be of the order of a few hundreds of microseconds. Thus, for short times—that is, $t \ll RC$, which is the time span of interest—Eq. 6.2 takes the form

$$V(t) = V_\infty (1 - e^{-t/T}) \tag{6.2a}$$

Notice that the rate at which the pulse rises (rise time) is determined by the decay time T. In certain measurements, for example, coincidence–anticoincidence measurements (Chapter 10), the timing characteristics of the pulse are extremely important.

6.2.3 Important Properties of Certain Inorganic Scintillators

NaI(Tl). NaI(Tl) is the most commonly used scintillator for gamma rays. It has been produced in single crystals of up to 0.75 m (~30 in.) in diameter and of considerable thickness (0.25 m ≈ 10 in.). Its relatively high density (3.67×10^3 kg/m³) and high atomic number combined with the large volume make it a γ-ray detector with very high efficiency. Although semiconductor detectors (Chapters 7 and 12) have better energy resolution, they cannot replace the NaI(Tl) in experiments where large detector volumes are needed.

The emission spectrum of NaI(Tl) peaks at 410 nm, and the light-conversion efficiency is the highest of all the inorganic scintillators (Table 6.1). As a material, NaI(Tl) has many undesirable properties. It is brittle and sensitive to temperature gradients and thermal shocks. It is also so hygroscopic that it should be kept encapsulated at all times. NaI always contains a small amount of potassium, which creates a certain background because of the radioactive ^{40}K.

CsI(Tl). CsI(Tl) has a higher density (4.51×10^3 kg/m³) and higher atomic number than NaI; therefore, its efficiency for gamma detection is higher. The light-conversion efficiency of CsI(Tl) is about 45% of that for NaI(Tl) at room temperature. At liquid nitrogen temperatures (77K), pure CsI has a light output equal to that of NaI(Tl) at room temperature and a decay constant equal to 10^{-8} s.[6] The emission spectrum of CsI(Tl) extends from 420 to about 600 nm.

CsI is not hygroscopic. Being softer and more plastic than NaI, it can withstand severe shocks, acceleration, and vibration, as well as large temperature gradients and sudden temperature changes. These properties make it suitable for space experiments. Finally, CsI does not contain potassium.

CsI(Na). The density and atomic number of CsI(Na) are the same as those of CsI(Tl). The light-conversion efficiency is about 85% of that for NaI(Tl). Its emission spectrum extends from 320 to 540 nm (see Figure 6.3). CsI(Na) is slightly hygroscopic.

CaF$_2$(Eu). CaF$_2$(Eu) consists of low-atomic-number materials, and for this reason makes an efficient detector for β particles[7] and X-rays[8] with low gamma sensitivity. It is similar to Pyrex and can be shaped to any geometry by grinding and polishing. Its insolubility and inertness make it suitable for measurements involving liquid radioisotopes. The light-conversion efficiency of CaF$_2$(Eu) is about 50% of that for NaI(Tl). The emission spectrum extends from about 405 to 490 nm.

LiI(Eu). LiI(Eu) is an efficient thermal-neutron detector through the reaction $^6_3\text{Li}(n, \alpha)^3_1\text{H}$. The alpha particle and the triton, both charged particles, produce the scintillations. LiI has a density of 4.06×10^3 kg/m³, decay time of about 1.1 μs, and emission spectrum peaking at 470 nm. Its conversion efficiency is about one-third of that for NaI. It is very hygroscopic and is subject to radiation damage as a result of exposure to neutrons.

Other inorganic scintillators. Many other scintillators have been developed for special applications. Examples are Bi$_4$Ge$_3$O$_{12}$, CdWO$_4$, and others such as[9] MF$_2$:UF$_4$:CeF$_3$, where M stands for one of the following: Ca, Sr, Ba. This last scintillator, containing 2% UF$_4$ and using Ce as the fluorescing agent, has been used for detection of fission fragments.

6.2.4 Applications of Inorganic Scintillators

During the past two decades there has been an increasing usage of inorganic scintillators in a wide variety of materials research. The following is a small sample of the many new applications: multienergy neutron detector for counting thermal neutrons, high energy neutrons, and photons separately,[10] solid scintiallation, counting as a new technique for measuring radiolabeled compounds,[11] the use of HgI$_2$ photodetectors combined with scintillators for gamma-ray spectroscopy,[12] application of a cerium fluoride scintillator for high couth rates,[13] high-position-resolution neutron imaging detector with crossed wavelength shifting fiber readout using two ZnS/6LiF scintillator sheets[14] the growth and scintillation properties of CsCe$_2$Cl$_7$ crystal,[15] slow neutron beam diagnostics with a scintillating fiber detector,[16] spectroscopy of high rate events during active interrogation,[17] and large area plastic scintillator detector array for fast neutron measurements.[18]

6.3 ORGANIC SCINTILLATORS

The materials that are efficient *organic scintillators* belong to the class of aromatic compounds. They consist of planar molecules made up of benzenoid rings. Two examples are toluene and anthracene, having the structures shown in Figure 6.6.

Organic scintillators are formed by combining appropriate compounds. They are classified as unitary, binary, ternary, and so on, depending on the number of compounds in the mixture. The substance with the highest concentration is called the *solvent*. The others are called *solutes*. A binary scintillator consists of a solvent and a solute, while a ternary scintillator is made of a solvent, a primary solute, and a secondary solute. Table 6.2 lists the most common compounds used.

6.3.1 The Mechanism of the Scintillation Process

The production of light in organic scintillators is the result of molecular transitions. Consider the energy-level diagram of Figure 6.7, which shows how the potential energy of a molecule changes

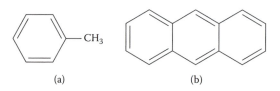

(a)   (b)

Figure 6.6   Molecular structure of (a) toluene and (b) anthracene.

Table 6.2
Organic Scintillator Compounds

Compound	Formula	Application[a]
Benzene	C_6H_6	S
Toluene	$C_6H_5CH_3$	S
p-Xylene	$C_6H_4(CH_3)_2$	S
1,2,4-Trimethylbenzene (pseudocumene)	$C_6H_3(CH_3)_3$	S
Hexamethylbenzene	$C_6(CH_3)_6$	S
Styrene monomer	$C_6H_5C_2H_3$	S
Vinyltoluene monomer	$C_6H_4CH_3C_2H_3$	S
Naphthalene	$C_{10}H_8$	S', C
Anthracene	$C_{14}H_{10}$	C
Biphenyl	$C_{12}H_{10}$	S'
p-Terphenyl	$C_{18}H_{14}$	C, PS
p-Quaterphenyl	$C_{24}H_{18}$	C
trans-Stilbene	$C_{14}H_{12}$	C
Diphenylacetylene	$C_{14}H_{10}$	C
1,1',4,4'-Tetraphenylbutadiene	$C_{28}H_{22}$	SS
Diphenylstilbene	$C_{26}H_{20}$	SS
PPO (2,5-diphenyloxazole)	$C_{15}H_{11}NO$	PS
α-NPO [2-(1-Naphthyl)-5-phenyloxazole]	$C_{19}H_{13}NO$	PS
PBD [2-Phenyl,5-(4-biphenylyl)-1,3,4-oxadiazole]	$C_{20}H_{14}N_2O$	PS
BBO [2,5-Di(4-biphenylyl)-oxazole]	$C_{27}H_{19}NO$	SS
POPOP {1,4-Bis[2-(5-phenyloxazolyl)]-benzene}	$C_{24}H_{16}N_2O_2$	SS
TOPOT {1,4-Di-[2-(5-p-tolyloxazolyl)]-benzene}	$C_{26}H_{20}N_2O_2$	SS
DiMePOPOP {1,4-Di[2-(4-methyl-5-phenyloxazolyl)]-benzene}	$C_{26}H_{20}N_2O_2$	SS

Source: From Brooks, F. D., *Nucl Instrum. Meth.*, 162, 477, 1979.

[a] S—primary solvent; S'—secondary solvent; PS—primary solute; SS—secondary solute; C—crystal scintillator.

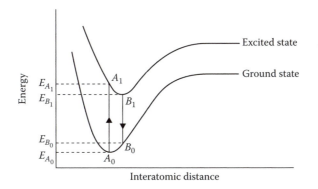

Figure 6.7 A typical (simplified) energy diagram of a molecule.

Table 6.3
Properties of Certain Organic Scintillators

Material	Wavelength of Maximum Emission (nm)	Relative Scintillation Efficiency (%)	Decay Time (ns)	Density (10^3 kg/m³)
Anthracene	445	100	~30	1.25
trans-Stilbene	385	~60	4–8	1.16
NE 102	350–450	~65	2	1.06
NE 110	350–450	60	3	1.06
NE 213 (liquid)	350–450	~60	2	0.867
PILOT B	350–450	68	2	1.06
PILOT Y	350–450	64	~3	1.06

with interatomic distance. The ground state of the molecule is at point A_0, which coincides with the minimum of the potential energy. Ionizing radiation passing through the scintillator may give energy to the molecule and raise it to an excited state, that is, the transition $A_0 \rightarrow A_1$ may occur. The position A_1 is not the point of minimum energy. The molecule will release energy through lattice vibrations (that energy is eventually dissipated as heat) and move to point B_1. The point B_1 is still an excited state and, in some cases, the molecule will undergo the transition $B_1 \rightarrow B_0$ accompanied by the emission of the photon with energy equal to $E_{B_1} - E_{B_0}$. This transition, if allowed, takes place at times of the order of 10^{-8} s. It should be noted that the energy of the emitted photon ($E_{B_1} - E_{B_0}$) is less than the energy that caused the excitation ($E_{A_1} - E_{A_0}$). This difference is very important because otherwise the emission spectrum of the scintillator would completely coincide with its absorption spectrum and no scintillations would be produced. A more detailed description of the scintillation process is given in the references (see Birks, 1964; and Ref. 19).

One of the important differences between inorganic and organic scintillators is in the response time, which is less than 10 ns for the latter (response time of inorganic scintillators is ~1 μs; see Table 6.1) and makes them suitable for fast timing measurements (see Chapter 10). Table 6.3 lists important properties of some organic scintillators.

6.3.2 Organic Crystal Scintillators

No activator is needed to enhance the luminescence of organic crystals. In fact, any impurities are undesirable because their presence reduces the light output, and for this reason, the material used to make the crystal is purified. Two of the most common organic crystal scintillators are anthracene and *trans*-stilbene.

Anthracene has a density of 1.25×10^3 kg/m³ and the highest light conversion efficiency of all organic scintillators (see Table 6.3)—which is still only about one-third of the light conversion efficiency of NaI(Tl). Its decay time (~30 ns) is much shorter than that of inorganic crystals. Anthracene can be obtained in different shapes and sizes.

trans-Stilbene has a density of 1.15×10^3 kg/m³ and a short decay time (4–8 ns). Its light conversion efficiency is about half of that for anthracene. It can be obtained as a clear, colorless, single crystal with a size up to several millimeters. Stilbene crystals are sensitive to thermal and mechanical shock.

6.3.3 Organic Liquid Scintillators

Liquid scintillator counting was codiscovered in 1950 by Kellman[20] and Reynolds *et al.*[21] It light has been extensively used with radioactive tracers research in the life sciences. The organic liquid scintillators consist of a mixture of a solvent with one or more solutes. Compounds that have been used successfully as solvents include xylene, toluene, and hexamethylbenzene (see Table 6.2). Satisfactory solutes include *p*-terphenyl, PBD, and POPOP.

In a binary scintillator, the incident radiation deposits almost all of its energy in the solvent but the luminescence is due almost entirely to the solute. Thus, as in the case of inorganic scintillators, an efficient energy transfer is taking place from the bulk of the phosphor to the material with the small concentration (activator in inorganic scintillators, solute in organic ones). If a second solute is added, it acts as a *wavelength shifter*, that is, it increases the wavelength of the light emitted by the first solute, so that the emitted radiation is better matched with the characteristics of the cathode of the photomultiplier tube.

Liquid scintillators are very useful for measurements where a detector with large volume is needed to increase efficiency. Examples are counting of low-activity β-emitters (^{3}H and ^{14}C in particular), detection of cosmic rays, and measurement of the energy spectrum of neutrons in the MeV range (see Chapter 14) using the scintillator NE 213. The liquid scintillators are well suited for such measurements because they can be obtained and used in large quantities (kiloliters) and can form a detector of desirable size and shape by utilizing a proper container.

In certain cases, the radioisotope to be counted is dissolved in the scintillator, thus providing 4π geometry (see Chapter 8) and, therefore, high detection efficiency. In others, an extra element or compound is added to the scintillator to enhance its detection efficiency without causing significant deterioration of the luminescence. Boron, cadmium, or gadolinium,[22–24] used as additives, cause an increase in neutron detection efficiency. On the other hand, fluorine-loaded scintillators consist of compounds in which fluorine has replaced hydrogen, thus producing a phosphor with low neutron sensitivity.

Some of the more recent applications of liquid scintillation counting includes digital discrimination of neutrons and gamma-rays in liquid scintillators using wavelets,[25] classical vs. evolved quenching parameters and procedures in scintillation measurements,[26] speciation of ^{129}I in the environmental and biological samples[27] and ^{210}Po determination in urines of people living in Central Italy.[28]

6.3.4 Plastic Scintillators

The plastic scintillators may be considered as solid solutions of organic scintillators. They have properties similar to those of liquid organic scintillators (Table 6.3), but they have the added advantage, compared to liquids, that they do not need a container. Plastic scintillators can be machined into almost any desirable shape and size, ranging from thin fibers to thin sheets. They are inert to water, air, and many chemicals, and for this reason they can be used in direct contact with the radioactive sample.

Plastic scintillators are also mixtures of a solvent and one or more solutes. The most frequently used solvents are polysterene and polyvinyltoluene. Satisfactory solutes include *p*-terphenyl and POPOP. The exact compositions of some plastic scintillators are given in Reference 29.

Plastic scintillators have a density of about 10^3 kg/m^3. Their light output is lower than that of anthracene (Table 6.3). Their decay time is short, and the wavelength corresponding to the maximum intensity of their emission spectrum is between 350 and 450 nm. Trade names of commonly used plastic scintillators are Pilot B, Pilot Y, NE 102, and NE 110. The characteristics of these phosphors are discussed in References 30–32. Plastic scintillators loaded with tin and lead have been tried as X-ray detectors in the 5–100 keV range.[33,34] Thin plastic scintillator films (as thin as 20×10^{-5} kg/m^2 = 20 μg cm^2) have proven to be useful detectors in time-of-flight measurements[35–37] (see Chapter 13).

Some recent applications of plastic scintillators include a large area detector array for fast neutron measurements,[38] energy calibration of gamma spectra using Compton kinematics,[39] design of a detector for γ-ray bursts,[40] and a γ-ray large area space telescope.[41]

6.4 GASEOUS SCINTILLATORS

Gaseous scintillators are mixtures of noble gases.[42,43] The scintillations are produced as a result of atomic transitions. Since the light emitted by noble gases belongs to the ultraviolet region, other gases, such as nitrogen, are added to the main gas to act as wavelength shifters. Thin layers of fluorescent materials used for coating the inner walls of the gas container achieve the same effect.

Gaseous scintillators exhibit the following features:

1. Very short decay time

2. Light output per MeV deposited in the gas depending very little on the charge and mass of the particle being detected

3. Very low efficiency for gamma detection.

These properties make the gaseous scintillators suitable for the energy measurement of heavy charged particles (alphas, fission fragments, other heavy ions). A recent book on noble gas detectors including those used in scintillators was published by Aprile *et al.* in 2006.

6.5 THE RELATIONSHIP BETWEEN PULSE HEIGHT AND ENERGY AND TYPE OF INCIDENT PARTICLE

To measure the energy of the incident particle with a scintillator, the relationship between the pulse height and the energy deposited in the scintillator must be known. Because the pulse height is proportional to the output of the photomultiplier, which output is in turn proportional to the light produced by the scintillator, it is necessary to know the light-conversion efficiency of the scintillator as a function of type and energy of incident radiation. The rest of this section presents experimental results for several cases of interest.

6.5.1 The Response of Inorganic Scintillators

Photons. The response of NaI(Tl) to gammas is linear, except for energies below 400 keV, where a slight nonlinearity is present. Experimental results are shown in Figure 6.8.[44] More details about the NaI(Tl) response to gammas are given in Chapter 12.

Charged particles. For protons and deuterons, the response of the scintillator is proportional to the particle energy, at least for $E > 1$ MeV. For alpha particles, the proportionality begins at about 15 MeV (Figure 6.9).[45] Theoretical aspects of the response have been studied extensively.[46–49] Today, inorganic scintillators are seldom used for detection of charged particles.

Neutrons. Because neutrons are detected indirectly through charged particles produced as a result of nuclear reactions, to find the response to neutrons, one looks at the response to alphas and protons. LiI(Eu), which is the crystal used for neutron detection, has essentially the same response as NaI(Tl) (Figure 6.9).

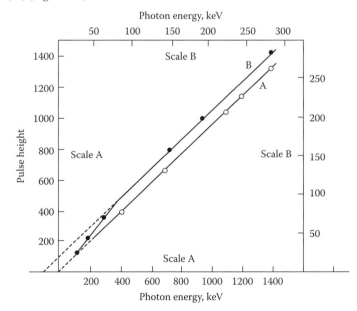

Figure 6.8 Pulse height versus energy for a NaI(Tl) crystal. The region below 300 keV has been expanded in curve B to show the nonlinearity. (From Engelkemeir, B., *Rev. Sci. Instrum.*, 27, 989, 1956.)

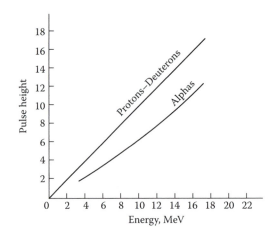

Figure 6.9 Pulse height versus energy for a NaI(Tl) crystal resulting from charged particles. (From Eby, F. S., and Jentschke, W. K., *Phys. Rev.*, 96, 911, 1954.)

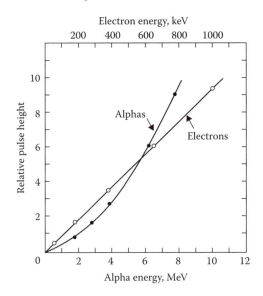

Figure 6.10 Pulse height versus energy for a liquid scintillator resulting from alphas and electrons. (From Flynn, K. F., et al., *Nucl. Instrum. Meth.*, 27, 13, 1964.)

6.5.2 The Response of Organic Scintillators

Charged particles. Experiments have shown that organic crystal scintillators (e.g., anthracene) exhibit a direction-dependent response to alphas[50] and protons.[51] An adequate explanation of the direction-dependent characteristics of the response does not exist at present. The user should be aware of the phenomenon to avoid errors.

The response of plastic and liquid scintillators to electrons, protons, and alphas is shown in Figures 6.10, 6.11, and 6.12.[52-54] Notice that the response is not linear, especially for heavier ions. The response has been studied theoretically by many investigators (Birks, 1964; and References 55–58).

Photons and neutrons. Organic scintillators are not normally used for detection of gammas because of their low efficiency. The liquid scintillators NE 213* is being used for γ detection in

---

* NE 213 consists of xylene, activators, and POPOP as the wavelength shifter. Naphthalene is added to enhance the slow components of light emission. The composition of NE 213 is given as $CH_{1,21}$ and its density as 0.867×10^3 kg/cm³.

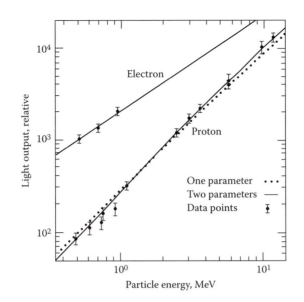

Figure 6.11 Plastic scintillator (NE 102) response to electrons and protons. (From Craun, R. L., and Smith, D. L., *Nucl. Instrum. Meth.*, 80, 239, 1970.)

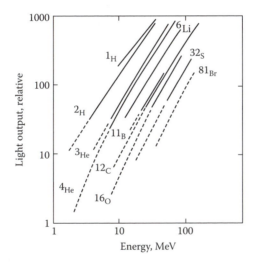

Figure 6.12 Plastic scintillator (NE 102) response to heavy ions. (From Becchetti, F. D., et al., *Nucl. Instrum. Meth.*, 138, 93, 1976.)

mixed neutron-gamma fields[59] because of its ability to discriminate against neutrons. Neutrons are detected by NE 213 through the proton-recoil method. More details about the use of the NE 213 scintillator and its response function are given in Chapters 12 and 14.

6.6 THE PHOTOMULTIPLIER TUBE

6.6.1 General Description

The photomultiplier tube or phototube is an integral part of a scintillation detector system. Without the amplification produced by the photomultiplier, a scintillator is useless as a radiation detector. The photomultiplier is essentially a fast amplifier, which in times of 10^{-9} s amplifies an incident pulse of visible light by a factor of 10^6 or more.

A photomultiplier consists of an evacuated glass tube with a photocathode at its entrance and several dynodes in the interior (Figure 6.13). The anode, located at the end of a series of dynodes

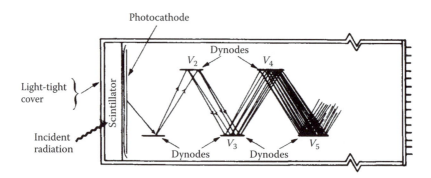

Figure 6.13 Schematic diagram of the interior of a photomultiplier tube.

serves as the collector of electrons. The photons produced in the scintillator enter the phototube and hit the photocathode, which is made of a material that emits electrons when light strikes it. The electrons emitted by the photocathode are guided, with the help of an electric field, toward the first dynode, which is coated with a substance that emits secondary electrons, if electrons impinge upon it. The secondary electrons from the first dynode move toward the second, from there toward the third, and so on. Typical commercial phototubes may have up to 15 dynodes. The production of secondary electrons by the successive dynodes results in a final amplification of the number of electrons as shown in the next section.

The electric field between dynodes is established by applying a successively increasing positive high voltage to each dynode. The voltage difference between two successive dynodes is of the order of 80–120 V (see Section 6.6.2).

The photocathode material used in most commercial phototubes is a compound of cesium and antimony (Cs–Sb). The material used to coat the dynodes is either Cs–Sb or silver–magnesium (Ag–Mg). The secondary emission rate of the dynodes depends not only on the type of surface but also on the voltage applied.

A very important parameter of every photomultiplier tube is the spectral sensitivity of its photocathode. For best results, the spectrum of the scintillator should match the sensitivity of the photocathode. The Cs–Sb surface has a maximum sensitivity at 440 nm, which agrees well with the spectral response of most scintillators (Tables 6.1 and 6.3). Such a response, called S-11, is shown in Figure 6.3. Other responses of commercial phototubes are known as S-13, S-20, etc.

Another important parameter of a phototube is the magnitude of its *dark current*. The dark current consists mainly of electrons emitted by the cathode after thermal energy is absorbed. This process is called *thermionic emission*, and a 50 mm diameter photocathode may release in the dark as many as 10^5 electrons/s at room temperature. Cooling of the cathode reduces this source of noise by a factor of about 2 per 10–15°C reduction in temperature. Thermionic emission may also take place from the dynodes and the glass wall of the tube, but this contribution is small. Electrons may be released from the photocathode as a result of its bombardment by positive ions coming from ionization of the residual gas in the tube. Finally, light emitted as a result of ion recombination may release electrons upon hitting the cathode or the dynodes. Obviously, the magnitude of the dark current is important in cases where the radiation source is very weak. Both the dark current and the spectral response should be considered when a phototube is to be purchased.

Recall that the electrons are guided from one dynode to the next by an electric field. If a magnetic field is present, it may deflect the electrons in such a way that not all of them hit the next dynode, and the amplification is reduced. Even the earth's weak magnetic field may sometimes cause this undesirable effect. The influence of the magnetic field may be minimized by surrounding the photomultiplier tube with a cylindrical sheet of metal, called *μ-metal*. The μ-metal is commercially available in various shapes and sizes.

Commercial photomultiplier tubes are made with the variety of geometrical arrangements of photocathode and dynodes. In general, the photocathode is deposited as a semitransparent layer on the inner surface of the end window of the phototube (Figure 6.14). The external surface of the window is, in most phototubes, flat for easier optical coupling with the scintillator (see Section 6.7). Two different geometries for the dynodes are shown in Figure 6.14.

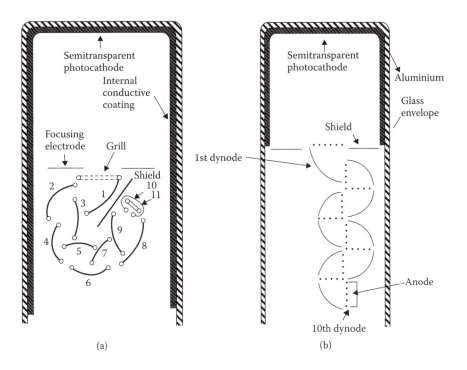

Figure 6.14 Two dynode arrangements in commercial phototubes: (a) Model 6342 RCA, 1–10 are dynodes, 11 is anode; (b) model 6292 DuMont.

6.6.2 Electron Multiplication in a Photomultiplier

The electron multiplication M in a photomultiplier can be written as

$$M = (\theta_1 \varepsilon_1)(\theta_2 \varepsilon_2) \cdots (\theta_n \varepsilon_n) \tag{6.3}$$

where

n = number of dynodes

$$\varepsilon_i = \frac{\text{number of electrons collected by } i\text{th dynode}}{\text{number of electrons emitted by } (i-1)\text{th dynode}}$$

$$\theta_i = \frac{\text{number of electrons collected by } i\text{th dynode}}{\text{number of electrons impinging upon } i\text{th dynode}}$$

If θ_i and ε_i are constant for all dynodes, then

$$M = (\theta \varepsilon)^n \tag{6.4}$$

The quantity ε depends on the geometry. The quantity θ depends on the voltage between two successive dynodes and on the material of which the dynode is made. The dependence of θ on voltage is of the form

$$\theta = kV^a \tag{6.5}$$

where $V = V_i - V_{i-1}$ = potential difference between two successive dynodes, assumed the same for all dynode pairs and k, a = constants (the value of a is about 0.7).

Using Eq. 6.5, the multiplication M becomes

$$M = \varepsilon^n (kV^a)^n = CV^{an} \tag{6.6}$$

where $C = (\varepsilon k)^n$ = constant, independent of the voltage.

Equation 6.6 indicates that the value of M increases with the voltage V and the number of stages n. The number of dynodes is limited, because as n increases, the charge density between two dynodes distorts the electric field and hinders the emission of electrons from the previous dynode with the lower voltage. In commercial photomultipliers, the number of dynodes is 10 or more. If one takes $n = 10$ and $\varepsilon\theta = 4$, typical value, the value of M becomes equal to 10^6.

To apply the electric field to the dynodes, a power supply provides a voltage adequate for all the dynodes. A voltage divider, usually an integral part of the preamplifier, distributes the voltage to the individual dynodes. When reference is made to *phototube voltage*, one means the total voltage applied. For example, if 1100 V is applied to a phototube with 10 dynodes, the voltage between any two dynodes is 100 V.

6.7 ASSEMBLY OF A SCINTILLATION DETECTOR AND THE ROLE OF LIGHT PIPES

A scintillation detector consists of the scintillator and the photomultiplier tube. It is extremely important that these two components be coupled in such a way that a maximum amount of light enters the phototube and strikes the photocathode. This section presents a brief discussion of the problems encountered during the assembly of a scintillation detector, with some of the methods used to solve them.

A solid scintillator is coupled to the photomultiplier through the end window of the tube (Figure 6.15). During the transfer from the scintillator to the photocathode, light may be lost by leaving through the sides and front face of the scintillator, or by being reflected back to the scintillator when it hits the window of the phototube.

To avoid loss of light through the sides and front face, the scintillator is painted with a material that reflects toward the crystal the light that would otherwise escape. Examples of reflecting materials commercially available are alpha alumina and Al_2O_3.

To avoid reflection of light from the end window of the phototube, a transparent viscous fluid (such as Dow–Corning 200 Silicone fluid) is placed between the scintillator and the phototube (Figure 6.15). The optical fluid minimizes reflection because it reduces the change of the index of refraction during the passage of light from the scintillator to the phototube. A sharp change in the index of refraction results in a small critical angle of reflection, which in turn increases total reflection.

In certain experiments, the scintillator has to be a certain distance away from the photocathode. Such is the case if the phototube should be protected from the radiation impinging upon the scintillator or from a magnetic field. Then a *light pipe* is interposed between the scintillator and the phototube. The light pipe is made of a material transparent to the light of the scintillator. Lucite, quartz, plexiglas, and glass have been used in many applications to form light pipes of different lengths and shapes. Light pipes of several feet—sometimes with bends—have been used with success. The optical coupling of the light pipe at both ends is accomplished by the same methods used to couple the scintillator directly to the phototube.

One of the major reasons for using scintillators is their availability in large sizes. In fact, commercially available scintillators are larger than the biggest commercial photomultipliers. In cases where the scintillator is too large, multiple phototubes are coupled to the same crystal. Figure 6.16 shows a NaI(Tl) crystal coupled to six photomultipliers.

When a liquid scintillator is used, the phototube is optically coupled to the scintillator through a window of the vessel containing the liquid scintillator. The efficiency of such a counting system increases by using a large volume of liquid and more than one photomultiplier tube (Figure 6.17).

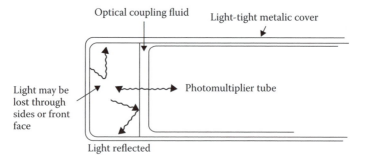

Figure 6.15 Assembly of a scintillation detector.

Figure 6.16 A special 30-in (0.762-m) diameter scintillator crystal coupled to six photomultiplier tubes (from Harshaw Chemical Company).

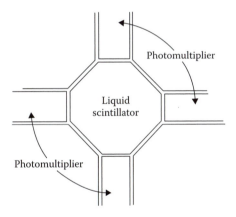

Figure 6.17 A counting system using a liquid scintillator and four photomultiplier tubes.

6.8 DEAD TIME OF SCINTILLATION DETECTORS

The *dead time* or *resolving time* is the minimum time that can elapse after the arrival of two successive particles and still result in two separate pulses (see Section 2.2.1).

For a scintillation detector this time is equal to the sum of three time intervals:

1. Time it takes to produce the scintillation, essentially equal to the decay time of the scintillator (see Eq. 6.1 and Tables 6.1 and 6.3).

2. Time it takes for electron multiplication in the phototube, of the order of 20–40 ns.

3. Time it takes to amplify the signal and record it by a scaler. The resolving time of commercial scalers is of the order of 1 μs. The time taken for amplification and discrimination is negligible.

By adding the three above components, the resulting dead time of a scintillation detector is of the order of 1–5 μs. This is much shorter than the dead time of gas-filled detectors, which is of the order of tens to hundreds of microseconds.

Scintillators are detectors with fast response. As seen in Tables 6.1 and 6.3, the risetime of the pulse is very short for all of them. Short risetime is important in measurements that depend on the time of arrival of the particle (see Chapter 10).

6.9 SOURCES OF BACKGROUND IN A SCINTILLATION DETECTOR

One of the major sources of background in a scintillation detector is the dark current of the phototube (see Section 6.6.1). Other background sources are naturally occurring radioisotopes, cosmic rays, and phosphorescing substances.

The holder of a liquid scintillator may contain small amounts of naturally occurring isotopes. In particular, ^{40}K is always present (isotopic abundance of ^{40}K is 0.01%). Another isotope, ^{14}C, is a constituent of contemporary organic materials. Solvents, however, may be obtained from petroleum, consisting of hydrocarbons without ^{14}C.

The term *phosphorescence* refers to delayed emission of light as a result of deexcitation of atoms or molecules. Phosphorescent half-lives may extend to hours. This source of background may originate in phosphorescent substances contained in the glass of the phototube, the walls of the sample holder, or the sample itself.

Cosmic rays, which are highly energetic charged particles, produce background in all types of detectors, and scintillators are no exception. The effect of cosmic-ray background, as well as that of the other sources mentioned earlier, will be reduced if two detectors are used in coincidence or anticoincidence.

6.10 THE PHOSWICH DETECTOR

The phoswich detector is used for the detection of low-level radiation in the presence of considerable background. It consists of two different scintillators coupled together and mounted on a single photomultiplier tube.[1] By using the difference in the decay constants of the two phosphors, differentiation between events taking place in the two detectors is possible. The combination of crystals used depends on the types of particles present in the radiation field under investigation.[60,61]

The basic structure of a phoswich detector is shown in Figure 6.18. A thin scintillator (scintillator A) is coupled to a larger crystal (scintillator B), which in turn is coupled to the cathode of a single phototube. Two examples of scintillators used are the following:

1. NaI(Tl) is the thin scintillator (A) and CsI(Tl) is the thick one (B). Pulses originating in the two crystals are differentiated based on the difference between the 0.25-μs decay constant of the NaI(Tl) and the 1-μs decay constant of the CsI(Tl). Slow pulses come from particles losing energy in the CsI(Tl) or in both crystals simultaneously. In a mixed low-energy–high-energy photon field, the relatively fast pulses of the NaI(Tl) will come from the soft component of the radiation. [Soft photons will not reach the CsI(Tl).] Phoswich detectors of this type have been used in X-ray and γ-ray astronomy, in detection of plutonium in the environment, and in other cases of mixed-radiation fields.

2. CaF$_2$(Eu) is the thin scintillator (A) and NaI(Tl) is the thick one (B). This combination is used for measurements of low-energy beta particles in the presence of a gamma background. The thin (0.1 mm) CaF$_2$(Eu) crystal detects the betas, but is essentially transparent to gammas because

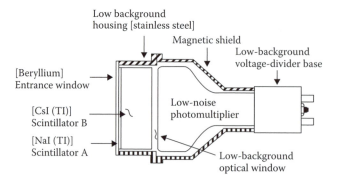

Figure 6.18 A phoswich detector (from Harshaw Chemical Company).

of its relatively low atomic number and thickness. A quartz window is usually placed between the two scintillators to stop the betas that did not deposit all their energy in the $CaF_2(Eu)$. The fast pulses of the NaI(Tl), which are due to gammas, are time-discriminated against the slower pulses from the $CaF_2(Eu)$ ($T = 0.94$ μs). Thus, the background due to gammas is reduced.

Some recent applications include a detector composed of an inner NaI(Tl) crystal and surrounding a NE102A plastic scintillator for neutron spectrometry[62] state-of-the-art of PET, SPECT and CT for small animal imaging[63], coincidence measurements in $\alpha/\beta/\gamma$ spectrometry with phoswich detectors using digital pulse shape discrimination analysis[64] and a well detector for radioxenon monitoring.[65]

PROBLEMS

6.1 If the dead time of a detection system using a scintillator is 1 μs, what is the gross counting rate that will result in a loss of 2% of the counts?

6.2 A typical dead time for a scintillation detector is 5 μs. For a gas counter, the corresponding number is 200 μs. If a sample counted with a gas counter results in 8% loss of gross counts due to dead time, what is the corresponding loss in a scintillation detector that records the same gross counting rate?

6.3 A parallel beam of 1.5-MeV gammas strikes a 25-mm-thick NaI crystal. What fraction of these gammas will have at least one interaction in the crystal ($\mu = 0.0047$ m^2/kg)?

6.4 A 1.75 MeV gamma hits a 25-mm-thick NaI crystal. (a) What fraction of these photons will have at least one interaction in the crystal? (b) What is the average distance traveled before the first interaction in the crystal? (c) What fraction of the interactions is photoelectric? [$\tau = 1.34 \times 10^{-3}$ cm^2/g]

6.5 What is the range of 2-MeV electrons in a plastic scintillator? Assume that the composition of the scintillator is $C_{10}H_{11}$ ($\rho = 1.02 \times 10^3$ kg/m^3).

6.6 Consider two electrons, one with kinetic energy 1 MeV, the other with 10 MeV. Which electron will lose more energy going through a 1-mm-thick plastic scintillator? Consider both ionization and radiation loss. Composition of the scintillator is given in Problem 6.4. For radiation loss, use

$$Z_{eff} = \frac{N_H Z_H^2 + N_C Z_C^2}{N_H Z_H + N_C Z_C}$$

6.7 A phoswich detector consists of a 1-mm-thick NaI(Tl) scintillator coupled to a 25-mm-thick CsI(Tl) scintillator. A 0.1 mm thick beryllium window protects the NaI(Tl) crystal. If the detector is exposed to a thin parallel beam of 150-keV X-rays and 1.5-MeV γ-rays, what are the fractions of interactions of each type of photon in each scintillator?

BIBLIOGRAPHY

Aprile, E., Bolotnikov, A. E., Bolozdynya, A. I., and Doke, T., *Noble Gas Detectors*, Wiley-VCH, 2006.

Birks, J. B., *The Theory and Practice of Scintillation Counting*, McMillan Co., New York, 1964.

Birowosuto, M. D., *Novel Gamma-Ray and Thermal-Neutron Scintillators: Search for High-Light-Yield and Fast-Response Materials*, IOS Press, Amsterdam, 2007.

Dorenbos, P., and Van Eijk, C. W. (eds.), *Inorganic Scintillators and Their Applications: Proceedings of the International Conference, Scint 95*, Coronet Books, Philadelphia, PA, 1996.

Eichholz, G. G., and Poston, J. W., *Nuclear Radiation Detection*, Lewis Publishers, Chelsea, Michigan, 1985.

Knoll, G. F., *Radiation Detection and Measurement*, 3rd ed., Wiley, New York, 2000.

L'Annunziata, M. F., *Handbook of Radioactivity Analysis*, Academic Press, San Diego, CA, 1998.

Lecoq, P., Annenkov, A., Gektin, A., and Korzhik M., *Inorganic Scintillators for Detector Systems: Physical Principles and Crystal Engineering (Particle Acceleration and Detection)*, Springer, Berlin and Heidelberg, 2006.

Price, W. J., *Nuclear Radiation Detection*, McGraw-Hill, New York, 1964.

Rodnyi, P. A., *Physical Processes in Inorganic Scintillators (Laser & Optical Science & Technology)*, CRC Press, Boca Raton and New York, 1997.

Ross, H., *Liquid Scintillation Counting and Organic Scintillators*, CRC Press, Boca Raton, 1991.

Ross, H., Noakes, J. E., and Spaulding, J. D., *Liquid Scintillation Counting and Organic Scintillators*, Lewis Publishers, Chelsea, Michigan, 1991.

Snell, A. H. (ed.), *Nuclear Instruments and Their Uses*, Wiley, New York, 1962.

Tait, W. H., *Radiation Detection*, Butterworth, London, 1980.

Tan, S. R., Donangelo, W. P., Gelbke, R., Lynch, C. K., Tsang, W. G., and Souza, M. B., *Energy Resolution and Energy-Light Response of CsI(Tl) Scintillators for Charged Particle Detection*, IOS Press, Philadelphia, 2000.

REFERENCES

1. Kolar, Z. I., and den Hollander, W., *J. Appl. Radiat. Isotop.* **61**:261 (2004).

2. Hofstatder, R., *Phys. Rev.* **74**:100 (1948).

3. Hofstatder, R., *Nucleonics* **6**:72 (1950).

4. Heath, R. L., Hofstadter, R., and Hughes, E. B., *Nucl. Instrum. Meth.* **162**:431 (1979). (Review article listing 127 references.)

5. Lempicki, A., *Appl. Spectrosc.* **62**:209 (1995).

6. Aliaga-Kelly, D., and Nicoll, D. R., *Nucl. Instrum. Meth.* **43**:110 (1966).

7. Colmenares, C., Shapiro, E. G., Barry, P. E., and Prevo, C. T., *Nucl. Instrum. Meth.* **114**:277 (1974).

8. Campbell, M., Ledingham, K. W. D., Baillie, A. D., and Lynch, J. G., *Nucl. Instrum. Meth.* **137**:235 (1976).

9. Catalano, E., and Czirr, J. B., *Nucl. Instrum. Meth.* **143**:61 (1977).

10. Chiles, M. M., Bauer, M. L., and McElhaney, S. A., *IEEE Trans. Nucl. Sci.* **37**:1348 (1990).

11. Wunderly, S. W., *Appl. Radiat. Isotop.* **40**:569 (1989).

12. Wang, Y. J., Patt, B. E., and Iwanczyk, J. S., *Nucl. Instrum. Meth. Phys. Res. A* **353**:50 (1994).

13. Anderson, D. F., *Nucl. Instrum. Meth. Phys. Res. A* **287**:606 (1990).

14. Katagiri, M., Nakamuraa, T., Ebinea, M., Birumachia, A., Satob, S., Shooneveld, E. M., and Rhodes, N. J., *Nucl. Instrum. Meth. Phys. Res. A* **573**:149 (2007).

15. Rooh, G., Kang, H. J., Kim, H. J., Park, H., Doh, S-H., and Kim, S., *Crys. Growth*, **311**:128 (2008).

16. Ottonello, P., Palestini, V., Rottigni, G. A., Zanella, G., and Zannoni, R., *Nucl. Instrum. Meth. Phys. Res. A* **366**:248 (1995).

17. Yang, H., Wehe, D. K., and Bartels, D. M., *Nucl. Instrum. Meth. Phys. Res. A* **598**:779 (2009).

18. Rout, P. C., Chakrabarty, D. R., Datar, V. M., Kumar, S., Mirgule, E. T., Mitra, A., Nanal, V., and Kujur, R., *Nucl. Instrum. Meth. Phys. Res. A* **598**:526 (2009).

19. Brooks, F. D., *Nucl Instrum. Meth. Phys. Res. A* **162**:477 (1979). (Review article listing 274 references.)

20. Kellman, H., *Phys. Rev.* **78**:621 (1950).

21. Reynolds, G. T., Harrison, F. B., and Salvine, G., *Phys. Rev.* **78**:488 (1950).

22. Bollinger, L. M., and Thomas, G. E., *Rev. Sci. Instrum.* **28**:489 (1957).

23. Hellstrom, J., and Beshai, S., *Nucl. Instrum. Meth.* **101**:267 (1972).

24. Bergere, R., Beil, H., and Veyssiere, A., *Nucl. Phys. A* **121**:463 (1968).

25. Yousefi, S., Lucchese, L., and Aspinall, M. D., *Nucl. Instrum. Meth. Phys. Res. A* **598**:551 (2009).

26. Bagán, H., Tarancón, A., Rauret, G., García, J. F., *Nucl. Instrum. Meth. Phys. Res. A* **592**:361 (2008).

27. Hou, X., Hansen, V., Aldahan, A., Pssnert, G., Lind, O. C., and Lujaniene, G., *Analytica Chimica Acta* **632**:18 (2009).

28. Meli, M. A., Desideri, D., Roselli, C., and Feduzi, L., *J. Environ. Radioactiv.* **100**:84 (2009).

29. Swank, R. K., *Annu. Rev. Nucl. Sci.* **4**:111 (1954).

30. Walker, J. K., *Nucl. Instrum. Meth.* **68**:131 (1969).

31. Moszynski, M., and Bengtson, B., *Nucl. Instrum. Meth.* **142**:417 (1977).

32. Moszynski, M., and Bengtson, B., *Nucl. Instrum. Meth.* **158**:1 (1979).

33. Eriksson, L. A., Tsai, C. M., Cho, Z. H., and Hurlbut, C. R., *Nucl Instrum. Meth.* **122**:373 (1974).

34. Becker, J., Eriksson, L., Monberg, L. C., and Cho, Z. H., *Nucl. Instrum. Meth.* **123**:199 (1975).

35. Muga, M. L., Burnsed, D. J., Steeger, W. E., and Taylor, H. E., *Nucl. Instrum. Meth.* **83**:135 (1970).

36. Muga, M. L., *Nucl. Instrum. Meth.* **95**:349 (1971).

37. Batra, R. K., and Shotter, A. C., *Nucl. Instrum. Meth.* **124**:101 (1975).

38. Rout, P. C., Chakrabarty, D. R., Datar, V. M., Kumar, S., Mirgule, E. T., Mitra, A., Nanal, V., and Kujur, R., *Nucl. Instrum. Meth. Phys. Res. A* **598**:526 (2009).

39. Siciliano, E. R., Ely, J. H., Kouzes, R. T., Schweppe, J. E., Strachan, D. M., and Yokuda, S. T., *Nucl. Instrum. Meth. Phys. Res. A* **594**:232 (2008).

40. Gierlik, M., Batsch, T., Marcinkowski, R., Moszyński, M., and Sworobowicz, T., *Nucl. Instrum. Meth. Phys. Res. A*, **593**:426 (2008).

41. Moissey, A., *Nucl. Instrum. Meth. Phys. Res. A*, **588**:41 (2008).

42. Policarpo, A. J. P. L., Conde, C. A. N., and Alves, M. A. F., *Nucl. Instrum. Meth.* **58**:151 (1968).

43. Morgan, G. L., and Walter, R. L., *Nucl. Instrum. Meth.* **58**:277 (1968).

44. Engelkemeir, B., *Rev. Sci. Instrum.* **27**:989 (1956).

45. Eby, F. S., and Jentschke, W. K., *Phys. Rev.* **96**:911 (1954).

46. Murray, R. B., and Meyer, A., *Phys. Rev.* **122**:815 (1961).

47. Meyer, A., and Murray, R. B., *Phys. Rev.* **128**:98 (1962).

48. Prescott, J. R., and Narayan, G. H., *Nucl. Instrum. Meth.* **75**:51 (1969).

49. Hill, R., and Collinson, A. J. L., *Nucl. Instrum. Meth.* **44**:245 (1966).

50. Brand, W., Dobrin, R., Jack, H., Aubert, R. L., and Roth, S., *Con. J. Phys.* **46**:537 (1968).

51. Brooks, F. D., and Jones, D. T. L., *Nucl. Instrum. Meth.* **121**:69 (1974).

52. Flynn, K. F., Glendenin, C. E., Steinberg, E. P., and Wright, P. M., *Nucl. Instrum. Meth.* **27**:13 (1964).

53. Craun, R. L., and Smith, D. L., *Nucl. Instrum. Meth.* **80**:239 (1970).

54. Becchetti, F. D., Thorn, C. E., and Levine, M. S., *Nucl. Instrum. Meth.* **138**:93 (1976).

55. Chou, C. N., *Phys. Rev.* **87**:376, 904 (1952).

56. Wright, G. T., *Phys. Rev.* **91**:1282 (1953).

57. Voltz, R., Lopes da Silva, J., Laustriat, G., and Coche, A., *J. Chem. Phys.* **45**:3306 (1966).

58. Voltz, R., du Pont, H., and Laustriat, G., *J. Physique.* **29**:297 (1968).

59. Ingersoll, D. T., and Wehring, B. W., *Nucl. Instrum. Meth.* **147**:551 (1977).

60. Pastor, C., Benrachi, F., Chambon, B., Cheynis, B., Drain, D., Dauchy, A., Giorni, A., and Morand, C., *Nucl. Instrum. Meth.* **227**:87 (1984).

61. Pouliot, J., Chan, Y., Dacal, A., Harmon, A., Knop, R., Ortiz, M. E., Plagnol, E., and Stokstad, R. G., *Nucl. Instrum. Meth.* **270**:69 (1988).

62. Watanabe, T., Arakawa, H., Kajimoto, T., Iwamoto, Y., Satoh, D., Kunieda, S., Noda, S., *et al.*, *Nucl. Instrum. Meth. Phys. Res. A* **587**:20 (2008).

63. Del Guerra, A., and Belcari, N., *Nucl. Instrum. Meth. Phys. Res. A* **583**:119 (2007).

64. De Celis, B., de la Fuente, R., Williart, A., and de Celis Alonso, B., *Nucl. Instrum. Meth. Phys. Res. Sec. A* **580**:206 (2007).

65. Hennig, W., Tan, H., Fallu-Labruyere, A., Warburton, W. K., McIntyre, J. I., and Gleyzer, A., *Nucl. Instrum. Meth. Phys. Res. A* **579**:431 (2007).

7

Semiconductor Detectors

7.1 INTRODUCTION

Semiconductor detectors are solid-state devices that operate essentially like ionization chambers. The charge carriers in semiconductors are not electrons and ions, as in the gas counters, but electrons and "holes."[1,2] At present, the most successful semiconductor detectors are made of silicon and germanium. However, starting in the 1990s different materials have been tried, with some very good successes, for example, cadmium telluride (CdTe), cadmium zinc telluride ([CdZnTe] often named CZT), cesium iodide (CsI), mercuric iodide (HgI_2), and others,

The most important advantage of the semiconductor detectors, compared to other types of radiation counters, is their superior *energy resolution*: the ability to resolve the energy of particles out of a polyenergetic energy spectrum (energy resolution and its importance are discussed in Chapters 9 and 12–14). Other advantages are as follows:

1. Linear response (pulse height versus particle energy) over a wide energy range

2. Higher efficiency for a given size, because of the high density of a solid relative to that of a gas

3. Possibility for special geometric configurations

4. Fast pulse rise time (relative to gas counters)

5. Ability to operate in vacuum

6. Insensitivity to magnetic fields.

The characteristics of a semiconductor detector depend not only on the type of material used—for example, Si or Ge—but also on the way the semiconductor is shaped and treated. The type, size, shape, and treatment of the crystal play a role in the operation and performance of a semiconductor detector.

This chapter first discusses the fundamentals of energy states in crystals, a subject necessary for understanding the creation and movement of electrons and holes in a solid. The properties of semiconductors are discussed next, with special emphasis given to the properties of silicon and germanium. The principle of construction and operation is accompanied by a description of the different types of detectors available in the market.

7.2 ELECTRICAL CLASSIFICATION OF SOLIDS

Solids are divided according to their electrical conductivity into three groups: conductors, insulators, and semiconductors. If a piece of solid material is placed in an electric field, whether or not current will flow depends on the type of material. If current flows, the material is a *conductor*. If current is zero at low temperatures but larger than zero at higher temperatures, the material is a *semiconductor*. If current is zero at all temperatures, the material is an *insulator*.

Conductivity and *electric current* mean motion of electrons, and according to the results of this simple experiment, we conclude that:

1. In conductors, electrons can move freely at any voltage other than zero.

2. In insulators, electrons cannot move under any voltage (except, of course, when the voltage is so high that an electrical discharge occurs).

3. In semiconductors, electrons cannot move at low temperatures (close to absolute zero) under any voltage. As the temperature of a semiconductor increases, however, electrons can move and electric current will flow at moderate voltages.

These properties can be explained by examining the electronic structure of crystals.

7.2.1 Electronic States in Solids—The Fermi Distribution Function

In a free atom, the electrons are allowed to exist only in certain discrete energy states (Figure 7.1a). In solids, the energy states widen into energy bands. Electrons can exist only in bands 1, 3, and 5,

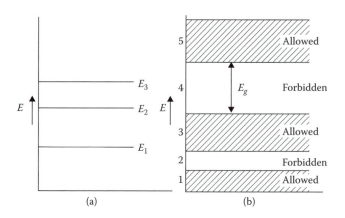

Figure 7.1 (a) The atomic energy levels are discrete lines, (b) in a solid, the allowed energy states become energy bands.

but not in bands 2 and 4 (Figure 7.1b). An electron can move from band 1 to band 3 if

1. The electron acquires the energy E_g necessary to cross the forbidden gap and

2. There is an empty state in band 3, which the jumping electron can occupy.*

The energy distribution of electronic states is described in terms of the following quantities:

$N(E)dE$ = number of electrons per unit volume with energy between E and $E + dE$
$S(E)dE$ = number of allowed electronic energy states, per unit volume, in the energy
        interval between E and $E + dE$
$P(E)$ = probability that a state of energy E is occupied = *Fermi distribution function*.

Then

$$N(E)dE = P(E)\big[S(E)\,dE\big] \tag{7.1}$$

The form of $P(E)$ is given by

$$P(E) = \frac{1}{1 + e^{(E - E_f)/kT}} \tag{7.2}$$

where E_f = Fermi energy
     k = Boltzmann constant
     T = temperature, Kelvin

The *Fermi energy* E_f is a constant that does not depend on temperature but it does depend on the purity of the solid. The function $P(E)$ is a universal function applying to all solids and having these properties (Figure 7.2):

1. At $T = 0$,

$$P(E) = 1, \quad E < E_f$$
$$P(E) = 0, \quad E > E_f$$

2. At any T,

$$P(E_f) = \tfrac{1}{2}$$

---

* This constraint is due to the Pauli principle, which forbids two or more electrons to be in the same state.

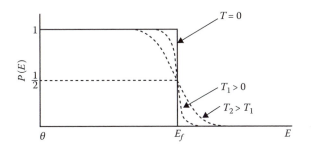

Figure 7.2 The Fermi distribution function.

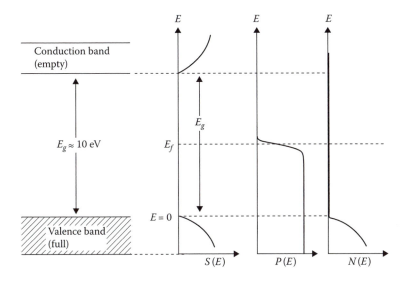

Figure 7.3 All the energy states in the conduction band of an insulator are empty. Since there are no charge carriers, the conductivity is zero.

3. For $T > 0$, the function $P(E)$ extends beyond E_f. If $E - E_f \gg kT$, $P(E)$ takes the form

$$P(E) = \frac{1}{1 + e^{(E-E_f)/kT}} \sim \frac{1}{e^{(E-E_f)/kT}} = \exp\left(-\frac{E-E_f}{kT}\right) \qquad (7.3)$$

which resembles the classical Boltzmann distribution.

Notice that at $T = 0$ (Figure 7.2), all the states are occupied for $E < E_f$ but all the states are empty for $E > E_f$.

7.2.2 Insulators

In insulators, the highest allowed band, called the *valence band*, is completely occupied (Figure 7.3). The next allowed band, called the *conduction band*, is completely empty. As Figure 7.3 shows, the gap is so wide that the number of occupied states in the conduction band is always zero. No electric field or temperature rise can provide enough energy for electrons to cross the gap and reach the conduction band. Thus, insulators *are* insulators because it is impossible for electrons to be found in the conduction band, where under the influence of an electric field, they would move and generate an electric current.

7.2.3 Conductors

In conductors, the conduction band is partially occupied (Figure 7.4). An electron close to the top of the filled part of this band (point A, Figure 7.4) will be able to move to the empty part (part B) under the influence of any electric field other than zero. Thus, because of the lack of a forbidden

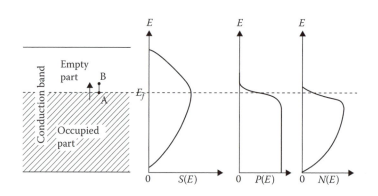

Figure 7.4 In conductors, the conduction band is partially occupied. If an electric field is applied, the electrons move and conductivity is not zero.

gap, there is no threshold of electric field intensity below which electrons cannot move. Motion of the charge carriers and, consequently, conductivity are always possible for any voltage applied, no matter how small.

7.3 SEMICONDUCTORS

In semiconductors, the valence band is full and the conduction band is empty, but the energy gap, E_g as shown, between these two bands is very small. At very low temperatures, close to $T = 0$, the conductivity of the semiconductors is zero and the energy-band picture looks like that of an insulator (Figure 7.3). As temperature increases, however, the "tail" of the Fermi distribution brings some electrons into the conduction band and conductivity increases (Figure 7.5). That is, as temperature increases, some electrons obtain enough energy to cross over to the conduction band. Once there, they will move under the influence of an electric field for the same reason that electrons of conductors move.

When an electron moves to the conduction band, an empty state is left in the valence band. This is called a *hole*. A hole is the absence of an electron. When the electron moves in one direction, the hole moves in the opposite direction (Figure 7.6). Holes are treated as particles with positive charges: $-(-e) = +e$. They contribute to the conductivity in the same way electrons do (see Section 7.3.2). In a pure and electrically neutral semiconductor, the number of electrons is always equal to the number of holes.

Heat—that is, temperature increase—is not the only way energy may be given to an electron. Absorption of radiation or collision with an energetic charged particle may produce the same effect. The interaction of ionizing radiation with a semiconductor is a complex process and there is no agreement upon a common model explaining it. One simplified model is the following.

An energetic incident charged particle collides with electrons of the semiconductor and lifts them, not only from the valence to the conduction band but also from deeper lying occupied bands to the conduction band, as shown in Figure 7.7a. Electrons appear in normally empty bands and holes appear in normally fully occupied bands. However, this configuration does not last long. In times of the order of 10^{-12} s, the interaction between electrons and holes makes the electrons concentrate at the bottom of the lowest lying unoccupied (conduction) band. The holes, on the other hand, concentrate near the top of the highest full (valence) band. During this deexcitation process, many more electrons and holes are generated. Because of this multistep process, the average energy necessary for the creation of one electron-hole pair is much larger than the energy gap E_g. For example, for silicon at room temperature, $E_g = 1.106$ eV, and the average energy for the production of one electron–hole pair is 3.66 eV.

In the absence of an electric field, the final step of the deexcitation process is the recombination of electrons and holes and the return of the crystal to its neutral state.

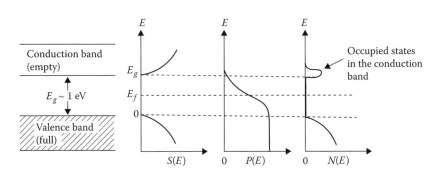

Figure 7.5 In semiconductors, the energy gap, E_g as shown, is relatively narrow. As temperature increases, some electrons have enough energy to be able to move to the conduction band and conductivity appears.

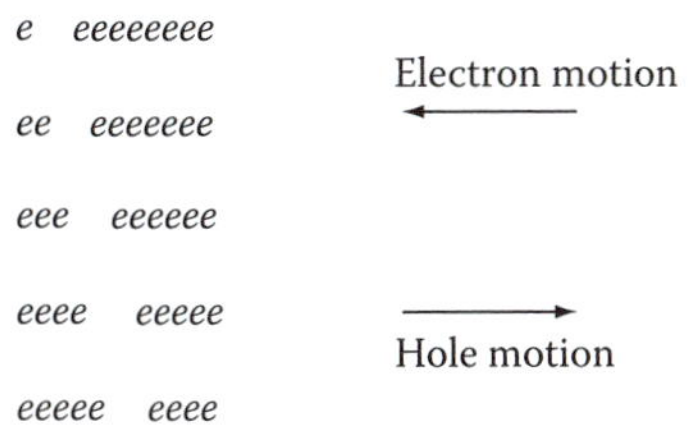

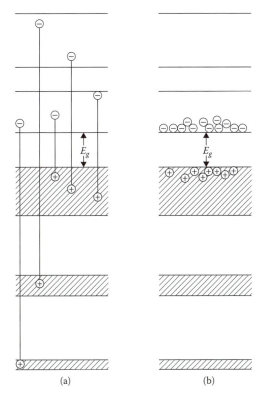

Figure 7.6 Electrons and holes move in opposite directions. A hole behaves like a positively charged carrier.

Figure 7.7 (a) Collisions with an energetic charged particle raise electrons to the conduction bands, (b) after times of the order of 10^{-12} s, electrons and holes tend to deexcite to the upper part of the valence band and lower part of the conduction band, respectively.

7.3.1 The Change of the Energy Gap with Temperature

The value of the energy gap E_g (Figure 7.5) is not constant, but it changes with temperature as shown in Figure 7.8. For silicon and germanium, E_g initially increases linearly as temperature decreases; but at very low temperatures, E_g reaches a constant value.

The average energy needed to create an electron-hole pair follows a similar change with temperature (Figure 7.9).

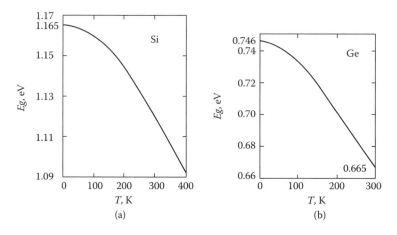

(a)   (b)

Figure 7.8   The variation of E_g with temperature: (a) for silicon; (b) for germanium. (From Bertolini, G., and Coche, A., *Semiconductor Detectors*, North-Holland Publishing Co., Amsterdam, Chapter 1.1.1, 1968.).

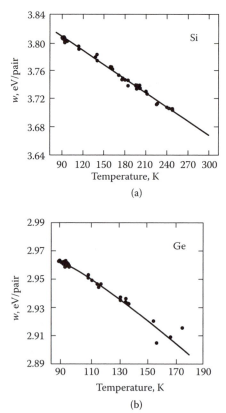

Figure 7.9   Energy needed to produce an electron-hole pair in (a) silicon and (b) germanium, as a function of temperature. (From Pehl, R. H., et al., *Nucl. Instrum. Meth.*, 59, 45, 1968.)

7.3.2 Conductivity of Semiconductors

Conductivity σ is the inverse of resistivity and is defined by

$$j = \sigma E \tag{7.4}$$

where j = current density (A/m^2)
σ = conductivity [A/(V m)] (conductivity is the reverse of resistivity)
E = electric field (V/m)

Another expression for the current density is

$$j = eN\upsilon \tag{7.5}$$

where N = number of charge carriers/m^3
υ = speed of carriers

Using Eqs. 7.4 and 7.5, one obtains the following equation:

$$\sigma = eN\frac{\upsilon}{E} \tag{7.6}$$

The ratio υ/E is given a new name, *mobility of the carrier*:

$$\mu = \frac{\upsilon}{E} \tag{7.7}$$

All the types of charge carriers present in a medium contribute to the conductivity. In the case of semiconductors, both electrons and holes should be taken into account when conductivity is calculated, and the expression for the conductivity becomes (using Eqs. 7.6 and 7.7).

$$\sigma = e(N_e\mu_e + N_p\mu_p) \tag{7.8}$$

where N_e and N_p are charge carrier concentrations and μ_e and μ_p are mobilities of electrons and holes, respectively. According to Eq. 7.8, the conductivity changes if the mobility of the carriers or their concentration or both change.

The mobilities of electrons and holes are independent of the electric field over a wide range of carrier velocities, but they change with temperature. If the temperature decreases, the mobility of both carriers increases. The mobility of electrons and holes in pure germanium as a function of temperature is shown in Figure 7.10.[4] The mobility changes at $\mu \sim T^{-\alpha}$ with $\alpha \approx 1.5$, for $T < 80$ K. For $T > 80$ K, the value of α is somewhat larger. It is worth noting that for $T < 80$ K, $\mu_n \approx \mu_p$.

In a pure semiconductor, $N_e = N_p$ and each one of these quantities is given by

$$N_e = N_p = AT^{1.5}\exp\left(-\frac{E_g}{2kT}\right) \tag{7.9}$$

where A is a constant independent of T.

The motion of the carriers in a semiconductor is also affected by the presence of impurities and defects of the crystal. A small amount of impurities is always present, although impurities are usually introduced deliberately to make the properties of the crystal more appropriate for radiation detection (see Section 7.3.3). Crystal defects are present too. Even if one starts with a perfect crystal, defects are produced by the incident particles (this is called *radiation damage*). In the language of energy bands, impurities and defects create new energy states that may trap the carriers. Trapping is, of course, undesirable because it means loss of part of the charge generated by the incident particle.

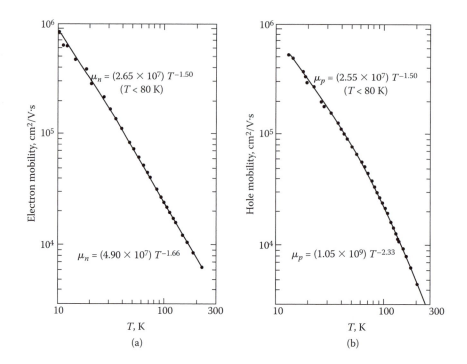

Figure 7.10 (a) Electron mobility versus temperature for *n*-type germanium, (b) hole mobility versus temperature for *p*-type germanium. (From Delaet, L. H., et al., *Nucl. Instrum. Meth.*, 101, 11, 1972.)

For semiconductors, the probability that an electron will move from the valence to the conduction level is proportional to the factor (Eq. 7.3)

$$\exp\left(-\frac{E_g}{2kT}\right) \tag{7.10}$$

(E_f is located in the middle of the gap; thus $E - E_f = E_g/2$.) Because of the exponential form of Eq. 7.10, there are always some electrons in the conduction band. These electrons produce a leakage current. Obviously, a successful detector should have as low a leakage current as possible to be able to detect the ionization produced by the incident radiation. The leakage current decreases as temperature decreases, and for two different materials it will be smaller for the material with the larger energy gap.

7.3.3 Extrinsic and Intrinsic Semiconductors—The Role of Impurities

The properties of a pure semiconductor change if impurities are introduced. With impurities present, new states are created and the semiconductor obtains extra electrons or extra holes, which increase the conductivity of the material.

Actually, pure semiconductors are not available. All materials contain some impurities and for this reason they are called *impure* or *extrinsic* in contrast to a pure semiconductor, which is called *intrinsic*. In most cases, controlled amounts of impurities are introduced purposely by a process called *doping*, which increases the conductivity of the material by orders of magnitude.

Doping works in the following way. Consider silicon (Si), which has four valence electrons. In a pure Si crystal, every valence electron makes a covalent bond with a neighboring atom (Figure 7.11a). Assume now that one of the atoms is replaced by an atom of arsenic (As), which has five valence electrons (Figure 7.11b). Four of the valence electrons form covalent bonds with four neighboring Si atoms, but the fifth electron does not belong to any chemical bond. It is bound very weakly and only a small amount of energy is necessary to free it, that is, to move it to the conduction band. In terms of the energy-band model, this fifth electron belongs to an energy state located very close to the conduction band. Such states are called *donor* states (Figure 7.12), and impurity

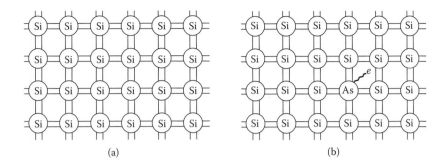

Figure 7.11 (a) Pure (intrinsic) silicon, (b) silicon doped with arsenic. The fifth electron of the arsenic atom is not tightly bound, and little energy is needed to move it to the conduction band.

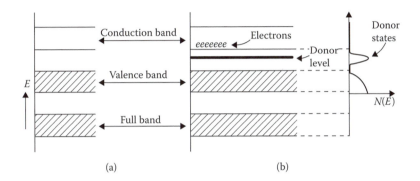

Figure 7.12 (a) Intrinsic and (b) n-type semiconductor. New electron states (donor states) are created close to the conduction band.

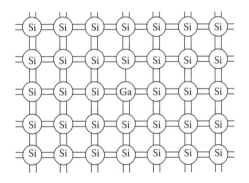

Figure 7.13 Silicon doped with gallium. One of the covalent bonds is not matched.

atoms that create them are called *donor* atoms. The semiconductor with donor atoms has a large number of electrons and a small number of holes. Its conductivity will be due mainly to electrons, and it is called an *n-type* semiconductor (*n* is for negative).

If a gallium atom is the impurity, three valence electrons are available; thus only three Si bonds will be matched (Figure 7.13). Electrons from other Si atoms can attach themselves to the gallium atom, leaving behind a hole. The gallium atom will behave like a negative ion after it accepts the extra electron. In terms of the energy-band theory, the presence of the gallium atom creates new states very close to the valence band (Figure 7.14). These are called *acceptor* states. The impurity is called an *acceptor* atom. For every electron that moves to the acceptor states, a hole is left behind. The acceptor impurity atoms create holes. The charge carriers are essentially positive, and the semiconductor is called *p-type*.

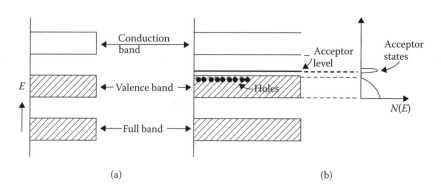

(a) (b)

Figure 7.14 (a) Intrinsic and (b) p-type semiconductor. New hole states (acceptor states) are created close to the top of the valence band.

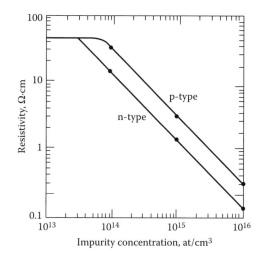

Figure 7.15 Resistivity as a function of impurity concentration in germanium. (From Bertolini, G., and Coche, A., *Semiconductor Detectors*, North-Holland Publishing Co., Amsterdam, Chapter 1.1.3, 1968.)

Interstitial atoms can act as donors or acceptors. Lithium, as an interstitial in either silicon or germanium, creates donor states very close to the conduction band. Copper and nickel introduce donor states midway between the valence and conduction bands. Gold may act as either an acceptor or donor, depending on its position on the lattice.

For every atom of n- or p-type impurity, an electron or hole is located at the donor or acceptor state, respectively. The material is still neutral, but when conductivity appears,

Electrons are the major carriers for n-type semiconductors.
Holes are the major carriers for p-type semiconductors.

Since the addition of impurities creates new states that facilitate the movement of the carriers, it should be expected that the conductivity of a semiconductor increases with impurity concentration. Figures 7.15 and 7.16 show how the resistivity of germanium and silicon changes with impurity concentration.

The energy gap E_g depends on temperature, as shown in Figure 7.8, and on the number of impurities and defects of the crystal. With increasing temperatures, if E_g is small as in germanium, the electrical conduction is dominated by electron-hole pairs created by thermal excitation and not by the presence of the impurity atoms. Therefore, at high enough temperatures, any semiconductor can be considered as intrinsic.

Table 7.1 presents the most important physical and electrical properties of silicon and germanium, the two most widely used semiconductors.

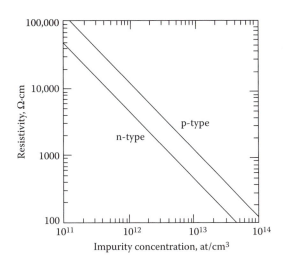

Figure 7.16 Resistivity as a function of impurity concentration in silicon. (From Bertolini, G., and Coche, A., *Semiconductor Detectors*, North-Holland Publishing Co., Amsterdam, Chapter 1.1.3, 1968.)

Table 7.1
Properties of Si and Ge

Property	Si	Ge
Atomic number	14	32
Atomic weight	28.1	72.6
Density (300 K)	2.33×10^3 kg/m^3	5.33×10^3 kg/m^3
Energy gap (E_g), 300 K	1.106 eV	0.67 eV
Energy gap (E_g), 0 K	1.165 eV	0.75 eV
Average energy per electron–hole pair, 77 K	3.7 eV	2.96 eV
Average energy per electron–hole pair, 300 K	3.65 eV	–
Diffusion voltage (V_0)	0.7 V	0.4 V
Atomic concentration	5×10^{28} m^{-3}	4.5×10^{28} m^{-3}
Intrinsic carrier concentration (300 K)	1.5×10^{16} m^{-3}	2.4×10^{19} m^{-3}
Intrinsic resistivity (300 K)	2.3×10^3 Ω·m	0.47 Ω·m
Intrinsic resistivity (77 K)	∞	5×10^2 Ω·m
Electron mobility (300 K)	0.1350 m^2/V·s	0.3900 m^2/V·s
Hole mobility (300 K)	0.0480 m^2/V·s	0.1900 m^2/V·s
Electron mobility (77 K)	4.0–7.0 m^2/V·s	3.5–5.5 m^2/V·s
Hole mobility (77 K)	2.0–3.5 m^2/V·s	4.0–7.0 m^2/V·s
Dielectric constant	12	16

Source: From Fenyves, E., and Haiman, O., *The Physical Principles of Nuclear Radiation Measurements*, Academic Press, New York, 1969; Knoll, G. F., *Radiation Detection and Measurement*, 3rd ed., Wiley, New York, 2000; Ewan, G. T., *Nucl. Instrum. Meth.*, 162, 75, 1979.

7.4 THE p–n JUNCTION

7.4.1 The Formation of a p–n Junction

As stated in the introduction to this chapter, semiconductor detectors operate like ionization detectors. In ionization chambers (see Chapter 5), the charges produced by the incident radiation are collected with the help of an electric field from an external voltage. In semiconductor detectors, the electric field is established by a process more complicated than in gas-filled detectors, a process that depends on the properties of n- and p-type semiconductors. The phenomena involved will be better understood with a brief discussion of the so-called p–n junction.

An n-type semiconductor has an excess of electron carriers. A p-type has excess holes. If a p-type and an n-type semiconductor join together, electrons and holes move for two reasons:

1. Both electrons and holes will move from areas of high concentration to areas of low concentration. This is simply diffusion, the same as neutron diffusion or diffusion of gas molecules.

2. Under the influence of an electric field, both electrons and holes will move, but in opposite directions because their charge is negative and positive, respectively.

Consider two semiconductors, one p-type, the other n-type, in contact, without an external electric field (Figure 7.17). The n-type semiconductor has a high electron concentration; the p-type has a high hole concentration. Electrons will diffuse from the n- to the p-type; holes will diffuse in the opposite direction. This diffusion will produce an equilibrium of electron and hole concentrations, but it will upset the original charge equilibrium. Originally, both p- and n-type semiconductors were electrically neutral, but as a result of the diffusion, the n-type region will be positively charged, while the p-type region will be negatively charged. After equilibrium is established, a potential difference exists between the two regions. This combination of p- and n-type semiconductor with a potential difference between the two types constitutes a p–n junction.

The potential V_0 (Figure 7.17a) depends on electron–hole concentrations and is of the order of 0.5 V. If an external voltage V_b is applied with the positive pole connected to the n side, the total potential across the junction becomes $V_0 + V_b$. This is called *reverse bias*. Such external voltage tends to make the motion of both electrons and holes more difficult. In the region of the changing potential, there is an electric field $E = -\partial V / \partial x$. The length X_0 of the region where the potential and the electric field exist increases with reverse bias. Calculation shows that

$$X_0 \approx \sqrt{\mu_p \rho (V_0 + V_b)} \quad \text{for p-type semiconductor} \tag{7.11a}$$

and

$$X_0 \approx \sqrt{\mu_n \rho (V_0 + V_b)} \quad \text{for n-type semiconductor} \tag{7.11b}$$

where $\rho(\Omega \cdot m)$ is the resistivity of the crystal. Application of a negative potential on the n side will have the opposite effect. The total potential difference will be $V_0 - V_b$. This is called *forward bias*. For a successful detector, reverse bias is applied. Since, usually, $V_b \gg V_0, X_0 \sim \sqrt{V_b}$.

In practice, a p–n junction is not made by bringing two pieces of semiconductor into contact. Instead, one starts with a semiconductor of one type (say, n-type) and then transforms one end of it into the other type (p-type).

7.4.2 The p–n Junction Operating as a Detector

The operation of a semiconductor detector is based, essentially, on the properties of the p–n junction with reverse bias (Figure 7.18). Radiation incident upon the junction produces electron–hole

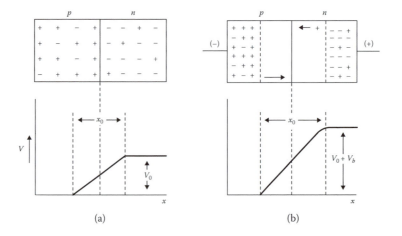

(a) (b)

Figure 7.17 (a) A p–n junction without external voltage, (b) If a reverse voltage is applied externally, the potential across the junction increases, and so does the depth x_0 along which an electric field exists.

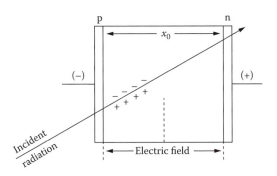

Figure 7.18 A p–n junction with reverse bias operating as a detector.

pairs as it passes through it. For example, if a 5-MeV alpha particle impinges upon the detector and deposits all its energy there, it will create about

$$\frac{5 \times 10^6 \, \text{eV}}{3 \, \text{eV/pair}} \sim 1.7 \times 10^6 \, \text{electron–hole pairs}$$

Electrons and holes are swept away under the influence of the electric field and, with proper electronics, the charge collected produces a pulse that can be recorded.

The performance of a semiconductor detector depends on the region of the p–n junction where the electric field exists (region of width X_0, Figure 7.18). Electrons and holes produced in that region find themselves in an environment similar to what electrons and ions see in a plate ionization chamber (see Section 5.4). There are some differences, however, between these two types of detectors.

In a gas-filled detector, the electron mobility is thousands of times bigger than that of the ions. In semiconductors, the electron mobility is only about two to three times bigger than that of the holes. The time it takes to collect all the charge produced in a gas counter is of the order of milliseconds. In semiconductors, the sensitive region of the counter is only a few millimeters, and the speed of electrons and holes is such that the charge carriers can traverse the sensitive region and be collected in times of the order of 10^{-7} s.

It is always the objective in either an ionization or a semiconductor detector to collect all the charges produced by the incident particle. This is achieved by establishing an electric field in the detector such that there is zero recombination of electrons and ions (or holes) before they are collected. In a semiconductor detector, even if recombination is zero, some charge carriers may be lost in "trapping" centers of the crystal, such as lattice imperfections, vacancies, and dislocations. The incident radiation creates crystal defects that cause deterioration of the detector performance and, thus, reduce its lifetime (see Section 7.6).

The capacitance of p–n junction is important because it affects the energy resolution of the detector. For a detector such as that shown in Figure 7.18, the *capacitance C* is given by

$$C = \varepsilon \frac{A}{4\pi X_0} \tag{7.12}$$

where ε = dielectric constant of the material
A = surface area of the detector
X_0 = depletion depth (detector thickness)

Combining Eqs. 7.11 and 7.12,

$$C \sim \frac{1}{\sqrt{V_b}}$$

To summarize, a material that will be used for the construction of a detector should have certain properties, the most important of which are the following:

1. *High resistivity.* This is essential, since otherwise current will flow under the influence of the electric field, and the charge produced by the particles will result in a pulse that may be masked by the steadily flowing current.

2. *High carrier mobility.* Electrons and holes should be able to move quickly and be collected before they have a chance to recombine or be trapped. High mobility is in conflict with property (1) because in high-resistivity materials, carrier mobility is low. Semiconductor materials doped with impurities have proven to have the proper resistivity-carrier mobility combination.

3. *Capability of supporting strong electric fields.* This property is related to property (1). Its importance stems from the fact that the stronger the field, the better and faster the charge collection becomes. Also, as the electric field increases, so does the depth of the sensitive region (Eq. 7.11a) for certain detectors.

4. *Perfect crystal lattice. Apart from externally* injected impurities, the semiconductor detector material should consist of a perfect crystal lattice without any defects, missing atoms, or interstitial atoms. Any such defect may act as a "trap" for the moving charges.

7.5 THE DIFFERENT TYPES OF SEMICONDUCTOR DETECTORS

The *several types of semico*nductor detectors that exist today differ from one another because of the material used for their construction or the method by which that material is treated. The rest of this section describes briefly the method of construction and the characteristics of the most successful detectors—made of silicon or germanium. Section 7.5.6 discusses room-temperature detectors such as CZT, CdTe, HgI_2, and CSI. These newer detectors aspire to rival the commonly used hyper-pure Ge detectors.

7.5.1 Surface-Barrier Detectors

Silicon of high purity, usually n-type, is cut, ground, polished, and etched until a thin wafer with a high-grade surface is obtained. The silicon is then left exposed to air or to another oxidizing agent for several days. Because of surface oxidization, surface energy states are produced that induce a high density of holes and form, essentially, a p-type layer on the surface (Figure 7.19). A very thin layer of gold evaporated on the surface serves as the electrical contact that will lead the signal to the preamplifier. In Figure 7.19, X_0 is the depth of the sensitive region, t is the total silicon thickness, and D is the diameter of the detector. The size of the detector is the length (or depth) X_0. The primary use of surface barrier detectors is for the detection and measurement of charged particles. These measurements may include alpha, beta and heavier ions in fundamental atomic[5] and nuclear physics,[6] depth profiling,[7] and environmental radioactivity[8] (particularly for alpha particles).

7.5.2 Diffused-Junction Detectors

Silicon of high purity, normally p-type, is the basic material for this detector type. As with surface-barrier detectors, the silicon piece has the shape of a thin wafer. A thin layer of n-type silicon is formed on the front face of the wafer by applying a phosphorus compound to the surface and then heating the assembly to temperatures as high as 800–1000°C for less than an hour. The phosphorus diffuses into the silicon and "dopes" it with donors (Figure 7.20). The n-type silicon in front and the p-type behind it form the p–n junction.

Both surface-barrier and diffused-junction detectors are used for the detection of charged particles. To be able to measure the energy of the incident radiation, the size X_0 of the detector should be at least equal to the range of the incident particle in silicon. The value of X_0 depends on the resistivity of the material (which in turn, depends on impurity concentration) and on the applied voltage, as shown by Eq. 7.11. Blankenship and Borkowski have designed a nomogram relating all these quantities (Figure 7.21).[9]

7.5.3 Silicon Lithium–Drifted [Si(Li)] Detectors

For surface-barrier and diffused-junction detectors, the sensitive region—that is, the actual size of the detector—has an upper limit of about 2000 μm. This limitation affects the maximum energy of a charged particle that can be measured. For electrons in Si, the range of 2000 μm corresponds to energy of about 1.2 MeV; for protons, the corresponding number is about 18 MeV; for

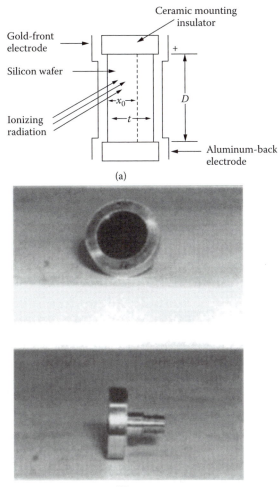

(a)

(b)

Figure 7.19 A typical surface barrier detector: (a) a schematic representation; (b) photograph of a commercial detector. (Reproduced from *Instruments for Research and Applied Science* by permission of EG & GORTEC, Oak Ridge, Tennessee.)

alphas, it is about 72 MeV. The length of the sensitive region can be increased if lithium ions are left to diffuse from the surface of the detector toward the other side. This process has been used successfully with silicon and germanium and has produced the so-called Si(Li) (pronounced silly) and Ge(Li) (pronounced jelly) semiconductor detectors. Lithium-drifted detectors have been produced with depth up to 5 mm in the case of Si(Li) detectors and up to 12 mm in the case of Ge(Li) detectors.

The lithium drifting process, developed by Pell,[10] and Mayer[11] consists of two major steps: (1) formation of an n–p junction by lithium diffusion, and (2) increase of the depletion depth by ion drifting.

The n–p junction is formed by letting lithium diffuse into a p-type silicon. The diffusion can be accomplished by several methods.[12–15] Probably the simplest method consists of painting a lithium-in-oil suspension onto the surface from which drifting is to begin. Other methods are lithium deposition under vacuum, or electrodeposition. After the lithium is applied on the surface, the silicon wafer is heated at 250–400° C for 3–10 min in an inert atmosphere, such as argon or helium.

Lithium is an n-type impurity (donor atom) with high mobility in silicon (and germanium; see next section). When the diffusion begins, the acceptor concentration (N_p) is constant throughout the silicon crystal (Figure 7.22a), while the donor concentration (N_n) is high on the surface and

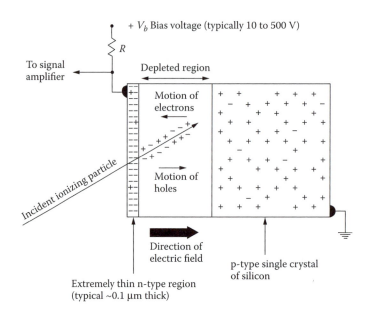

Figure 7.20 A diffused-junction detector.

zero everywhere else. As the diffusion proceeds, the donor concentration changes with depth, as shown in Figure 7.22a. At the depth x_j where

$$N_n(x_j) = N_p$$

and n–p junction has been formed (Figure 7.22b).

After the diffusion is completed, the crystal is left to cool, the excess lithium is removed, and ohmic contacts are put on the n and p sides of the junction. The contact on the p side is usually formed by evaporating aluminum or gold doped with boron. The contact on the n side can be formed by using pure gold or antimony-doped gold.

Drifting is accomplished by heating the junction to 120–150°C while applying a reverse bias that may range from 25 V up to about 1000 V. In general, the higher the temperature and the voltage are, the faster the drifting proceeds. Depending on the special method used, the semiconductor may be under vacuum or in air or be placed in a liquid bath (e.g., silicon oil or fluorocarbon). The electric field established by the reverse bias tends to move the n-type atoms (lithium) toward the p side of the junction. As a result, the concentration of lithium atoms becomes lower for $x < x_j$ (Figure 7.22a) and higher for $x > x_j$. For $x < x_j$, N_n cannot become less than N_p because then a local electric field would appear pushing the lithium atoms toward the η side. Similarly, for $x > x_j$, N_n cannot increase very much because the local electric field works against such a concentration. Thus, a region is created that looks like an intrinsic semiconductor because $N_n \approx N_p$. For long drifting times, the thickness of the intrinsic region $X_0(t)$ as a function of time is given by

$$X_0(t) = \sqrt{2V\mu_{Li}t} \tag{7.13}$$

where V = applied voltage

μ_{Li} = mobility of Li ions in silicon at the drifting temperature

The mobility of lithium, which increases with temperature,[16] has a value of about 5×10^{-14} m²/Vs at $T = 150$°C. Drifting is a long process. Depending on the desired thickness, drifting may take days and sometimes weeks.

Example 7.1 How long will it take to obtain an intrinsic region of 1.5 mm in a silicon wafer drifted at 150°C under a reverse bias of 500 V?

Answer Using Eq. 7.13 with $\mu_{Li} = 5 \times 10^{-14}$ m²/Vs, one obtains

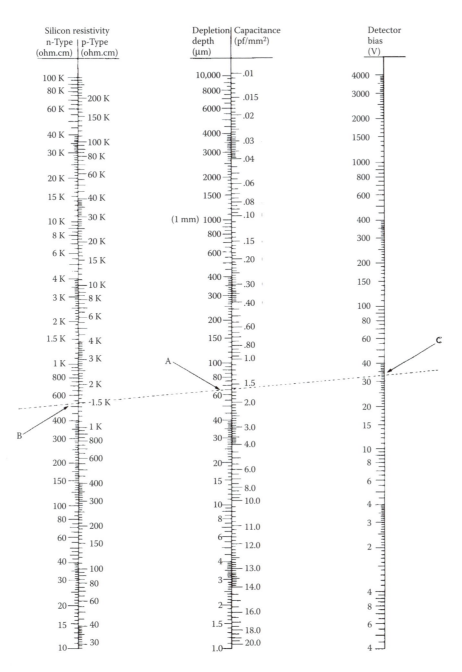

Figure 7.21 The Blankenship and Borkowski nomogram that relates resistivity, detector thickness and detector bias. The detector capacitance as a function of detector thickness is also given.

$$t = \frac{X_0^2(t)}{2V\mu_{Li}} = \frac{(1.5\times10^{-3})^2 m^2}{2(500\ V)[5\times10^{-14}\ m^2\ /(Vs)]} = 4.5\times10^4\ s = 12.5\ h$$

After drifting is completed, the Si(Li) detector is mounted on a cryostat, since the best results are obtained if the detector is operated at a very low temperature. Usually, this temperature is 77 K, the temperature of liquid nitrogen. Si(Li) detectors may be stored at room temperature for a short period without catastrophic results, but for longer periods it is advisable to keep the detector

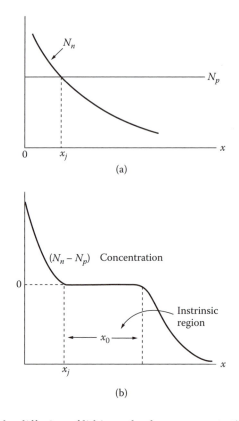

Figure 7.22 (a) During the diffusion of lithium, the donor concentration changes with depth as shown, (b) during drifting (at elevated temperature and under reverse bias), and almost intrinsic region is formed with thickness x_0.

cooled at all times. The low temperature is necessary to keep the lithium drifting at a "frozen" stage. At room temperature, the mobility of lithium is such that its continuous diffusion and precipitation[12] will ruin the detector.

Si(Li) detectors are used for detection of charged particles and especially X-rays. Their characteristics with respect to energy measurements are described in Chapters 12 and 13.

7.5.4 Germanium Lithium–Drifted [Ge(Li)] Detectors

Ge(Li) detectors are not made anymore; they have been replaced by pure germanium crystals. Historically, Ge(Li) detectors dominated the gamma detection field until about 1985. While there are hardly any Ge(Li) detectors still in use, a brief discussion is presented in this section for historical purposes. Also described are the technical evolutionary upgrades germanium detectors underwent.

Ge(Li) detectors are made from horizontally grown or pulled single crystals of germanium. As the crystal is grown, it is doped with acceptor impurities such as indium, gallium, or boron; and it becomes a p-type semiconductor. Germanium crystals may be cut to length and shaped by a variety of means, including the use of diamond wheels or band saws. In these mechanical operations, great care must be taken not to fracture the brittle material.

Lithium drifting in germanium follows the same approach as in silicon. The deposition and diffusion of lithium are accomplished by one of the methods discussed in the previous section. The ohmic contacts are made by electrolytic deposition of gold,[17] by using gallium-indium,[18] mercury-indium,[19] or by ion implantation.[20] The drifting process itself takes place at a lower temperature (<60°C) than for silicon, with the germanium diode in air[21] or immersed in a liquid maintained at its boiling point.[22]

After the drifting process has been completed, the detector is mounted on a cryostat and is always kept at a low temperature (liquid nitrogen temperature ~77 K). Keeping the Ge(Li) detector at a low temperature is much more critical than for a Si(Li) detector. The mobility of the lithium atoms in germanium is so high at room temperature that the detector will be ruined if brought to room temperature even for a short period. If this happens, the detector may be redrifted, but at a considerable cost.

7.5.5 Germanium (Ge) Detectors

The production of high-purity germanium (HPGe) with an impurity concentration of 10^{16} atoms/cm³ or less has made possible the construction of detectors without lithium drifting.[23–25] These detectors are now designated as Ge, not HPGe, and are simply formed by applying a voltage across a piece of germanium. The sensitive depth of the detector depends on the impurity concentration and the voltage applied, as shown in Figure 7.23.

The major advantage of Ge versus Ge(Li) detectors is that the former can be stored at room temperature and cooled to liquid nitrogen temperature (77 K) *only* when in use. Cooling the detector, when in use, is necessary because germanium has a relatively narrow energy gap, and at room or higher temperatures a leakage current due to thermally generated charge carriers induces such noise that the energy resolution of the device is destroyed.

Germanium detectors are fabricated in various geometries, thus offering devices that can be tailored to the specific needs of the measurement. Two examples, the coaxial and the well-type detector, are shown in Figure 7.24.

Efficiencies of germanium detectors are calibrated against 7.6 cm × 7.6 cm (3 in. × 3 in.) sodium iodide (NaI) with a ⁶⁰Co source. In the past, these efficiencies ranged between 10% and 50%. However, improvements in detector technology have resulted in germanium detectors to exceed 100%. In fact, detectors with efficiencies up to 200% can be readily manufactured (keep in mind that efficiency numbers quoted for Ge detectors are always values relative to the efficiency of a 7.62 × 7.62 NaI for ⁶⁰Co photons). This has greatly enhanced measurements in environmental radioactivity studies. Excellent overviews of germanium detector characteristics can be seen in manufacturers' websites (http://www.canberra.com, http://www.ortec-online.com, http://www.pgt.com, etc.).

Another great improvement in detector technology has been the introduction of a new cooling method of germanium detectors. These new mechanical coolers have great advantages over those using liquid nitrogen cooling. An excellent overview of this technology has been published.[26]

More details about these detectors are presented in Chapter 12 in connection with γ-ray spectroscopy.

7.5.6 CdTe, CdZnTe, and HgI₂ Detectors

The major disadvantage of germanium detectors is the requirement for continuous cooling. Cooling requires a cryostat, which makes the counter bulky and thus impossible to use in cases where only a small space is available; another disadvantage is the cost of continuously buying liquid nitrogen. There is a great incentive, therefore, to develop semiconductor detectors that can be stored and operated at room temperature. Over the last two decades, there have

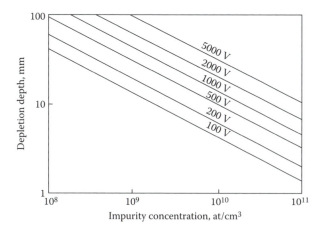

Figure 7.23 Depletion depth as a function of impurity concentration and applied voltage for planar diodes of high-purity germanium. (From Pehl, R. H., *Physics Today*, 30, 11, 50, 53, 1977.)

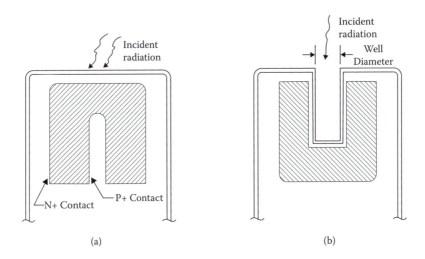

Figure 7.24 Two examples of geometries used for Ge Detectors; (a) coaxial; (b) well type. (From Canberra Nuclear, *Edition Nine Instruments Catalog.*)

Table 7.2
Properties of Si, Ge, CdTe, HgI$_2$, and CdZnTe

Material	Atomic Number	Energy Gap (eV)	Energy Needed to Form the Pair (eV)
Si	14	1.106 (300 K)	3.65 (300 K)
Ge	32	0.67 (77 K)	2.96 (77 K)
CdTe	48 and 52	1.47 (300 K)	4.43 (300 K)
HgI$_2$	80 and 53	2.13 (300 K)	4.22 (300 K)
CdZnTe	48, 30, and 52	1.64 (300 K)	5.0 (300 K)

been great strides in manufacturing such detectors. Earlier on two materials that have been studied and show great promise for the construction of such detectors are CdTe and HgI$_2$.[27–40] A comprehensive review of the state-of-the-art (until 1978) for both materials can be found in Reference 41.

Successful detectors using CdTe or HgI$_2$ have been constructed with thickness up to 0.7 mm and area 100 mm^2 (as of 1978).[36] These detectors are small in size, compared to Si(Li) or Ge(Li) detectors, but the required detector volume depends on the application. For CdTe and HgI$_2$, the favored applications are those that require a small detector volume: monitoring in space,[42] measurement of activity in nuclear power plants,[43] medical portable scanning,[44] or medical imaging devices.[45] Although the detector volume is small, efficiency is considerable because of the high atomic number of the elements involved (Table 7.2). The energy needed for the production of an electron-hole pair is larger for CdTe and HgI$_2$ than it is for Si and Ge; as a result, the energy resolution of the former is inferior to that of the latter (see also Chapter 12). However, CdTe and HgI$_2$ detectors are used in measurements where their energy resolution is adequate while, at the same time, their small volume and, in particular, their room-temperature operation offers a distinct advantage over Si(Li) and Ge detectors.

CdTe detectors have been used for Mossbauer spectroscopy[46] and induced radiation damage.[47] Afterwards CdZnTe (often named CZT) detectors have appeared on the market used in a variety of research topics such as ion beam induced charge collection,[48] implementation in ion beam facilites,[49] time resolved ion beam induced charge,[50] and the effect of boron ion implantation on the structural and optical properties of polycrystalline Cd$_{0.96}$Zn$_{0.04}$Te thin films.[51]

Other detectors include HgI$_2$, gallium arsenide (GaAs), CsI, and lanthanum trichloride (LaCl$_3$). There was a performance comparison for efficiency, resolution, and peak shape of four compact room-temperature detectors, two CZT semiconductor detectors, a LaCl$_3$(Ce) scintillator, and an NaI(Tl) scintillator.[52]

Table 7.3
Particle Fluence That Causes Significant Radiation Damage

	Heavy Ions (Particles/m²)	Alphas (α/m²)	Fast Neutrons (n/m²)
Junction detectors	10^{12}	10^{14}	
Si(Li)		10^{12}	10^{14}
Ge(Li) or Ge			10^{13}–10^{14}

7.6 RADIATION DAMAGE TO SEMICONDUCTOR DETECTORS

The fabrication and operation of a semiconductor detector are based on the premise that one starts with a perfect crystal containing a known amount of impurities. Even if this is true at the beginning, a semiconductor detector will suffer damage after being exposed to radiation. The principal type of radiation damage is caused by the collision of an incident particle with an atom. As a result of the collision, the atom may be displaced into an interstitial position, thus creating an interstitial-vacancy pair known as the *Frenkel defect*. A recoiling atom may have enough energy to displace other atoms; therefore, an incident particle may produce many Frenkel defects.

Crystal defects affect the performance of the detector because they may act as trapping centers for electrons and holes or they may create new donor or acceptor states. New trapping centers and new energy states change the charge collection efficiency, the leakage current, the pulse rise time, the energy resolution, and other properties of the detector. The changes are gradual, but the final result is shortening of the detector lifetime.

Electrons and photons cause negligible radiation damage compared to charged particles and neutrons. Heavier and more energetic charged particles cause more damage than lighter and less energetic particles.[2,25] Also, the damage is not the same for all detector types. Table 7.3 gives the fluences that cause considerable radiation damage for different detectors and bombarding particles. Ge detectors are not affected by gammas, but they are damaged by the neutrons in a mixed n–γ field.

There has been a multitude of radiation damage studies on various types of detectors. These include investigation of radiation damage in a Si pin photodiode for particle detection,[53] mechanisms of damage formation in semiconductors,[54] radiation damage mechanisms in CsI(Tl) studied by ion beam induced luminescence,[55] frequency dependence of ac conductance of neutron irradiated silicon detectors to fluences up to 10^{16} n cm^{-2},[56] radiation damage induced by 2 MeV protons in CdTe and CdZnTe semiconductor detectors,[57] effect of neutron damage on energy and position resolution of the GRETINA germanium detector,[58] neutron damage tests of a highly segmented germanium crystal,[59] radiation damage in p-type silicon irradiated with neutrons and protons,[60] annealing studies of silicon microstrip detectors irradiated at high neutron fluences,[61] an assessment of radiation damage in space-based germanium detectors due to solar proton events,[62] solar proton damage in high-purity germanium detectors,[63] numerical simulation of radiation damage effects in p-type silicon detectors,[64] radiation damage in silicon detectors,[65] radiation damage measurements in room-temperature semiconductor radiation detectors,[66] radiation damage measurements on CZT drift strip detectors,[67] and radiation damage study of GaAs detectors irradiated by fast neutrons.[68]

PROBLEMS

7.1 What is the probability that an electron energy state in Ge will be occupied at temperature $T = 300$ K if the energy state is greater than the Fermi energy by 2 eV?

7.2 Repeat Problem 7.1 for $T = 11$ K.

7.3 The energy gap for diamond is 7 eV. What temperature will provide thermal energy (kT) equal to that amount?

7.4 What should be the maximum thickness of the gold layer covering the front face of a surface barrier detector used for the measurement of 10-MeV alphas, if the energy loss of the alphas traversing the layer should be less than 0.1% of the kinetic energy?

7.5 Repeat Problem 7.4 for 6-MeV electrons.

7.6 A 6 MeV alpha particle strike a Si wafer with thickness equal to 0.8 R, where R = range of this particle in Si. What is the total energy loss of this particle as it traverses this Si wafer?

7.7 The thickness of the gold layer covering the front face of a semiconductor detector may be measured by detecting particles entering the detector at two different angles. Calculate that thickness if alphas that enter in a direction perpendicular to the front face register as having energy 4.98 MeV, but those that enter at a 45° angle register as having energy 4.92 MeV.

7.8 What is the average distance traveled in Si by a 50-KeV gamma before it has an interaction? What is the corresponding distance in Ge?

7.9 Lithium has been drifted in germanium at 50°C under a reverse bias of 500 V for 2 weeks. What is your estimate of the drifting depth? $[\mu_{Li} = 1.5 \times 10^{-13} \, \text{m}^2 / (\text{V s})]$

7.10 A parallel beam of 0.5-MeV gammas is normally incident upon 2 mm thick crystals of Si, Ge, CdTe, and HgI_2. What fraction of photons will interact at least once in each crystal?

BIBLIOGRAPHY

Ahmed, S. N., *Physics and Engineering of Radiation Detection*, Academic Press, Elsevier, 2007.

Bertolini, G., and Coche, A., *Semiconductor Detectors*, North-Holland Publishing Co., Amsterdam, 1968.

Brown, W. L., Higinbotham, W. A., Miller, G. L., and Chace, R. L. (eds.), "Semiconductor Nuclear Particle Detectors and Circuits," Proceedings of a conference conducted by the Committee of Nuclear Science of the National Academy of Sciences, NAS Publication 1593, Washington, D.C. (1969).

Dearnaley, G., and Northrop, D. C., *Semiconductor Counters for Nuclear Radiations*, E & F.N. Spon. Ltd., London, 1964.

Eichholz, G. G., and Poston, J. W., *Nuclear Radiation Detection*, Lewis Publishers, Chelsea, Michigan, 1985.

Fenyves, E., and Haiman, O., *The Physical Principles of Nuclear Radiation Measurements*, Academic Press, New York, 1969.

Knoll, G. F., *Radiation Detection and Measurement*, 3rd ed., Wiley, New York, 2000.

http://www.canberra.com

http://www.ortec-online.com

http://www.pgt.com

Price, W. J., *Nuclear Radiation Detection*, McGraw-Hill, New York, 1964.

Tait, W. H., *Radiation Detection*, Butterworth, London, 1980.

REFERENCES

1. McKenzie, J. M., *Nucl. Instrum. Meth.* **162**:49 (1979).

2. Ewan, G. T., *Nucl. Instrum. Meth.* **162**:75 (1979).

3. Pehl, R. H., Goulding, F. S., Landis, D. A., and Lenzlinger, M., *Nucl. Instrum. Meth.* **59**:45 (1968).

4. Delaet, L. H., Schoenmaekers, W. K., and Guislain, H. J., *Nucl. Instrum. Meth.* **101**:11 (1972).

5. Abdesselam, M., Ouichaoui, S., Azzouz, M., Chami, A. C., and Siad, M., **B 266**:3899 (2008).

6. Büschera, J., Ponsaers, J., Raabe, R., Huyse, M., Van Duppen, P., Aksouh, F., Smirnov, D., Fynbo, H. O. U., Hyldegaard, S., and Diget, C. A., *Nucl. Instrum. Meth. Phys. Res.* **B 266**:4652 (2008).

7. Yasuda, K., Ishigami, R., Sasase, M., and Ito, Y., *Nucl. Instrum. Meth. Phys. Res.* **B 266**:1416 (2008).

8. Aguado, J. L., Bolivar, J. P., and García-Tenorio, R., *J. Radioanal. Nucl. Chem.* **278**:191 (2008).

9. Friedland, S. S., Mayer, J. W., and Wiggins, J. S., *IRE Trans. Nucl. Sci.* **NS-7**(2–3):181 (1960).

10. Pell, E. M., *J. Appl. Phys.* **31**:291 (1960).

11. Mayer, J. W., *J. Appl. Phys.* **33**:2894 (1962).

12. Elliott, J. H., *Nucl. Instrum. Meth.* **12**:60 (1961).

13. Baily, N. A., and Mayer, J. W., *Radiology* **76**:116 (1961).

14. Siffert, P., and Coche, A., *Compt. Rend.* **256**:3277 (1963).

15. Dearnaley, G., and Lewis, J. C., *Nucl. Instrum. Meth.* **25**:237 (1964).

16. Siffert, P., and Coche, A., "Behavior of Lithium in Silicon and Germanium," in G. Bertolini and A. Coche (eds.), *Semiconductor Detectors*, North-Holland Publishing Co., Amsterdam, 1968.

17. Janarek, F. J., Helenberg, H. W., and Mann, H. M., *Rev. Sci. Instrum.* **36**:1501 (1965).

18. Mooney, J. B., *Nucl. Instrum. Meth.* **50**:242 (1967).

19. Hansen, W. L., and Jarrett, B. V., *Nucl. Instrum. Meth.* **31**:301 (1964).

20. Meyer, O., and Haushahn, G., *Nucl. Instrum. Meth.* **56**:177 (1967).

21. Ewan, G. T., and Tavendale, A. J., *Can. J. Phys.* **42**:3386 (1964).

22. Cappellani, E., Fumagulli, W., and Restelli, G., *Nucl. Instrum. Meth.* **37**:352 (1965).

23. Hall, R. N., and Soltys, T. J., *IEEE Trans. Nucl. Sci.* **NS-18**:160 (1971).

24. Hansen, W. L., *Nucl. Instrum. Meth.* **94**:377 (1971).

25. Pehl, R. H., *Physics Today* **30**:11, 50, 53 (1977).

26. Upp, D. L., Keyser, R. M., and Twomey, T. R., *J. Radioanal. Nucl. Chem.* **264**:121 (2005).

27. Siffert, I., Gonidec, J. P., and Cornet, A., *Nucl. Instrum. Meth.* **115**:13 (1974).

28. Eichinger, P., Haider, N., and Kemmer, J., *Nucl. Instrum. Meth.* **117**:305 (1974).

29. Jones, L. T., and Woollam, P. B., *Nucl. Instrum. Meth.* **124**:591 (1975).

30. Iwanczyk, J., and Dabrowski, A. J., *Nucl. Instrum. Meth.* **134**:505 (1976).

31. Siffert, P., *Nucl. Instrum. Meth.* **150**:1 (1978).

32. Dabrowski, A. J., Iwanczyk, J., and Szymczyk, W. M., *Nucl. Instrum. Meth.* **150**:25 (1978).

33. Schieber, M., Beinglass, I., Dishon, G., Holzer, A., and Yaron, G., *Nucl. Instrum. Meth.* **150**:71 (1978).

34. Shalev, S., *Nucl Instrum. Meth.* **150**:79 (1978).

35. Caine, S., Holzer, A., Beinglass, I., Dishon, G., Lowenthal, E., and Schieber, M., *Nucl. Instrum. Meth.* **150**:83 (1978).

36. Whited, R. C., and Schieber, M. M., *Nucl. Instrum. Meth.* **162**:113 (1979).

37. Ristinen, R. A., Peterson, R. J., Hamill, J. J., and Becchetti, F. D., *Nucl. Instrum. Meth.* **188**:445 (1981).

38. Markakis, J. M., *Nucl. Instrum. Meth.* **263**:499 (1988).

39. Courat, B., Fourrier, J. P., Silga, M., and Omaly, J., *Nucl. Instrum. Meth.* **269**:213 (1988).

40. McKee, B. T. A., Goetz, T., Hazlett, T., and Forkert, L., *Nucl. Instrum. Meth.* **272**:825 (1988).

41. "International Workshop on Mercuric Iodide and Cadmium Telluride Nuclear Detectors," *Nucl. Instrum. Meth.* **150**:1–112 (1978).

42. Lyons, R. B., *Rev. Phys. Appl.* **12**:385 (1977).

43. Jones, L. T., *Rev. Phys. Appl.* **12**:379 (1977).

44. Vogel, J., Ullman, J., and Entine, G., *Rev. Phys. Appl.* **12**:375 (1977).

45. Canali, C., Gutti, E., Kozlov, S. F., Manfredi, P. F., Manfredotti, C., Nava, F., and Quirini, A., *Nucl. Instrum. Meth.* **160**:73 (1979).

46. Bargholtz, Chr, J., Blomquist, Fumero, E., Mårtensson, L., Einarsson, L., and Wäppling, R., *Nucl. Instrum. Meth. Phys. Res. B,* **170**:239 (2000).

47. Pastuović, Ž., and Jakšić, M., *Nucl. Instrum. Meth. Phys. Res. B,* **181**:344 (2001).

48. Doyle, B. L., Vízkelethy, G., and Walsh, D. S., *Nucl. Instrum. Meth. Phys. Res. B,* **161**:457 (2000).

49. Simon, A., Jeynes, C., Webb, R. P., Finnis, Tabatabaian, R. Z., Sellin, P. J., Breese, M. B. H., Fellows, D. F., van den Broek, R., and Gwilliam, R. M., *Nucl. Instrum. Meth. Phys. Res. B,* **219–220**:405 (2004).

50. Medunić, Z., Jakšić, M., Pastuović, Ž., and Skukan, N., *Nucl. Instrum. Meth. Phys. Res. B,* **210**:237 (2003).

51. Sridharan, M., Narayandass, Sa. K., Mangalaraj, D., and Lee, H. C., *Nucl. Instrum. Meth. Phys. Res B,* **201**:465 (2003).

52. Hartwell, J. K., Gehrke, R. J., and Ilwain, M. E. M., *Nucl. Sci. IEEE Trans.* **52**:1813 (2005).

53. Simon, A., Kalinka, G., Jakšić, M., Pastuović, Ž., Novák, M., and Kiss, Á. Z., *Nucl. Instrum. Meth. Phys. Res B,* **260**:304 (2007).

54. Wendler, E., *Nucl. Instrum. Meth. Phys. Res.* In Press, Corrected Proof, Available online 2 June 2009.

55. Quaranta, A., Gramegna, F., Kravchuk, V., and Carlo Scian, C., *Nucl. Instrum. Meth. Phys. Res B*, **266**:2723 (2008).

56. Croitoru, N., David, G., Rancoita, P. G., Rattaggi, M., and Seidman, A., *Nucl. Instrum. Meth. Phys. Res B*, **134**:209 (1998).

57. Zanarini, M., Chirco, P., Dusi, W., Auricchio, N., Cavallini, A., Fraboni, B., Siffert, P., and Bianconi, M., *Nucl. Instrum. Meth. Phys. Res B*, **213**:320 (2004).

58. Descovich, M., Lee, I. Y., Cromaz, M., Clark, R. M., Deleplanque, M. A., Diamond, R. M., Fallon, P., *et al.*, *Nucl. Instrum. Meth. Phys. Res B*, **241**:931(2005).

59. Ross, T. J., Beausang, C. W., Lee, I. Y., Macchiavelli, A. O., Gros, S., Cromaz, M., Clark, R. M., Fallon, P., Jeppesen, H., and Allmond, J. M., *Nucl. Instrum. Meth. Phys. Res A*, **606**:533 (2009).

60. Cindro, V., Kramberger, G., Lozano, M., Mandić, I., Mikuž, M., Pellegrini, G., Pulko, J., Ullan, M., and Zavrtanik, M. *Nucl. Instrum. Meth. Phys. Res A*, **599**:60 (2009).

61. Miñano, M., Balbuena, J. P., García, C., González, S., Lacasta, C., Lacuesta, V., Lozano, M., Martí i Garcia, S., Pellegrini, G., and Ullán, M., *Nucl. Instrum. Meth. Phys. Res A*, **591**:181 (2008).

62. Owens, A., Brandenburg, S., Buis, E-J., Kiewiet, H., Kraft, S., Ostendorf, R. W., Peacock, A., Quarati, F., and Quirin, P., *Nucl. Instrum. Meth. Phys. Res A*, **583**:285 (2007).

63. Pirard, B., Cabrera, J., d'Uston, C., Thocaven, J. J., Gasnault, O., Leleux, P., and Brückner, J., *Nucl. Instrum. Meth. Phys. Res A*, **572**:698 (2007).

64. Petasecca, M., Moscatelli, F., Passeri, D., Pignatel, G. U., and Scarpello, C., *Nucl. Instrum. Meth. Phys. Res A*, **563**:192 (2006).

65. Lindström, G., *Nucl. Instrum. Meth. Phys. Res A*, **512**:30 (2003).

66. Franks, L. A., Brunett, B. A., Olsen, R. W., Walsh, D. S., Vizkelethy, G., Trombka, J. I., Doyle, B. L., and James, R. B., *Nucl. Instrum. Meth. Phys. Res A*, **428**:95 (1999).

67. Kuvvetli, I., Budtz-Jørgensen, C., Korsbech, U., and Jensen, H. J., *Nucl. Instrum. Meth. Phys. Res. A*, **512**:98 (2003).

68. Linhart, V., Bém, P., Götz, M., Honusek, M., Mareš, J. J., Slavíček, T., Sopko, B., and Šimečková, E., *Nucl. Instrum. Meth. Phys. Res. A*, **563**:66 (2006).

8

Relative and Absolute Measurements

8.1 INTRODUCTION

An *absolute measurement* is one in which the exact number of particles emitted or the exact number of events taking place is determined, for example,

1. Determination of the activity of a radioactive source, that is, measurement of the number of particles emitted by the source per second

2. Determination of the neutron flux (neutrons per square meter per second) at a certain point in a reactor

3. Measurement of the number of neutrons emitted per fission

4. Measurement of the first cross section for a nuclear interaction

A *relative measurement* is one in which the exact number of particles emitted or the exact number of events taking place is not determined. Instead, a "relative" number of particles or events is measured, a number that has a fixed, but not necessarily known, relationship to the exact number, for example,

1. Determination of the G–M plateau. The relative change of the number of particles counted versus HV is measured. The exact number of particles emitted by the source is not determined; in fact, it is not needed.

2. Determination of half-life by counting the decaying activity of an isotope. The relative change of the number of atoms versus time is measured. The exact number of nuclei decaying per second is not needed.

3. Measurement of the fission cross section for ^{239}Pu, based on the known fission cross section for ^{235}U.

4. Determination of the variation of the neutron flux along the axis of a cylindrical reactor. The relative change of the flux from point to point along the axis of the reactor is measured, and not the exact number of neutrons per square meter per second.

Relative measurements are, in most cases, easier than absolute measurements. For this reason, investigators tend to perform the very minimum of absolute measurements and use their results in subsequent relative measurements. One of the most characteristic examples is the determination of the value of nuclear cross sections. Absolute measurements have been performed for very few cross sections. After certain cross sections have been measured, most of the others may be determined relative to the known ones.

This chapter discusses the factors that should be taken into account in performing relative and absolute measurements. Assume that there is a source of particles placed a certain distance away from a detector (Figure 8.1) and that the detector is connected to a pulse-type counting system. The source may be located outside the detector as shown in Figure 8.1, or it may be inside the detector (e.g., liquid-scintillation counting and internal-gas counting), and may be isotropic (e.g., particles emitted with equal probability in all directions) or anisotropic (e.g., parallel beam of particles). Both cases will be examined. Let

S = number of particles per second emitted by the source
r = number of particles per second recorded by the scaler

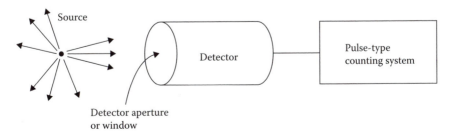

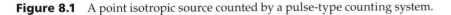

Figure 8.1 A point isotropic source counted by a pulse-type counting system.

It is assumed that the counting rate r has been corrected for dead time and background, if such corrections are necessary. The measured rate r is related to S by

$$r = f_1 f_2 f_3 \cdots f_n S \tag{8.1}$$

where the f factors represent the effects of the experimental setup on the measurement. These factors may be grouped into three categories, to be discussed in detail in the following sections:

1. *Geometry effects.* The term *geometry* refers to size and shape of source (point, parallel beam, disk, rectangular), size and shape of detector aperture (cylindrical, rectangular, etc.), and distance between source and detector.

2. *Source effects.* The size and, in particular, the way the source is made may have an effect on the measurement. Whether the source is a solid material or a thin deposit evaporated on a metal foil may make a difference. The effect of source thickness is different on charged particles, gammas, and neutrons.

3. *Detector effects.* The detector may affect the measurement in two ways. First, the size and thickness of the detector window (Figure 8.1) determine how many particles enter the detector and how much energy they lose, as they traverse the window. Second, particles entering the detector will not necessarily be counted. The fraction of particles that is recorded depends on the efficiency of the detector (see Section 8.4.2).

8.2 GEOMETRY EFFECTS

The geometry may affect the measurement in two ways. First, the medium between the source and the detector may scatter and may also absorb some particles. Second, the size and shape of the source and the detector and the distance between them determine what fraction of particles will enter the detector and have a chance to be counted.

8.2.1 The Effect of the Medium between Source and Detector

Consider a source and a detector separated by a distance d (Figure 8.2). Normally, the medium between the source and detector is air, a medium of low density. For measurements of photons and neutrons, the air has no effect. If the source emits charged particles, however, all the particles suffer some energy loss, and some of them may be scattered in or out of the detector (Figure 8.2). If this effect is important for the measurement, it can be eliminated by placing the source and the detector inside an evacuated chamber. If the use of an evacuated chamber is precluded by the conditions of the measurement, then appropriate corrections should be applied to the results.

8.2.2 The Solid Angle—General Definition

To illustrate the concept of solid angle, consider a point isotropic source at a certain distance from a detector as shown in Figure 8.3. Since the particles are emitted by the source with equal probability in every direction, only some of the particles have a chance to enter the detector. That

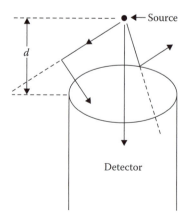

Figure 8.2 The medium between the source and the detector may scatter and/or absorb particles emitted by the source.

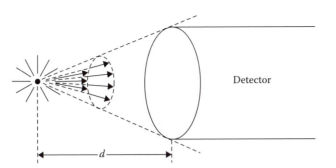

Figure 8.3 The fraction of particles emitted by a point isotropic source and entering the detector is defined by the solid angle subtended by the detector at the location of the source.

portion is equal to the fractional solid angle subtended by the detector at the location of the source. In the general case of an extended source, the *solid angle* Ω is defined by

$$\Omega = \frac{\text{number of particles per second emitted inside the space defined by the contours of the source and the detector aperture}}{\text{number of particles per second emitted by the source}} \tag{8.2}$$

The mathematical expression for Ω is derived as follows (Figure 8.4). A plane source of area A_s emitting S_0 particles/(m^2 s), isotropically, is located a distance d away from a detector with an aperture equal to A_d. Applying the definition given by Eq. 8.2 for the two differential areas dA_s and dA_d and integrating, one obtains*

$$\Omega = \frac{\int_{A_s} \int_{A_d} (S_0 dA_s / 4\pi r^2) dA_d (\hat{\mathbf{n}} \cdot \mathbf{r} / r)}{S_0 A_s} \tag{8.3}$$

where $\hat{\mathbf{n}}$ is a unit vector normal to the surface of the detector aperture. Since $\hat{\mathbf{n}} \cdot \mathbf{r}/r = \cos \omega$, Eq. 8.3 takes the form

$$\Omega = \frac{1}{4\pi A_s} \int_{A_s} dA_s \int_{A_d} dA_d \frac{\cos \omega}{r^2} \tag{8.4}$$

Equation 8.4 is valid for any shape of source and detector. In practice, one deals with plane sources and detectors having regular shapes, examples of which are given in the following sections.

As stated earlier, Ω is equal to the fractional solid angle ($0 \le \Omega \le 1$). In radiation measurements, it is called either *solid angle* or *geometry factor*. In this text it will be called the solid angle.

8.2.3 The Solid Angle for a Point Isotropic Source and a Detector with a Circular Aperture

The most frequently encountered case of obtaining a solid angle is that of a point isotropic source at a certain distance away from a detector with a circular aperture (Figure 8.5). In Eq. 8.4, cos $\omega = d/r$, and the integration gives

$$\Omega = \frac{1}{2}\left(1 - \frac{d}{\sqrt{d^2 + R^2}}\right) \tag{8.5}$$

From Figure 8.5,

$$\cos \theta_0 = \frac{d}{\sqrt{d^2 + R^2}} \tag{8.6}$$

---

* Equation 8.3 applies to isotropic sources: nonisotropic sources, seldom encountered in practice, need special treatment.

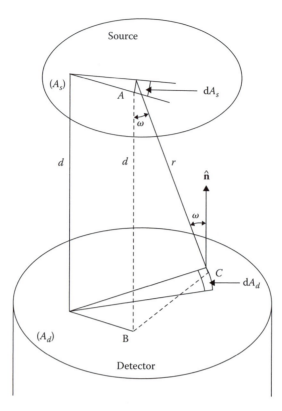

Figure 8.4 Definition of the solid angle for a plane source and a plane detector parallel to the source.

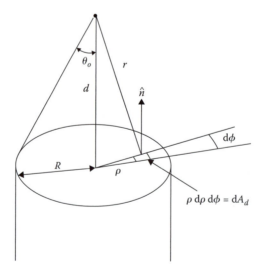

Figure 8.5 The solid angle between a point isotropic source and a detector with a circular aperture.

Therefore, an equation equivalent to Eq. 8.5 is

$$\Omega = \frac{1}{2}(1 - \cos\theta_0)$$ (8.7)

It is instructive to rederive Eq. 8.7, not by using Eq. 8.4, but by a method that gives more insight into the relationship between detector size and source-detector distance.

Consider the point isotropic source of strength S_0 particles per second located a distance d away from the detector, as shown in Figure 8.6. It is assumed that all space outside the detectors is void. If one draws a sphere centered at the source position and having a radius R_s greater than d, the number of particles/(m^2 s) on the surface of the sphere is $S_0/4\pi R_s^2$. The particles that will hit the detector are those emitted within a cone defined by the location of the source and the detector aperture. If the lines that define this cone are extended up to the surface of the sphere, an area A_s is defined there. A_s is a nonplanar area on the surface of the sphere. The number of particles per second entering the detector is $A_s(S_0/4\pi R_s^2)$ and, using Eq. 8.2, the solid angle becomes

$$\Omega = \frac{A_s(S_0/4\pi R_s^2)}{S_0} = \frac{A_s}{4\pi R_s^2}$$

The area A_s is given by (Figure 8.7):

$$A_s = \int dA_s = \int (R_s d\theta)(R_s \sin\theta d\phi) = R_s^2 \int_0^{2\pi} d\phi \int_0^{\theta_0} d\theta \sin\theta$$
$$= 2\pi R_s^2 (1 - \cos\theta_0)$$

Therefore, the expression for the solid angle becomes

$$\Omega = \frac{A_s}{4\pi R_s^2} = \frac{2\pi R_s^2(1 - \cos\theta_0)}{4\pi R_s^2} = \frac{1}{2}(1 - \cos\theta_0)$$ (8.7a)

which is, of course, Eq. 8.7.

If $R \ll d$, Eq. 8.5 takes the form [after expanding the square root (Eq. 8.6) and keeping only the first two terms]

$$\Omega = \frac{R_2}{4d^2} = \frac{\pi R^2}{4\pi d^2} = \frac{\det ector\ aperture}{4\pi d^2}$$ (8.8)

Equation 8.8 is valid even for a noncylindrical detector *if the source-detector distance is much larger than any of the linear dimensions of the detector aperture.*

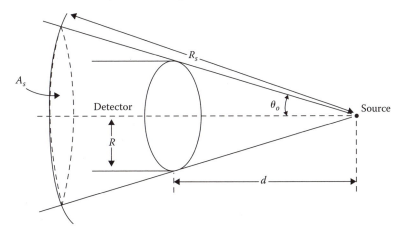

Figure 8.6 Diagram used for the calculation of the solid angle between a point isotropic source and a detector with a circular aperture.

Example 8.1 A typical Geiger–Muller counter is a cylindrical detector with an aperture 50 mm in diameter. What is the solid angle if a point isotropic source is located 0.10 m away from the detector?
Answer Using Eq. 8.5 with $d = 0.10$ m and $R = 25$ mm,

$$\Omega = \frac{1}{2}\left[1 - \frac{0.10}{\sqrt{0.10^2 + (25\times10^{-3})^2}}\right] = 0.015$$

If $\Omega = 1$, the setup is called a 4π geometry because the detector sees the full 4π solid angle around the source. A spherical detector represents such a case (Figure 8.8a). If $\Omega = 1/2$, the setup is called a 2π geometry. Then half of the particles emitted by the source enter the detector (Figure 8.8b).

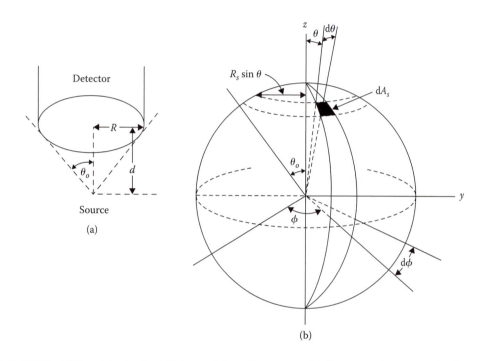

Figure 8.7 (a) The detector is at distance d from the source, (b) the source is assumed to be at the center of the sphere. The cone defined by the angle θ_0 determines the area A_s (differential area dA) on the surface of the sphere.

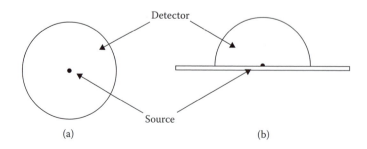

Figure 8.8 (a) 4π Geometry and (b) 2π geometry.

8.2.4 The Solid Angle for a Disk Source Parallel to a Detector with a Circular Aperture

Consider a disk source parallel to a detector with a circular aperture (Figure 8.9). Starting with Eq. 8.4, one may obtain an expression involving elliptic integrals[1,2] or the following equation in terms of Bessel functions[3,4]:

$$\Omega = s \int_0^\infty dx e^{-xz} \frac{J_1(x)}{x} J_1(xs)$$

(8.9)

where $s = R_d/R_s$, $z = d/R_s$, and $J_1(x)$ = Bessel function of the first kind. If R_d/d and R_s/d are less than 1, the following algebraic expression is obtained for the solid angle (see Prob. 8.1):

$$\Omega = \frac{\omega^2}{4} \left\{ 1 - \frac{3}{4}(\psi^2 + \omega^2) + \frac{15}{8}\left(\frac{\psi^4 + \omega^4}{3} + \psi^2\omega^2 \right) \right.$$
$$\left. - \frac{35}{16}\left[\frac{\psi^6 + \omega^6}{4} + \frac{3}{2}\psi^2\omega^2(\psi^2 + \omega^2) \right] \right\}$$

(8.10)

where $\psi = R_s/d$
$\quad\quad \omega = R_d/d$

The accuracy of Eq. 8.10 increases as ψ and ω decrease. If $\psi < 0.2$ and $\omega < 0.5$, the error is less than 1%.

8.2.5 The Solid Angle for a Point Isotropic Source and a Detector with a Rectangular Aperture

Consider the geometry of Figure 8.10 with a point isotropic source located a distance d away from a detector having a rectangular aperture with area equal to ab. The solid angle is given by[5]

$$\Omega = \frac{1}{4\pi} \arctan \frac{ab}{d\sqrt{a^2 + b^2 + d^2}}$$

(8.11)

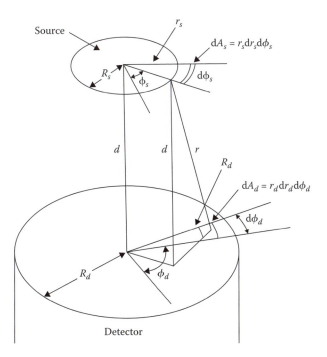

Figure 8.9 A disk source and a detector with a circular aperture.

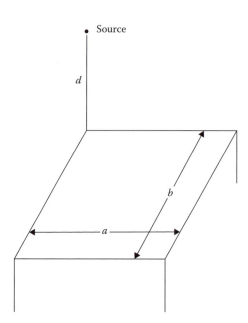

Figure 8.10 The solid angle between a point isotropic source and a detector with a rectangular aperture. Source is located directly above one corner of the detector.

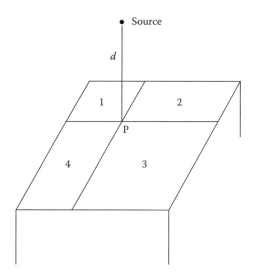

Figure 8.11 A point isotropic source located at an arbitrary point above a detector with a rectangular aperture. The solid angle is equal to four terms, each given by Eq. 8.11.

If the source is located at an arbitrary point above the detector, the solid angle is the sum of four terms (Figure 8.11), each of them similar to Eq. 8.11. As Figure 8.11 shows, the detector is divided into four rectangles by the lines that determine the coordinates of the point P. The solid angle is then

$$\Omega = \Omega_1 + \Omega_2 + \Omega_3 + \Omega_4$$

where Ω_i for $i = 1, \ldots, 4$ is given by Eq. 8.11 for the corresponding rectangles.

8.2.6 The Solid Angle for a Disk Source and a Detector with a Rectangular Aperture

Consider the geometry shown in Figure 8.12. A disk source is located at a distance d above a detector having a rectangular aperture with an area equal to ab. It is assumed that the center of

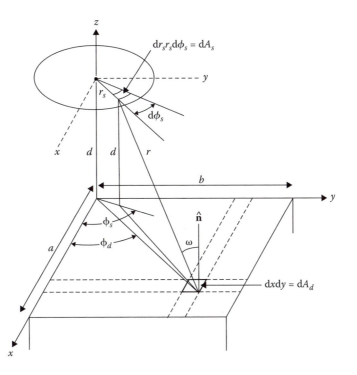

Figure 8.12 A disk source and a detector with a rectangular aperture.

the source is directly above one corner of the aperture, as shown in Figure 8.12. The expression of the solid angle in the general case of the arbitrary position is derived in the manner shown in Section 8.2.5.

The distance r (Figure 8.12) is equal to

$$r^2 = d^2 + r_s^2 + x^2 + y^2 - 2r_s\sqrt{x^2 + y^2}\cos\left[\phi_s - \cos^{-1}\left(\frac{x}{\sqrt{x^2 + y^2}}\right)\right]$$

Equation 8.4 is then written as

$$\Omega = \frac{1}{4\pi A_s d^2}\int_0^{R_s} dr_s r_s \int_0^{2\pi} d\phi_s \int_0^a dx \int_0^b dy \left\{1 + \frac{r_s^2}{d^2} + \frac{x^2}{d^2}\right.$$

$$\left. + \frac{y^2}{d^2} - 2\frac{r_s}{d}\sqrt{\frac{x^2 + y^2}{d^2}}\cos\left[\phi_s - \cos^{-1}\left(\frac{x}{\sqrt{x^2 + y^2}}\right)\right]\right\}^{-3/2}$$

(8.12)

As in Section 8.2.4, if the ratios R_s/d, a/d, and b/d are less than 1, the expression in the braces may be expanded in a series. If only the first four terms are kept, the result of the integration is

$$\Omega = \frac{\omega_1\omega_2}{4\pi}\left[1 - \frac{3}{4}\psi^2 - \frac{1}{2}(\omega_1^2 + \omega_2^2) + \frac{1}{8}(5\psi^4 + 3\omega_1^4 + 3\omega_2^4)\right.$$

$$+ \frac{5}{4}\psi^2(\omega_1^2 + \omega_2^2) - \frac{35}{64}\psi^6 + \frac{5}{12}\omega_1^2\omega_2^2 - \frac{35}{16}\psi^4(\omega_1^2 + \omega_2^2)$$

$$- \frac{7}{32}\psi^2(9\omega_1^4 + 9\omega_2^4 + 10\omega_1^2\omega_2^2) - \frac{7}{16}\omega_1^2\omega_2^2(\omega_1^2 + \omega_2^2)$$

$$\left. - \frac{5}{16}(\omega_1^6 + \omega_2^6)\right]$$

(8.13)

where $\omega_1 = a/d$

$$\omega_2 = \frac{b}{d}$$

$$\psi = \frac{R_\varepsilon}{d}$$

8.2.7 The Use of the Monte Carlo Method for the Calculation of the Solid Angle

The basic equation defining the geometry factor (Eq. 8.4) can be solved analytically in very few cases. Approximate solutions can be obtained either by a series expansion (Eqs. 8.10 and 8.13 are such results) or by a numerical integration or using other approximations.[1,6,7]

A general method that can be used with any geometry is based on a Monte Carlo calculation,[8-11] which simulates, in a computer, the emission and detection of particles. A computer program is written based on a model of the source-detector geometry. Using random numbers, the particle position of birth and the direction of emission are determined. The program then checks whether the randomly selected direction intersects the detector volume. By definition, the ratio of particles hitting the detector to those emitted by the source is equal to the solid angle.

The advantage of a Monte Carlo calculation is the ability to study complicated geometries. The result has an error associated with it that decreases as the number of particles studied increases.

8.3 SOURCE EFFECTS

Two source effects are discussed in this section: absorption of particles in the source, and the effect of the backing material that supports the source. Both effects are always important in measurements of charged particles. In some cases, however, they may also be significant in X-ray or thermal-neutron measurements.

8.3.1 Source Self-Absorption Factor (f_a)

Radioactive substances are deposited on a backing material in thin deposits. But no matter how thin, the deposit has a finite thickness and may cause absorption of some particles emitted by the source. Consider the source of thickness t shown in Figure 8.13. Particle 1 traverses the source deposit and enters the detector. Particle 2 is absorbed inside the source so that it will not be counted. Therefore, source self-absorption will result in a decrease of the counting rate r.

Source self-absorption may be reduced to an insignificant amount but it cannot be eliminated completely. It is always important for charged particles and generally more crucial for heavier particles (p, α, d, heavy ions) than for electrons.

Source self-absorption, in addition to altering the number of particles leaving the source, may also change the energy of the particles escaping from it. Particle 1 in Figure 8.13 successfully leaves

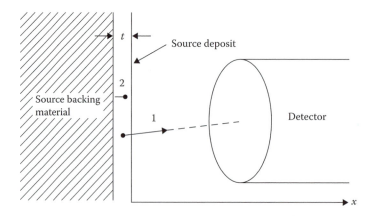

Figure 8.13 Source self-absorption. Particles may be absorbed in the source deposit.

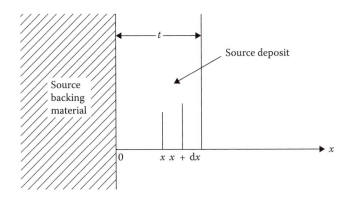

Figure 8.14 Diagram used for the calculation of the source self-absorption factor for betas.

the deposit, but it loses some energy as it goes through the deposit. This energy loss is important when the energy of the particle is measured.

An approximate correction for self-absorption can be obtained if the source emits particles following a known attenuation law. As an example, consider a source with thickness t (Figure 8.14) that has a uniform deposit of a radioisotope emitting β particles. Assume that the source gives S betas per second in the direction of the positive x axis. If self-absorption is absent, S betas per second will leave the source (toward positive x). Because of the source thickness, betas produced in dx, around x, have to successfully penetrate the thickness $(t - x)$ to escape. The probability of escape is $e^{-\mu(t-x)}$ where μ is the attenuation coefficient for the betas in the material of which the deposit is made. The total number of betas escaping is

$$\int_0^t \frac{dx}{t} Se^{-\mu(t-x)} = \frac{S}{t\mu}(1-e^{-\mu t})$$

A self-absorption factor f_a is defined by

$$f_a = \frac{\text{number of particles leaving source with self-absorption}}{\text{number of particles leaving source without self-absorption}}$$

Using the result obtained above,

$$f_a = \frac{(S/t\mu)(1-e^{-\mu t})}{S} = \frac{1}{t\mu}(1-e^{-\mu t}), \quad 0 \le f_a \le 1 \tag{8.14}$$

Example 8.2 Assume that ^{137}Cs was deposited on a certain material. The thickness of the deposit is $t = 0.1$ mm. ^{137}Cs emits betas with $E_{max} = 0.514$ MeV. What is the value of f_a for such a source? The density of cesium is 1.6×10^3 kg/m^3.

Answer For betas of $E_{max} = 0.514$ MeV, the attenuation coefficient is (from Chapter 4)

$$\mu = 1.7E_{max}^{-1.14}, \quad \mu = 1.7(0.514)^{-1.14} = 2.14\,\text{m}^2/\text{kg}$$
$$\mu t = (2.14\,\text{m}^2/\text{kg})(0.1\times 10^{-3}\text{m})(1.6\times 10^3\,\text{kg/m}^3) = 0.34$$

Using Eq. 8.14,

$$f_a = \frac{1}{0.34}(1 - e^{-0.34}) = 0.85$$

Therefore, only 85% of the betas escape this source. Or, if this effect is not taken into account, the source strength will have an error of 15%.*

---

* A similar calculation of f_a may be repeated for an X-ray or a neutron source. For X-rays the probability of escape $e^{-\mu t}$; for neutrons it is $e^{-\Sigma t}$.

If the source emits monoenergetic charged particles, essentially all the particles leave the source deposit as long as $t < R$, where R = range of the particles. In practice, the sources for monoenergetic charged particles are such that $t \ll R$, in which case, $f_a \approx 1$. Then the only effect of the source deposit is an energy loss for the particles that traverse it (see also Chapter 13).

8.3.2 Source Backscattering Factor (f_b)

A source cannot be placed in midair. It is always deposited on a material that is called *source backing* or *source support*. The source backing is usually a very thin material, but no matter how thin, it may backscatter particles emitted in a direction away from the detector (Figure 8.15). To understand the effect of backscattering, assume that the solid angle in Figure 8.15 is $\Omega = 10^{-2}$. Also assume that all the particles entering the detector are counted, self-absorption is zero, and there is no other medium that might absorb or scatter the particles except the source backing.

Particle 1 in Figure 8.15 is emitted toward the detector. Particle 2 is emitted in the opposite direction. Without the source backing, particle 2 would not turn back. With the backing material present, there is a possibility that particle 2 will have scattering interactions there, have its direction of motion changed, and enter the detector. If the counting rate is $r = 100$ counts per minute and there is no backscattering of particles toward the detector, the strength of the source will be correctly determined as

$$S = \frac{r}{\Omega} = \frac{100}{10^{-2}} = 10{,}000 \text{ part.} / \text{min}$$

If the source backing backscatters 5% of the particles, the counting rate will become 105 counts/min, even though it is still the same source as before. If source backscattering is not taken into account, the source strength will be erroneously determined as

$$S = \frac{r}{\Omega} = \frac{105}{10^{-2}} = 10{,}500 \text{ part.} / \text{min}$$

To correct properly for this effect, a source backscattering factor (f_b) is defined by

$$f_b = \frac{\text{number of particles counted with source backing}}{\text{number of particles counted without source backing}} \tag{8.15}$$

From the definition it is obvious that

$$2 > f_b \geq 1$$

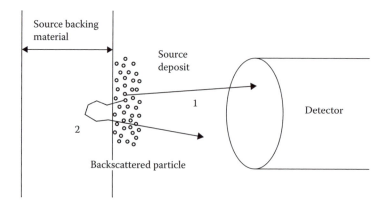

Figure 8.15 The source backing material backscatters particles and necessitates the use of a backscattering factor f_b.

In the example discussed above, $f_b = 1.05$, and the correct strength of the source is

$$S = \frac{r}{\Omega f_b} = \frac{105}{10^{-2} \times 1.05} = 10,000 \, \text{part./min}$$

The backscattering factor is important, in most cases, only for charged particles. It depends on three variables:

1. Thickness (b) of the backing material
2. Particle kinetic energy (T)
3. Atomic number of the backing material (Z)

The dependence of f_b on thickness b is shown in Figure 8.16. As $b \to 0, f_b \to 1$, which should be expected. For large thicknesses, f_b reaches a saturation value, which should also be expected. Since charged particles have a definite range, there is a maximum distance they can travel in the backing material, be backscattered, and traverse the material again in the opposite direction. Therefore an upper limit for that thickness is $b = R/2$, where R is the range of the particles. Experiments have shown that

$$b_s = b(\text{saturation}) \approx 0.2R$$

The dependence of the saturation backscattering factor of electrons on the kinetic energy and the atomic number of the backing material is given by the following empirical equation,[12,13] based on a least-squares fit of experimental results:

$$f_b(\text{sat}) = 1 + \frac{b_1 \exp(-b_2 Z^{-b_3})}{1 + (b_4 + b_5 Z^{-b_6}) \alpha^{(b_7 - b_8 / Z)}} \tag{8.16}$$

where the constants b_i for $i = 1, 2, \ldots, 8$ have these values:

$$b_1 = 1.15 \pm 0.06 \qquad b_5 = 15.7 \pm 3.1 \quad \alpha = \frac{T}{mc^2}$$
$$b_2 = 8.35 \pm 0.25 \qquad b_6 = 1.59 \pm 0.07$$
$$b_3 = 0.525 \pm 0.02 \qquad b_7 = 1.56 \pm 0.02$$
$$b_4 = 0.0185 \pm 0.0019 \quad b_8 = 4.42 \pm 0.18$$

Figure 8.17 shows the change of $f_b(\text{sat})$ versus kinetic energy T for four elements.

A backscattering correction should be applied to alpha counting in 2π counters (Figure 8.8b). It has been determined[14–16] that the number of backscattered alphas is between 0% and 5%, depending on the energy of the alphas, the uniformity of the source, and the atomic number of the material forming the base of the counter.

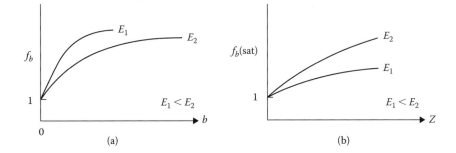

Figure 8.16 (a) The backscattering factor f_b as a function of thickness b of the backing material, (b) the saturation backscattering factor as a function of the atomic number Z of the backing material.

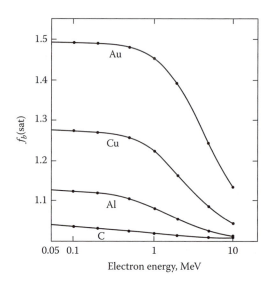

Figure 8.17 Backscattering factor as a function of energy for C, Al, Cu, and Au. Curves were obtained using Eq. 8.16 for electrons only.

Correction for source backscattering is accomplished in two ways:

1. An extremely thin backing material is used for which $f_b \approx 1$. In general, a low-Z material is used, for example, plastic, if possible.

2. A thick backing material is used, for which the saturation backscattering factor should be employed for correction of the data.

For accurate results, the backscattering factor should be measured for the actual geometry of the experiment.

8.4 DETECTOR EFFECTS

The detector may affect the measurement in two ways. First, if the source is located outside the detector (which is usually the case), the particles may be scattered or absorbed by the detector window. Second, some particles may enter the detector and not produce a signal, or they may produce a signal lower than the discriminator threshold.

8.4.1 Scattering and Absorption due to the Window of the Detector

In most measurements the source is located outside the detector (Figure 8.18). The radiation must penetrate the detector window to have a chance to be counted. Interactions between the radiation and the material of which the detector window is made may scatter and/or absorb particles. This is particularly important for low-energy β particles.

Figure 8.18 shows a gas-filled counter and a source of radiation placed outside it. Usually the particles enter the detector through a *window* made of a very thin material (such as glass, mica, or thin metal). Looking at Figure 8.18, most of the particles, like particle 1, traverse the window and enter the counter. But, there is a possibility that a particle, like particle 2, may be scattered at the window and never enter the counter. Or, it may be absorbed by the material of the window (particle 3).

In the case of scintillation counters, the window consists of the material that covers the scintillator and makes it light-tight. In some applications the source and the scintillator are placed in a light-tight chamber, thus eliminating the effects of a window.

In semiconductor detectors, the window consists of the metallic layer covering the front face of the detector necessary to provide an ohmic contact. That layer is extremely thin, but may still affect measurements of alphas and heavier charged particles because of energy loss there.

There is no direct way to correct for the effect of the window. Commercial detectors are made with very thin windows, but the investigator should examine the importance of the window effect for the particular measurement performed. If there is a need for an energy-loss correction, it is

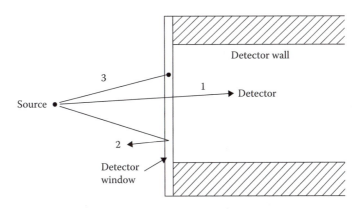

Figure 8.18 The window of the detector may scatter and/or absorb some of the particles emitted by the source.

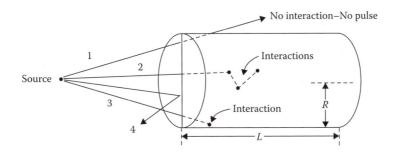

Figure 8.19 Particles detected are those that interact inside the detector and produce a pulse higher than the discriminator level.

applied separately to the energy spectrum. If, however, there is a need to correct for the number of particles stopped by the window, that correction is incorporated into the detector efficiency.

8.4.2 Detector Efficiency (ε)

It is not certain that a particle will be counted when it enters a detector. It may, depending on the type and energy of the particle and type and size of detector, go through without having an interaction (particle 1 in Figure 8.19); it may produce a signal so small it is impossible to record with the available electronic instruments (particle 3); or, it may be prevented from entering the detector by the window (particle 4). In Figure 8.19, the particle with the best chance of being detected is particle 2.

The quantity that gives the fraction of particles being detected is called the *detector efficiency* ε, given by

$$\varepsilon = \frac{\text{number of particles recorded per unit time}}{\text{number of particles impinging upon the detector per unit time}} \tag{8.17}$$

The detector efficiency depends upon the following*:

1. Density and size of detector material

2. Type and energy of radiation

3. Electronics

Effect of density and size of detector material. The efficiency of a detector will increase if the probability of an interaction between the incident radiation and the material of which the detector is made increases. That probability increases with detector size. But larger size is of limited usefulness because the background increases proportionally with the size of the detector, and because

---

* In gamma spectroscopy, several other efficiencies are being used in addition to this one (see Chapter 12).

in some cases it is practically impossible to make large detectors. (Semiconductor detectors are a prime example.)

The probability of interaction per unit distance traveled is proportional to the density of the material. The density of solids and liquids is about a thousand times greater than the density of gases at normal pressure and temperature. Therefore, other things being equal, detectors made of solid or liquid material are more efficient than those using gas.

Effect of type and energy of radiation. Charged particles moving through matter will always have Coulomb interactions with the electrons and nuclei of that medium. Since the probability of interaction is almost a certainty, the efficiency for charged particles will be close to 100%. Indeed, detectors for charged particles have an efficiency that is practically 100%, regardless of their size or the density of the material of which they are made. For charged particles, the detector efficiency is practically independent of particle energy except for very low energies, when the particles may be stopped by the detector window.

Charged particles have a definite range. Therefore, it is possible to make a detector with a length L such that all the particles will stop and deposit their energy in the counter. Obviously, the length L should be greater than R, where R is the range of the particles in the material of which the detector is made.

Photons and neutrons traversing a medium show an exponential attenuation (see Chapter 4), which means that there is always a nonzero probability for a photon or a neutron to traverse any thickness of material without an interaction. As a result of this property, detectors for photons or neutrons have efficiency less than 100% regardless of detector size and energy of the particle.

Effect of electronics. The electronics of a counting setup affects the counter efficiency indirectly. If a particle interacts in the detector and produces a signal, that particle will be recorded only if the signal is recorded. The signal will be registered if it is higher than the discriminator level, which is, of course, determined by the electronic noise of the counting system. Thus, the counting efficiency may increase if the level of electronic noise is decreased.

As an example, consider a counting system with electronic noise such that the discriminator level is at 1 mV. In this case, only pulses higher than 1 mV will be counted; therefore, particles that produce pulses lower than 1 mV will not be recorded. Assume next that the preamplifier or the amplifier or both are replaced by quieter ones, and the new noise level is such that the discriminator level can be set at 0.8 mV. Now, pulses as low as 0.8 mV will be registered, more particles will be recorded, and hence the efficiency of the counting system increases.

If electronics is included in the discussion, it is the efficiency of the *system* (detector plus electronics) that is considered rather than the efficiency of the counter.

8.4.3 Determination of Detector Efficiency

The efficiency of a detector can be determined either by measurement or by calculation. Many methods have been used for the measurement of detection efficiency,[17–19] but the simplest and probably the most accurate is the method of using a calibrated source, that is, a source of known strength. In Figure 8.19, assume that the source is a monoenergetic point isotropic source emitting S particles per second. If the true net counting rate is r counts per second, the solid angle is Ω, and the efficiency is ε, the equation giving the efficiency is

$$\varepsilon(E) = \frac{r}{\Omega F(E)S} \tag{8.18}$$

where $F = f_d f_b \cdots$ is a combination of all the correction factors that may have to be applied to the results. Note that the correction factors and the efficiency depend on the energy of the particle.

Accurate absolute measurements rely on measured rather than calculated efficiencies. Nevertheless, an efficiency calculation is instructive because it brings forward the parameters that are important for this concept. For this reason, two cases of efficiency calculation for a photon detector are presented below.

Consider first a parallel beam of photons of energy E impinging upon a detector of thickness L (Figure 8.20). The probability that a photon will have at least one interaction in the detector is $1 - e^{-\mu(E)L}$, where $\mu(E)$ is the total linear attenuation coefficient of photons with energy E in the material of which the detector is made. If one interaction is enough to produce a detectable pulse, the efficiency is

$$\varepsilon(E) = 1 - e^{-\mu(E)L} \tag{8.19}$$

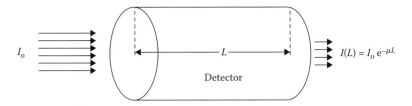

Figure 8.20 A parallel beam of photons going through a detector of length (thickness) L.

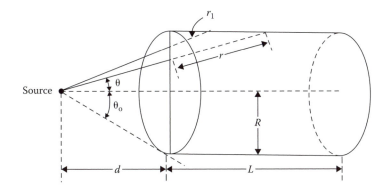

Figure 8.21 A point isotropic photon source at a distance d away from a cylindrical detector.

Equation 8.19 shows the dependence of $\varepsilon(E)$ on

1. The size L of the detector
2. The photon energy (through μ)
3. The density of the material (through μ)

Example 8.3 What is the efficiency of a 50 mm long NaI(Tl) crystal for a parallel beam of (a) 2-MeV gammas or (b) 0.5-MeV gammas?
Answer (a) From the table in Appendix D, the total mass attenuation coefficient for 2-MeV gammas in NaI(Tl) is $\mu = 0.00412$ m²/kg. The density of the scintillator is 3.67×10^3 kg/m³. Therefore, Eq. 8.19 gives

$$\varepsilon = 1 - \exp[-\,0.00412 \text{ m}^2/\text{kg}(3.67 \times 10^3 \text{ kg/m}^3)0.05 \text{ m}] = 1 - \exp(-0.756) = 0.53 = 53\%$$

(b) For 0.50-MeV gammas, $\mu = 0.00921$ m²/kg. Therefore,

$$\varepsilon = 1 - \exp[-0.00921 \text{ m}^2/\text{kg } (3.67 \times 10^3 \text{ kg/m}^3)0.05 \text{ m}] = 1 - \exp(-1.69) = 0.81 = 81\%$$

The next case to consider is that of a point isotropic monoenergetic source, at a distance d away from a cylindrical detector of length L and radius R (see Figure 8.21). For photons emitted at an angle θ, measured from the axis of the detector, the probability of interaction is $1 - \exp[-\mu(E)r(\theta)]$ and the probability of emission between angles θ and $\theta + d\theta$ is $(1/2)\sin\theta\,d\theta$. Assuming, as before, that one interaction is enough to produce a detectable pulse, the efficiency is given by

$$\varepsilon(E) = \frac{\int_0^{\theta_0} S\{1 - \exp[-\mu(E)r(\theta)]\}(1/2)\sin\theta\,d\theta}{(S/2)\int_0^{\theta_0}\sin\theta\,d\theta} \tag{8.20}$$

233

Equation 8.20 shows that the efficiency depends, in this case, not only on L, μ, and E, but also on the source-detector distance and the radius of the detector. Results obtained by numerically integrating Eq. 8.20 are given in Section 12.4.1, where the efficiency of gamma detectors is discussed in greater detail. Many graphs and tables based on Eq. 8.20 can be found in Reference 20.

Equations 8.19 and 8.20 probably overestimate efficiency, because their derivation was based on the assumption that a single interaction of the incident photon in the detector will produce a detectable pulse. This is not necessarily the case. A better way to calculate efficiency is by determining the energy deposited in the detector as a result of all the interactions of an incident particle. Then one can compute the number of recorded particles based on the minimum energy that has to be deposited in the detector in order that a pulse higher than the discriminator level may be produced. The Monte Carlo method, which is ideal for such calculations, has been used by many investigators[11,21] for that purpose.

Efficiencies of neutron detectors are calculated by methods similar to those used for gammas. Neutrons are detected indirectly through gammas or charged particles produced by reactions of nuclei with neutrons. Thus, the neutron detector efficiency is essentially the product of the probability of a neutron interaction, with the probability to detect the products of that interaction (see Chapter 14).

8.5 RELATIONSHIP BETWEEN COUNTING RATE AND SOURCE STRENGTH

Equation 8.18 rewritten in terms of the true net counting rate r gives the relationship between r and the source strength:

$$r = \Omega F \varepsilon S \tag{8.21}$$

In terms of gross counts G obtained over time t_G and background count B obtained over time t_B, the true net counting rate (Eq. 2.113) is

$$r = \frac{G/t_G}{1 - (G/t_G)\tau} - \frac{B}{t_B} \tag{8.22}$$

where τ is the counter dead time. Usually the objective of the measurement is to obtain the source strength S using a detection system of known Ω, F, and ε. Combining Eqs. 8.21 and 8.22, the source strength becomes

$$S - \frac{r}{\Omega F \varepsilon} = \frac{1}{\Omega F \varepsilon} \left(\frac{G/t_G}{1 - (G/t_G)\tau} - \frac{B}{t_B} \right) \tag{8.23}$$

The error in the value of S is due to errors in the values of Ω, F, ε, and the statistical error of r. In many cases encountered in practice, the predominant error is that of r. Then one obtains, from Eq. 8.23,

$$\frac{\sigma_S}{S} = \frac{\sigma_r}{r}$$

That is, the percent error of S is equal to the percent error of the true net counting rate r.

Example 8.4 The geometric setup shown in Figure 8.22 was used for the measurement of the strength of the radioactive source. The following data were obtained:

$$G = 6000, \quad B = 400, \quad \tau = 100 \text{ μs}, \quad \varepsilon = 0.60 \pm 0.005$$
$$t_G = 10 \text{ min}, \quad t_B = 10 \text{ min}, \quad F = 1 \pm 0.001$$

What is the strength S and its standard error?
Answer The true net counting rate r is

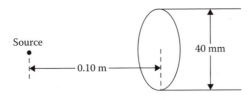

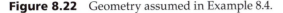

Figure 8.22 Geometry assumed in Example 8.4.

$$r = \frac{6000/10}{1-\left[6000\times10\times\left(10^{-6}/10\times60\right)\right]} - \frac{400}{10} = \frac{6000/10}{0.999} - 40$$

$$= 600.6 - 40 \approx 561 \text{ counts/min}$$

The standard error of r is (Eq. 2.114)

$$\sigma_\gamma = \sqrt{\left(\frac{1}{1-(G/t_G)\tau}\right)^4 \frac{G_2}{t_G^2} + \frac{B}{t_B^2}} = \sqrt{\left(\frac{1}{0.999}\right)^4 \frac{6000}{100} + \frac{400}{100}} = 8 \text{ counts/min}$$

Therefore

$$\frac{\sigma_r}{r} = \frac{8}{561} = 1.4\%$$

The solid angle is

$$\Omega = \frac{1}{2}\left(1 - \frac{d}{\sqrt{d^2 + r^2}}\right) = 0.0097$$

Using Eq. 8.23,

$$S = \frac{r}{\Omega F \varepsilon} = \frac{561}{(0.0097)(0.60)} = 96{,}392 \text{ part./min}$$

The standard error of S is (Section 2.15.1)

$$\frac{\sigma_S}{S} = \sqrt{\left(\frac{\sigma_r}{r}\right)^2 + \left(\frac{\sigma_F}{F}\right)^2 + \left(\frac{\sigma_\varepsilon}{\varepsilon}\right)^2}$$

$$= \sqrt{(1.4\times10^{-2})^2 + \left(\frac{0.001}{1}\right)^2 + \left(\frac{0.005}{0.60}\right)^2} = 1.7\times10^{-2} = 1.7\%$$

8.6 REFERENCE MATERIALS FOR RELATIVE AND ABSOLUTE MEASUREMENTS

While mathematical descriptions of relative and absolute measurements are very useful in understanding geometrical source and detector effects for various radiation sources, the use of standards is imperative to establish the credentials of any laboratory. Nowadays no analytical laboratory can distribute or publish results without stringent quality control (QC) procedures. QC is the methodology that each laboratory undertakes to ascertain that the measurements obtained are accurate within acceptable errors for any particular procedure. In essence it is a blind test to assure accuracy. The International Organization for Standardization has been at the forefront of setting guidelines for laboratories to formalize procedures to attain the highest possible quality control. In the past two decades there has been a lot of research in the development of standard and certified reference materials to ensure quality control in both relative and absolute measurements. Institutions such as the USA National Institute of Science and Technology (NIST), USA New Brunswick Laboratory, British Nuclear Physics Laboratory (NPL), European Institute for Reference Materials and Measurements (IRMM), International Atomic Energy Agency, (IAEA) and the Japan National Institute of Technology and Evaluation (NITE) have a very wide range of natural and artificially produced radioactivity standards for alpha, beta and gamma measurements.

There are many published works on the intercomparisons of ionizing radiation standards,[22] preparation and analysis of ^{226}Ra-^{222}Rn emanation standards for calibrating passive radon detectors,[23] accreditation for the radioactivity metrology,[24] a procedure for the standardization of gamma reference sources for quality assurance in activity measurements of radiopharmaceuticals,[25] and the preparation of a soil reference material for the determination of radionuclides.[26] A more practical application of such reference sources is presented in Chapter 12 in the section on efficiency of X-ray and γ-ray detectors.

PROBLEMS

8.1 Show that if $R_d/d < 1$ and $R_s/d < 1$, the solid angle between two parallel disks with radii R_d and R_s a distance d apart is given to a good approximation by

$$\Omega = \frac{\omega^2}{4}\left\{1 - \frac{3}{4}(\psi^2 + \omega^2) + \frac{15}{8}\left(\frac{\psi^4 + \omega^4}{3} + \psi^2\omega^2\right)\right.$$
$$\left. - \frac{35}{16}\left[\frac{\psi^6 + \omega^6}{4} + \frac{3}{2}\psi^2\omega^2(\psi^2 + \omega^2)\right]\right\}$$

where $\psi = R_s/d$ and $\omega = R_d/d$.

8.2 Show that an approximate expression for the solid angle between two nonparallel disks is

$$\Omega = \frac{\omega^2}{4}\left\{\begin{array}{l} 1 - \frac{3}{4}\left[(\psi^2 + \omega^2)(1 + \sin^2\theta)\right] \\ + \frac{15}{8}\left[\frac{\omega^4 + \psi^4}{3} + \psi^2\omega^2 + \psi^2\left(1 + \frac{1}{4}\omega^2 + \frac{2}{3}\psi^2\right)\sin^2\theta\right]\end{array}\right\}$$

where θ is the angle between the planes of the two disks, and ψ and ω are defined as in Prob. 8.1.

8.3 Show that the solid angle between a disk source and a detector with a rectangular aperture is given, approximately, by Eq. 8.13 under the conditions given in Section 8.2.6.

8.4 A 1-mCi point isotropic gamma source is located 0.10 m away from a 60° spherical shell of a NaI detector, as shown in the figure below. Assuming that all the pulses at the output of the photomultiplier tube are counted, what is the counting rate of the sealer? The gamma energy is 1.25 MeV.

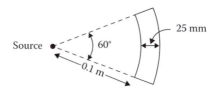

8.5 A hemispherical detector is used to count betas with an efficiency of 95%. The saturation backscattering factor has been determined to be 1.5 and the background is known to be 35 ± 2 counts/min. If with the source present 4200 counts are recorded in 2 min, what is the strength of this source in Bq and Ci? The dead time of the detector is estimated to be 50 μs.

8.6 Calculate the counting rate for the case shown in the figure below. The source has the shape of a ring and emits 10^6 part./s isotropically. The background is zero. The detector efficiency is 80%, and $F = 1$.

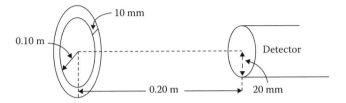

8.7 Calculate the self-absorption factor for a ^{14}C source that has a thickness of 10 μg/cm^2 (10^{-4} kg/m^2); $E_{max} = 156$ keV.

8.8 An attempt was made to measure the backscattering factor by placing foils of continuously increasing thickness behind the source and observing the change in the counting rate. The foils were of the same material as the source backing. The results of the measurements are given in the table below. Calculate the saturation backscattering factor and the source backscattering factor.

Thickness Behind Source (mm)	Counting Rate (Counts/min)
0.1 (Source backing only)	3015
0.15	3155
0.2	3365
0.25	3400
0.3	3420
0.35	3430
0.4	3430

8.9 What is the counting rate in a detector with a rectangular aperture measuring 1 mm × 40 mm, if a 1-mCi gamma-ray point isotropic source is 0.10 m away? The efficiency of the detector for these gammas is 65%.

8.10 A radioactive source emits electrons isotropically at the rate of 10^4 electrons/s. A plastic scintillator having the shape of a cylindrical disk with a 25 mm radius is located 120 mm away from the source. The efficiency of the detector for these electrons is 95%. The backscattering factor is 1.02, and the source self-absorption factor is 0.98. Dead time of the counting system is 5 μs. How long should one count, under these conditions, to obtain the strength of the source with a standard error of 5%? Background is negligible. The only error involved is that due to counting statistics.

8.11 How would the result of Problem 8.9 change if the backscattering factor was known with an error of ±1%, the efficiency with an error of ± 0.5%, and the source self-absorption factor with an error of ±1%?

8.12 Calculate the strength of a point isotropic radioactive source if it is given that the gross counting rate is 200 counts/min, the background counting rate is 25 counts/min, the counter efficiency is 0.90, the source-detector distance is 0.15 m, and the detector aperture has a radius of 20 mm ($F = 1$). What is the standard error of the results if the error of the gross counting rate is known with an accuracy of ± 5% and the background with ± 3%? Dead time is 1 μs.

8.13 A point isotropic source is located at the center of a hemispherical 2π counter. The efficiency of this detector for the particles emitted by the source is 85%. The saturation backscattering factor is 1.5. The background is 25 ± 1 counts/min. What is the strength of the source if 3000 counts are recorded in 1 min? What is the standard error of this measurement?

REFERENCES

1. Masket, A. V., *Rev. Sci. Instrum.* **28**:191 (1957).

2. Ruffle, M. P., *Nucl. Instrum. Meth.* **52**:354 (1967).

3. Ruby, L., and Rechen, J. B., *Nucl. Instrum. Meth.* **58**:345 (1968).

4. Ruby, L., *Nucl. Instrum. Meth.* **337**:531 (1994).

5. Gotoh, H., and Yagi, H., *Nucl. Instrum. Meth.* **96**:485 (1971).

6. Gardner, R. P., and Verghese, K., *Nucl. Instrum. Meth.* **93**:163 (1971).

7. Verghese, K., Gardner, R. P., and Felder, R. M., *Nucl. Instrum. Meth.* **101**:391 (1972).

8. Williams, I. R., *Nucl. Instrum. Meth.* **44**:160 (1966).

9. Green, M. V., Aamodt, R. L., and Johnston, G. S., *Nucl. Instrum. Meth.* **117**:409 (1974).

10. Wielopolski, L., *Nucl. Instrum. Meth.* **143**:577 (1977).

11. Beam, G. B., Wielopolski, L., Gardner, R. P., and Verghese, K., *Nucl. Instrum. Meth.* **154**:501 (1978).

12. Tabata, T., Ito, R., and Okabe, S., *Nucl. Instrum. Meth.* **94**:509 (1971).

13. Kuzminikh, V. A., and Vorobiev, S. A., *Nucl. Instrum. Meth.* **129**:561 (1975).

14. Deruytter, A. J., *Nucl. Instrum. Meth.* **15**:164 (1962).

15. Walker, D. H., *Int. J. Appl. Rad. Isot.* **16**:183 (1965).

16. Hutchinson, J. M. R., Nass, C. R., Walker, D. H., and Mann, W. B., *Int. J. Appl. Rad. Isot.* **19**:517 (1968).

17. Waibel, E., *Nucl. Instrum. Meth.* **74**:236 (1969).

18. Waibel, E., *Nucl. Instrum. Meth.* **86**:29 (1970).

19. Waibel, E., *Nucl. Instrum. Meth.* **131**:133 (1975).

20. Heath, R. L., "Scintillation Spectrometry—Gamma-Ray Spectrum Catalogue," IDO-16880–1, 1964.

21. Nakamura, T., *Nucl. Instrum. Meth.* **86**:163 (1970).

22. Jerome, S. N., and Woods, M. J., *J. Radioanal. Nuc. Chem.* **233**:25 (1998).

23. Lavi, N., and Alfassi, Z. B., *J. Radioanal. Nuc. Chem.* **265**:123 (2005).

24. Jerome, S. M., and Judge, S. M., *J. Radioanal. Nuc. Chem.* **276**:353 (2008).

25. Oropesa, P., Serra, R., Gutierrez, S., and Hernandez, A. T., *Appl. Rad. Isotop.* **56**:787 (2002).

26. Llaurado, M., Torres, J. M., Tent, J., Sahuquillo, A., Muntau, H., and Rauret, G., *Anal. Chim. Acta*, **445**:99 (2001).

27. Bell, P. R., in K. Siegbahn (ed.), *Beta and Gamma-Ray Spectrometry*, Interscience Publishers, New York, 1955, Chapter 5.

9

Introduction to Spectroscopy

9.1 INTRODUCTION

Spectroscopy is the aspect of radiation measurements that deals with measuring the energy distribution of particles emitted by a radioactive source or produced by a nuclear reaction.

This introduction to spectroscopy is complemented by Chapters 11–14, which present details on spectroscopy of photons, charged particles, and neutrons. This chapter discusses the following broad subjects:

1. Definition of differential and integral spectra

2. Energy resolution of the detector

3. The function of a multichannel analyzer (MCA).

9.2 DEFINITION OF ENERGY SPECTRA

A particle energy spectrum is a function giving the distribution of particles in terms of their energy. There are two kinds of energy spectra, differential and integral.

The *differential energy spectrum*, the most commonly studied distribution, is also known as an energy spectrum. It is a function $n(E)$ with the following meaning:

$n(E)\,dE$ = number of particles with energies between E and $E + dE$

or

$n(E)$ = number of particles with energy E per unit energy interval

The quantity $n(E)\,dE$ is represented by the cross-hatched area of Figure 9.1.

The *integral energy spectrum* is a function $N(E)$, where $N(E)$ is the number of particles with energy greater than or equal to E. The quantity $N(E)$ is represented by the hatched area of Figure 9.1. The integral energy spectrum $N(E)$ and the differential energy spectrum $n(E)$ are related by

$$N(E) = \int_E^\infty n(E)\,dE \tag{9.1}$$

The two examples that follow illustrate the relationship between a differential spectrum and an integral spectrum.

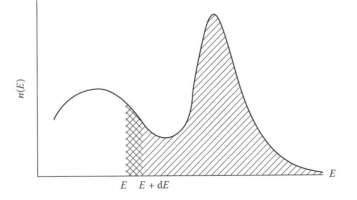

Figure 9.1 A differential energy spectrum. The quantity $n(E)\,dE$ is equal to the number of particles between E and $E + dE$ (cross-hatched area).

Example 9.1 Consider a monoenergetic source emitting particles with energy E_0. The differential energy spectrum $n(E)$ is shown in Figure 9.2. Since there are no particles with energy different from E_0, the value of $n(E)$ is equal to zero for any energy other than $E = E_0$.

The corresponding integral spectrum $N(E)$ is shown in Figure 9.3. It indicates that there are no particles with $E > E_0$. Furthermore, the value of $N(E)$ is constant for $E \leq E_0$, since all the particles have energy E_0 and only those particles exist. In other words,

$N(E_0)$ = number of particles with energy greater than or equal to E_0 $N(E_1)$

= number of particles with energies greater than or equal to E_1 (Figure 9.3)

Example 9.2 Consider the energy spectrum shown in Figure 9.4. According to this spectrum, there are 10 particles per MeV at 11, 12, and 13 MeV. The total number of particles is 30. The integral spectrum is shown in Figure 9.5. Its values at different energies are

$N(14) = 0$	no particles above $E = 14$ MeV
$N(13) = 10$	10 particles at $E = 13$ MeV and above
$N(12) = 20$	20 particles at $E = 12$ MeV and above
$N(11) = 30$	30 particles at $E = 11$ MeV and above
$N(10) = 30$	30 particles at $E = 10$ MeV and above
$N(0) = 30$	30 particles above $E = 0$

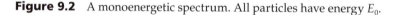

Figure 9.2 A monoenergetic spectrum. All particles have energy E_0.

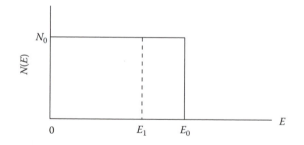

Figure 9.3 The integral spectrum of a monoenergetic source.

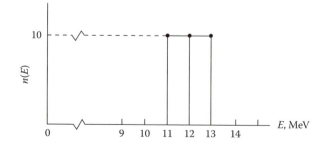

Figure 9.4 The energy spectrum considered in Example 9.2.

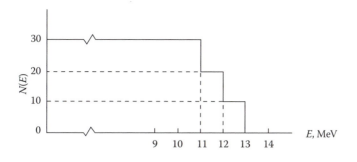

Figure 9.5 The integral spectrum corresponding to that of Figure 9.4.

The determination of energy spectra is based on the measurement of pulse-height spectra, as shown in the following sections. Therefore, the definitions of differential and integral spectra given in this section in terms of energy could be expressed equivalently in terms of pulse height. The relationship between particle energy and pulse height is discussed in Section 9.5.

9.3 MEASUREMENT OF AN INTEGRAL SPECTRUM WITH A SINGLE-CHANNEL ANALYZER

Measurement of an integral spectrum means to count all particles that have energy greater than or equal to a certain energy E or, equivalently, to record all particles that produce pulse height greater than or equal to a certain pulse height V. A device is needed that can sort out pulses according to height. Such a device is a *single-channel analyzer* (SCA) operating as a discriminator (integral mode). If the discriminator is set at V_0 volts, all pulses with height less than V_0 will be rejected, while all pulses with heights above V_0 will be recorded. Therefore, a single discriminator can measure an integral energy spectrum. The measurement proceeds as follows.

Consider the differential pulse spectrum shown in Figure 9.6 for which all pulses have exactly the same height V_0. To record this spectrum, one starts with the discriminator threshold set very high (higher than V_0) and then lowers the threshold by a certain amount ΔV (or ΔE) in successive steps. Table 9.1 shows the results of this measurement, where $N(V)$ is the number of pulses higher than or equal to V. A plot of these results is shown in Figure 9.7.

9.4 MEASUREMENT OF A DIFFERENTIAL SPECTRUM WITH A SINGLE-CHANNEL ANALYZER

Measurement of a differential energy spectrum amounts to the determination of the number of particles within a certain energy interval ΔE for several values of energy; or, equivalently, it amounts to the determination of the number of pulses within a certain interval ΔV, for several pulse heights. A SCA operating in the differential mode is the device that is used for such a measurement.

If the lower threshold of the SCA is set at V_1 (or E_1) and the window has a width ΔV (or ΔE), then only pulses with height between V_1 and $V_1 + \Delta V$ are recorded. All pulses outside this range are rejected. To measure the pulse spectrum of Figure 9.6, one starts by setting the lower threshold at V_1, where $V_1 > V_0$, with a certain window ΔV (e.g., $\Delta V = 0.1$ V) and then keeps lowering the lower threshold of the SCA. Table 9.2 shows the results of the measurement, where $n(V) \Delta V$ is the number of pulses with height between V and $V + \Delta V$. Figure 9.8 shows these results. It is assumed that the width is $\Delta V = V_i - V_{i+1}$, where V_i are the successive settings of the lower threshold of the SCA. It is important to note that one never measures the value of $n(V)$ but only the product $n(V) \Delta V$.

9.5 THE RELATIONSHIP BETWEEN PULSE-HEIGHT DISTRIBUTION AND ENERGY SPECTRUM

To determine the energy spectrum of particles emitted by a source, one measures, with the help of a detector and appropriate electronics, the pulse-height distribution produced by these particles. Fundamental requirements for the detector and the electronics are as follows:

1. The particle should deposit all its energy or a known constant fraction of it in the detector.

2. The voltage pulse produced by the detector should be proportional to the particles energy dissipated in it, or a known relationship should exist between energy dissipated and pulse height.

3. The electronic amplification should be the same for all pulse heights.

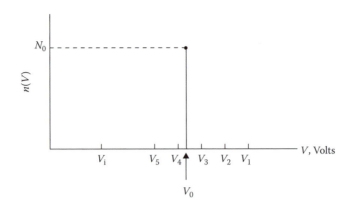

Figure 9.6 A differential pulse spectrum consisting of pulses with the same height V_0.

Table 9.1
Measurement of Integral Spectrum

Discriminator Threshold	$N(V)$
$V_1 > V_0$	0
V_2	0
V_3	0
$V_4 < V_0$	N_0
V_5	N_0
V_6	N_0
$V_i < V_0$	N_0

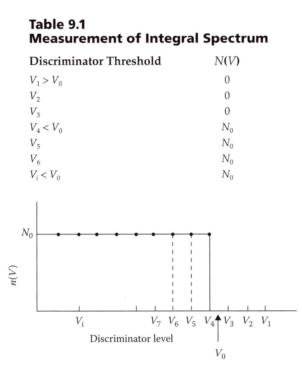

Figure 9.7 The integral spectrum corresponding to the pulse spectrum of Figure 9.6.

Since the relationship between pulse-height distribution and energy spectrum depends on these three requirements, it is important to discuss them in some detail.

Charged particles deposit all their energy in the detector, as long as their range is shorter than the size of the detector. Gammas do not necessarily deposit all their energy in the detector, regardless of detector size. Neutrons are detected indirectly through other particles produced by nuclear reactions. The energy deposited in the detector depends not only on the energy of the neutron but also on the energy and type of the reaction products.

The events that transform the particle energy into a voltage pulse are statistical in nature. As a result, even if all the particles deposit exactly the same energy in the detector, and the amplification is the same for all pulses, the output pulses will not be the same but they will have a certain distribution.

The state of commercial electronics is such that the amplification is essentially the same for all pulse heights (see also Section 10.11).

242

Table 9.2
Measurement of
Differential Spectrum

SCA Threshold	$n(V) \, \Delta V$
$V_1 > V_0$	0
V_2	0
V_3	0
$V_4 < V_0$	N_0
V_5	0
$V_i < V_5$	0

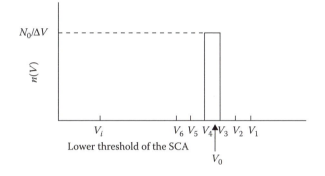

Figure 9.8 A differential energy spectrum measured with an SCA.

As a result of incomplete energy deposition and the statistical nature of the events that take place in the detector, the shape of the pulse-height distribution is different from that of the source energy spectrum. In other words, two spectra are involved in every measurement as follows:

1. The *source spectrum*: This is the energy spectrum of particles emitted by the source.

2. The *measured spectrum*: This is the measured pulse-height distribution.

Consider, for example, the measured pulse-height distribution shown in Figure 9.9b produced by a monoenergetic gamma source. This distribution (or spectrum) is obtained using a scintillation counter. The observer records the data shown in Figure 9.9b, which is not identical to that of the source, Figure 9.9a. The objective of the measurement is to obtain the spectrum of Figure 9.9a, but the observer actually measures the distribution shown in Figure 9.9b. The task of the observer is, therefore, to apply appropriate corrections to the measured spectrum to finally obtain the source spectrum.

9.6 ENERGY RESOLUTION OF A DETECTION SYSTEM

The quality of the performance of a detection system used for energy measurements is characterized by the width of the pulse-height distribution obtained with particles of the same energy (monoenergetic source). Even in the case where each particle deposits exactly the same energy in the detector, the pulse-height distribution will not be a single line (like that shown in Figure 9.9a); instead, it will have a certain finite width (Figure 9.10) due to the following:

1. Statistical fluctuations in the number of charge carriers produced in the detector

2. Electronic noise in detector itself, the preamplifier, and the amplifier

3. Incomplete collection of the charge produced in the detector.

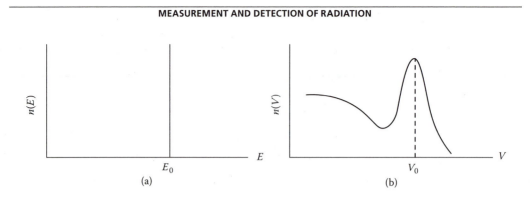

Figure 9.9 (a) The source energy spectrum of a monoenergetic gamma source, (b) the pulse-height distribution obtained with a NaI(Tl) scintillation counter.

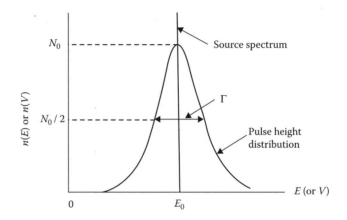

Figure 9.10 The energy resolution of the detector is given by the width Γ or the ratio Γ/E_0.

The *width*, measured at half of the maximum of the bell-shaped curve, is indicated by Γ or by FWHM, the initials of full width at half maximum. The ability of a detector to identify particles of different energies, called the *energy resolution*, is given either in terms of Γ or in terms of the ratio $R(E_0)$, where

$$R(E_0) = \frac{\Gamma}{E_0} \tag{9.2}$$

The width Γ is given in energy units, while the ratio $R(E_0)$ is given as a percentage.

The most important elements affecting the energy resolution of a radiation detection system are the three statistical factors mentioned above in relation to the width Γ. It is worth repeating that in energy measurements it is the energy resolution of the counting system (detector-preamplifier-amplifier) that is the important quantity and not the energy resolution of just the detector.

9.6.1 The Effect of Statistical Fluctuations: The Fano Factor

To discuss the effect of the statistical fluctuations on energy resolution, consider a monoenergetic source of charged particles being detected by a silicon semiconductor detector. (The discussion would apply to a gas-filled counter as well.) The average energy w needed to produce one electron–hole pair in silicon is 3.66 eV, although the energy gap (E_g) is 1.1 eV. This difference between w and E_g shows that part of the energy of the incident particles is dissipated into processes that do not generate charge carriers. Any process that consumes energy without producing electron-hole pairs is, of course, useless to the generation of the detector signal. If the energy deposited in the detector is E, the average number of charge carriers is E/w. If the process of the electron-hole generation were purely statistical, Poisson statistics would apply and the standard deviation of the number of pairs would be

$$\sigma = \sqrt{\frac{E}{w}} \tag{9.3}$$

Experience has shown that the fluctuations are smaller than what Eq. 9.3 gives. The observed statistical fluctuations are expressed in terms of the *Fano factor F*,[1] where

$$F = \frac{(\text{standard deviation of the number of pairs produced})^2}{\text{number of pairs produced}}$$

or, using Eq. 9.3,

$$\sigma = \sqrt{\frac{FE}{w}} \tag{9.4}$$

The two extreme values of F are 0 and 1.

$F = 0$ means that there are no statistical fluctuations in the number of pairs produced. That would be the case if all the energy was used for production of charge carriers. $F = 1$ means that the number of pairs produced is governed by Poisson statistics.

Fano factors have been calculated and also measured.[2–7] For semiconductor detectors, F values as low as 0.06 have been reported.[8] For gas-filled counters, reported F values lie between 0.2 and 0.5. Values of $F < 1$ mean that the generation of electron-hole pairs does not exactly follow Poisson statistics. Since Poisson statistics applies to outcomes that are independent, it seems that the ionization events in a counter are interdependent.

The width Γ of a Gaussian distribution, such as that shown in Figure 9.10, is related to the standard deviation σ by

$$\Gamma_f = 2\sqrt{2\ln 2}\ w\sigma \approx 2.355 w\sigma \tag{9.5}$$

Combining Eqs. 9.4 and 9.5,

$$\Gamma_f = 2\sqrt{2(\ln 2)wFE} \tag{9.6}$$

Equation 9.5 shows that the width Γ_f, which is due to the statistical fluctuations, is roughly proportional to the square root of the energy (the Fano factor is a weak function of energy).

To compare the contribution of the statistical fluctuations to the resolution of different types of detectors at a certain energy, one can use Eqs. 9.2 and 9.6 and write for detectors 1 and 2

$$\frac{R_1}{R_2} = \frac{\Gamma_1/E}{\Gamma_2/E} = \sqrt{\frac{w_1 F_1}{w_2 F_2}} \tag{9.7}$$

It can be seen from Eq. 9.7 that the resolution is better for the detector with the smaller average energy needed for the creation of a charge carrier pair (and smaller Fano factor). Thus, the energy resolution of a semiconductor detector ($w \sim 3$ eV, $F < 0.1$) should be expected to be much better than the resolution of a gas-filled counter ($w \approx 30$ eV, $F \approx 0.2$), and indeed it is (see Chapters 12 and 13). Some recent work in the determination of the Fano factor involving signal summation, Monte Carlo simulation, etc., can be found in References 9–13.

9.6.2 The Effect of Electronic Noise on Energy Resolution

The electronic noise consists of a small voltage variation around the zero line (Figure 9.11), with average voltage $\bar{v}_n \neq 0$. To see the effect of the noise on the energy resolution, consider pulses of constant height V. In the absence of noise, the FWHM of the distribution of these pulses is zero. If noise is present, the pulses will be superimposed on the noise with the results that the pulses are not of equal height any more (Figure 9.12), and that the pulses form a Gaussian distribution centered at V and having a width equal to $\Gamma_n = 2\sqrt{2\ln 2}\ \sigma_n$. The width Γ_n is due to the noise only and has nothing to do with statistical effects in the detector.

Figure 9.11 The electronic noise.

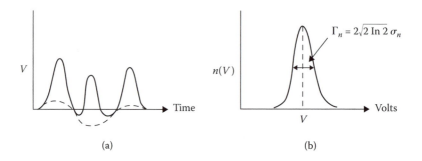

(a) (b)

Figure 9.12 (a) The pulses are superimposed on the noise, as a result of which (b) they show a distribution with a width that depends on the standard deviation of the noise.

The *signal-to-noise ratio* is frequently the quantity used to indicate the magnitude of the noise. It is defined by

$$\text{Signal-to-noise ratio} = \frac{\text{mean pulse height}}{\text{noise standard deviation}} = \frac{V}{\sigma_n}$$

Or, one can write

$$\frac{V}{\sigma_n} = 2\sqrt{2\ln 2}\,\frac{V}{\Gamma_n} = \frac{2\sqrt{2\ln 2}}{R} \tag{9.8}$$

where R is given by Eq. 9.2. This last equation may be rewritten as

$$R = \frac{2\sqrt{2\ln 2}}{V/\sigma_n} \tag{9.8a}$$

to show that the higher the signal-to-noise ratio is, the better the resolution becomes (other things being equal, of course).

9.6.3 The Effect of Incomplete Charge Collection

The effect of incomplete charge collection in gas-filled counters is small compared to the effect of the statistical fluctuations. In semiconductor detectors, incomplete charge collection is due to trapping of carriers. The amount of charge trapped is approximately proportional to the energy deposited in the detector, which in turn, is proportional to the energy of the incident particles.[14] For this reason, the resolution is affected by trapping effects more at high energy than at low energy. As discussed in Chapter 7, trapping effects depend on the material of which the detector is made and on radiation damage suffered by the semiconductor.

Usually, the effect of incomplete charge collection is included in the statistical fluctuations.

9.6.4 The Total Width Γ

The total width Γ (or the total energy resolution) is obtained by adding in quadrature the contributions from the statistical effects (Γ_f) and from the noise and incomplete charge collection (Γ_n). Thus,

$$\Gamma = \sqrt{\Gamma_f^2 + \Gamma_n^2} \tag{9.9}$$

For gas and scintillator counters, the main contribution comes from the statistical fluctuations. For semiconductor detectors at low energies, measurements have shown that $\Gamma_n \geq \Gamma_f$. At higher energies this is reversed, since Γ_n is essentially independent of energy while Γ_f increases with it (see Eq. 9.6).

9.7 DETERMINATION OF THE ENERGY RESOLUTION—THE RESPONSE FUNCTION

Depending on the type and energy of the incident particle and the type of the detector, a monoenergetic source produces a pulse-height distribution that may be a Gaussian (Figure 9.10) or a more complicated function (Figure 9.9). In either case, one concludes that although all the particles start at the source with the same energy, there is a probability that they may be recorded within a range of energies. That probability is given by the *response function* or *energy resolution function* $R(E, E')$ of the detection system, defined as

$R(E, E')\, dE$ = probability that a particle emitted by the source with energy E' will be recorded with energy between E and $E + dE$.

One measures, of course, a pulse-height distribution, but the energy calibration of the system provides a one-to-one correspondence between energy and pulse height. If one defines

$S(E)\, dE$ = source spectrum = number of particles emitted by the source with energy between E and $E + dE$

and

$M(E)\, dE$ = measured spectrum = number of particles recorded as having energy between E and $E + dE$

then the three functions $R(E, E')$, $S(E)$ and $M(E)$ are related by

$$M(E) = \int_0^\infty R(E, E') S(E')\, dE' \qquad (9.10)$$

Equation 9.10 is an integral equation with the source spectrum $S(E)$ being the unknown. The procedure by which $S(E)$ is obtained, after $R(E, E')$ and $M(E)$ have been determined, is called *unfolding* of the measured spectrum. Methods of unfolding are discussed in Chapters 11–14.

To determine the response function of a detection system at energy E, the energy spectrum of a monoenergetic source emitting particles with that energy is recorded. Since the resolution changes with energy, the measurement is repeated using several sources spanning the energy range of interest. The response function can also be calculated, as shown in Chapters 12–14. Figure 9.13 shows response functions for several commonly encountered cases.

9.8 THE IMPORTANCE OF GOOD ENERGY RESOLUTION

Regardless of the type of particle or photon interaction the importance of good energy resolution becomes obvious if the energy spectrum to be measured consists of several discrete energies. Consider as an example the source spectrum of Figure 9.14, consisting of two energies E_1 and E_2. Assume that this spectrum is measured with a system having energy resolution equal to Γ,* and examine the following cases:

Case I: $E_2 - E_1 > 2\Gamma$. The measured spectrum is shown in Figure 9.15 for this case. The system can *resolve* the two peaks—that is, the two peaks can be identified as two separate energies.

Case II: $E_2 - E_1 = 2\Gamma$. This case is shown in Figure 9.16. The peaks can still be resolved.

Case III: $E_2 - E_1 = \Gamma$. This case is shown in Figure 9.17. The solid line shows how the measured spectrum will look as the sum of the two peaks (dashed lines).

It is obvious that it is difficult to identify two distinct peaks if $E_2 - E_1 = \Gamma$, and the situation will be worse if $E_2 - E_1 < \Gamma$.

The three cases examined above intend to show how important good energy resolution is for the measurement of spectra with many energy peaks. If the response function of the detector is not known and the measured spectrum shows no well-identified peaks, the following criterion is used for the energy resolution required to identify the peaks of about equal magnitude. To be

---

* Γ may be different at E_1 and E_2. However, the difference is very small since E_1 and E_2 are close. For the present discussion, the same Γ will be used at E_1 and E_2.

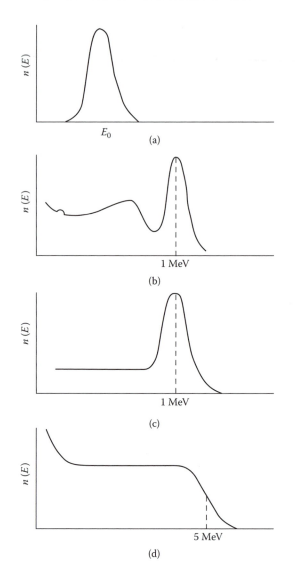

Figure 9.13 Four examples of response functions: (a) 5-MeV Alpha particles detected by a silicon surface barrier detector (Chapter 12), or 20-keV X-rays detected by a Si(Li) reactor (Chapter 11). (b) 1-MeV Gamma ray detected by a NaI(Tl) crystal (Chapter 11). (c) 1-MeV Electrons detected by a plastic scintillator (Chapter 12). (d) 5-MeV Neutrons detected by an NE 213 organic scintillator (Chapter 13).

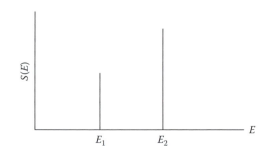

Figure 9.14 Source spectrum consisting of two distinct energies.

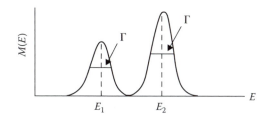

Figure 9.15 Measured spectrum for Case I: $2\Gamma < E_2 - E_1$.

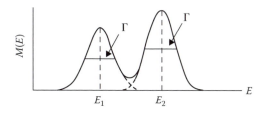

Figure 9.16 Measured spectrum for Case II: $2\Gamma = E_2 - E_1$.

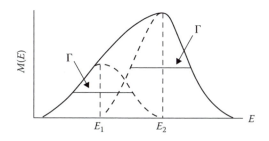

Figure 9.17 Measured spectrum for Case III: $\Gamma = E_2 - E_1$.

able to resolve two energy peaks at E_1 and E_2, the resolution of the system should be such that $\Gamma \leq |E_1 - E_2|$.

9.9 BRIEF DESCRIPTION OF A MULTICHANNEL ANALYZER

To measure an energy spectrum of a radioactive source means to record the pulse-height distribution produced by the particles emitted from the source, which is achieved with the use of an instrument called the multichannel analyzer (MCA). MCAs are used in either of two different modes: the *pulse-height analysis* (PHA) mode or the *multichannel scaling* (MCS) mode.

The MCS mode is used to count events as a function of time. The individual channels of the memory count all incoming pulses for a preset time width Δt. After time Δt, the counting operation is switched automatically to the next channel in the memory, thus producing in the end a time sequence of the radiation being detected. For example, if the radiation source is a short-lived isotope, the MCS mode will provide the exponential decay curve that can be used for the measurement of the half-life of this isotope. The MCS mode is also useful for Mossbauer experiments.

The PHA mode is the traditional function of an MCA and is used to sort out incoming pulses according to their height and store the number of pulses of a particular height in a corresponding address of the MCA memory called the *channel number*.

In the PHA mode, an MCA performs the function of a series of SCAs placed adjacent to one another. When only one SCA with width ΔE is used, the experimenter has to sweep the spectrum by moving the lower threshold of the SCA manually (see Section 9.4). On the other hand, if one had many SCAs, all counting simultaneously, the whole spectrum would be recorded simultaneously.

This is exactly what the MCA does, although its principle of operation is not based on a series of SCAs.

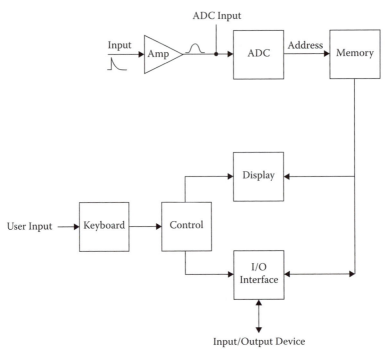

Figure 9.18 Block diagram showing the components of a typical MCA counting system.

Figure 9.18 shows a simplified block diagram of an MCA. In the PHA mode, the incoming pulse enters into a unit called the *analog-to-digital converter* (ADC). The ADC *digitizes* the pulse amplitude: it produces a number proportional to the height of the pulse, a number that determines the channel where the pulse will be stored. The size of the ADC, given in terms of channels, defines the absolute resolution of the system. Actually, the ADC determines the number of discrete parts into which the pulse height can be subdivided. Commercial ADCs have at the present time a size up to 16,384 channels, with the full scale adjustable in steps of 256, 512, 1024, etc., channels.

The number of discrete parts (channels) into which the input pulse range (0 to +10 V) is divided is called the *conversion gain*. The conversion gain is set by a stepwise control knob located on the front of the instrument. As an example, if the conversion gain is set at 2048 channels, it means that the maximum pulse height (10 V) is divided into that many parts. Therefore, the resolution of the MCA at this setting is 10 V/2048 = 4.88 mV/channel.

More details about the operation and characteristics of ADCs are given in Section 10.12.

The memory of the MCA is a data-storage unit arranged in a series of channels. Every channel is capable of storing up to $2^{20} - 1$ data (pulses), in most cases. Normally, the MCA provides for selection and use of the full memory, only half of it, or one-fourth of it. Transfer of data from one fraction of the memory to another is also possible.

In the PHA mode, the first channel of the region used is called *channel zero* and records, in almost all late model MCAs, the live time of the analysis, in seconds. If the full memory or the first half or first quarter of the memory is used, channel zero is the address 0000. If the second half of 4096 memory is used, channel zero is address 2048; if the second quarter is used, channel zero is address 1024; and so on.

How does one determine the size of the MCA memory needed for a specific experiment? The decision is made based on the requirements for the PHA mode. One equation frequently used is

$$\text{Number of channels} = h\frac{\text{energy range of interest (keV)}}{\Gamma(\text{keV})} \qquad (9.11)$$

where Γ is the FWHM of the detector used. The factor h is equal to the number of channels at or above the FWHM of the peak. Its value is between 3 and 5.

As an example, assume that the energy range of interest is 0–2.0 MeV and consider a NaI(Tl) and a Ge detector. The resolution of the NaI(Tl) detector is about 50 keV. Therefore, the minimum number of channels is ($h = 5$)

$$5\left(\frac{2000}{50}\right) \approx 200 \text{ channels}$$

The resolution of a Ge detector is about 2 keV. Now, the number of channels is

$$5\left(\frac{2000}{2}\right) \approx 5000 \text{ channels}$$

The user should remember that the ADC, not the memory, determines the absolute resolution of an MCA. An MCA with an ADC of 1000 channels and a memory of 2000 channels has an actual resolution of only 1000 channels.

In using an MCA to record a spectrum, there is "dead time" involved, which is, essentially, the time it takes to store the pulse in the appropriate channel. That time depends on and increases with the channel number. More details about the MCA dead time are given in Section 10.12 in connection with the discussion of ADCs, which also includes the newer zero dead-time correction modules. A more in depth description of such modules is given in Chapter 10.

Commercial MCAs have a meter that shows, during counting, the percentage of dead time. They also have timers that determine the counting period in *live time* or *clock time*. In clock time mode, the counting continues for as long as the clock is set up. In live time mode, an automatic correction for dead time is performed. In this case, the percent dead time indication can be used to determine the approximate amount of actual time the counting will take. For example if the clock is set to count for 5 min (in live mode) and the dead time indicator shows 25%, the approximate actual time of this measurement is going to be

$$\text{Actual time} = \frac{\text{Live time}}{1 - \left(\text{Dead time fraction}\right)} = \frac{300 \text{ s}}{1 - 0.25} = 400 \text{ s}$$

Modern MCAs can do much more than just store pulses in memory. They are computers that may, depending on the hardware and software available, be able to

1. Perform the energy calibration of the system

2. Determine the energy of a peak

3. Integrate the area over a desired range of channels

4. Identify an isotope, based on the energy peaks recorded, and so forth.

9.10 CALIBRATION OF A MULTICHANNEL ANALYZER

The calibration of an MCA follows these steps:

1. Determination of range of energies involved. Assume this is $0 \leq E \leq E_m$ (MeV).

2. Determination of preamplifier-amplifier setting. Using a source that emits particles of known energy, one observes the signal generated on the screen of the oscilloscope. It should be kept in mind that the maximum possible signal at the output of the amplifier is 10 V. In energy spectrum measurements, one should try to stay in the range 0–9 V.

Assume that the particle energy E_1 results in pulse height V_1. Is this amplification proper for obtaining a pulse height $V_m \leq 10$ V for energy E_m? To find this out, the observer should use the fact that pulse height and particle energy are proportional. Therefore,

$$\frac{V_m}{E_m} = \frac{V_1}{E_1} \rightarrow V_m = \frac{E_m}{E_1} V_1$$

If $V_m < 10$ V, then the amplification setting is proper. If $V_m > 10$ V, the amplification should be reduced. (If $V_m < 2$ V, amplification should be increased. It is good practice, but not necessary, to use the full range of allowed voltage pulses.) The maximum pulse V_m can be changed by changing the amplifier setting.

3. Determination of MCA settings. One first decides the part of the MCA memory to be used, that is, how many channels. Calibration of energy spectra is highly dependent on the range of energies considered, investigated or studied For instance in alpha spectrometry 1024 channels are usually adequate, while for prompt gamma activation analysis the full 16K channels are needed. Assume that the MCA has a 1024-channel memory and it has been decided to use 256 channels, one-fourth of the memory. Also assume that a spectrum of a known source with energy E_1 is recorded and that the peak is registered in channel C_1. Will the energy E_m be registered in $C_m < 256$, or will it be out of scale?

The channel number and energy are almost proportional,* that is, $E_i \sim C_t$. Therefore

$$\frac{C_m}{E_m} \approx \frac{C_1}{E_1} \rightarrow C_m \approx \frac{E_m}{E_1}C_1$$

If $C_m < 256$, the setting is proper and may be used. If $C_m > 256$, a new setting should be employed. This can be done in one of two ways or a combination of the two:

1. The fraction of the memory selected may be changed. One may use 512 channels of 1024, instead of 256.

2. The conversion gain may be changed. In the example discussed here, if a peak is recorded in channel 300 with conversion gain of 1024, that same peak will be recorded in channel 150 if the conversion gain is switched to 512.

There are analyzer models that do not allow change of conversion gain. For such an MCA, if C_m is greater than the total memory of the instrument, one should return to step 2 and decrease V_m by reducing the gain of the amplifier.

4. Determination of the energy-channel relationship. Calibration of the MCA means finding the expression that relates particle energy to the channel where a particular energy is stored. That equation is written in the form

$$E = a_1 + a_2 C + a_3 C^2 + \cdots \tag{9.12}$$

where C is the channel number and $a_1, a_2, a_3, \ldots$ are constants.

The constants $a_1, a_2, a_3, \ldots$ are determined by recording spectra of sources with known energy. In principle, one needs as many energies as there are constants. In practice, a large number of sources is recorded with energies covering the whole range of interest, and the constants are then determined by a least-squares fitting process (see Chapter 11).

Most detection systems are essentially linear, which means that Eq. 9.12 takes the form

$$E = a_1 + a_2 C \tag{9.13}$$

Example 9.3 Obtain the calibration constants for an MCA based on the spectrum shown in Figure 9.19. The peaks correspond to the following three energies:

$$E_1 = 0.662\,\text{MeV}, \quad C_1 = 160$$
$$E_2 = 1.173\,\text{MeV}, \quad C_2 = 282.5$$
$$E_3 = 1.332\,\text{MeV}, \quad C_3 = 320$$

Answer Plotting energy versus channel on linear graph paper, one obtains the line shown in Figure 9.20, which indicates that the linear equation, Eq. 9.13, applies, and one can determine the constants a_1 and a_2 from

---

* The correct equation is $E = a + bC$, but a is small and for this argument it may be neglected; proper evalua-tion of a and b is given in step 4 of the calibration procedure.

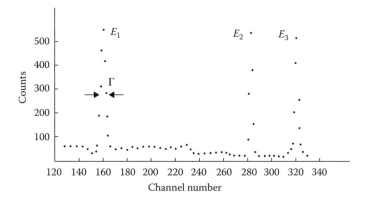

Figure 9.19 A gamma spectrum used for calibration of an MCA.

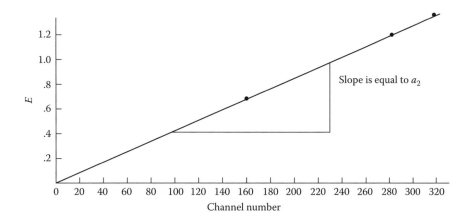

Figure 9.20 Plot of energy versus channel number. In this case, the relationship is linear.

the slope and the zero intercept of the straight line. From Figure 9.20, the value of a_2 is

$$a_2 = \frac{950 - 400}{230 - 97.5} = 4.15 \, \text{keV/channel}$$

The constant a_1 is equal to the zero-intercept of the line.* In the present case, it is almost zero. Based on these results, the calibration equation of this MCA is $E = 4.15C$.

5. Calculation of the energy resolution. By definition, the energy resolution is $R = \Gamma/E$, where Γ is the FWHM of the peak of energy E. Therefore, using Eq. 9.13,

$$R = \frac{\Gamma}{E} = \frac{(a_1 + a_2 C_R) - (a_1 - a_2 C_L)}{E} = \frac{a_2 (C_R - C_L)}{E} \tag{9.14}$$

where C_R and C_L are the channel numbers on either side of the peak at half of its maximum. If a_1 is zero, the resolution is given by

$$R = \frac{a_2 (C_R - C_L)}{a_2 C_{\text{peak}}} = \frac{C_R - C_L}{C_{\text{peak}}} \tag{9.15}$$

---

* Most commercial MCAs have a hand-screw adjustment that makes a_1 equal to zero.

For peak E_1 (Figure 9.19),

$$C_L = 158, \quad C_{peak} = 160, \quad C_R = 162$$

Therefore

$$R = \frac{162 - 158}{160} = 2.5\%$$

Or $\Gamma = a_2 (C_R - C_L) = 4.15(4) = 16.6 \text{ keV}.$

Typically, all MCA's in conjunction with ADCs in the market now have multi-energy calibration software with automated resolution measurements. A more in depth description of the digital signal processors is given in Chapter 10.

PROBLEMS

9.1 Sketch the integral spectrum for the differential spectrum shown in the figure below.

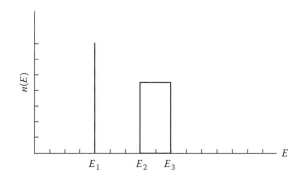

9.2 Sketch the differential energy spectrum for the integral spectrum shown in the figure below.

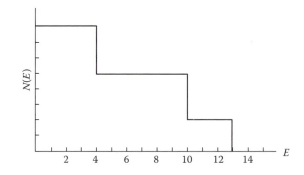

9.3 Sketch the integral spectrum for the differential spectrum shown in the figure below.

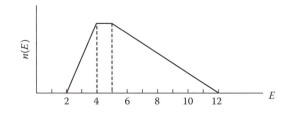

9.4 If the energy resolution of a NaI(Tl) scintillator system is 11% at 600 keV, what is the width Γ of a peak at that energy?

9.5 What is the maximum energy resolution necessary to resolve two peaks at 720 and 755 keV?

9.6 Prove that if a detection system is known to be linear, the calibration constants are given by

$$a_1 = \frac{E_2 C_1 - E_1 C_2}{C_1 - C_2}, \quad a_2 = \frac{E_1 - E_2}{C_1 - C_2}$$

where E_1 and E_2 are two energies recorded in channels C_1 and C_2, respectively.

9.7 Shown in the following figure is the spectrum of ^{22}Na, with its decay scheme. Determine the calibration constants of the MCA that recorded this spectrum, based on the two peaks of the ^{22}Na spectrum.

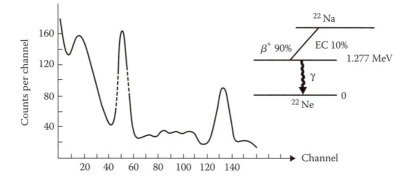

9.8 In Problem 9.7, the channel number cannot be read exactly. What is the uncertainty of the calibration constants a_1 and a_2 if the uncertainty in reading the channel is one channel for either peak?

9.9 Assume that the energy resolution of a scintillation counter is 9% and that of a semiconductor detector is 1% at energies around 900 keV. If a source emits gammas at 0.870 and 0.980 MeV, can these peaks be resolved with a scintillator or a semiconductor detector?

9.10 Consider the two peaks shown in the accompanying figure. How does the peak at E_2 affect the width of the peak at E_1 and vice versa? What is the width Γ for either peak?

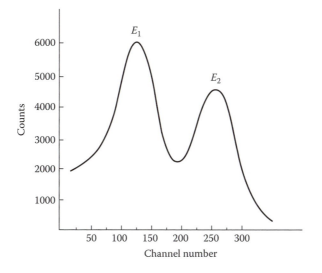

REFERENCES

1. Fano, U., *Phys. Rev.* **72**:26 (1947).

2. van Roosbroeck, W., *Phys. Rev.* **139A**:1702 (1965).

3. Goulding, F. S., *Nucl. Instrum. Meth.* **43**:1 (1966).

4. Mann, H. M., Bilger, H. R., and Sherman, I. S., *IEEE* **NS-13**(3):352 (1966).

5. Deshpande, R. Y., *Nucl. Instrum. Meth.* **57**:125 (1967).

6. Hashiba, A., Masuda, K., and Doke, T., *Nucl. Instrum. Meth.* **227**:305 (1984).

7. Kase, M., Akioka, T., Mamyoda, H., Kikuchi, J., and Doke, T., *Nucl. Instrum. Meth.* **227**:311 (1984).

8. Pehl, R. H., and Goulding, F. S., *Nucl. Instrum. Meth.* **81**:329 (1970).

9. Doke, T., *Nucl. Instrum. Meth. in Phys. Res. Section B* **234**:203 (2005).

10. Gao, F., Campbell, L. W., Devanathan, R., Xie, Y. L., Zhang, Y., Peurrung, A. J., and Weber, W. J., *Nucl. Instrum. Meth. Phys. Res. Section B* **255**:286 (2007).

11. Owens, A., Fraser, G. W., and McCarthy, K. J., *Nucl. Instrum. Meth. Phys. Res. Section A* **491**:437 (2002).

12. Mazziotta, M. N., *Nucl. Instrum. Meth. Phys. Res. Section A* **584**:436 (2008).

13. Jordan, D. V., Renholds, A. S., Jaffe, J. E., Anderson, K. K., Corrales, L. R., and Peurrung, A. J., *Nucl. Instrum. Meth. Phys. Res. Section A* **585**:146 (2008).

14. Ewan, G. T., *Nucl. Instrum. Meth.* **162**:75 (1979).

10

Electronics

10.1 INTRODUCTION

This chapter presents a brief and general description of electronic units used in radiation measurements. The subject is approached from the viewpoint of "input-output"—that is, the input and output signals of every component unit or instrument are presented with a minimum of discussion on circuitry. The objective is to make the reader aware of the capabilities and limitations of the different types of units and, at the same time, create the capacity to choose the right component for a specific counting system.

As in other 21st century technologies the advancement of smaller and faster electronic parts has been integrated into nuclear instrumentation. The increased speed of information processing of personal computers coupled with modern electronics has allowed for the miniaturization of many systems. Details about construction and operation of electronic components and systems are given in books specializing on that subject. A few such texts are listed in the bibliography at the end of the chapter. Also, the vendors of nuclear instruments provide manuals for their products with useful information about their operation.

10.2 RESISTANCE, CAPACITANCE, INDUCTANCE, AND IMPEDANCE

To understand what factors affect the formation, transmission, amplification, and detection of a detector signal, it is important to comprehend the function of resistance, capacitance, inductance, and impedance, which are the basic constituents of any electronic circuit. For this reason, a brief review of these concepts is offered.

The *resistance* R is a measure of how difficult (or easy) it is for an electric current to flow through a conductor. The resistance is defined by Ohm's law as the ratio of a voltage to current flowing through a conductor (Figure 10.1a). The resistance is measured in ohms (Ω). If a potential difference of 1 V generates a current of 1 A, the resistance is 1 Ω; that is,

$$R = \frac{V}{i} \tag{10.1}$$

Capacitance C is the ability to store electrical charge. A capacitor usually consists of two conductors separated by an insulator or a dielectric (Figure 10.1b). Every conductor, for example, a simple metal wire, has a certain capacitance. The capacitance is measured in farads (F). If a charge of 1 coulomb produces a potential difference of 1 V between the two conductors forming the capacitor, then its capacitance is 1 F; that is,

$$C = \frac{q}{V} \tag{10.2}$$

If the voltage across the capacitor is constant, no current flows through it; that is, a capacitor acts as an open circuit to dc voltage. If, however, the voltage changes, a current flows through the capacitor equal to

$$i = \frac{dq}{dt} = C\frac{dV}{dt} \tag{10.3}$$

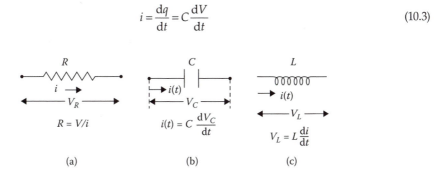

(a) (b) (c)

Figure 10.1 (a) A resistor, (b) a capacitor, (c) an inductor (coil).

Inductance refers to the property of conductors to try to resist a change in a magnetic field. If the current flowing through a conductor changes with time (in which case the magnetic field produced by the current also changes), a potential difference is induced that opposes the change. The induced potential difference V_L is given by (Figure 10.1c)

$$V_L = L\frac{di}{dt} \qquad (10.4)$$

where L is called the inductance of the conductor and is measured in henrys (H). If a current change of 1 A/s induces a potential difference of 1 V, the inductance is 1 H. An inductor is usually indicated as a coil (Figure 10.1c), but any conductor, for example, a metal wire, has a certain inductance. No pure inductor exists because there is always some ohmic resistance and some capacitance in the wires making the coils.

Capacitance and inductance are important for time-varying signals. To be able to introduce and discuss the pertinent concepts, consider a sinusoidal voltage signal with maximum voltage V_m and frequency ω applied to an RC circuit, as shown in Figure 10.2:

$$V(t) = V_m \sin\omega t \qquad (10.5)$$

Kirchhoff's second law applied to this circuit gives

$$V_m \sin\omega t - \frac{q}{C} = Ri \qquad (10.6)$$

Differentiating Eq. 10.6 with respect to time,

$$V_m \,\omega\cos\omega t - \frac{i}{C} = R\frac{di}{dt} \qquad (10.7)$$

The current flowing through the circuit is sinusoidal with frequency ω but with a phase difference relative to the input voltage. Let us call this phase difference φ. Then, we can write for the current

$$i(t) = i_m \sin(\omega t + \varphi) \qquad (10.8)$$

To evaluate the phase difference φ, substitute Eq. 10.8 into Eq. 10.7 and compute the resulting expression at some convenient value of the time t (in this case, $t = \pi/2\omega$; see Prob. 10.1). The result is

$$\tan\varphi = \frac{1}{RC\omega} \qquad (10.9)$$

Note that as $R \to 0$, $\tan\varphi \to \infty$ and $\varphi \to \pi/2$. The voltage across the capacitor is given by

$$V_c(t) = \frac{q}{C} = \frac{1}{C}\int i_m \sin(\omega t + \varphi)dt = -\frac{i_m}{C\omega}\cos(\omega t + \varphi)$$

$$= \frac{i_m}{C\omega}\sin\left(\omega t + \varphi - \frac{\pi}{2}\right)$$

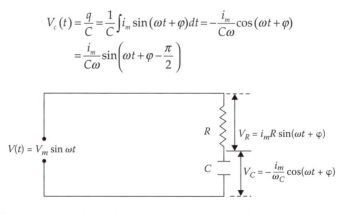

Figure 10.2 The interaction of an RC circuit with a sinusoidal input.

At every time t, the instantaneous potentials $V(t)$, $V_R(t)$, and $V_C(t)$ satisfy the equation

$$V(t) = V_R(t) + V_C(t) \tag{10.10}$$

The peak potentials, however, are not additive linearly because their maxima do not occur at the same time. To find the correct relationship, apply Eq. 10.7 at time $t = -(\varphi/\omega)$ with $i(t)$ from Eq. 10.8. The result is

$$i_m = \frac{V_m}{R}\cos\varphi = \frac{V_m}{R}\frac{1}{\sqrt{1+\tan^2\varphi}} = \frac{V_m}{\sqrt{R^2+(1/\omega^2C^2)}} \tag{10.11}$$

Equation 10.11 is the analog of Ohm's law for the RC circuit. The "resistance" of the circuit is called the impedance Z and is given by

$$Z = \sqrt{R^2 + \left(\frac{1}{\omega^2C^2}\right)} \tag{10.12}$$

Pictorially, the relationships expressed by Eqs. 10.11 and 10.12 are shown in Figure 10.3. The quantity $R_c = 1/\omega C$ is called the capacitive resistance. The impedance Z, as well as R and R_c, are measured in ohms.

Consider now an LR circuit as shown in Figure 10.4. If the voltage given by Eq. 10.5 is applied at the input, Kirchhoff's second law gives, in this case,

$$V_m \sin\omega t - L\frac{di}{dt} = iR \tag{10.13}$$

As in the RC circuit, the current will have the frequency ω and a phase difference φ, relative to the voltage. The current may be written as

$$i(i) = i_m \sin(\omega t - \varphi) \tag{10.14}$$

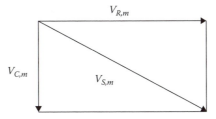

Figure 10.3 The addition of the peak potentials in an RC circuit.

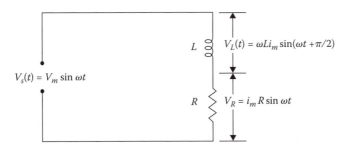

Figure 10.4 The interaction of an LR circuit with a sinusoidal input.

Substituting the value of the current from Eq. 10.14 into Eq. 10.13 and evaluating the resulting expression at $t = 0$ (see Prob. 10.2), one obtains

$$\tan\varphi = \frac{L\omega}{R} \tag{10.15}$$

which is the phase difference in the LR circuit (equivalent to Eq. 10.9). To find the relationship between the peak values, Eq. 10.13 is evaluated at time $t = \varphi/\omega$. The result is

$$i_m = \frac{V_m}{L\omega}\sin\varphi = \frac{V_m}{L\omega}\frac{\tan\varphi}{\sqrt{1+\tan^2\varphi}} = \frac{V_m}{\sqrt{R^2 + L^2\omega^2}} \tag{10.16}$$

Thus, the impedance of an LC circuit is

$$Z = \sqrt{R^2 + L^2\omega^2} \tag{10.17}$$

The quantity $R_L = L\omega$ is called the inductive reactance. If a circuit contains all three elements, R, L, C, it can be shown (see Prob. 10.3) that the phase difference and impedance are given by

$$\tan\varphi = \frac{\omega L - (1/\omega C)}{R} \tag{10.18}$$

$$Z = \sqrt{R^2 + \left[\omega L - \left(\frac{1}{\omega C}\right)\right]^2} \tag{10.19}$$

Every electronic component has a characteristic impedance. When a signal is transmitted from a unit with a high-impedance output to a low-impedance input, there is going to be a loss in the signal unless an impedance-matching device is used to couple the two units. Manufacturers of preamplifiers and amplifiers quote the impedance of the input and output for their products. Coaxial cables have an impedance between 90 and 100 Ω.

10.3 A DIFFERENTIATING CIRCUIT

A *differentiating circuit* consists of a capacitor and a resistor (Figure 10.5). If a time-dependent voltage $V_i(t)$ is applied at the input, Eq. 10.10 relating the instantaneous values of the three voltages involved (Figure 10.5) becomes

$$\frac{q(t)}{C} + Ri = \frac{q(t)}{C} + R\frac{dq(t)}{dt} = V_i(t) \tag{10.20}$$

where $q(t)$ is the charge of the capacitor at time t. If the input signal is a step function (Figure 10.6), the output voltage is given (after solving Eq. 10.20) by

$$V_0(t) = R\frac{dq(t)}{dt} = V_{i,m}e^{-t/RC} \tag{10.21}$$

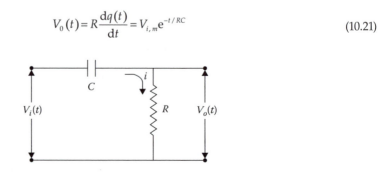

Figure 10.5   A CR shaping circuit (differentiator).

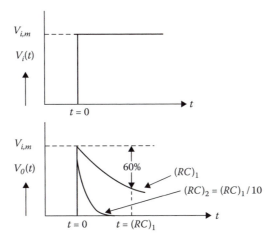

Figure 10.6 The output signal of a CR shaping circuit for a step input.

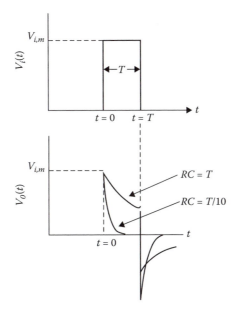

Figure 10.7 The response of a CR circuit to a rectangular pulse.

Figure 10.7 shows the output voltage if the input signal is a rectangular pulse of height $V_{i,m}$ and duration T. Notice that if $RC \ll T$, the output signal represents the derivative of the input. Indeed, from Eq. 10.20, if, $RC \ll T$, then

$$\frac{V_i(t)}{R} = \frac{q}{RC} + \frac{dq}{dt} \approx \frac{q}{RC}$$

or

$$\frac{1}{R}\frac{dV_i}{dt} = \frac{1}{RC}\frac{dq}{dt} = \frac{1}{RC}i$$

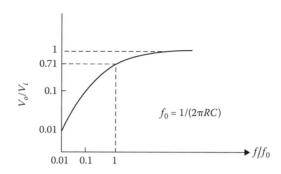

Figure 10.8 The output of a differentiator (high-pass) filter as a function of input frequency.

Thus,

$$V_0(t) = iR = R\frac{dq}{dt} = RC\frac{dV_i}{dt}$$

(10.22)

and for this reason, this circuit is called a *differentiator*. If $V_i(t)$ is the pulse from a detector, the effect of differentiation is to force the pulse to decay faster.

As shown in Section 10.2, for a sinusoidal signal the peak value of the potential across the resistor of an RC circuit is related to the peak of the input signal by

$$\frac{V_{0,m}}{V_{i,m}} = \frac{R}{\sqrt{R^2 + R_C^2}} = \frac{R}{\sqrt{R^2 + 1/\omega^2 C^2}}$$

(10.23)

where $\omega = 2\pi f$ and f is the frequency of input signal. According to Eq. 10.23, as the frequency decreases, the fraction of the signal appearing at the output of the differentiator also decreases, approaching zero for very low frequencies. For this reason, this circuit is called a *high-pass filter*. The output of the filter as a function of frequency is shown in Figure 10.8. If the signal is not purely sinusoidal, it may be decomposed into a series of sine components with frequencies that are multiples of a fundamental one (this is called Fourier analysis). Going through the high-pass filter, the lower frequencies will be attenuated more than the higher ones.

10.4 AN INTEGRATING CIRCUIT

An *integrating circuit* also consists of a resistor and a capacitor, but now the output signal is taken across the capacitor (Figure 10.9). Equation 10.20 applies in such a case too, and the output signal as a result of a step input is given by

$$V_0(t) = \frac{q(t)}{C} = V_{i,m}(1 - e^{-t/RC})$$

(10.24)

Input and output signals are shown in Figure 10.10. Figure 10.11 shows the output voltage if the input signal is a rectangular pulse of height $V_{i,m}$ and duration T. If $RC \gg T$, the output signal looks like the integral of the input. Indeed, from Eq. 10.20,

$$\frac{V_i(t)}{R} = \frac{q(t)}{CR} + \frac{dq(t)}{dt}$$

which gives, if $RC \gg T$,

$$\frac{V_i(t)}{R} \approx \frac{dq}{dt}$$

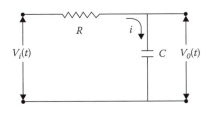

Figure 10.9 An *RC* shaping circuit (integrator).

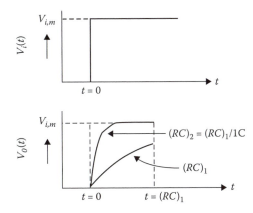

Figure 10.10 The output signal of an *RC* shaping circuit for a step input (top).

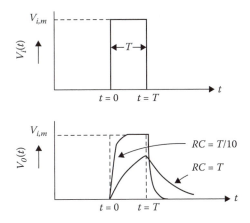

Figure 10.11 The response of an *RC* circuit to a rectangular pulse.

Then, from Eq. 10.24,

$$V_0(t) = \frac{q(t)}{C} = \frac{1}{RC}\int V_i(t)\,dt \tag{10.25}$$

and for this reason, this circuit is called an *integrator*.

For a sinusoidal input signal, the peak value of the voltage across the capacitor is related to the peak value of the input signal by

$$\frac{V_{0,m}}{V_{i,m}} = \frac{R_C}{\sqrt{R^2 + R_C^2}} = \frac{1/\omega C}{\sqrt{R^2 + 1/\omega^2 C^2}} \tag{10.26}$$

As the frequency increases, the ratio given by Eq. 10.26 decreases, and for this reason this circuit is called a *low-pass filter*. Going through the filter, lower frequencies fare better than the higher ones. The output of the filter as a function of frequency is shown in Figure 10.12.

10.5 DELAY LINES

Any signal transmitted through a coaxial cable is delayed by a time $T = \sqrt{LC}$ seconds per unit length, where L is the inductance per unit length and C is the capacitance per unit length. For ordinary coaxial cables, the delay is about 5 ns/m. For larger delays, the central conductor of the cable is spiraled to increase the inductance per unit length.

Commercial delay lines are a little more complicated than a simple cable. They are used not only to delay a signal, but also to produce a rectangular pulse for subsequent pulse shaping or for triggering another electronic unit (e.g., a scaler). The formation of the rectangular pulse is achieved by reflecting the delayed signal at the end of the delay line, bringing it back to the input and adding it to the original signal (Figure 10.13). A double delay line produces the double rectangular pulse shown in Figure 10.14.

Figure 10.12   The output of an integrator (low-pass) filter as a function of input frequency.

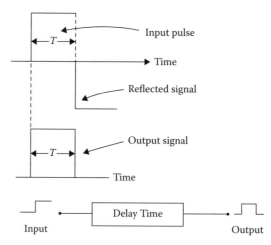

Figure 10.13   The use of a delay line to form a rectangular pulse.

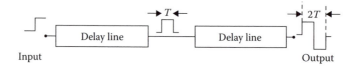

Figure 10.14   The effect of a double delay line.

264

10.6 PULSE SHAPING

The pulse produced at the output of a radiation detector has to be modified or *shaped* for better performance of the counting system. There are three reasons that necessitate pulse shaping:

1. *To prevent overlap.* Each pulse should last for as short a period of time as possible, and then its effect should be abolished so that the system may be ready for the next pulse. Without pulse shaping, the detector signal lasts so long that pulses overlap. If only the number of particles is counted, pulse overlap leads to loss of counts (dead time loss). In spectroscopy measurements, pulse overlap worsens the resolution.

2. *To improve the signal-to-noise ratio.* Noise created in the detector and the early amplification stages accompanies the detector signal. Appropriate pulse shaping can enhance the signal while at the same time reduce the noise. Thus, the signal-to-noise ratio will improve, which in turn, leads to better energy resolution.

3. *For special pulse manipulation.* The detector pulse may, in certain applications, need special pulse shaping to satisfy the needs of certain units of the counting system. As an example, the signal at the output of the amplifier needs to be stretched before it is recorded in the memory of a multichannel analyzer (see Section 10.12).

The pulse-shaping methods used today are based on combinations of RC circuits and delay lines. For example, the use of a CR–RC circuit combination produces the pulse shown in Figure 10.15. The exact shape and size of the output pulse depends on the relative magnitudes of the time constants C_1R_1 and C_2R_2. The use of the CR–RC circuit combination provides, in addition to pulse shaping, a better signal-to-noise ratio by acting as high-pass and low-pass filter for undesired frequencies.

If one adds more RC integrating circuits, the pulse will approach a Gaussian shape (Figure 10.16).

If one applies a CR–RC–CR combination, the result is a doubly differentiated pulse as shown in Figure 10.17. Commercial amplifiers usually provide either singly or doubly differentiated pulses.

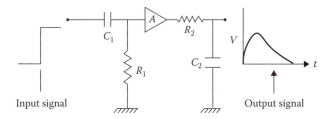

Figure 10.15   An example of CR–RC shaping. The triangle indicates the amplification unit (A) that isolates the two shaping circuits.

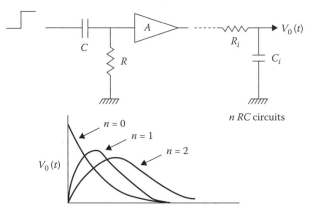

Figure 10.16   The output pulse after using many shaping circuits. The triangle indicates the amplification unit that isolates any two consecutive shaping circuits.

265

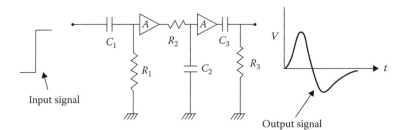

Figure 10.17 A doubly differentiated pulse.

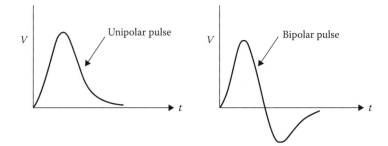

Figure 10.18 A unipolar and bipolar pulse as a result of the application of many CR and RC circuits.

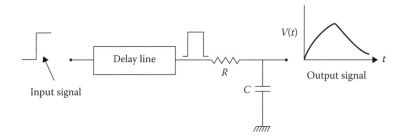

Figure 10.19 Pulse shaping using a single delay line and an RC circuit.

In all cases, the final pulse is the result of repeated application of CR–CR circuits. Figure 10.18 shows such pulses produced by the application of many RC and CR circuits, called unipolar and bipolar pulses, respectively. Pulse shaping using a delay line and an RC circuit is shown in Figure 10.19.

10.7 TIMING

The term *timing* refers to the determination of the time of arrival of a pulse. Timing experiments are used in measurement of the time development of an event (e.g., measurement of the decay of a radioactive species), measurement of true coincident events out of a large group of events, and discrimination of different types of particles based on the different time characteristics of their pulse (pulse-shape discrimination).

Timing methods are characterized as "slow" or "fast" depending on the way the signal is derived. Slow timing signals are generated by an integral discriminator or a timing single-channel analyzer. In either case, timing is obtained by using a shaped signal at the output of an amplifier. Fast timing signals are based on the unshaped pulse at the output of the detector or on a signal shaped specifically for timing.

Many timing methods have been developed over the years. All the methods pick the time based on a certain point in the "time development" of the pulse, but they differ in the way that point is selected. Three methods are discussed here.[1–5]

10.7.1 The Leading-Edge Timing Method

The leading-edge timing method determines the time of arrival of a pulse with the help of a discriminator, as shown in Figure 10.20. A discriminator threshold is set and the time of arrival of the pulse is determined from the point where the pulse crosses the discriminator threshold.*

The leading-edge timing method is simple, but it introduces uncertainties because of "jitter" and "walk" (Figure 10.21). *Jitter* is another name for electronic noise. The timing uncertainty due to jitter depends on the amplitude of the noise and the slope of the signal close to the discriminator threshold. *Walk* originates when differences in the rate of pulse-rise time cause pulses starting at the same point in time to cross the discriminator level at different positions. Walk can be reduced by setting the discriminator level as low as possible or by restricting the amplitude range of the acceptable pulses. Both of these corrective measures, however, introduce new difficulties. Setting the discriminator level too close to the noise level may allow part of the random noise to be counted. Limiting the range of acceptable pulses reduces the counting rate.

10.7.2 The Zero-Crossing Timing Method

The zero-crossing method reduces the errors due to jitter and walk by picking the time from the zero crossing of a bipolar pulse (Figure 10.22). Ideally, all the pulses cross the zero at the same point, and the system is walk free. In practice, there is some walk because the position of zero crossing depends on pulse rise time.† The dependence on pulse rise time is particularly important for Ge detectors because the pulses produced by Ge detectors exhibit considerable variations in their time characteristics. To reduce the uncertainties still present with the zero-crossing method, the constant-fraction method has been developed specifically for Ge detectors.

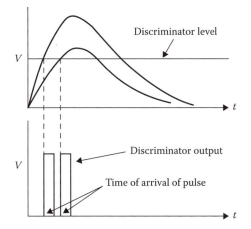

Figure 10.20 The time of arrival of the pulse is determined from the instant at which the pulse crosses the discriminator threshold.

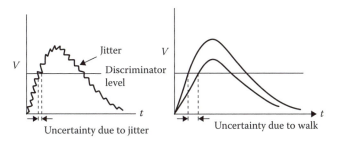

Figure 10.21 Timing uncertainty due to jitter and walk.

---

* Rectangular pulses such as the ones shown in Figure 10.20 are called logical pulses.
† Pulse rise time is taken as the time it takes the pulse to increase from 10% to 90% of its value.

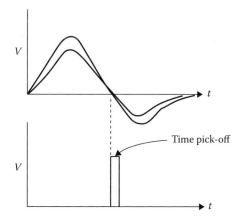

Figure 10.22 Timing by the zero-crossing method.

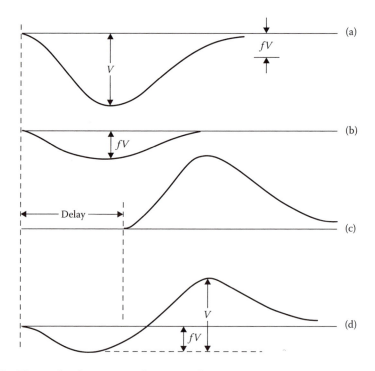

Figure 10.23 Timing by the constant-fraction technique.

10.7.3 The Constant-Fraction Timing Method

The principle of constant-fraction timing is shown in Figure 10.23. First, the original pulse (Figure 10.23a) is attenuated by a factor f equal to the fraction of the pulse height on which the timing will be based (Figure 10.23b). The original pulse is inverted and delayed (Figure 10.23c) for a time longer than its rise time. Finally, the signals in Figures 10.23c and 10.23b are added to give the signal in Figure 10.23d. The time pick-off, taken as the zero-crossing point, is thus defined by the preselected fraction of the pulse height and is independent of the pulse amplitude. It can be shown that pulses with the same rise time always give the same zero-crossing time.

10.7.4 Applications of Novel Timing Methods

With the increased technological advances of electronic components there appears to be an endless quantity of novel applications of timing methods in nuclear instrumentation. While it is not possible to list all the specific ever changing advancements in timing methods, here are some representative applications. A high-throughput, multichannel photon-counting detector with picosecond timing,[6] timing resolution of the scintillation detectors,[7] time resolution improvement of CdTe detectors using digital signal processing,[8] a simple method for rise-time discrimination of slow pulses from charge-sensitive preamplifiers,[9] single photon timing resolution and detection efficiency of silicon photo-multipliers,[10] photon counting detector with picoseconds timing resolution for X to visible range on the basis of GaP,[11] time resolution studies using digital constant fraction discrimination,[12] and double-layer silicon PIN photodiode X-ray detector.

10.8 COINCIDENCE–ANTICOINCIDENCE MEASUREMENTS

There are times in radiation measurements when it is desirable or necessary to discard the pulses due to certain types of radiation and accept only the pulses from a single type of particle or from a particle or particles coming from a specific direction. Here are two examples of such measurements.

1. *Detection of pair-production events.* When pair production occurs, two 0.511-MeV gammas are emitted back-to-back. To insure that only annihilation photon are counted, two detectors are placed 180° apart, and only events that register simultaneously (coincident events) in both detectors are recorded.

2. *Detection of internal conversion electrons.* Radioisotopes emitting internal conversion (IC) electrons also emit gammas and X-rays. The use of a single detector to count electrons will record not only IC electrons but also Compton electrons produced in the detector by the gammas. To eliminate the Compton electrons, one can utilize the X-rays that are emitted simultaneously with the IC electrons. Thus, a second detector is added for X-rays and the counting system is required to record only events that are coincident in these two detectors. This technique excludes the detection of Compton electrons.

Elimination of undesirable events is achieved by using a coincidence (or anticoincidence) unit. Consider the counting system shown in Figure 10.24. The source emits particles detected by detectors 1 and 2. After amplification, the detector signals are fed into a timing circuit, which in turn generates a pulse signifying the time of occurrence in the detector of the corresponding event (1) or (2). The timing signals are fed into a coincidence unit, so constructed that it produces an output signal only when the two timing pulses are coincident. If the objective is to count only the number of coincident events, the output of the coincidence unit is fed into a scaler. If, on the other hand, the objective is to measure the energy spectrum of particles counted by detector 1 in coincidence with particles counted by detector 2, the output signal of the coincidence unit is used to "gate" a multichannel analyzer (MCA) that accepts the energy pulses from detector 1. The gating signal permits the MCA to store only those pulses from detector 1 that are coincident with events in detector 2.

In theory, a true coincidence is the result of the arrival of two pulses at exactly the same time. In practice, this "exact coincidence" seldom occurs, and for this reason a coincidence unit is designed to register as a coincident event those pulses arriving within a finite but short time interval τ. The interval τ, called the *resolving time* or the *width* of the coincidence, is set by the observer. Typical

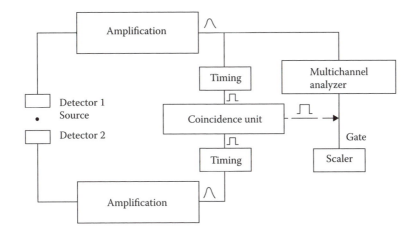

Figure 10.24 A simple coincidence measurement setup.

values of τ are 1–5 μs for "slow" coincidence and 1–10 ns for "fast" coincidence measurements. By introducing the width τ, the practical definition of coincidence is

Two or more events are coincident if they occur within the time period τ.

According to this definition, events (1) and (2) or (2) and (3) in Figure 10.25 are coincident, but events (1) and (3) are not.

As stated earlier, the coincidence unit is an electronic device that accepts pulses (events) in two or more input channels and provides an output signal *only* if the input pulses arrive within the time period τ. The logic of a coincidence unit is shown in Figure 10.26.

An anticoincidence unit is an electronic device that accepts pulses (events) in two input channels and provides an output signals *only* if the two events *do not* arrive within the time period τ. The logic of an anticoincidence unit is shown in Figure 10.27.

Figures 10.26 and 10.27 both show a "coincidence" unit as the instrument used because, commercially, a single component is available that, with the flip of a switch, is used in the coincidence or anticoincidence mode.

For a successful coincidence or anticoincidence measurement, the detector signals should not be delayed by any factors other than the time of arrival of the particles at the detector. If it is known that it takes longer to generate the signal in one detector than in another, the signal from the fast detector should be delayed accordingly to compensate for this difference. This compensation is accomplished by passing the signal through a delay line before it enters the coincidence unit. A delay line is always needed if the detectors used in the coincident measurement are not identical. The value of the relative delay needed is determined as follows.

The simplest type of coincidence circuit is probably the additive type shown in Figure 10.28. The coincidence unit is summing the input pulses. When two pulses overlap, their sum exceeds a discriminator threshold and the unit produces an output pulse. If the width of the input pulse is T, the resolving time is essentially $\tau = 2T$. Assume now that a system has been set up as shown in Figure 10.24, with the addition of a delay line in channel 2 between the timing and the coincidence units. If one measures the number of coincidences as a function of the delay between the two signals, the result will be the delay or resolving-time curve shown in Figure 10.29. The proper relative delay is the value corresponding to the center of the flat region. The ideal (rectangular) curve will be obtained if the time jitter is zero.

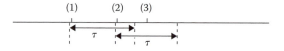

Figure 10.25 Events (1) and (2) or events (2) and (3) are coincident. Events (1) and (3) are not.

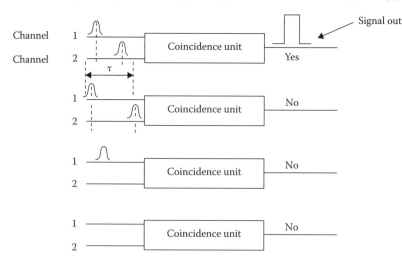

Figure 10.26 The logic of a coincidence unit with two input channels.

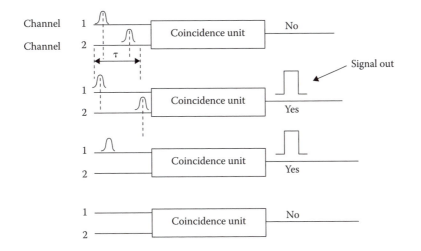

Figure 10.27 The logic of an anticoincidence unit.

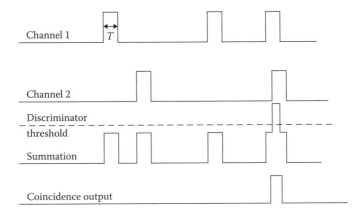

Figure 10.28 The coincidence circuit of the additive type.

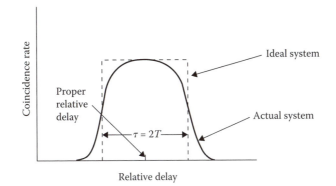

Figure 10.29 The delay or resolving-time curve.

Since the pulses from the two detectors arrive randomly, a certain number of accidental (or chance) coincidences will always be recorded. Let

r_1 = counting rate of detector 1
r_2 = counting rate of detector 2
r_a = accidental coincidence rate

Consider a single pulse in channel 1. If a pulse occurs in channel 2 within the time period τ, then a coincidence will be registered. Since the number of pulses in channel 2 during time τ is $r_2\tau$, the rate of accidental coincidences is

$$r_a = r_1 r_2 \tau \qquad (10.27)$$

Equation 10.27 gives accidental coincidences of first order. Corrections for multiple coincidences of higher order have also been calculated.[14,15]

If S is the strength of the source, ε_1 and ε_2 the efficiencies, Ω_1 and Ω_2 the solid angle factors, and F_1 and F_2 any other factors that affect the measurement of particles counted by detectors 1 and 2 (see Chapter 8), then the true coincidence rate r_t is given by

$$r_t = S\varepsilon_1\varepsilon_2\Omega_1\Omega_2 F_1 F_2 \qquad (10.28)$$

and from Eq. 10.27 the accidental coincidence rate is

$$r_a = S^2\varepsilon_1\varepsilon_2\Omega_1\Omega_2 F_1 F_2 \tau \qquad (10.29)$$

The *figure of merit* in a coincidence experiment is the ratio

$$Q = \frac{r_t}{r_a} = \frac{1}{S\tau} \qquad (10.30)$$

which should be as high as possible. Equation 10.30 shows that this ratio improves when S and τ decrease. Unfortunately, the values for both of these quantities have constraints. The value of τ is limited by the performance of the detector and by the electronics. The source strength S has to be of a certain value for meaningful counting statistics to be obtained in a reasonable time. It is interesting to note that when the source strength increases, both true and accidental coincidence rates increase but the ratio Q (Eq. 10.30) decreases, because $r_t \propto S$ but $r_a \propto S^2$.

Another coincidence technique involves the use of a *time-to-amplitude converter* (TAC). A TAC is an electronic unit that converts the time difference between two pulses into a voltage pulse between 0 and 10 V. The height of the pulse is proportional to the time difference between the two events. The time spectrum of the two detectors is stored directly in the MCA. A "time" window is set around the coincidence peak (Figure 10.30). A second window of equal width is set outside the peak to record accidental coincidences only.

The advantages of using a TAC are as follows:

1. No resolving curve need be taken.

2. No resolving time is involved.

3. The number of channels and the range of time intervals analyzed may be changed over a wide range by simply changing the conversion gain of the MCA.

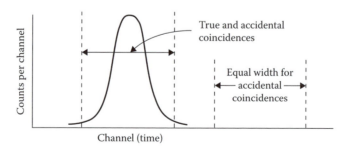

Figure 10.30 An MCA spectrum taken with a TAC.

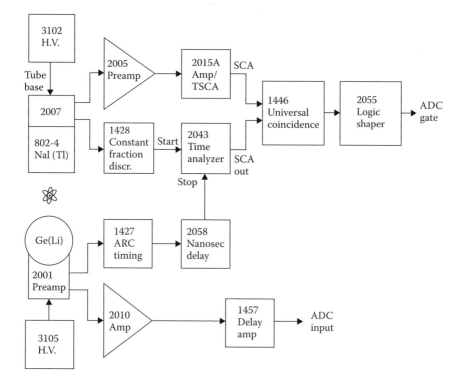

Figure 10.31 A NaI–Ge(Li) γ–γ coincidence system (numbers indicate Canberra models).

Most TACs cannot distinguish the sequence of events—that is, they cannot tell if a pulse from channel 1 precedes a pulse from channel 2 and vice versa. To avoid this ambiguity and also to create a measurable difference between the pulses, the signal from one detector is usually shifted by a fixed delay.

Figure 10.31 shows what is now a common counting system for γ–γ coincidence measurements using a NaI(Tl) detector and a Ge(Li) detector. The initials ADC stand for analog-to-digital converter (see Section 10.12).

10.9 PULSE-SHAPE DISCRIMINATION

Pulse-shape discrimination (PSD) is the name given to a process that differentiates pulses produced by different types of particles in the same detector. Although PSD has found many applications, its most common use is to discriminate between pulses generated by neutrons and gammas in organic scintillators (see also Chapter 13), and it is this type of PSD that will be discussed.

Measurement of the amount of light produced in organic scintillators by neutrons and gammas shows that both the differential and integral light intensities are different as functions of time. Figure 10.32, presenting the results of Kuchnir and Lynch,[16] illustrates this point. It is obvious that the pulses from neutrons and gammas have different time characteristics, and it is this property that is used as the basis for PSD.

Many different methods have been proposed and used for successful PSD.[17–22] One method doubly differentiates the detector pulse, either using CR circuits or a delay line, and bases the PSD on the time interval between the beginning of the pulse and the zero crossing point. This time interval, which is essentially independent of the pulse amplitude but depends on the pulse shape, is usually converted into a pulse by means of a TAC. The pulse from the TAC may be used to gate the counting system. Figure 10.33 shows a block diagram for such a counting system. The result of n–γ discrimination is usually a spectrum that resembles Figure 10.34. Actually, the γ peak is due to electrons produced by the gammas, and the neutron peak is due to protons recoiling after collisions with the incident neutrons. More details of this method of neutron detection are given in Chapter 13.

A second method, introduced by Brooks[20] integrates the charge from the early part of the pulse and compares it to the total charge. A third method, introduced by Kinbara and Kumahara,[21]

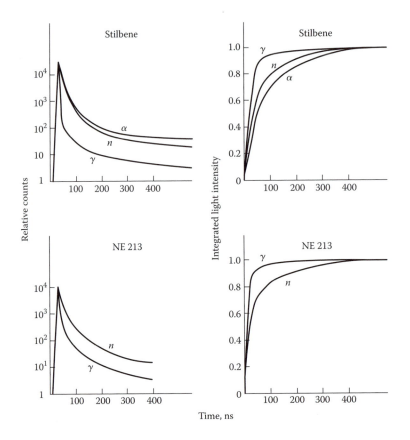

Figure 10.32   The light produced by neutrons and gammas in stilbene and NE 213. Light intensity is shown at left: integrated light intensity is shown at right. (Based on data of Nakhostin, M. et al., *Nucl. Instrum. Meth. Phys. Res. A*, 606, 681, 2009.)

differentiates $n–\gamma$ pulses by a measurement of the rise time of the pulse. A final example of a PSD technique is that used by Burrus and Verbinski,[22] and Verbinski *et al.*[23] based on a design by Forté.[24] Details of the circuitry are given in References 22 and 23. This PSD method produces a large positive pulse output for neutrons and a small positive or a large negative pulse for gammas.

10.10 PREAMPLIFIERS

In Section 1.5.5, a few general comments were made about preamplifiers. It was stated that the primary purpose of the preamplifier is to provide an optimum coupling between the detector and the rest of the counting system. A secondary purpose of the preamplifier is to minimize any sources of noise, which will be transmitted along with the pulse and thus may degrade the energy resolution of the system. This second objective, low noise, is particularly important with semiconductor detectors, which are the counters offering the best energy resolution.

There are three basic types of preamplifiers: charge-sensitive, current-sensitive, and voltage-sensitive. The voltage-sensitive preamplifier is not used in spectroscopy because its gain depends on the detector capacitance, which in turn depends on the detector bias. The charge-sensitive preamplifier is the most commonly used in spectroscopic measurements and the only type used with semiconductor detectors.

To understand the basic features of a charge-sensitive preamplifier, consider the basic circuit associated with a semiconductor detector, shown in Figure 10.35. The high-voltage (HV) bias applied to the detector is usually connected through the detector to the first stage of the charge-sensitive preamplifier. In Figure 10.35, C_f is the feedback capacitor (~1 pF) and R_f is the feedback resistor (~1000 MΩ).[25] The triangle with the letter A indicates the first stage of the preamplifier, which today is usually a field-effect transistor (FET). The FET is a p–n junction with reverse bias, exhibiting extremely low noise. The type of coupling shown in Figure 10.35 is called dc coupling.

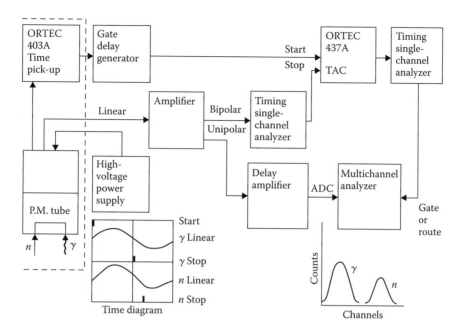

Figure 10.33 A block diagram for a PSD system. (Reproduced from *Instruments for Research and Applied Sciences* by permission of EG & G ORTEC, Oak Ridge, Tennessee.)

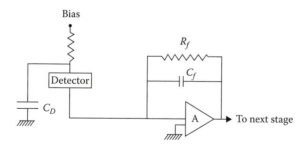

Figure 10.34 The result of γ–n discrimination using PSD.

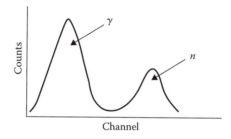

Figure 10.35 The first stage of a charge-sensitive preamplifier dc-coupled to the bias circuit.

There is an ac coupling, too, in which the detector is coupled to the FET through a coupling capacitor (see Nicholson, 1974, p. 110).

The detector sees the FET stage as a large capacitor of magnitude AC_f. As long as $AC_f \gg C_i$, where C_i is the total input capacitance consisting of the detector capacitance C_D, the cable capacitance, etc., the voltage at the output of the preamplifier is equal to

$$V_0 = \frac{Q}{C_f}$$

(10.31)

where Q, the charge produced in the detector, is given by

$$Q = \frac{Ee}{w} \qquad (10.32)$$

where E = energy of the particle
e = electronic charge = 1.6×10^{-19} coulombs
w = average energy required to produce one electron–hole pair

The major components of C_i are the detector capacitance C_D and that of the cables between the detector and the preamplifier. Both of these components are controlled by the user to a certain extent.

The noise of the charge-sensitive preamplifier depends on three parameters: the noise of the input FET, the input capacitance C_i, and the resistance connected to the input. The noise can be determined by injecting a charge Q, equivalent to E, into the preamplifier and measuring the amplitude of the generated pulse. Commercial preamplifiers are provided with a test input for that purpose. In general, the noise expressed as the width (keV) of a Gaussian distribution increases as input capacitance increases (Figure 10.36).

The output pulse of the preamplifier has a fast rise time (of the order of nanoseconds) followed by a slow exponential decay, ~100 μs (Figure 10.37). The useful information in the pulse is its amplitude and its rise time. The rise time is particularly important when the signal is going to be used for timing. The observer should be aware that the rise time increases with external capacitance. The preamplifier pulse is shaped in the amplifier by the methods described in Section 10.6.

The *sensitivity* (or *gain*) of a charge-sensitive preamplifier is expressed by the ratio V/E, where V is given by Eq. 10.31. For a 1-MeV particle in a germanium detector, the sensitivity is (using $C_f \approx 5$ pF)

$$\frac{V}{E} = \frac{Q}{EC_f} = \frac{Ee}{EC_f w} = \frac{e}{C_f w} = \frac{1.6 \times 10^{-19}}{(5 \times 10^{-12})(3 \times 10^{-6})} \sim 10 \, \text{mV/MeV}$$

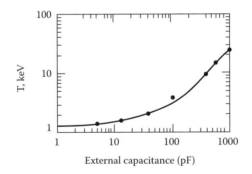

Figure 10.36 The dependence of noise on external capacitance for a typical charge-sensitive preamplifier. (From *Instruments for Research and Applied Sciences* by permission of EG & G ORTEC, Oak Ridge, Tennessee.)

Figure 10.37 Typical pulse from a charge-sensitive preamplifier.

A current-sensitive preamplifier is used to transform fast current pulses produced by a photomultiplier into a voltage pulse. The current-sensitive preamplifier is an amplifying instrument. The sensitivity (or gain) of such a unit is expressed as V_{out}/I_{in}, that is, in mV/mA with typical values of the order of 500 mV/mA. The rise time of the pulse is ~1 ns.

10.11 AMPLIFIERS

As explained in Section 1.5.6, the amplifier plays the two roles of amplifying and shaping the signal. The need for amplification is obvious. The output signal of the preamplifier, being in the range of a few millivolts, cannot travel very far or be manipulated in any substantial way without losing the information it carries or being itself lost in the noise. Commercial amplifiers consisting of many amplification stages increase the amplitude of the input signal by as many as 2000 times, in certain models.

The need for shaping the signal was explained in Section 10.6. The type of shaping that is applied depends on the requirements of the measurement. For spectroscopy measurements where good energy resolution is the important parameter, pulse shaping should not decrease the signal-to-noise ratio. For timing measurements, depending on the method of time pick-off, the signal may be singly or doubly differentiated or be shaped by a single or double delay line.

For a good measurement, the amplifier should satisfy many requirements.[26–31] Not all types of measurements, however, require the same level of performance. For example, if one measures only the number of particles and not their energy, the precision and stability of the amplification process can be relatively poor. It is in spectroscopy measurements, particularly measurements using semiconductor detectors, that the requirements for precision and stability are extremely stringent. Since the energy resolution of Ge detectors is of the order of 0.1%, the dispersion of the pulses due to the amplification process should be much less, about 0.01%.

An ideal spectroscopy amplifier should have a constant amplification for pulses of all amplitudes without distorting any of them. Unfortunately, some pulse distortion is always present because of electronic noise, gain drift due to temperature, pulse pile-up, and limitations on the linearity of the amplifier.

The effect of electronic noise on energy resolution was discussed in Section 9.6.2. Random electronic noise added to equal pulses makes them unequal (see Figure 9.12). Gain drift of an amplifier is caused by small changes in the characteristics of resistors, capacitors, transistors, etc., as a result of temperature changes. The value of the gain drift, always quoted by the manufacturer of the instrument, is for commercial amplifiers of the order of 0.005% per °C or less.

Since the time of arrival of pulses is random, it is inevitable that a pulse may arrive at a time when the previous one did not fully decay. Then the incoming pulse "piles up" on the tail of the earlier one and appears to have a height different from its true one. Pulse pile-up depends on the counting rate.

The linearity of an amplifier is expressed as differential and integral.

Differential nonlinearity is a measure of the change in amplifier gain as a function of amplifier input signal. Referring to Figure 10.38, the differential nonlinearity in percent is given by

$$\frac{(\Delta V_0/\Delta V_i)_A}{(\Delta V_0/\Delta V_i)_B}(100) \tag{10.33}$$

In Eq. 10.33, the numerator is the slope of the amplifier gain curve at the point where the nonlinearity is measured, and the denominator is the slope of the straight line as shown in Figure 10.38.

Integral nonlinearity is defined as the maximum vertical deviation between the straight line shown in Figure 10.38 and the actual amplifier gain curve, divided by the maximum rated output of the amplifier. Referring to Figure 10.38, the integral nonlinearity in percent is given by

$$\frac{V_m - V_L}{V_{max}}(100) \tag{10.34}$$

The integral nonlinearity is one of the specifications of commercial amplifiers and has a value of about 0.05% or less over the range 0–10 V.

There are many types of commercial amplifiers designed to fit the specific needs of spectroscopic or timing measurements. Companies like Canberra, EG & G ORTEC, etc., offer a wide selection of such instruments.

10.12 ANALOG-TO-DIGITAL CONVERTERS

As discussed in Section 9.9, the backbone of an MCA is the analog-to-digital converter (ADC), the unit that digitizes the input pulse height and assigns it to a specific channel. Many types of ADCs have been developed, but the most frequently used is the Wilkinson type.[32]

The Wilkinson-type ADC operates as shown in Figure 10.39. When a pulse enters the MCA, two events are initiated:

1. A capacitor starts charging.

2. An input gate prevents the acceptance of another pulse until the previous one is fully processed and registered.

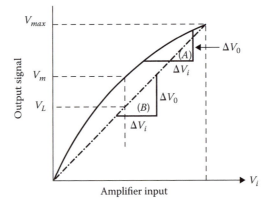

Figure 10.38 Diagram used for the definition of differential and integral linearity of an amplifier. The output signal of a perfect amplifier plotted versus input signal should give the straight line shown (–·–·–).

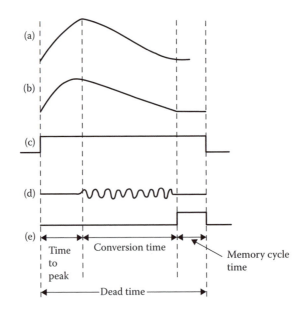

Figure 10.39 The processing of the pulse in a Wilkinson ADC. (a) Input pulse, (b) charge capacitor voltage, (c) gate stops other pulses, (d) oscillator clock, (e) pulse storage.

The capacitor keeps charging until the peak of the pulse is reached. At that point in time, two new events are initiated:

1. The voltage on the charged capacitor is discharged by a constant current.

2. An oscillator clock starts. The clock stops its oscillations when the capacitor is fully discharged.

The number of oscillations during this time—called *rundown* or *conversion time*—is proportional to the pulse height and constitutes the information that determines the channel number in which that pulse will be stored.

A variation of this method is shown in Figure 10.40. The steps followed in this case are as follows:

1. The input pulse is stretched in such a way that its flat portion is proportional to its height.

2. At the moment the pulse reaches its maximum (time t_1), a linear ramp generator is triggered, producing a voltage C.

3. At the same moment (t_1) a gate signal is produced and an oscillator clock is turned on.

When the voltage ramp signal reaches the flat part of the stretched pulse (P), the gate signal turns the clock off. Thus, the time interval ($t_2 - t_1$) and, therefore, the number of oscillations during ($t_2 - t_1$) are again proportional to the height of the pulse. This second method of ADC operation (Figure 10.40) is not favored because it is difficult to keep the pulse height constant for the time interval ($t_2 - t_1$).

Figure 10.39 shows, in addition to the principle of operation of the Wilkinson ADC, the reason for the dependence of the MCA dead time on the channel number. The dead time consists of three components:

1. Pulse rise time

2. Conversion time

3. Memory cycle time (time it takes to store the digitized signal)

Of the three components, the second is the most important because it depends on the channel number. One can reduce the size of the conversion time by using a clock with higher frequency. Today's ADCs use quartz-stabilized clocks with a frequency of up to 450 MHz. Obviously, for a Wilkinson ADC, the higher the clock frequency is, the shorter the dead time will be. The equation for dead time is written as

$$\tau(C) = a_1 + 0.01(C + X)\mu s \tag{10.35}$$

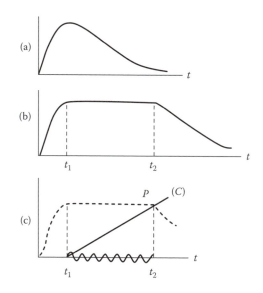

Figure 10.40 The processing of the pulse by the ADC. (a) Input pulse, (b) input pulse stretched; flat part proportional to pulse height, (c) ramp voltage and oscillator clock start at t_1.

where a typical value of a_1 is 1.5 μs, C = address (channel) count, and X = effective digital off-set.[33] The digital offset is a capability offered by modern ADCs of subtracting a certain number of channels from the converted channel number before the data are introduced into the memory. One application of digital offset is to enhance resolution in a measurement performed with a small MCA. For example, with a 1000-channel MCA and an 8000 channel ADC, a 7000 digital offset allows data to be recorded for the top eighth of the spectrum only. A fixed dead time (FDT) ADC has also been developed for certain applications.[33]

The resolution of an ADC is expressed in terms of channels. It represents the maximum number of discrete voltage increments into which the maximum input pulse can be subdivided. ADC reso-lutions range from 4096 to 16,384 channels. Since commercial amplifiers can provide a maximum 10 V pulse, an ADC with a resolution of 4096 channels may subdivide 10 V into 4096 increments. Another quantity used is the conversion gain of the ADC. The conversion gain may be considered as a subset of the resolution. An ADC with a resolution of 16,384 channels may be used, depend-ing on the application, with a conversion gain of 4096, or 8192, or 16,384 channels.

The accuracy of the ADC is expressed in terms of its differential and integral nonlinearity. The *differential nonlinearity* describes the uniformity of address widths over the entire range of the ADC. To make this point better understood, assume that a 1000-channel ADC is used to pro-cess pulses with maximum height of 10 V. Then the average address width is 10/1000 = 10mV/channel. The ideal ADC should provide a conversion of 10 mV/channel at any channel. Any deviation between this width and the actual one is expressed by the differential nonlinearity. Mathematically, if

$\overline{\Delta V}$ = average width
ΔV_{max} = maximum width
ΔV_{min} = minimum width

then the differential nonlinearity is given by

$$\% \text{ Differential nonlilnearity} = \frac{\Delta V_{max} - \Delta V_{min}}{\overline{\Delta V}}(100) \qquad (10.36)$$

Commercial ADCs have differential nonlinearity of the order of ±0.5% to ±1%.

The *integral nonlinearity* is defined as the maximum deviation of any address (ADC channel) from its nominal position, determined by a linear plot of address (ADC channel) versus input pulse amplitude (Figure 10.41). The maximum pulse height V_{max} corresponds to the maximum address N_{max}. If N is the address number with the maximum deviation between the actual and nominal pulse heights, the integral nonlinearity is given by

$$\% \text{ Integral nonlinearity} = \frac{V_{nom} - V_{act}}{V_{max}}(100) \qquad (10.37)$$

Modern commercial ADCs have integral nonlinearity of the order of ±0.05% over 98–99% of the full range.

The integral nonlinearity affects the centroid position of energy peaks, which in turn affects the calibration of the system as well as the identification of unknown energy peaks.

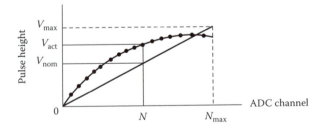

Figure 10.41 The definition of integral nonlinearity is based on a linear plot of ADC channel versus pulse amplitude.

10.13 MULTIPARAMETER ANALYZERS

The MCA is an instrument that stores events by a single parameter, namely, pulse height. When the need arises, however, there are many experiments for the study of events in terms of more than one parameter. Such requirements occur in the following:

1. Coincidence measurements where the energy spectrum from both detectors need be analyzed

2. Simultaneous measurement of energy and mass distribution of fission fragments

3. Study of energy and angular dependence of nuclear reactions involving many particles, etc.

The "direct" method of multiparameter analysis would be to use an arrangement such that all parameters but one are limited to a narrow range (by using a single-channel analyzer) and the remaining parameter is recorded by an MCA. After an adequate number of events have been recorded, the value of one of the fixed parameters is changed, and the measurement is repeated. This process continues until all values of all parameters are covered. Obviously, such an approach is cumbersome and time consuming.

A more efficient way of performing the measurement is by storing the information simultaneously for more than one parameter. For example, consider a coincidence measurement involving two detectors (Figure 10.42). The detector signals are fed into a coincidence unit, which then is used to gate the corresponding ADCs. The amplified detector pulses that are coincident are thus digitized by the ADCs, and the information is stored in the memory of the system. Any event that reaches the memory is defined like a point in a two-dimensional space. For example, if a pulse from ADC_1 has the value 65 (i.e., ADC channel 65) and one from ADC_2 has the value 18, the event is registered as 6518 (assuming 100 channels are available for each parameter). The measured data may be stored in the computer, for subsequent analysis, and may also be displayed on the screen of the monitor for an immediate preliminary assessment of the results. A dual-parameter system as described above is shown in Figure 10.42.

One of the difficulties with multiparameter measurements is to secure sufficient memory capacity to register all possible events. The necessary storage increases exponentially with the number of parameters. For a k parameter measurement with N channels per parameter, the capacity of the memory should be N^k. Thus a two-parameter system with 100 channels per parameter needs 10^4 memory locations. If both parameters are registered in 1000 channels, the requirements are 10^6 locations.

Some recent applications of multiparamter systems include: ^{7}Be analyses in seawater by low background gamma-spectroscopy,[34] a dual purpose Compton suppression spectrometer,[35] application of multidimensional spectrum analysis for neutron activation,[36] application of neutron well coincidence counting for plutonium determination in mixed oxide fuel fabrication plant,[37] and deconvolution of three-dimensional beta-gamma coincidence spectra from xenon sampling and measurement units.[38]

10.14 HIGH COUNT RATES

Since the middle 1990s there has been a concerted effort to develop hardware and software modules that can alleviate the problem of counting losses due to dead time that become significant when the counting rates are elevated (see Section 2.21). A detailed review article on this subject matter[39] has been published. While this effort has been done for germanium detectors, it has applications in all nuclear spectroscopy systems.[40] This technique can be employed for diminishing

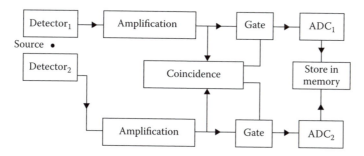

Figure 10.42 A two-parameter measurement.

high count rates, such as very short-lived isotopes or stable high count rates for longer-lived ones. The basic assumption of this technique is that either the dead-time system is extended (known as Loss Free Counter)[41] or by forcing the live-time counter to count backwards (known as Zero Deadtime Counting).[42] Both systems which are marketed by Canberra[33] and Ortec,[43] have been very successful. Pommé[44] has found that there is statistical control of both systems.

10.15 DIGITAL PROCESSING

The other major electronic innovation in nuclear spectroscopy has been the development of digital processing to replace the standard analog to digital converters (ADC). In the ORTEC[33] and Canberra[43] systems all the needed modules are assimilated into one simple box. This also includes the power supply and high voltage. Thus, there is no need for bulky nuclear instrumentation bins (NIM) to house the individual components. These systems can be integrated to all existing detector technologies. These systems are very well suited for most spectroscopy systems in teaching and research. Other specific references include a digital signal processing module for gamma-ray tracking detectors,[45] a comparison of pulsar based analog and digital spectrometers,[46] advanced compact accelerator neutron generator technology for active neutron interrogation field work,[47] automatic activation analysis,[48] and comparison of the precision and accuracy of digital versus analogue γ-ray spectrometric systems.[49]

10.16 DATA MANIPULATION

Raw experimental data seldom give the answer to the problem that is the objective of the measurement. In most cases, additional calculations or analysis of the raw data is necessary. The analysis of the raw data may consist of a simple division of the counts recorded in a scaler by the counting time to obtain counting rates, may require fitting an analytical function to the data, or may necessitate unfolding of a measured spectrum. Advances in software has been made in all permutations of data handling very easy to use; everything from least square fitting to unfolding of complicated spectra is now readily attainable.

PROBLEMS

10.1 Prove that the phase difference between voltage and current maximum values in an RC circuit is given by $\varphi = \tan^{-1}(1/RC\omega)$.

10.2 Prove that the phase difference between voltage and current maximum values in a LR circuit is given by $\varphi = \tan^{-1}(L\omega/R)$.

10.3 Prove that (a) the phase difference between voltage and current maximum values in an RCL circuit is given by

$$\varphi = \tan^{-1}\left[\frac{\omega L - (1/\omega C)}{R}\right]$$

and (b) the impedance is given by $Z = \sqrt{R^2 + \left[\omega L - (1/\omega C)\right]^2}$.

10.4 Prove that the output signal of a differentiating circuit is, for a step input, equal to

$$V_0(t) = V_i e^{-t/RC}$$

10.5 Show that the output signal of a differentiating circuit is given by

$$V_0(t) = \frac{V_i}{\tau} RC \left(1 - e^{-t/RC}\right)$$

when the input signal is given by $V_i(t) = V_i t/\tau$.

10.6 Show that the output signal of an integrating circuit is, for a step input, equal to

$$V_0(t) = V_i \left(1 - e^{-t/RC}\right)$$

10.7 Show that the output signal of a differentiating circuit is given by

$$V_0(t) = V_i \frac{1}{1 - \tau/RC} (e^{-t/RC} - e^{-t/\tau})$$

when the input signal is $V_i(1 - e^{-t/\tau})$.

10.8 A coincidence measurement has to be performed within a time T. Show that the standard deviation of the true coincidence rate is given by

$$\sigma_t = \frac{1}{\sqrt{T}} \left(\sqrt{r_a + r_t} + \sqrt{r_a} \right)$$

where r_a = accidental coincidence rate and r_t = true coincidence rate.

BIBLIOGRAPHY

http://www.canberra.com.

http://www.ortec-online.com.

http://www.pgt.com.

Kowalski, E., *Nuclear Electronics*, Springer–Verlag, New York, 1970.

Malmstadt, H. V., Enke, C. G., and Toren, E. C, *Electronics for Scientists*, W. A. Benjamin, New York, 1963.

Nicholson, P. W., *Nuclear Electronics*, Wiley, London, 1974.

REFERENCES

1. Chase, R. L., *Rev. Sci. Instrum.* **39**:1318 (1968).

2. Gedcke, D. A., and McDonald, W. J., *Nucl. Instrum. Meth.* **55**:377 (1967).

3. Maier, M. R., and Sperr, P., *Nucl. Instrum. Meth.* **87**:13 (1970).

4. Strauss, M. G., Larsen, R. N., and Sifter, L. L., *Nucl. Instrum. Meth.* **46**:45 (1967).

5. Graham, R. L., Mackenzie, I. K., and Ewan, G. T., *IEEE Trans. NS 13* **1**:72 (1966).

6. Lapington, J. S., Fraser, G. W., Miller, G. M., Ashton, T. J. R., Jarron, P., Despeisse, M., Powolny, F., Howorth, J., and Milnes, J., *Nucl. Instrum. Meth. in Phys. Res. A* **604**:199 (2009).

7. Dalena, B., D'Erasmo, G., Di Santo, D., Fiore, E. M., Palomba, M., Simonetti, G., Andronenkov, A., Pantaleo, A., Paticchio, V., and Faso, D., *Nucl. Instrum. Meth. Phys. Res. A* **603**:276 (2009).

8. Nakhostin, M., Ishii, K., Kikuchi, Y., Matsuyama, S., Yamazaki, H., and Torshabi, A. E., *Nucl. Instrum. Meth. Phys. Res. A* **606**:681 (2009).

9. Tõke, J., Quinlan, M. J., Gawlikowicz, W., and Udo Schröder, W. U., *Nucl. Instrum. Meth. Phys. Res. A* **595**:460 (2008).

10. Collazuol, G., Ambrosi, G., Boscardin, M., Corsi, F., Betta, G. F. D., Del Guerra, A. D., Dinu, N., *et al.*, *Nucl. Instrum. Meth. Phys. Res. A* **581**:461 (2007).

11. Ivan Prochazka, I., Hamal, K., Sopko, B., Blazej, and Chren, J. D., *Nucl. Instrum. Meth. Phys. Res. A* **568**:437 (2006).

12. Fallu-Labruyere, A., Tan, H., Hennig, W., and Warburton, W. K., *Nucl. Instrum. Meth. Phys. Res. A* **579**:247 (2007).

13. Feng, H., Kaaret, P., and Andersson, H., *Nucl. Instrum. Meth. Phys. Res. A* **564**:347 (2006).

14. Viencent, C. H., *Nucl. Instrum. Meth.* **127**:421 (1975).

15. Smith, D., *Nucl. Instrum. Meth.* **152**:505 (1978).

16. Kuchnir, F. T., and Lynch, F. J., *IEEE Trans. NS 15* 3:107 (1968).

17. Alexander, T. K., and Goulding, F. S., *Nucl. Instrum. Meth.* **13**:244 (1961).

18. Heistek, L. J., and Van der Zwan, L., *Nucl. Instrum. Meth.* **80**:213 (1970).

19. McBeth, G. W., Lutkin, J. E., and Winyard, R. A., *Nucl. Instrum. Meth.* **93**:99 (1971).

20. Brooks, F. D., *Nucl Instrum. Meth.* **4**:151 (1959).

21. Kinbara, S., and Kumahara, T., *Nucl. Instrum. Meth.* **70**:173 (1969).

22. Burrus, W. R., and Verbinski, V. V., *Nucl. Instrum. Meth.* **67**:181 (1969).

23. Verbinski, V. V., Burrus, W. R., Love, T. A., Zobel, W., and Hill, N. W., *Nucl. Instrum. Meth.* **65**:8 (1968).

24. Forté, M., Konsta, A., and Moranzana, C., *Electronic Methods for Discrimination of Scintillation Shapes* (IAEA Conf. Nucl. Electr. Belgrade, 1961) NE-59.

25. Brenner, R., *Rev. Sci. Instrum.* **40**:1011 (1969).

26. Nowlin, C. H., and Blankenship, J. L., *Rev. Sci. Instrum.* **36**:1830 (1965).

27. Fairstein, E., and Hahn, J., *Nucleonics 23* **7**:56 (1965).

28. Fairstein, E., and Hahn, J., *Nucleonics 23* **9**:81 (1965).

29. Fairstein, E., and Hahn, J., *Nucleonics 23* **11**:50 (1965).

30. Fairstein, E., and Hahn, J., *Nucleonics 24* **1**:54 (1966).

31. Fairstein, E., and Hahn, J., *Nucleonics 24* **3**:68 (1966).

32. Wilkinson, D. H., *Philos. Soc.* **46**:508 (1950).

33. http://www.canberra.com

34. Andrews, J. E., Hartin, C., and Buesseler, K. O., *J Radioanal. Nucl. Chem.* **277**:253 (2008).

35. Parus, J., Kierzek, J., Raab, W., and Donohue, D., *J. Radioanal. Nucl. Chem.* **258**:123 (2003).

36. Hatsukawa, Y., Oshima, M., Hayakawa, T., Toh, Y., and Shinohara, N., *J. Radioanal. Nucl. Chem.* **248**:121 (2001).

37. Kumar, P., and Ramakumar, K. L., *J. Radioanal. Nucl. Chem.* **277**:419 (2008).

38.   Foltz Biegalski, K. M., and Biegalski, S. R., *J. Radioanal. Nucl. Chem.* **263**:259 (2005).

39.   Westphal, G. P., *J. Radioanal. Nucl. Chem.* **275**:677 (2008).

40.   Hansen, K., Reckleben, C., Diehl, I., and Klär, H., *Nucl. Instrum. Meth. Phys. Res A* **589**:250 (2008).

41.   Zeisler, R., *J. Radioanal. Nucl. Chem.* **244**:507 (2000).

42.   Blaauw, M., *J. Radioanal. Nucl. Chem.* **257**:457 (2003).

43.   http://www.ortec-online.com

44.   Pommé, S., *J. Radioanal. Nucl. Chem.* **257**:463 (2003).

45.   Cromaz, M., Riot, V., Fallon, J. P., Gros, S., Holmes, B., Lee, I. Y., Macchiavelli, A. O., Vu, C., Yaver, H., and Zimmermann, S., *Nucl. Instrum. Meth. Phys. Res A* **597**:233 (2008).

46.   McGrath, C. A., and Gehrke, R. J., *J. Radioanal. Nucl. Chem.* **276**:669 (2008).

47.   Chichester, D. L., Simpson, J. D., and Lemchak, M., *J. Radioanal. Nucl. Chem.* **271**:629 (2007).

48.   Westphal, G. P., Grass, F., Lemmel, H., Sterba, J. P., and Bloch, Ch., *J. Radioanal. Nucl. Chem.* **271**:145 (2007).

49.   Makarewicz, M., and Burns, K., *J. Radioanal. Nucl. Chem.* **244**:649 (2000).

11

Data Analysis Methods

11.1 INTRODUCTION

Raw* experimental data seldom give the answer to the problem that is the objective of the measurement. In most cases, additional calculations or analysis of the raw data is necessary. The analysis of the raw data may consist of a simple division of the counts recorded in a scaler by the counting time to obtain counting rates, may require fitting an analytical function to the data, or may necessitate unfolding of a measured spectrum.

In the second addition of this book a significant amount of space was given to curve fitting, interpolation scheme, least squares fitting and data smoothing and folding and unfolding of energy spectra. Since the advent of very powerful desktop statistical packages such as EXCEL, SPSS (Statistical Packages for Social Scientists) and SAS (Statistical Analysis Software) to name but a few, many of these operations are now easily handled.

Much of the current work in data analysis in nuclear instrumentation revolves around three main areas. One is the folding and unfolding of energy spectra such in x-ray, gamma-ray and neutron spectroscopy. The second is the quality control of data acquisitions systems and lastly the assignment of errors and detection limits values. The latter can have important consequences within various regulatory agencies. For the sake of completeness and historical value we have left the original chapter unchanged with the exception of adding a few new references. However, we have included one new section on total quality management including quality assurance and quality control.

11.2 CURVE FITTING

The results of most experiments consist of a finite number of values (and their errors) of a dependent variable y measured as a function of the independent variable x (Figure 11.1). The objective of the measurement of $y = y(x)$ may be one of the following:

1. To find how y changes with x

2. To prove that $y = y(x)$ follows a theoretically derived function

3. To use the finite number of measurements of $y(x)$ for the evaluation of the same function at intermediate points or at values of x beyond those measured

These objectives could be immediately achieved if the function $y(x)$ were known. Since it is not, the observer tries to determine it with the help of the experimental data. The task of obtaining an analytic function that represents $y(x)$ is called *curve fitting*.

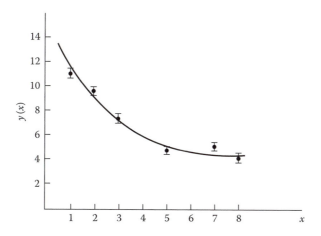

Figure 11.1 Experimental results consist of the values of the dependent variable $y(x)$ and their errors. The curve was drawn to help the eye.

---

* Raw data consist of the numbers obtained by the measuring device, for example, a scaler, a clock, or a voltmeter.

The first step in curve fitting is to plot the data (y versus x) on linear graph paper (Figure 11.1). A smooth curve is then drawn, following as closely as possible the general trend of the data and trying to have an equal number of points on either side of the curve. The experimental points always have an error associated with them, so the smooth curve is not expected to pass through all the measured (x, y) points. Obviously, there is no guarantee that the smooth curve so drawn is the "true" one. Criteria that may help the observer draw a curve with a certain degree of confidence are then needed. Such criteria exist and are described in Section 11.4.

After the data are plotted and a smooth curve is drawn, the observer has to answer the following two questions:

1. What type of function would represent the data best (e.g., exponential, straight line, parabola, logarithmic)?

2. After the type of function is decided upon, how can one determine the best values of the constants that define the function uniquely?

Since there exists an infinite number of functions, the observer would like to have certain criteria or rules that limit the number of possible functions. While no such formal set of criteria exists, the following suggestions have proved useful.

First, the observer should utilize any a priori knowledge about $y(x)$ and x. Examples are restrictions of x and y within a certain range (e.g., in counting experiments both x and y are positive) or information from theory that suggests a particular function (e.g., counting data follow Poisson statistics).

Second, the observer should try the three simple expressions listed next, before any complicated function is considered.

1. The linear relation (straight line)

$$y(x) = ax + b \qquad (11.1)$$

where a and b are constants to be determined based on the data. A linear relationship will be recognized immediately in a linear plot of $y(x)$ versus x.

2. The exponential relationship

$$y(x) = ae^{-bx} \qquad (11.2)$$

If the data can be represented by such a function, a plot on semi log paper—that is, a plot of $\ln y$ versus x—will give a straight line.

3. The power relationship

$$y(x) = ax^b \qquad (11.3)$$

If the data can be represented by this expression, a plot on log–log paper—that is, a plot of $\ln y$ versus $\ln x$—will give a straight line.

Third, the observer should know that a polynomial of degree N can always be fitted exactly to $N + 1$ pieces of data (see also Sections 11.3–11.5).

If no satisfactory fit can be obtained by using any of these suggestions, the analyst should try more complicated functions. Plotting the data on special kinds of graph paper, such as reciprocal or probability paper may be helpful. After the type of function is found, the constants associated with it are determined by a least-squares fit (see Section 11.4).

There is software now available that accepts a table of data points as input and tests possible fits of this data set to a large number of analytic functions. At the end of the operation, both the function representing the best fit and a degree of "confidence" are provided.

11.3 INTERPOLATION SCHEMES

It was mentioned in Section 11.2 that one of the reasons for curve fitting is to be able to evaluate the function $y(x)$ at values of x for which measurements do not exist. An alternative to curve fitting that can be used for the calculation of intermediate $y(x)$ values is the method of interpolation. This section presents one of the basic interpolation techniques—the *Lagrange formula*.

Many other formulas exist that the reader can find in the bibliography of this chapter (e.g., see Hildebrand, and Abramowitz and Stegan's *Handbook or Mathematical Functions*).

Assume that N values of the dependent variable $y(x)$ are known at the N points x_i, $x_a \leq x_i \leq x_b$ for $i = 1, \dots, N$. The pairs of data (y_i, x_i) for $i = 1, \dots, N$ where $y(x_i) = y_i$, may be the results of an experiment or tabulated values. Interpolation means to obtain a value $y(x)$ for $x_a < x < x_b$ based on the data (y_i, x_i), when the point x is not one of the N values for which $y(x)$ is known.

The Lagrange interpolation formula expresses the value $y(x)$ in terms of polynomials (up to degree $N - 1$ for N pairs of data). The general equation is

$$y(x) = \sum_{i=0}^{M} P_i(x) y(x_i), \quad M \leq N - 1 \tag{11.4}$$

where

$$P_i(x) = \frac{\prod_{j=0, j \neq i} (x - x_j)}{\prod_{j=0, j \neq i}^{M} (x_i - x_j)}, \quad M \leq N - 1 \tag{11.5}$$

The error associated with Eq. 11.4 is given by

$$\text{Error}[y(x)] = \prod_{j=0}^{M} (x - x_j) \frac{y^{M+1}(\xi)}{(M + 1)!}, \quad M \leq N - 1 \tag{11.6}$$

where $y^{M+1}(\xi)$ is the $(M + 1)$ derivative of $y(x)$ evaluated at the point ξ, $x_a < \xi < x_b$. Since $y(x)$ is not known analytically, the derivative in Eq. 11.6 has to be calculated numerically.

Equation 11.4 is the most general. It uses all the available points to calculate any new value of $y(x)$ for $x_a < x < x_b$. In practice, people use only a few points at a time, as the following two examples show.

Example 11.1   Derive the Lagrange formula for $M = 1$.
Answer   If $M = 1$, Eq. 11.4 takes the form (also using Eq. 11.5)

$$y(x) = \sum_{i=0}^{1} P_i(x) y(x_i) = \frac{x - x_1}{x_0 - x_1} y_0 + \frac{x - x_0}{x_1 - x_0} y_1 \tag{11.7}$$

where $y_i = y(x_i)$. The points x_0 and x_1 could be anywhere between x_a and x_b, but the point x should be $x_0 \leq x \leq x_1$.

To calculate $y(x)$ at any x, Eq. 11.7 uses two points, one on either side of x, and for this reason it is called the *Lagrange two-point interpolation formula*. Equation 11.7 may be written in the form

$$y(x) = y_0 + \frac{x - x_0}{x_1 - x_0} (y_1 - y_0) \tag{11.8}$$

which shows that the two-point formula amounts to a linear interpolation.

The error associated with the two-point formula is obtained from Eq. 11.6:

$$\text{Error}[y(x)] = (x - x_0)(x - x_1) \frac{y^{(2)}(\xi)}{2}$$

where $y^{(2)}(\xi)$ is the second derivation evaluated at ξ, $x_0 \leq \xi < x_1$.

Example 11.2 Derive the Lagrange formula for $M = 2$.
Answer If $M = 2$, Eq. 11.4 takes the form

$$y(x) = \sum_{i=0}^{2} P_i(x) y(x_i) = \frac{(x - x_1)(x - x_2)}{(x_0 - x_1)(x_0 - x_2)} y_0 + \frac{(x - x_0)(x - x_2)}{(x_1 - x_0)(x_1 - x_2)} y_1$$
$$+ \frac{(x - x_0)(x - x_1)}{(x_2 - x_0)(x_2 - x_1)} y_2 \tag{11.9}$$

To calculate $y(x)$ at any point x, Eq. 11.9 uses three points x_0, x_1, x_2 with $x_0 \le x \le x_2$, and is called the Lagrange three-point interpolation formula. The three-point formula amounts to a parabolic representation of the function $y(x)$ between any three points.

The error associated with the three-point formula is (applying again Eq. 11.6):

$$\text{Error}[y(x)] = (x - x_0)(x - x_1)(x - x_2) \frac{y^{(3)}(\xi)}{3!}$$

where $y^{(3)}(\xi)$ is the third derivative evaluated at ξ, $x_0 \le \xi < x_2$.

Example 11.3 Calculate the value of the function $f(x)$ at $x = 11.8$ and the associated interpolation error based on the data in the table below using the Lagrange two- and three-point interpolation formulas. The data are plotted in Figure 11.2.

x	f(x)
10	30.5
11	33.0
12	35.8
13	36.7
14	37.2

Answer (a) Using the two-point formula (Eq. 11.7), one has $x_0 = 11$, $x_1 = 12$, $y_0 = 33.0$, $y_1 = 35.8$.

$$y(x) = y(11.8) = \frac{11.8 - 12}{11 - 12}(33.0) + \frac{11.8 - 11}{12 - 11}(35.8) = 35.2$$

(b) Using the three-point formula (Eq. 11.9), one has $x_0 = 11$, $x_1 = 12$, $x_2 = 13$, $y_0 = 33$, $y_1 = 35.8$, $y_2 = 36.7$.

$$y(11.8) = \frac{(11.8 - 12)(11.8 - 13)}{(11 - 12)(11 - 13)}(33) + \frac{(11.8 - 11)(11.8 - 13)}{(12 - 11)(12 - 13)}(35.8)$$
$$+ \frac{(11.8 - 11)(11.8 - 12)}{(13 - 11)(13 - 12)}(36.7) = 3.96 + 34.37 - 2.94 = 35.4$$

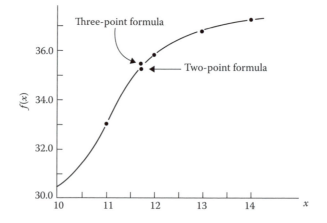

Figure 11.2 A plot of the data used in Example 11.3.

The error of $y(x)$ associated with the interpolation is, for the first case only,

$$\text{Error}[y(x)] = (x - x_0)(x - x_1)\frac{y^{(2)}(\xi)}{2}$$

The second derivative evaluated at $\xi = \frac{1}{2}(x_0 + x_1)$ is

$$y^{(2)}(\xi) = \frac{4(y_1 - 2y + y_0)}{(x_1 - x_0)^2}$$

Thus the error is

$$\text{Error}[y(x)] = (11.8 - 11)(11.8 - 12)\frac{4(35.8 - 2(35.2) + 33)}{2(1^2)} = 0.512$$

or $0.512/35.2 = 0.014 = 1.4\%$.

11.4 LEAST-SQUARES FITTING

Assume that an observer obtained the experimental data $y_i = y(x_i)|_{i=1;N}$ It is often desirable to find a function that can represent the data, that is, to find a function that can be fitted to the data. Let such a function be written as

$$f(x, a_1, a_2, \ldots, a_M), \quad M < N$$

where $a_m|_{m=1, M}$ are parameters to be determined. According to the method of least squares, the best values of the parameters a_M are those that minimize the quantity

$$Q = \sum_{i=1}^{N} w_i[y_i - f(x_i)]^2 \tag{11.10}$$

where $w_i|_{i=1, N}$ are weighting functions. Minimization of Q is achieved by requiring

$$\frac{\partial Q}{\partial a_m} = 0, \quad m = 1, \ldots, M \tag{11.11}$$

Before Eq. 11.11 is solved, it should be pointed out that the observer decides about the form of the function $f(x)$ and the weighting functions w_i. The form of $f(x)$ is obtained by the curve-fitting methods discussed in Section 11.2. The weighting functions are selected based on the type of data and the purpose of the fit. For example, if the data are the result of a counting experiment, $w_i = 1/\sigma_i^2$, where σ_i is the standard deviation of y_i.

There are two types of least-squares fit, linear and nonlinear. *Linear least-squares fit* is based on a function $f(x)$ of the form

$$f(x, a_1, a_2, \ldots, a_M) = \sum_{m=1}^{M} a_m \phi_m(x) \tag{11.12}$$

where the $\phi_m(x)$ are known functions of x.

Nonlinear least-squares fit is based on a function $f(x)$ nonlinear in a_m, such as

$$f(x, a_1, a_2, \ldots, a_M) = a_1 \cos(a_2 x)$$

The interested reader should consult the bibliography of this chapter (see Bevington, 1969) for further information on nonlinear least-squares fit.

For a linear least-squares fit, the parameters $a_m|_{m=1,M}$ are determined from Eq. 11.11, with Eq. 11.12 giving the form of $f(x)$. The result is

$$\frac{\partial Q}{\partial a_k} = \frac{\partial}{\partial a_k} \left\{ \sum_{i=1}^{N} w_i \left[y_i - \sum_{m=1}^{M} a_m \phi_m(x_i) \right]^2 \right\}, \quad k = 1,...,M \tag{11.13}$$

If one defines

$$A_{km} - \sum_{i=1}^{N} w_i \phi_k(x_i)\phi_m(x_i), \quad k, m = 1,...,M \tag{11.14}$$

and

$$B_k = \sum_{i=1}^{N} w_i y_i \phi_k(x_i), \quad k = 1,...,M \tag{11.15}$$

then Eq. 11.13 takes the form

$$\sum_{m=1}^{M} A_{km} a_m = B_k, \quad k = 1,...,M \tag{11.16}$$

Equation 11.16 forms a system of M linear nonhomogeneous equations for the M unknowns $a_m|_{m=1,M}$ and can be solved by using any of the standard methods (e.g., Kramer's rule). In matrix notation, the solution is

$$a_m = \mathbf{A}^{-1}\mathbf{B} \tag{11.17}$$

where $\mathbf{A}$ and $\mathbf{B}$ are matrices with elements given by Eqs. 11.14 and 11.15.

If the function $f(x)$ is a polynomial, then

$$f(x,a_1,a_2,...,a_m) = \sum_{m=1}^{M} a_m x^{m-1} \tag{11.18}$$

Equations 11.14 and 11.15 take the form [since $\phi_k(x) = x^{k-1}$]

$$A_{km} = \sum_i w_i x_i^{k-1} x_i^{m-1} \tag{11.19}$$

$$B_k = \sum_i w_i y_i x_i^{k-1} \tag{11.20}$$

The notation used in Eqs. 11.19 and 11.20 and in the next section is $\sum_{i=1}^{N} \rightarrow \sum_i$.

11.4.1 Least-Squares Fit of a Straight Line

If the function represented by Eq. 11.18 is a straight line, then

$$f(x_i, a_1, a_2) = a_1 + a_2 x_i \tag{11.21}$$

Thus, Eqs. 11.19 and 11.20 become

$$A_{11} = \sum_i w_i, \quad A_{12} = \sum_i w_i x_i = A_{21}, \quad A_{22} = \sum_i w_i x_i^2$$

$$B_1 = \sum_i w_i y_i, \quad B_2 = \sum_i w_i y_i x_i$$

Then, Eq. 11.16 takes the forms

$$A_{11}a_i + A_{12}a_2 = B_1$$
$$A_{21}a_1 + A_{22}a_2 = B_2$$

which are solved to give

$$a_1 = \frac{1}{D}\left[\left(\sum_i w_i y_i\right)\left(\sum_i w_i x_i^2\right) - \left(\sum_i w_i y_i x_i\right)\left(\sum_i w_i x_i\right)\right] \tag{11.22}$$

$$a_2 = \frac{1}{D}\left[\left(\sum_i w_i\right)\left(\sum_i w_i y_i x_i\right) - \left(\sum_i w_i x_i\right)\left(\sum_i w_i y_i\right)\right] \tag{11.23}$$

and

$$D = \left(\sum_i w_i\right)\left(\sum_i w_i x_i^2\right) - \left(\sum_i w_i x_i\right)^2 \tag{11.24}$$

The variance of a_1 and a_2 is obtained by using the principle of propagation of error presented in Chapter 2.

$$\sigma_{a_m}^2 = \sum_i \left(\frac{\partial a_m}{\partial y_i}\right)^2 \sigma_{y_i}^2, \quad m = 1, 2 \tag{11.25}$$

where σ_{y_i} = standard error of $y_i = \sigma_i$.

In many cases, the standard deviation of y_i defines the weighting functions, and specifically, analysts use

$$w_i = \frac{1}{\sigma_i^2}$$

Then, Eq. 11.25 gives

$$\sigma_{a_1}^2 = \frac{1}{D}\sum_i \frac{x_i^2}{\sigma_i^2} \tag{11.26}$$

$$\sigma_{a_2}^2 = \frac{1}{D}\sum_i \frac{1}{\sigma_i^2} \tag{11.27}$$

where D is given by Eq. 11.24 with $w_i = 1/\sigma_i^2$. Equations 11.22–11.27 are further simplified if all the σ_i have the same value.

11.4.2 Least-Squares Fit of General Functions

A straight-line least-squares fit is not limited to linear functions of x. It may be used with functions such as the exponential ($y = ae^{bx}$) or the power relationship ($y = ax^b$) after an appropriate transformation of variables. For example, the exponential function can be written as

$$\ln y = \ln a + bx \tag{11.28}$$

which is of the form given by Eq. 11.21 after setting

$$y' = \ln y, \qquad a'_1 = \ln a, \qquad a'_2 = b$$

When the variable is transformed, it is necessary to obtain the standard deviation of the new variable. In general, if one sets

$$y' = g(y)$$

then the standard deviation of $y'(x)$ is

$$\sigma'_i = \frac{\partial g(y_i)}{\partial y_i} \sigma_i \tag{11.29}$$

In the example given above, $y' = \ln y$ and

$$\sigma'_i = \frac{\partial(\ln y_i)}{\partial y_i} \sigma_i = \frac{\sigma_i}{y_i} \tag{11.30}$$

Therefore, if a transformation is applied to the function, all the σ_i in Eqs. 11.26 and 11.27 should be replaced by the values given by Eq. 11.30.

If the parameters a_m are transformed, the standard deviation of the new constant is again determined by Eq. 11.29. In the example given above, $a' = \ln a$ and

$$\sigma_{a'} = \frac{\sigma_a}{a} \tag{11.31}$$

Table 11.1 presents a number of functions that can be cast into a linear (or polynomial) form by a transformation of variables. It should be emphasized that although the functions shown in Table 11.1 are not linear in x, the least-squares fit is still linear. An example of a function $f(x, a_1, \ldots)$ that represents a nonlinear least-squares fit is

$$f(x, a_1, a_2, \ldots) = a_1 \cos(a_2 x)$$

Table 11.1
Functions That Can Be Changed into a Form Suitable for a Linear Least-Squares Fit

Function $y(x)$	Transformation	Function Used in the Least-Squares Fit
$y = ae^{bx}$	$y' = \ln y \quad a' = \ln a$	$y' = a' + bx$
$y = ax^b$	$y' = \ln y \quad a' = \ln a \quad x' = \ln x$	$y' = a' + bx'$
$y = a \exp[-(x-a)^2/2\sigma^2]$	$y' = \ln y$	$y' = a_1 + a_2 x + a_3 x^2$
$y = a_1 + a_2 x^b + a_3 x^{2b}$	$x' = x^b$	$y' = a_1 + a_2 x' + a_3 x'^2$
$y = a_1 x^{b_1} + a_2 x^{b_2}$	$y' = yx^{-b_1} \quad x' = x^{b_2 - b_1}$	$y' = a_1 + a_2 x'$

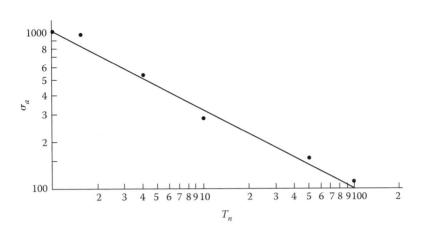

Figure 11.3 The least-squares fit to the data of Example 11.4.

Example 11.4 The following table gives neutron-absorption cross-section values and their errors as a function of neutron kinetic energy. Determine the analytic function that fits this data set.

T_n (eV)	σ_a (b)	Error (b)
1	1000	50
4	540	23
10	290	19
20	225	15
50	160	11
80	108	10
100	105	9

Answer If one plots the data on log–log paper, the result is very close to a straight line (see Figure 11.3). Therefore, the function to use is

$$\ln \sigma_a = a_1 + a_2 \ln T_n$$

If one takes $w_i = 1/\sigma_i^2$ and applies Eqs. 11.22–11.27, the result is

$$a_1 = 6.925, \qquad a_2 = -0.495, \qquad \frac{\sigma_{a_1}}{a_1} = 0.6\%$$

$$\sigma_{a_1} = 0.039, \qquad \sigma_{a_2} = 0.015, \qquad \frac{\sigma_{a_2}}{a_2} = 3\%$$

11.5 FOLDING AND UNFOLDING

To define the problems of folding and unfolding, consider the functions

$$S(x), \qquad 0 < x < \infty$$
$$M(x), \qquad 0 < x < \infty$$
$$R(x, x'), \qquad 0 < x, x' < \infty$$

where the function $R(x, x')$ is normalized to 1:

$$\int_0^\infty R(x, x')\, dx = 1 \tag{11.32}$$

Folding the function $S(x)$ with the function $R(x, x')$ to obtain the function $M(x)$ means to perform the integration:

$$M(x) = \int_0^\infty R(x, x') S(x') \, dx' \tag{11.33}$$

Unfolding means to obtain the function $S(x)$, knowing $M(x)$ and $R(x, x')$. Thus, folding is an integration, as shown by Eq. 11.33. Unfolding, on the other hand, entails solving the integral equation, Eq. 11.33—known as the *Fredholm equation*—for the unknown function $S(x)$.

In the field of radiation measurements, folding and (especially) unfolding are very important operations that have to be applied to the experimental data. In most radiation measurements, the variable x is the energy of the particle, and for this reason the discussion in this section will be based on that variable. The reader should be aware, however, that x may represent other quantities, such as time, velocity, or space variables. If x is the energy of the particle, the functions $S(x)$, $M(x)$, and $R(x, x')$ have the following meanings (also given in Section 9.7):

$S(E) \, dE$ = source spectrum = number of particles emitted by the source with energy
between E and $E + dE$

$M(E) \, dE$ = measured spectrum = number of particles recorded as having energy
between E and $E + dE$

$R(E, E') \, dE$ = response of the detector = probability that a particle emitted by the source
with energy E' will be recorded with energy between E and $E + dE$

As explained in Chapter 9, the response function is measured using monoenergetic sources. A monoenergetic source is represented mathematically by the *delta function* (δ function), which has these properties (Figure 11.4):

$$\delta(E - E_0) = 0, \quad E \neq E_0$$

$$\int_{E_1}^{E_2} \delta(E - E_0) \, dE = \begin{cases} 1 & \text{if } E_1 < E_0 < E_2 \\ 0 & \text{otherwise} \end{cases} \tag{11.34}$$

Thus, the δ function is equal to zero everywhere except at $E = E_0$, which is, of course, what the energy spectrum of a monoenergetic source represents. Because of the property expressed by Eq. 11.34, integrals involving the δ function are immediately evaluated. For any function $f(E)$, one obtains

$$\int_{E_1}^{E_2} f(E) \delta(E - E_0) \, dE = f(E_0), \quad E_1 < E_0 < E_2 \tag{11.35}$$

because there is no contribution to the integral except at $E = E_0$. For the same reason, if E_0 is outside the limits of integration, then

$$\int_{E_1}^{E_2} f(E) \delta(E - E_0) \, dE = 0, \quad \begin{cases} E_0 < E_1 \\ E_0 > E_2 \end{cases} \tag{11.36}$$

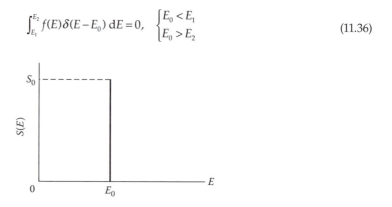

Figure 11.4 A monoenergetic source spectrum.

Assume that a monoenergetic source emitting S_0 particles per second (Figure 11.4) is used to measure the response function. If one substitutes the expression for this source,

$$S(E) = S_0 \delta(E - E_0)$$

(11.37)

into Eq. 11.33, the result is

$$M(E) = \int_0^\infty R(E, E') S_0 \delta(E' - E_0)\, dE' = S_0 R(E, E_0)$$

(11.38)

Equation 11.38 shows that the measured spectrum is indeed equal to the response function in the case of a monoenergetic source.

11.5.1 Examples of Folding

In radiation measurements, *folding* means to obtain the shape of the measured spectrum when the source and the detector response are known. Several examples of folding using a Gaussian distribution as the response function are presented next.

Example 11.5   The source spectrum is a step function:

$$S(E) = \begin{cases} S_0, & E \geq E_0 \\ 0, & E < E_0 \end{cases}$$

What is the measured spectrum?
Answer

$$M(E) = \int_0^\infty \frac{dE'}{\sqrt{2\pi}\sigma} \exp\left[-\frac{(E-E')^2}{2\sigma^2}\right] S(E')$$

$$= S_0 \int_{E_0}^\infty \frac{dE'}{\sqrt{2\pi}\sigma} \exp\left[-\frac{(E-E')^2}{2\sigma^2}\right]$$

$$M(E) = \frac{S_0}{2}\left[1 + \mathrm{erf}\left(\frac{E-E_0}{\sqrt{2}\sigma}\right)\right]$$

where

$$\mathrm{erf}\left(\frac{E-E_0}{\sqrt{2}\sigma}\right) = \text{error function} = \frac{1}{\sqrt{2\pi}} \int_0^{(E-E_0)/\sigma} e^{-t^2/2} dt$$

Figure 11.5 shows the three functions involved.

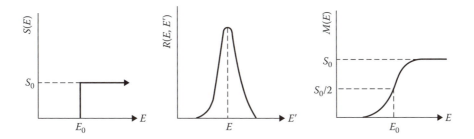

Figure 11.5   A step function folded with a Gaussian.

Example 11.6   The source spectrum is a square spectrum:

$$S(E) = \begin{cases} S_0, & E_1 \leq E \leq E_2 \\ 0, & \text{otherwise} \end{cases}$$

What is the measured spectrum?
Answer

$$M(E) = \int_0^{\infty} \frac{dE'}{\sqrt{2\pi}\sigma} \exp\left[-\frac{(E-E')^2}{2\sigma^2}\right] S(E')$$

$$= S_0 \int_{E_1}^{E_2} \frac{dE'}{\sqrt{2\pi}\sigma} \exp\left[-\frac{(E-E')^2}{2\sigma^2}\right]$$

$$= \frac{S_0}{2}\left[\text{erf}\left(\frac{E_2-E}{\sqrt{2}\sigma}\right) - \text{erf}\left(\frac{E_1-E}{\sqrt{2}\sigma}\right)\right]$$

Figure 11.6 shows the functions involved.

Example 11.7   The source spectrum is a Gaussian centered at $E = E_0$:

$$S(E) = \frac{S_0}{\sqrt{2\pi}\sigma_S} \exp\left[-\frac{(E-E_0)^2}{2\sigma_S^2}\right]$$

What is the measured spectrum?
Answer

$$M(E)^* = \int_{-\infty}^{\infty} \frac{dE'}{\sqrt{2\pi}\sigma} \exp\left[-\frac{(E-E')^2}{2\sigma^2}\right] \frac{S_0}{\sqrt{2\pi}\sigma} \exp\left[-\frac{(E-E_0)^2}{2\sigma_S^2}\right]$$

$$= \frac{S_0}{\sqrt{2\pi}(\sigma_S^2+\sigma^2)^{1/2}} \exp\left[-\frac{(E-E_0)^2}{2(\sigma^2+\sigma_S^2)}\right]$$

Figure 11.7 shows the three functions involved. It is worth noting that if a Gaussian is folded with another Gaussian, their standard deviations add in quadrature.

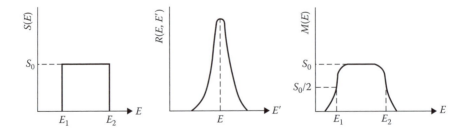

Figure 11.6   A "square" function folded with a Gaussian.

---

* The integral of Eq. 11.33 may be extended to $-\infty$ because the Gaussian drops off quickly to a negligible value for $E < 0$.

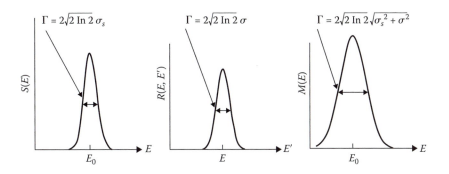

Figure 11.7 A Gaussian folded with a Gaussian gives a third Gaussian, which has a larger width.

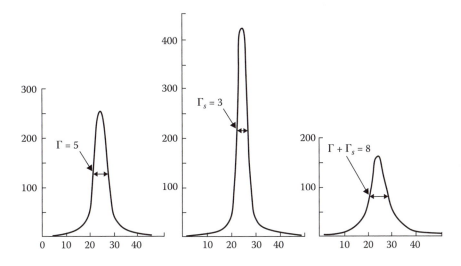

Figure 11.8 A Lorentzian folded with a Lorentzian gives a third Lorentzian with a width equal to the sum of the widths of the first two distributions.

Example 11.8 The source spectrum is a Lorentzian centered at E_0:

$$S(E) = \frac{\Gamma_s/2\pi}{(E-E_0)^2 + \Gamma_s^2/4}$$

and the response function is also a Lorentzian with width Γ. What is the measured spectrum?

Answer

$$M(E) = \int_{-\infty}^{\infty} \frac{\Gamma/2\pi}{(E-E')^2 + \Gamma^2/4} \frac{\Gamma_s/2\pi}{(E'-E_0)^2 + \Gamma_s^2/4} \, dE'$$

$$= \frac{(\Gamma+\Gamma_s)/2\pi}{(E-E_0)^2 + (\Gamma+\Gamma_s)^2/4}$$

Figure 11.8 shows the three functions involved. Notice that by folding a Lorentzian with a Lorentzian, the result is a third Lorentzian with width equal to the sum of the two widths.

11.5.2 The General Method of Unfolding

This section discusses methods of unfolding, assuming that an energy spectrum is measured with a multichannel analyzer or any other device that divides the measured spectrum into energy bins.

299

As stated at the beginning of Section 11.5, unfolding means to solve the Fredholm-type integral equation

$$M(E) = \int_0^\infty R(E,E')S(E')\,dE'$$

(11.39)

for the unknown function $S(E)$. Before possible methods of solution of Eq. 11.39 are discussed, it is important to note that no spectrometer measures $M(E)$. What is measured is the quantity

$$M_i = \int_{E_i}^{E_{i+1}} M(E)\,dE$$

(11.40)

where $E_{i+1} - E_i = \Delta E_i =$ energy "bin" of the spectrometer. For a multichannel analyzer, ΔE_i represents the width of one of the channels. Therefore, one never measures a continuous function $M(E)$ but obtains instead a histogram consisting of the quantities M_i (see Figure 11.9). As a first approximation, $M(E_i) \approx M_i/\Delta E_i$.

An analytic solution of Eq. 11.39 is immediately obtained if the detector response is a δ function. Indeed, if $R(E, E') = \delta(E - E')$, then

$$M(E) = \int_0^\infty \delta(E - E')S(E')\,dE' = S(E)$$

This case is not encountered in practice because there is no detector with such a response function; it indicates only that with perfect energy resolution there is no need for unfolding. In general, the more the detector response resembles a δ-function, the more the measured spectrum looks like the source spectrum.

A second type of response that gives an analytic solution, in principle, is a step function* (Figure 11.10). Let

$$R(E,E') = \begin{cases} \dfrac{C}{E'} & 0 < E \le E' \\ 0 & \text{otherwise} \end{cases}$$

(11.41)

where C is a normalization constant. Then Eq. 11.39 takes the form

$$M(E) = \int_E^\infty \frac{C}{E'}S(E')\,dE'$$

(11.42)

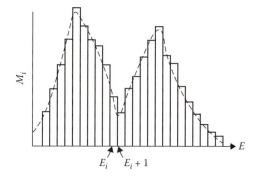

Figure 11.9 The spectrometer produces a histogram, that is, the quantities M_i, and not the continuous function $M(E)$ shown by the dashed line.

---

* The response of proton-recoil counters resembles a step function (see Chapter 14).

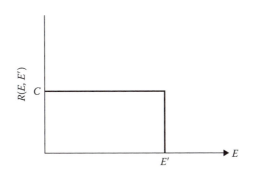

Figure 11.10   A step-function response.

The lower limit of the integral has been set equal to E because if the response function is that given by Eq. 11.41, no source particles with energy $E' < E$ can contribute to $M(E)$. Upon differentiation,[1–4] Eq. 11.42 gives

$$S(E) = \frac{E}{C} \left| \frac{dM(E)}{dE} \right|$$ (11.43)

which is the desired solution. This method of unfolding is known as the *differentiation method*.

Since only the quantities M_i (Eq. 11.40) are obtained, and not the function $M(E)$, the differentiation indicated by Eq. 11.43 must be performed numerically. There are several computer codes that perform such differentiation.

The most general method of unfolding is based on a transformation of Eq. 11.39 into a matrix equation. Equation 11.39 may be rewritten as

$$M(E) = \sum_{j}^{NR} \int_{E_j}^{E_{j+1}} R(E, E')S(E')\, dE'$$ (11.44)

where the integral over E' has been written as a sum of integrals over NR energy intervals $\Delta E_j = E_{j+1} - E_j$. Next, Eq. 11.44 is integrated over E to give (see Eq. 11.40)

$$M_i = \int_{E_i}^{E_{i+1}} dE\, M(E) = \sum_{j}^{NR} \int_{E_i}^{E_{i+1}} dE \int_{E_j}^{E_{j+1}} R(E, E')S(E')\, dE'$$ (11.45)

Equation 11.45 is still exact. To proceed further, one needs an approximation for the source spectrum $S(E)$. Two approximations and corresponding methods of solving the resulting matrix equation are presented in the next two sections.

11.5.3 An Iteration Method of Unfolding

There are several iteration methods. The method presented here is useful for slowly varying spectra and has been used successfully to unfold beta spectra.[5]

The source spectrum $S(E)$ is approximated over any interval ΔE_j by the expression

$$S(E') = \frac{S_j}{\Delta E_j}$$ (11.46)

Using Eq. 11.46 and defining

$$R_{ij} = \frac{1}{\Delta E_j} \int_{E_i}^{} dE \int_{E_j}^{} dE'\, R(E, E')$$ (11.47)

then Eq. 11.45 takes the form

$$M_i = \sum_i R_{ij}S_j, \quad i, j = 1, NR$$ (11.48)

301

or, in matrix notation,

$$\mathbf{M} = \mathbf{R} \cdot \mathbf{S} \tag{11.49}$$

A formal solution of Eq. 11.49 is

$$\mathbf{S} = \mathbf{R}^{-1}\, \mathbf{M} \tag{11.50}$$

where $\mathbf{R}^{-1}$ is the inverse of the matrix with elements given by Eq. 11.47. Although in principle, Eq. 11.50 represents a solution to the unfolding problem, in practice the matrix inversion is not always achieved or leads to a solution with a large error.

The iteration method to be discussed here starts with Eq. 11.48 and uses the measured spectrum as the first guess of the iteration procedure.[5]

$$S_j^{(1)} = M_j, \quad j = 1, NR$$

This source spectrum when substituted into Eq. 11.48 gives

$$M_j^{(1)} = \sum_i R_{ij} S_j^{(1)}, \quad j = 1, NR$$

The error of $S_j^{(1)}$ is taken to be

$$S_j - S_j^{(1)} = M_j - M_j^{(1)}, \quad j = 1, NR$$

and the new guess for the second iteration is

$$S_j^{(2)} = S_j^{(1)} + [M_j - M_j^{(1)}], \quad j = 1, NR$$

Substitution into Eq. 11.48 gives

$$M_j^{(2)} = \sum R_{ij} S_i^{(2)}, \quad i = 1, NR$$

and so on. The nth iteration uses

$$S_j^{(n)} = S_j^{(n-1)} + [M_j - M_j^{(n-1)}], \quad j = 1, NR$$

and is the solution to the problem if the difference $\left| M_j - M_j^{(n)} \right|$ for $j = 1, \ldots, NR$ is acceptably small, i.e., it satisfies a criterion set by the experiment. This iteration method converges in less than five iterations and gives good results.

11.5.4 Least-Squares Unfolding

A different approximation for the source spectrum, used with neutrons, assumes that $S(E)$ can be represented as a sum of NS discrete components.[6–8] Therefore, one can write

$$S(E') = \sum_{j=1}^{NS} X_j \delta(E' - E_j) \tag{11.51}$$

Using Eq. 11.51 and defining

$$A_{ij} = \frac{1}{E_i - E_{i-1}} \int_{E_{i-1}}^{E_i} R(E, E_j)\, \mathrm{d}E \tag{11.52}$$

then Eq. 11.45 takes the form

$$M_i = \sum_{j=1}^{NS} A_{ij} X_j \quad \begin{cases} i = 1, \ldots, NR \\ j = 1, \ldots, NS \end{cases} \tag{11.53}$$

or in matrix notation,

$$\mathbf{M} = \mathbf{AX} \tag{11.54}$$

If $NR = NS$, the formal solution of Eq. 11.54 is, as with Eq. 11.49,

$$\mathbf{X} = \mathbf{A}^{-1}\mathbf{M} \tag{11.55}$$

Because of the difficulties of matrix inversion, a least-squares solution has been attempted with $NR > NS$. If $NR < NS$, no unique solution exists, but an acceptable one has been obtained.

The least-squares unfolding starts with Eq. 11.53 and minimizes the quantity

$$Q = \sum_{i=1}^{NR} w_i \left(M_i - \sum_{j=1}^{NS} A_{ij} X_i \right)^2 \tag{11.56}$$

The weighting factors w_i are usually taken to be the inverse of the variance of M_i. The minimization is achieved by setting

$$\frac{\partial Q}{\partial X_k} = 0, \quad k = 1, \ldots, NS$$

which gives

$$\sum_{i=1}^{NR} w_i A_{ik} \left(M_i - \sum_{j=1}^{NS} A_{ij} X_j \right) = 0, \quad k = 1, \ldots, NS \tag{11.57}$$

and can be solved for X_j for $j = 1, NS$. Equation 11.57 may be written in matrix form[6]

$$\mathbf{X} = (\mathbf{A}^T \mathbf{WA})^{-1} \mathbf{A}^T \mathbf{WM} \tag{11.58}$$

where $\mathbf{A}^T$ is the transpose of $\mathbf{A}$.

Computer round-off errors in completing the matrix inversion shown by Eq. 11.58 lead to large oscillations in the solution $\mathbf{X}$. The oscillations can be reduced if the least-squares solution is "constrained." Details of least-squares unfolding with constraints are given in References 6 and 7.

There are a multitude of examples of unfolding algorithms that have appeared in the open literature. While many of these codes are probably of very high quality, they often lack rigorous documentation to be used by other researchers. Both Canberra[9] and Ortec[10] the two most popular nuclear instrumentation vendors, offer gamma ray unfolding codes and also offer a lot of service support and training sessions. Some of the unfolding techniques include a model for fitting peaks induced by fast neutrons in an HPGe detector[11] analysis of coincidence γ-ray spectra using advanced background elimination, unfolding and fitting algorithms,[12] an automated peak searching and fitting of data from γ-ray coincidence experiments,[13] application of robust fitting analysis techniques to low-resolution spectral data,[14] a comparative study of minimization methods in the fitting of α-particle spectra,[15] a genetic algorithm approach for deconvolution in γ-ray spectra,[16] a method for unfolding multiplet regions of X- and γ-ray spectra with a detection efficiency constraint avoiding inflecting the peak shapes for correction results,[17] and a gamma analysis code for ultra-low-level HPGe spectra.[18]

11.6 DATA SMOOTHING

The smoothing of raw experimental data is a controversial subject because it represents manipulation of the data without clear theoretical justification. However, smoothing is generally accepted as common practice, since experience has shown that it is beneficial in certain cases to the subsequent analysis of the data, for example, in identification of energy peaks in complex gamma energy spectra (Chapter 12) and unfolding of neutron energy spectra (Chapter 14). Data smoothing should be viewed as an attempt to filter out the statistical fluctuations without altering the significant features of the data.

To illustrate how data smoothing is performed, consider again N measurements $y_i|_{i=1, N}$, where $y_i = y(x_i)$. Smoothing, which is applied to the values of y_i, is an averaging process. In the simplest case, one adds a fixed odd number of y_i values, takes the arithmetic average of the sum, and sets the smoothed value of y_i at the center of the group equal to this average. Next, the first point of the group is dropped, the next point is added at the other end of the group, and the process is repeated for all y_i points. In general, the "smoothing" equation takes the form

$$y_i = \frac{1}{M} \sum_{j=-n}^{j=n} C_j y_{i+j} \tag{11.59}$$

where C_j = coefficients that depend on the method of smoothing (see below)
$\quad M$ = normalization constant
$\quad n$ = index showing the number of points used in the smoothing process
$\quad\quad$ (the index n means that $2n + 1$ points were used for smoothing)
$\quad y_i$ = smoothed value, replacing the old y_i in the middle of $2n + 1$ points

The coefficients C_j are determined by least-squares fitting a polynomial of order m to $2n + 1$ data points[19-21] and taking the smoothed value equal to the value of the polynomial in the middle point (Figure 11.11). To illustrate the method, a few examples are given below. The least-squares fit will be based on Eqs. 11.18–11.20.

Three-point zeroth-order smoothing. From Eq. 11.18,

$$f(x) = a_1$$

From Eq. 11.19,

$$A_{11} = \sum_{i=1}^{3} w_i = \sum_i 1 = 3$$

From Eq. 11.20,

$$B_1 = \sum_i^3 w_i y_i = \sum_{i=1}^{3} y_i$$

From Eq. 11.16,

$$a_1 = \frac{B_1}{A_{11}} = \frac{1}{3} \sum_{i=1}^{3} y_i$$

Therefore, if three-point zeroth-order smoothing is applied (Figure 11.12), the constants of Eq. 11.59 are

$$M = 3 \quad C_1 = C_2 = C_3 = 1$$

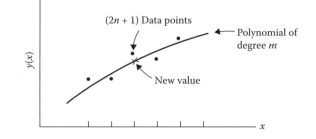

Figure 11.11 Data smoothing. A polynomial of degree m is fitted to $(2n + 1)$ data points, and the smoothed value is equal to the value of the polynomial in the middle point.

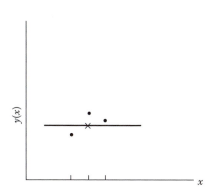

Figure 11.12 Three-point, zeroth-order smoothing.

Table 11.2
Coefficients for Second-Order Smoothing

n	5	7	9	11
M	35	21	231	429
C_{-5}				36
C_{-4}			−21	9
C_{-3}		−2	14	44
C_{-2}	−3	3	39	69
C_{-1}	12	6	54	84
C_0	17	7	59	89
C_1	12	6	54	84
C_2	−3	3	39	69
C_3		−2	14	44
C_4			−21	9
				−36

Source: From Savitzky, A., and Golay, M. J. E., *Anal. Chem.*, 36, 1627, 1964.

and

$$y_i = \frac{1}{3}(y_{i-1} + y_i + y_{i+1})$$

(11.60)

Five-point zeroth-order smoothing. Following the same steps as above, one obtains

$$M = 5, \quad C_i = 1, \quad i = 1, 5$$

and

$$y_i = \frac{1}{5}(y_{i-2} + y_{i-1} + y_i + y_{i+1} + y_{i+2})$$

(11.61)

Three-point first-order smoothing. From Eq. 11.18, $f(x) = a_1 + a_2 x$. Using Eqs. 11.19, 11.20, and 11.16, one can solve for the values of M and C_i. If the x_i points are equally spaced, the result is identical with three-point zeroth-order smoothing (Eq. 11.60). This is true, in general, for equally spaced x_i; that is, the result of smoothing with an even-order polynomial is the same as that with a polynomial of the next higher order. Table 11.2 gives the values of M and C_j for second-order smoothing.

As an example of using the various equations, Figure 11.13 shows results of three-point zeroth-order smoothing and five-point second-order smoothing, that is, using Eqs. 11.60 and 11.62:

$$y_i = \frac{1}{35}(-3y_{i-2} + 12y_{i-1} + 17y_i + 17y_i + 12y_{i+1} - 3y_{i+2})$$

(11.62)

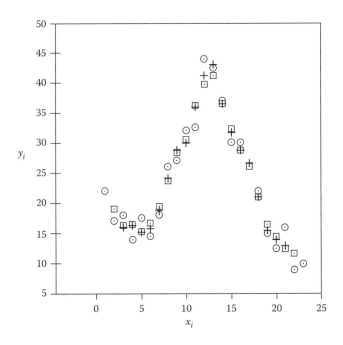

Figure 11.13 The same data smoothed with two different equations. Circles are original data, squares are data smoothed with Eq. 11.60, and crosses are data smoothed with Eq. 11.62.

If the total number of points is N, the number of smoothed points is $N - 2n$ for $(2n + 1)$-point smoothing because the first smoothed point is $i = n + 1$ and the last one is $N - n$. The smoothing process can be repeated if necessary—that is, one may smooth data that were previously smoothed.[19]

One of the difficulties in data smoothing is the choice of the correct value for n. Unfortunately, there are no strict criteria for the selection of n. The analyst should be guided mainly by experience and by the general effects of smoothing on the results. In particular, if the data represent energy spectra with many peaks, the smoothed spectrum may tend to flatten the peaks and fill the valleys.

The general smoothing equation may be written as

$$Z_i = \sum_j R_{i-j} Y_j \tag{11.63}$$

which has the same form as the folding-unfolding matrix equations (Eq. 11.48). Thus, smoothing may be considered as folding the data (y_i) with the weights (R_{i-j}) as the response function. Taking it one step further, one may perform smoothing by using a continuous function, that is, a Gaussian. Then

$$Z(x) = \int G(x, x') y(x') dx' \tag{11.64}$$

The operation indicated by Eq. 11.64 has been applied to neutron spectroscopic data.[22]

11.7 QUALITY ASSURANCE AND QUALITY CONTROL

Total quality management (TQM), which comprises of quality assurance (QA) and quality control (QC) is now well accepted vital component in any analytical laboratory. While TQM methodologies are similar in many cases, each different type of measurement has its own unique characteristics that needs to be specifically addressed.

Nuclear measurements span a very large cross-section of disciplines. Some of the more common examples include, health physics and dosimetry, neutron activation analysis, homeland security, radiochemistry in the nuclear fuel cycle, environmental radioactivity, production of radioactive certified reference materials, and nuclear medicine. Accreditation of any laboratory which includes a TQM procedure is very important not only for research purposes but also for potential legal procedures. The International Laboratory Accreditation Cooperation[23] has various documents on laboratory accreditation. Also the International Standards Organization[24] has a vast database of quality assurance and quality control procedures. Some examples include auditing quality systems, accuracy of measurement methods and results, terms and definitions used in connection with reference materials and proficiency testing by interlaboratory comparisons. All these aspects are very important in nuclear measurements.

Measurements in nuclear instrumentation require particular requirements which are not usually seen in traditional chemical methods. For instance, special care is needed in nuclear spectroscopy when analyzing radiation sources with elevated count rates (or high dead times) or measuring different geometries. The abundance of new measuring instruments on the market necessitates that personnel in each individual laboratory fully understand the implications of setting up the required parameters for measurement which requires knowledge of software used in unfolding energy spectra. This is particularly true when fitting complicated spectra in X-ray, γ-ray, or alpha spectroscopy.

The International Atomic Energy Agency has published technical documents entitled Quality System Implementation for Nuclear Analytical Techniques[25] and Quality Control Procedures Applied to Nuclear Instruments.[26] Both these manuscripts are of great value in any nuclear measurements laboratory.

The phenomena of detection limits, statistical uncertainty error analysis and propagation of errors have been discussed in many papers and review articles. The classic paper by Currie in 1968[27] was the first real attempt to quantify detection limits and error analysis in nuclear measurement. In Chapter 2 there is a very good treatise of statistical errors of radiation counting, as well as minimum detection limits. However, while statistical counting errors are the main sources of uncertainty, there are other sources of error uncertainity which may complicate the calculation of the overall error. For instance unfolding overlapping peaks, energy spectral interferences, and background radioactivity for low-level measurements are three prime examples. Poor energy calibration or increased full width half maximum (FWHM) due to deteriorating resolution can significantly add to the error. Thus; the proper operation of nuclear instruments is of prime importance as to not add any unintended errors to the measurement systems.

In conclusion, maintaining the instrument at a high level of confidence during operation will lead to a measurement with a reduced error. Also, all sources of error must be identified and propagated correctly to obtain a reliable uncertainity for the final measurement.

PROBLEMS

11.1 The table below shows radioactive decay data from a certain isotope. Using least-squares fit, determine the half-life of the isotope. What is the error of the half-life as determined by this set of data?

t (min)	Counts	t (min)	Counts
0	500	6	164
1	430	7	130
2	310	8	92
3	265	9	89
4	240	10	75
5	186		

11.2 The numbers below represent values of cosine for the corresponding angles.

Angle:	5°	10°	15°	20°	25°	30°
Cosine:	0.99619	0.98481	0.96593	0.93969	0.90631	0.86603

Obtain cosine values, by interpolation, for 22° using Lagrange's three-point interpolation formula.
Evaluate the error of your result. Compare the error with its correct value.

11.3 Prove Eqs. 11.22 to 11.24.

11.4 Obtain the least-squares fit equations for a quadratic fit.

11.5 Prove that the result of folding a step function with a Gaussian is

$$M(E) = \frac{S_0}{2}\left[1 + \mathrm{erf}\left(\frac{E - E_0}{\sqrt{2}\sigma}\right)\right]$$

where the source spectrum is

$$S(E) = \begin{cases} S_0, & E \geq E_0 \\ 0, & \text{otherwise} \end{cases}$$

11.6 Prove that the result of folding a Gaussian with a Gaussian is

$$M(E) = \frac{1}{\sqrt{2\pi}\sqrt{(\sigma^2 + \sigma_S^2)}} \exp\left[-\frac{(E - E_0)^2}{2(\sigma^2 + \sigma_S^2)} \right]$$

where the source spectrum is centered at E_0 and has a standard deviation σ_S.

11.7 Prove that the result of folding an exponential function e^{-aE} with a Gaussian is

$$M(E) = e^{a^2\sigma^2/2} e^{-aE}$$

11.8 What is the measured spectrum $M(E)$ if the detector response is a step function of the form $R(E, E') = C(E')/E'$ and the source emits two types of particles at energy E_1 and E_2?

11.9 What is the measured spectrum $M(E)$ if the detector response is a step function, as in Problem 11.8, and the source spectrum is

$$S(E) = \frac{S_0}{E_2 - E_1}, \quad E_1 \leq E \leq E_2$$

and is zero otherwise.

11.10 The following data represent results of counting an energy peak. How does the full width at half maximum of the peak change if one applies (a) three-point zeroth-order smoothing and (b) five-point second-order smoothing?

Channel	Counts	Channel	Counts
10	12	17	34
11	10	18	26
12	14	19	18
13	14	20	10
14	24	21	12
15	30	22	9
16	40		

BIBLIOGRAPHY

Abramowitz, M., and Stegan, I. A. (eds.), *Handbook of Mathematical Functions*, National Bureau of Standards, Applied Mathematics Series, 55, June 1964.

Bevington, R. P., *Data Reduction and Error Analysis for the Physical Sciences*, McGraw-Hill, New York, 1969.

Hildebrand, F. B., *Introduction to Numerical Analysis*, McGraw-Hill, New York, 1956.

Householder, A. S., *Principles of Numerical Analysis*, McGraw-Hill, New York, 1953.

Knoll, G. F., *Radiation Detection and Measurement*, 3rd ed., John Wiley and Sons, New York, 2000.

Lyons, L., *Statistics for Nuclear and Particle Physicists*, Cambridge University Press, 1989.

REFERENCES

1. Bennet, F. F., and Yule, T. J., ANL-7763, 1971.

2. Ciallela, C. M., and Devanney, J. A., *Nucl. Instrum. Meth.* **60**:269 (1968).

3. Toms, M. E., *Nucl. Instrum. Meth.* **92**:61 (1971).

4. Johnson, R. H., and Wehring, B. W., ORNL/RSIC-40, p. 33, 1976.

5. Tsoulfanidis, N., Wehring, B. W., and Wyman, M. E., *Nucl. Instrum. Meth.* **73**:98 (1969).

6. Burrus, W. R., and Verbinski, V. V., *Nucl. Instrum. Meth.* **67**:181 (1969).

7. Kendrick, H., and Sperling, S. M., GA-9882, 1970.

8. Johnson, R. H., Ingersoll, D. T., Wehring, B. W., and Doming, J. J., *Nucl. Instrum. Meth.* **145**:337 (1977).

9. http://www.canberra.com

10. http://www.ortec-online.com

11. Siiskonen, T., and Toivonen, H., *Nucl. Instrum. Meth. Phys. Res. A* **540**:403 (2005).

12. Morhác, M., Matousek, V., Kliman, J., Krupa, L., and Jandel, M., *Instrum. Meth. Phys. Res. A* **502**:784 (2003).

13. Smith, A. G., and Vermeer, W. J., *Instrum. Meth. Phys. Res. A* **350**:314 (1994).

14. Lasche, G. P., Coldwell, R. L., and Cray, C. J., *Instrum. Meth. Phys. Res. A* **458**:491 (2001).

15. García-Toraño, E., *Instrum. Meth. Phys. Res. A* **369**:608 (1996).

16. Garcia-Talavera, M., and Ulicny, B., *Instrum. Meth. Phys. Res. A* **512**:585 (2003).

17. Wang, S-G., Mao, Y-J., and Tang, P-J., *Instrum. Meth. Phys. Res. A* **600**:445 (2009).

18. Winn, W. G., *Instrum. Meth. Phys. Res. A* **450**:430 (2000).

19. Savitzky, A., and Golay, M. J. E., *Anal. Chem.* **36**:1627 (1964).

20. Yule, H. P., *Nucl. Instrum. Meth.* **54**:61 (1967).

21. Yule, H. P., *Anal. Chem.* **38**:103 (1966).

22. Johnson, R. H., Wehring, B. W., and Doming, J. J., *Nucl. Sci. Eng.* **73**:93 (1980).

23. http://www.ilac.org

24. http://www.iso.org

25. http://www-pub.iaea.org/MTCD/publications/PDF/TCS-24_web.pdf (2004).

26. http://www-pub.iaea.org/MTCD/publications/PDF/te_1599_web.pdf (2008).

27. Currie, L. A., *Anal. Chem.* **40**:586 (1968).

12

Photon (γ-Ray and X-Ray) Spectroscopy

12.1 INTRODUCTION

Photons, that is, gamma-rays and X-rays, may be treated either as electromagnetic waves or as particles. An *electromagnetic wave* is characterized by its wavelength λ or frequency v. A *photon* is a particle having zero charge and zero rest mass, traveling with the speed of light, and having an energy $E = hv$, where h = Planck's constant. The wave properties of a photon are used for low-energy measurements only. In all other cases, detection of photons is based on their interactions as particles.

This chapter first examines the mechanisms of detection in photon detectors and then discusses the spectroscopic characteristics of the different types of X-ray and γ-ray detectors.

12.2 MODES OF ENERGY DEPOSITION IN THE DETECTOR

Photons are detected by means of the electrons they produce when they interact in the material of which the detector is made. The main interactions are photoelectric effect, Compton scattering, and pair production. The electrons (or positrons) produced by these interactions deposit their energy in the counter and thus generate a voltage pulse that signifies the passage of the photon. The height of the voltage pulse is proportional to the energy deposited in the detector. Since the objective is to measure the energy of the incident photon, the question arises: Is this voltage pulse proportional to the energy of the incident particle? To provide an answer, one must examine how the photon interacts and what happens to its energy.

12.2.1 Energy Deposition by Photons with $E < 1.022$ MeV

A photon with $E < 1.022$ MeV can interact only through the photoelectric or the Compton effect. If a photoelectric interaction takes place, the photon disappears and an electron appears with energy equal to $E - B_e$, where B_e is the binding energy of that electron. The range of electrons in a solid, either a scintillator crystal or a semiconductor, is so short that it can be safely assumed that all the electron energy will be deposited in the detector (Figure 12.1a). If the interaction occurs very close to the wall, the electron may deposit only part of its energy in the counter (Figure 12.1b), but the probability of this happening is small. In practice, one assumes that all the photoelectrons deposit all their energy in the detector. This energy is less than the energy of the incident photon by the amount B_e, the binding energy of the electron. What happens to the energy B_e?

After a photoelectric effect takes place, an electron from one of the outer atomic shells drops into the empty inner state in about 10^{-8} s. This electronic transition is followed by an X-ray or by an Auger electron (see Chapter 4). The Auger electron will also deposit its energy in the detector. The X-ray with energy in the low keV range (~100 keV or less) interacts again photoelectrically and generates another electron.* The net result of these successive interactions is that the part B_e of the incident photon energy is also deposited in the counter. All these events take place within a time of the order of 10^{-8} s. Since the formation of the voltage pulse takes about 10^{-6} s, both parts of the

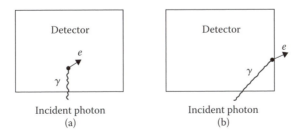

Figure 12.1 As a result of a photoelectric interaction, the photon disappears, (a) all the energy of the electron is deposited in the detector, (b) part of the energy is deposited in the wall.

---

* For thin detectors, or detectors made of high-Z material—for example, CdTe or HgI_2—some X-rays may escape, thus forming the so-called "escape peaks" (see the section CdTe and HgI_2 Detectors as Gamma Spectrometers).

energy—namely, $E - B_e$ = energy of photoelectron and B_e = energy of the X-ray—contribute to the same pulse, the height of which is proportional to $(E - B_e) + B_e = E$ = incident photon energy. The conclusion is, therefore, that if the photon interacts via photoelectric effect, the resulting pulse has a height proportional to the incident particle energy.

If Compton scattering takes place, only a fraction of the photon energy is given to an electron. A scattered photon still exists carrying the rest of the energy. The energy of the electron is deposited in the detector. But what happens to the energy of the scattered photon?

The scattered photon may or may not interact again inside the detector. The probability of a second interaction depends on the size of the counter (Figure 12.2), on the position of the first interaction, on the energy of the scattered photon, and on the material of which the detector is made. Unless the detector is infinite in size, there is always a chance that the scattered photon may escape, in which case a pulse will be formed with height proportional to an energy that is *less* than the energy of the incident photon.

From the study of the Compton effect (Chapter 4), it is known that Compton electrons have an energy range from zero up to a maximum energy T_{max}, which is

$$T_{max} = E - \frac{E}{1 + 2E/mc^2} \tag{12.1}$$

where $mc^2 = 0.511$ MeV, the rest mass energy of the electron. Therefore, if the interaction is Compton scattering, pulses are produced from Compton electrons with heights distributed from $V = 0$ V, corresponding to $T_{min} = 0$, up to a maximum height V_{max} volts corresponding to the maximum energy T_{max}. Figures 12.3 to 12.5 illustrate how a monoenergetic photon spectrum is recorded as a result of photoelectric and Compton interactions.

Figure 12.3 shows the source spectrum. In the case of perfect energy resolution, this monoenergetic source produces in an MCA the measured spectrum shown by Figure 12.4. Some photons produce pulses that register in channel C_0, corresponding to the source energy E_0, and thus contribute to the main peak of the spectrum, which is called the full-energy peak. The Compton electrons are responsible for the continuous part of the spectrum, extending from zero channel up to channel CC and called the *Compton continuum*. The end of the Compton continuum, called the *Compton edge*, corresponds to the energy given by Eq. 12.1. Since no detector exists with perfect energy resolution, the measured spectrum looks like that of Figure 12.5.

Sometimes the Compton interaction occurs very close to the surface of the detector or in the material of the protective cover surrounding the detector (Figure 12.6).* Then there is a high probability that the electron escapes and only the energy of the scattered photon is deposited in the detector. The minimum energy E_{min} of the scattered photon is given by

$$E_{min} = \frac{E}{1 + 2E/mc^2} \tag{12.2}$$

Occasionally, a rather broad peak, corresponding to the energy given by Eq. 12.2, is observed in γ-ray spectra. This peak is called the *backscatter peak* (Figure 12.5).

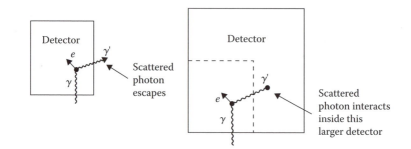

Figure 12.2 As a result of Compton scattering, part of the photon energy may escape.

* Backscattering may also take place in the source itself, or in the shield surrounding the detector.

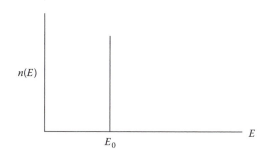

Figure 12.3 A monoenergetic gamma spectrum (source spectrum).

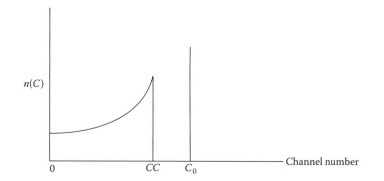

Figure 12.4 The pulse height spectrum obtained from the source spectrum of Figure 12.3, in the absence of statistical effects in the detector (perfect energy resolution).

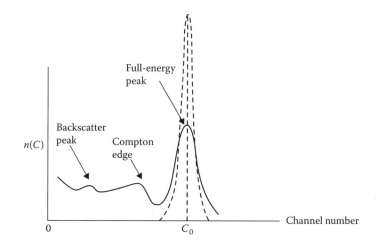

Figure 12.5 The measured pulse height spectrum for the source spectrum of Figure 12.3. The statistical effects in the detector broaden both the peak and the Compton continuum part of the spectrum. The dashed line shows the spectrum that would have been recorded in the absence of the Compton continuum.

The fraction of counts recorded outside the full-energy peak depends on the energy of the gamma and on the size of the detector. The energy of the photon determines the ratio σ/μ of the Compton scattering coefficient to the total attenuation coefficient. The lower the gamma energy is, the smaller this ratio becomes. Then a greater fraction of photons interacts photoelectrically and is recorded in the full-energy peak, thus reducing the Compton continuum part of the spectrum.

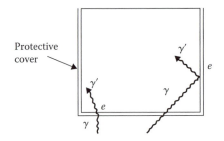

Figure 12.6 If Compton scattering occurs close to the surface of the detector, the only energy deposited may be that of the scattered photon.

As an example, consider gammas with energy 100 keV and 1 MeV, and a Ge detector. For 100-keV gammas in germanium, the ratio σ/μ is 0.9/3.6 ≈ 0.25 (Figure 12.31), which indicates that 25% of the interactions are Compton and 75% photoelectric. The number of pulses in the Compton continuum should be equal to or less than one-third the number recorded under the full energy peak. At 1 MeV, the ratio σ/μ is about 0.4/0.42 ≈ 0.95, which means that about 95% of the interactions are Compton and only 5% photoelectric. Thus, the Compton continuum due to 1-MeV photons is the largest part of the spectrum.

The magnitude of the Compton continuum is also affected by the size of the detector (Figure 12.2). The larger the detector is, the greater the probability of a second Compton interaction. If the detector size could become infinite, the Compton continuum would disappear.

12.2.2 Energy Deposition by Photons with $E > 1.022$ MeV

If $E > 1.022$ MeV, pair production is possible, in addition to photoelectric effect and Compton scattering. As a result of pair production, the photon disappears and an electron–positron pair appears, at the expense of 1.022 MeV transformed into the pair's rest masses. The total kinetic energy of the electron–positron pair is

$$T_{e^-} + T_{e^+} = T = (E - 1.022)\text{MeV}$$

The kinetic energy of the pair is deposited in the counter (the arguments are the same as for photoelectrons or Compton electrons). Therefore, pulses proportional to the energy $T = E - 1.022$ MeV are certainly produced, but what happens to the energy of 1.022 MeV?

The positron slows down and reaches the end of its range in a very short time, shorter than the time needed for pulse formation. Sometimes while in flight, but most of the time at the end of its track, it combines with an atomic electron, the two annihilate, and two gammas are emitted, each with energy 0.511 MeV.* There are several possibilities for the fate of these annihilation gammas:

1. The energy of both annihilation gammas is deposited in the detector. Then, a pulse height proportional to energy

$$(E - 1.022)\,\text{MeV} + 1.022\,\text{MeV} = E$$

is produced.

2. Both annihilation photons escape. A pulse height proportional to energy $(E - 1.022)$ MeV is formed.

3. One annihilation photon escapes. A pulse height proportional to energy

$$(E - 1.022)\,\text{MeV} + 0.511\,\text{MeV} = (E - 0.511)\,\text{MeV}$$

is formed.

---

* There is a small probability that three gamma may be emitted. This event has a negligible effect on spectroscopy measurements.

If the pair production event takes place on or close to the surface of the detector, it is possible that only one of the annihilation photons enters the counter. In such a case, a pulse height proportional to energy 0.511 MeV is formed.

Peaks corresponding to these energies could be identified, but this does not mean that they are observed in every γ-ray spectrum. The number, energy, and intensity of peaks depend on the size of the detector, the geometry of the source (is it collimated or not?), and the energies of the gammas in the spectrum. If a source emits only one gamma, the measured spectrum will certainly show the following:

1. The full energy peak, corresponding to E (this is the highest energy peak).

2. The Compton edge, corresponding to energy

$$E - \frac{E}{1 + 2E/mc^2}$$

In addition, other peaks that may be observed are as given below.

3. Backscatter peak, with energy

$$\frac{E}{1 + 2E/mc^2}$$

4. The *single-escape peak* with energy (E – 0.511) MeV.

5. The *double-escape peak* with energy (E – 1.022) MeV.

Figure 12.7 presents the spectrum of ^{24}Na. The single- and double-escape peaks due to the 2.754-MeV gamma are clearly shown. The single- and double-escape peaks are very important when complex gamma spectra are recorded. The observer should be extremely careful to avoid identifying them falsely as peaks produced by gammas emitted from the source.

If the source is a positron emitter, a peak at 0.511 MeV is always present. The positron-emitting isotope ^{22}Na is such an example. It emits only one gamma with energy 1.274 MeV, yet its spectrum shows two peaks. The second peak is produced by 0.511-MeV annihilation photons emitted after a positron annihilates (Figure 12.8).

The Compton continuum, present in gamma energy spectra recorded either by a NaI(Tl) scintillator or by a Ge detector, is a nuisance that impedes the analysis of complex spectra. It is therefore desirable to eliminate or at least reduce that part of the spectrum relative to the gamma energy peak. One way to achieve this is to use two detectors and operate them in anticoincidence. Such an arrangement, known as the *Compton-suppression spectrometer*, is shown in Figure 12.9. A large

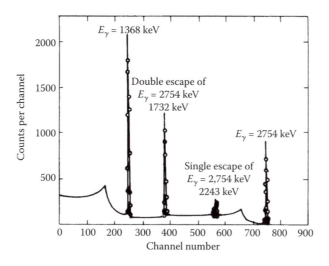

Figure 12.7 A gamma spectrum showing single- and double-escape peaks from ^{24}Na.

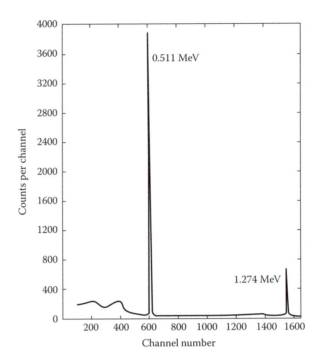

Figure 12.8 The ^{22}Na spectrum showing the 1.274-MeV peak and the 0.511-MeV peak that is due to annihilation gammas.

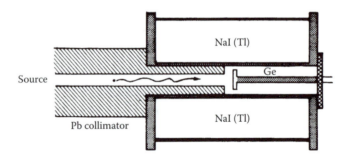

Figure 12.9 Diagram of a Compton suppression spectrometer using a NaI(Tl) and a Ge detector. The two detectors are operated in anticoincidence, with the Ge recording the energy spectrum.

NaI(Tl) scintillator surrounds a Ge detector, and the two detectors are operated in anticoincidence. The energy spectrum of the central detector (the Ge in this case) will consist of pulses that result from total energy absorption in that detector. Figure 12.10 shows the ^{60}Co spectrum obtained with and without Compton suppression.

12.3 EFFICIENCY OF X-RAY AND γ-RAY DETECTORS: DEFINITIONS

There are four types of efficiency reported in the literature:

1. Total detector efficiency

2. Full-energy peak efficiency

3. Double-escape peak efficiency

4. Single-escape peak efficiency

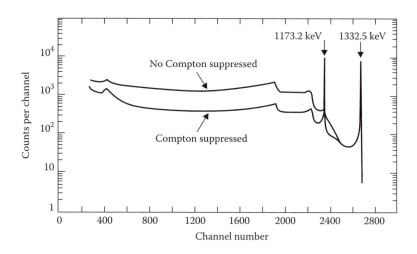

Figure 12.10 The ^{60}Co spectrum recorded with and without Compton suppression. Notice that the ordinate is in logarithmic scale.

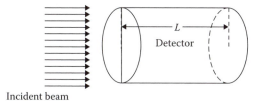

Figure 12.11 The geometry assumed in the definition of intrinsic efficiency.

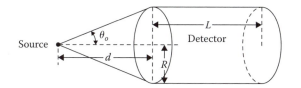

Figure 12.12 The geometry assumed in the definition of absolute efficiency.

The first two are much more frequently used than the last two. All four efficiencies may be intrinsic, absolute, or relative. The individual definitions are as follows.

Intrinsic total detector efficiency is the probability that a gamma of a given energy which strikes the detector will be recorded. The geometry assumed for the calculation or measurement of this efficiency is shown in Figure 12.11.

Absolute total detector efficiency is the probability that a gamma emitted from a specific source will be recorded in the detector. The geometry assumed for the absolute efficiency is shown in Figure 12.12. The intrinsic efficiency (Figure 12.11) depends on the energy of the gamma E and the size of the detector L. The absolute total efficiency (Figure 12.12) depends on, in addition to E and L, the radius of the detector R and the source-detector distance d. Therefore the absolute total efficiency, as defined here, is the product of intrinsic efficiency times the solid angle fraction (see also Chapter 8).

Full-energy peak efficiency is defined as follows:

$$\begin{pmatrix} \text{full-energy peak} \\ \text{efficiency} \end{pmatrix} = \begin{pmatrix} \text{total detector} \\ \text{efficiency} \end{pmatrix} \times \frac{\begin{pmatrix} \text{counts in full-} \\ \text{energy peak} \end{pmatrix}}{\begin{pmatrix} \text{total counts in} \\ \text{spectrum} \end{pmatrix}} \tag{12.3}$$

317

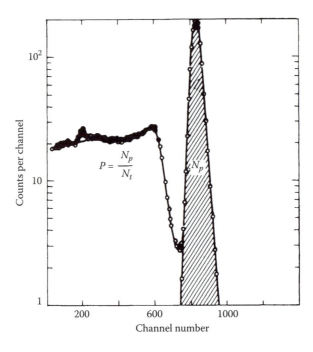

Figure 12.13 The peak-to-total ratio is equal to the number of counts under the peak (N_p) divided by the total number of counts (N_t).

The ratio by which the total detector efficiency is multiplied in Eq. 12.3 is called the *peak-to-total ratio* (P). Figure 12.13 shows how P is measured.

The *double-escape peak efficiency* is important if the energy of the gamma E is greater than about 1.5 MeV, in which case pair production becomes important. The energy of the double-escape peak, equal to $E - 1.022$ MeV, is used for identification of certain isotopes. This kind of efficiency is defined by

$$\begin{pmatrix} \text{double-escape} \\ \text{peak efficiency} \end{pmatrix} = \begin{pmatrix} \text{total detector} \\ \text{efficiency} \end{pmatrix} \times \dfrac{\begin{pmatrix} \text{counts in double-} \\ \text{escape peak} \end{pmatrix}}{\begin{pmatrix} \text{total counts in} \\ \text{spectrum} \end{pmatrix}} \tag{12.4}$$

The *single-escape peak efficiency* is important also for $E > 1.5$ MeV, and its definition is analogous to that of the double-escape peak:

$$\begin{pmatrix} \text{single-escape} \\ \text{peak efficiency} \end{pmatrix} = \begin{pmatrix} \text{total detector} \\ \text{efficiency} \end{pmatrix} \times \dfrac{\begin{pmatrix} \text{counts in single-} \\ \text{escape peak} \end{pmatrix}}{\begin{pmatrix} \text{total counts in} \\ \text{spectrum} \end{pmatrix}} \tag{12.5}$$

The double- and single-escape peak efficiencies are used with semiconductor detectors only. In the above definitions, if the total detector efficiency is replaced by intrinsic, the corresponding full-energy, single-, and double-escape peak efficiencies are also considered intrinsic.

Relative efficiency may be obtained for all the cases discussed above. In general,

$$(\text{relative efficiency})_i = \dfrac{(\text{absolute efficiency})_i}{\text{efficiency of a standard}} \tag{12.6}$$

where the subscript *i* refers to any one of the efficiencies defined earlier.

Depending on the type of detector and measurement, the user selects the efficiency to be used. For quantitative measurements, the absolute total efficiency of the detector has to be used at some stage of the analysis of the experimental data.

12.4 DETECTION OF PHOTONS WITH NaI(Tl) SCINTILLATION DETECTORS

Of all the scintillators existing in the market, the NaI crystal activated with thallium, NaI(Tl), is the most widely used for the detection of γ-rays. NaI(Tl) scintillation counters are used when the energy resolution is not the most important factor of the measurement. They have the following advantages over Ge and Si(Li) detectors:

1. They can be obtained in almost any shape and size. NaI(Tl) crystals with size 0.20 m (8 in.) diameter by 0.20 m (8 in.) thickness are commercially available.

2. They have rather high efficiency (see Section 12.4.1).

3. They cost less than semiconductor detectors.

A disadvantage of all scintillation counters, in addition to their inferior energy resolution relative to Si(Li) and Ge detectors, is the necessary coupling to a photomultiplier tube.

NaI(Tl) detectors are offered in the market today either as crystals that may be ordered to size or as integral assemblies mounted to an appropriate photo-multiplier tube.[1-3] The integral assemblies are hermetically sealed by an aluminum housing. Often, the housing is chrome-plated for easier cleaning. The phototube itself is covered by an antimagnetic μ-metal that reduces gain perturbations caused by electric and magnetic fields surrounding the unit.

The front face of the assembly is usually the "window" through which the photons pass before they enter into the crystal. The window should be as thin as possible to minimize the number of interactions of the incident photons in the materials of the window. Commercially available NaI(Tl) counters used for γ-ray detection have an aluminum window, which may be as thin as 0.5 mm (0.02 in.). X-ray scintillation counters usually have a beryllium window, which may be as thin as 0.13 mm (0.005 in.). Beryllium is an excellent material because it allows less absorption thanks to its low atomic number ($Z = 4$).

12.4.1 Efficiency of NaI(Tl) Detectors

The intrinsic efficiency of NaI(Tl) detectors (see Figure 12.11) is essentially equal to $1 - \exp[-\mu(E)L]$, where

$\mu(E)$ = total attenuation coefficient in NaI for photons with energy E
L = length of the crystal

A plot of $\mu(E)$ for NaI as a function of photon energy is shown in Figure 12.14.

Obviously, the efficiency increases with crystal size. The user should be aware, however, that when the detector volume increases, the background counting rate increases too. In fact, the background is roughly proportional to the crystal volume, while the efficiency increases with size at a slower than linear rate. Thus, there may be a practical upper limit to a useful detector size for a given experiment.

Calculated absolute total efficiencies of a NaI crystal are given in Figure 12.15 for several source-detector distances. They have been obtained by integrating Eq. 8.20, which is repeated here (refer to Figures 8.21 and 12.12 for notation):

$$\varepsilon(E) = \frac{\int_0^{\theta_0} S\{1 - \exp[-\mu(E)r(\theta)]\}\frac{1}{2}\sin\theta\, d\theta}{(S/2)\int_0^{\theta_0} \sin\theta\, d\theta} \tag{8.20}$$

or

$$\varepsilon(E) = \frac{\int_0^{\theta_1}\{1 - \exp[-\mu(E)L/\cos\theta]\}\sin\theta\, d\theta}{1 - \cos\theta_0} + \frac{\int_{\theta_1}^{\theta_0}(1 - \exp\{-\mu(E)[(R/\sin\theta) - (d/\cos\theta)]\})\sin\theta\, d\theta}{1 - \cos\theta_0}$$

where $\theta_1 = \tan^{-1}[R/(d+L)]$.
$\theta_0 = \tan^{-1}(R\,|\,d)$

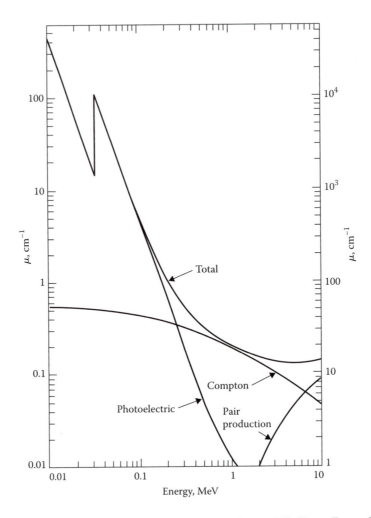

Figure 12.14 The photon linear attenuation coefficients for NaI(Tl). (From Evans, R. D., *The Atomic Nucleus*, McGraw-Hill, New York, 1972.)

The inherent approximation of Eq. 8.20 is that it considers detected every photon that interacted at least once inside the detector.

In Figure 12.15, note that the efficiency decreases with energy up to about 5 MeV. Beyond that point, it starts increasing because of the increase in the pair production probability. Figure 12.16 shows how the peak-to-total ratio (see Figure 12.13) changes with energy for a source located 0.10 m from detectors of different sizes.

The energy resolution of NaI(Tl) detectors is quoted in terms of the % resolution for the 0.662-MeV gamma of ^{137}Cs. Using the best electronics available, this resolution is about 7% and the FWHM is about 46 keV. As mentioned in Chapter 9, the FWHM is roughly proportional to the square root of the photon energy. For this reason, the resolution in % deteriorates as the energy decreases. For 10-keV X-rays, the best resolution achieved is about 40%, which makes the FWHM about 4 keV.

12.5 DETECTION OF GAMMAS WITH Gε DETECTORS

As mentioned in Section 7.5.5, the Ge(Li) detectors have been replaced by Ge detectors, which are devices that use hyper pure germanium (impurity concentration 10^{16} atoms/m^3 or less). The main advantage of Ge over Ge(Li) detectors is that the former should be kept at low temperatures only when in use; the latter must be kept cool at all times.

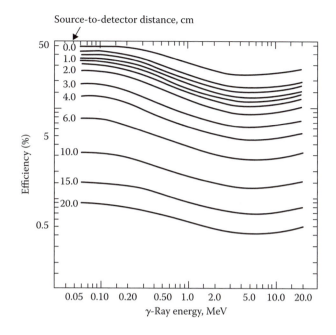

Figure 12.15 Calculated absolute total efficiencies of a 3 in. × 3 in. (7.6.2 cm × 7.6.2 cm) NaI(Tl) scintillator as a function of energy for different source-detector distances. (From http://www.thermo.com. Last accessed September 14, 2010.)

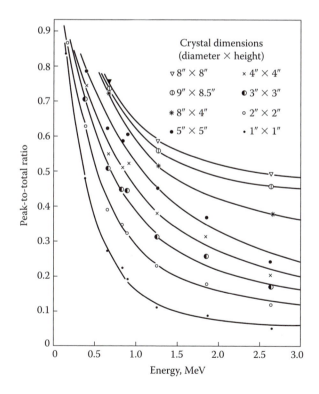

Figure 12.16 Peak-to-total ratio as a function of energy for NaI(Tl) scintillators of different sizes. The source-to-detector distance is 0.10 m. (From http://www.thermo.com. Last accessed September 14, 2010.)

The cooling of Ge detectors is achieved by permanently mounting the detector on a cryostat. The cryostat consists of a reservoir or Dewar containing the cooling medium and a vacuum chamber housing the detector. The Dewar is made of two concentric metal containers (Figure 12.17) with the space between the two containers evacuated for thermal insulation. In one design, called the "dipstick" (Figure 12.17a), the detector is housed in a separate vacuum chamber and the cooling rod is made of copper. The cooling medium is usually liquid nitrogen (boiling temperature –196°C, or 77 K). In another design, called the integral cryostat (Figure 12.17b), there is a common vacuum chamber for both the Dewar and the detector. One version of the integral cryostat is provided with a rotary vacuum seal, which allows the detector chamber to be rotated 180°. With respect to cooling, one manufacturer (Canberra) has designed and is offering a cryoelectric cryostat that uses a commercial refrigerator with helium gas as the refrigerant.

The vacuum chamber that contains the detector is made of stainless steel. The chamber protects the detector from dirt and, by being evacuated, prevents condensation of vapor on the detector surface or electrical discharge when high voltage is applied to the detector. A metal envelope, with a very thin window at its end for the passage of the incident photons, surrounds the detector. The window is made of beryllium, aluminum, or a carbon composite fiber. Transmission characteristics of several window thicknesses are shown in Figure 12.18. Most commercial cryostats include the preamplifier as a standard component.

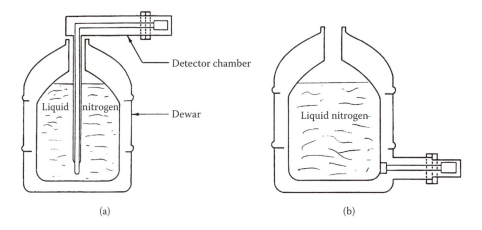

Figure 12.17 (a) A dipstick cryostat (cross section), (b) an integral cryostat (cross section) (courtesy of Canberra Nuclear).

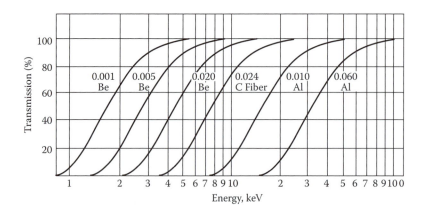

Figure 12.18 Transmission characteristics of various detector windows. The thicknesses shown are in mils of inches (courtesy of Canberra Nuclear).

Reduction of background in any measurement is very important. It becomes absolutely necessary in cases when the sample to be counted is a very weak radiation source and its activity barely exceeds the background. Complete elimination of the background radiation is impossible, but reduction of it is feasible by using special shields. Common shields are made of lead or steel and are 0.10–0.15 m thick. Figure 12.19 shows a typical arrangement of the cryostat, the detector, and the shield. Photographs of two commercial detectors, cryostats, and multichannel analyzers are shown in Figure 12.20.

12.5.1 Efficiency of Ge Detectors

The efficiency of Ge detectors quoted in the list of specifications by the manufacturer may be a relative full-energy peak efficiency or an absolute efficiency. Relative efficiencies are referenced in terms of the absolute full-energy peak efficiency of a 7.6 cm × 7.6 cm (3 in. × 3 in.) NaI(Tl) crystal. The measurement (or calculation) is based on the 1.33-MeV peak of ^{60}Co. It is assumed that a ^{60}Co source of known strength is positioned 0.25 m away from the face of the detector. A count is taken for a period of time, and the absolute full-energy peak efficiency of the Ge detector is determined by dividing the total number of counts under the 1.33-MeV peak (shaded area, Figure 12.21) by the number of photons emitted from the source during the same time period. This absolute efficiency is divided by 1.2×10^{-3}, which is the absolute efficiency of a 7.6 cm × 7.6 cm (3 in. × 3 in.) NaI(Tl) crystal 0.25 m from the source, to give the relative efficiency quoted in the specifications.

Absolute efficiencies as a function of energy for four types of Ge detectors are shown in Figures 12.22–12.25. In Figure 12.22, a Ge wafer is used to make what the manufacturer calls a low-energy Ge (LEGe) detector. In this detector, a p + contact is fabricated on the front face and the cylindrical surface with implanted boron; on the rear face, an n + contact is formed with lithium diffused along a spot that is smaller than the full rear area of the device. The efficiency of this detector is dropping for energies below 5 keV because of absorption in the Be window; at the other end of the graph, the efficiency drops for $E > 100$ keV because of a corresponding decrease in the value of the total linear attenuation coefficient of gamma rays in Ge (Figure 12.26). A coaxial Ge detector and its efficiency are shown in Figure 12.23. The contacts of this detector are formed by diffused lithium (n contact) and by implanted boron (p contact). The diffused-lithium n contact is given by the manufacturer as 0.5 mm thick. A variation of the coaxial detector, called the reverse-electrode (REGe) detector and its efficiency are shown in Figure 12.24. In the REGe detector the electrodes are opposite to those of the "normal" coaxial: the p-type electrode (formed by ion-implanted boron) is on the outside, and the η-type contact (formed by diffused lithium) is on the inside. This electrode arrangement leads to decreased window thickness (the p contact may be as thin as 0.3 μm; the Be window is ~0.5 mm), which, in turn, results in higher efficiency at lower energy (compare efficiency curves of Figures 12.2 and 12.23). Finally, in Figure 12.25 a Ge well-type detector is shown. The special characteristic of this device is its increased efficiency due to its particular geometry. The solid angle approaches 4π, resulting in close to 100% efficiency in a certain energy range.

In practice, however, efficiency is determined with the use of a calibrated source or multiple sources such as those provided by various national agencies or private companies.

For Ge detectors other than the well-type, the efficiency is low, relative to Na(Tl) scintillation counters. This statement holds true for Si(Li) detectors as well (see Section 12.9). Lower efficiency, however,

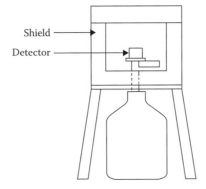

Figure 12.19 A typical arrangement of the detector, the shield, and the cryostat.

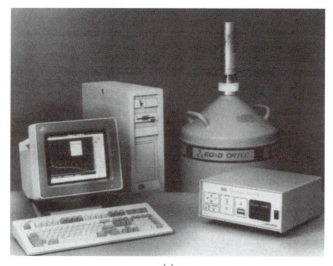

(a)

(b)

Figure 12.20 (a) Ge detector with its cryostat and multichannel analyzer (reproduced from Instruments for Research and Applied Sciences by permission of EG & G ORTEC, Oak Ridge, Tennessee), (b) portable Ge detector system with its cryostat and multichannel analyzer (courtesy of Canberra Nuclear).

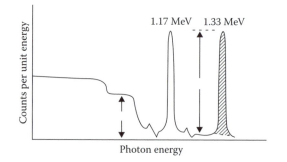

Figure 12.21 A sketch of the ^{60}Co spectrum, indicating how it is used for efficiency and peak-to-Compton ratio determination.

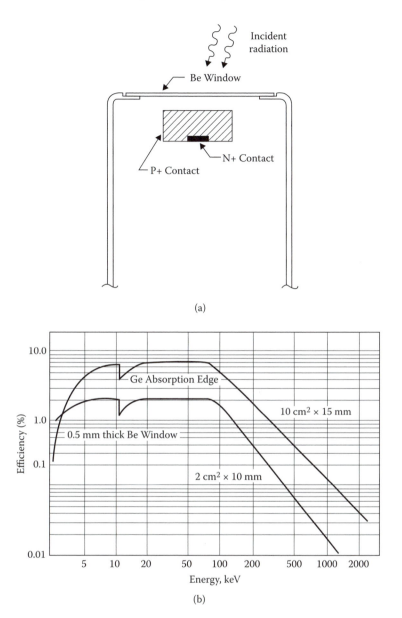

(a)

(b)

Figure 12.22 (a) LEGe detector, (b) absolute efficiency as a function of energy for a detector size shown on the graph and 2.5-cm distance assumed between source and detector. (From http://www.canberra.com. Last accessed September 14, 2010.)

is more than compensated for by the better energy resolution of the semiconductor detector. Figure 12.27 illustrates the outstanding resolution characteristics of a semiconductor detector by showing the same spectrum obtained with a NaI(Tl) and a Ge(Li) detector. Notice the tremendous difference in the FWHM. The Ge(Li) gives a FWHM = 1.9 keV, while the NaI(Tl) gives FWHM ≈ 70 keV.

Consider a case of 10,000 counts being recorded by a 7.6 cm × 7.6 cm (3 in. × 3 in.) NaI(Tl) detector under the 1.33-MeV peak of ^{60}Co. A Ge detector with 10% relative efficiency will record only 1000 counts. The FWHM of the NaI(Tl) peak is 70 keV; the corresponding width of the Ge peak is about 2 keV. Since the total number of counts under the peak is proportional to the product of the FWHM times the peak, the heights of the two peaks are related by

$$\frac{\text{Height of Ge peak}}{\text{Height of NaI(Tl) peak}} \approx \frac{(1000/2)}{(10,000/70)} = \frac{70}{20} = 3.5$$

325

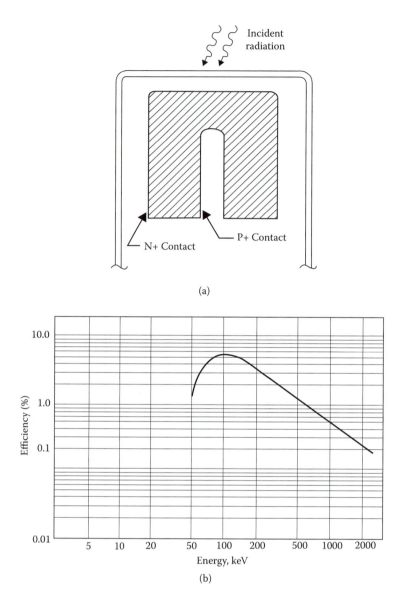

Figure 12.23 (a) Coaxial Ge detector, (b) absolute efficiency as a function of energy for a distance of 2.5 cm between source and detector.

Thus, even though the Ge detector is only 10% efficient, relative to the NaI(Tl) crystal, it produces a peak that is 3.5 times higher.

Another parameter specified by the manufacturer of Ge detectors is the *peak-to-Compton ratio* (PCR). Looking at Figure 12.21, the PCR is defined by

$$PCR = \frac{\text{height of 1.33-MeV peak}}{\left(\text{average height of Compton plateau of 1.33-MeV peak}\right)}$$

The average of the plateau is taken between 1040 and 1096 keV, in accordance with IEEE Standard No. 325–1971.[4] The PCR is important because it indicates the capability of the detector to identify low-energy peaks in the presence of stronger peaks of higher energy. PCR values of 70:1 are common, but higher values have also been reported.

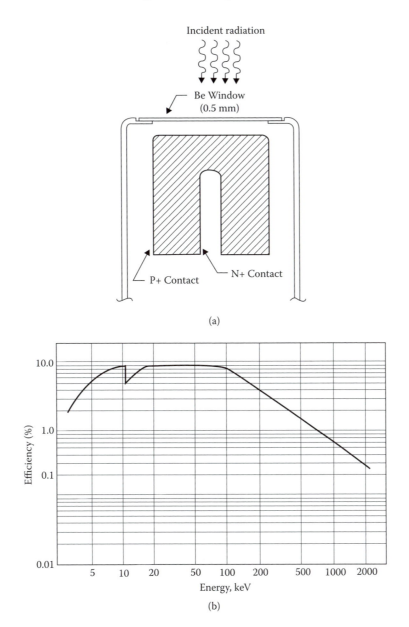

Figure 12.24 (a) REGe coaxial detector, (b) absolute efficiency as a function of energy for a distance of 2.5 cm between source and detector.

For the analysis of complex gamma spectra, it is helpful to have an analytic function that represents the efficiency of the detector as a function of energy. Many semiempirical equations have been developed to fit the efficiency of germanium detectors.[5-7] Three examples are given here.

The Freeman–Jenkin equation,[5]

$$\varepsilon = 1 - \exp(-\tau x) + \sigma A \exp(-BE) \tag{12.7}$$

where τ = photoelectric coefficient
x = thickness of the detector
σ = Compton coefficient
A, B = constants to be determined from measurement

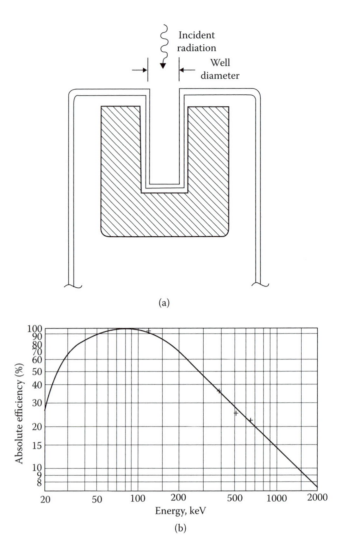

(a)

(b)

Figure 12.25 (a) A Ge well-type detector, (b) absolute detector efficiency for a detector with 40-mm depth and 10-mm-diameter well. Source assumed placed at the bottom of the well.

Equation 12.7 was used to determine the *relative* efficiency of a cylindrical and a trapezoidal detector over the range 500–1500 keV, with an accuracy of about 1%.

The Mowatt equation[6]

$$\varepsilon = a_1 F' \exp\left(-\mu_{Ge} a_2\right) \left[\frac{\tau + \sigma a_3 \exp\left(-a_4 E\right)}{\tau + \sigma}\right]\left[1 - \exp\left(-\mu_{Ge} a_5\right)\right] \tag{12.8}$$

where $F' = \Pi_i \exp\left(-\mu_i x_i\right)$ = product of attenuation factors outside the intrinsic region
a_1 = normalization factor a_2 = the thickness of the germanium front dead layer
a_3, a_4 = constants to be determined from measurement
a_5 = effective detector depth

Equation 12.8 is an improvement over Eq. 12.7 because it takes into account absorption in the window (through the factor F') and in the dead layer of the detector [through the factor $\exp\left(-\mu_{Ge} a_2\right)$] Mowatt's equation, developed for planar detectors, gives the efficiency with an accuracy of 1.5% over the range 100–1400 keV.

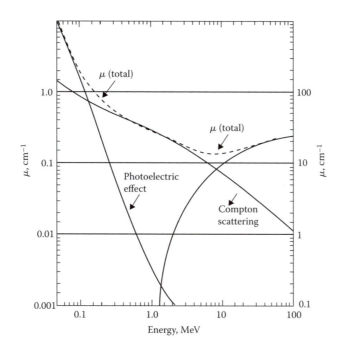

Figure 12.26 Photon attenuation coefficients for germanium. The dashed line is the approximate total linear attenuation coefficient.

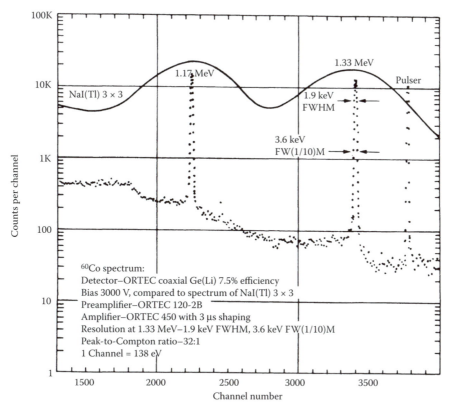

Figure 12.27 The ^{60}Co spectrum obtained with a NaI(Tl) scintillator and a Ge(Li) detector. (Reproduced from *Instruments for Research and Applied Sciences* by permission of EG & G ORTEC, Oak Ridge, Tennessee.)

McNelles–Campbell equation[7]

$$\varepsilon = \left(\frac{a_1}{E}\right)^{a_2} + a_3 \exp(-a_4 E) + a_5 \exp(-a_6 E) + a_7 \exp(-a_8 E) \tag{12.9}$$

where a_1 through a_8 are constants to be determined from measured gamma spectra. Equation 12.9, developed for coaxial detectors, predicts the efficiency with an accuracy of 0.2% over the energy range 160–1333 keV. Further testing of 12.9 showed[8] that the last term involving the constants a_7 and a_8 has a negligible effect on the result.

Equations 12.7–12.9 are given just as examples that indicate the general energy dependence of the efficiency curve. What is done in practice is to measure the efficiency as a function of energy using calibrated sources emitting gammas of known energies and intensities. A table of gamma energies used for calibration is given in Appendix C.

Applications for efficiency calculations include efficiency corrections for variable sample height in well-type germanium gamma detectors,[8] use of the Monte Carlo method to the analysis of measurement geometries for the calibration of a HP Ge detector in an environmental radioactivity laboratory,[9] a method for assessing and correcting coincidence summing effects for germanium detector efficiency calibrations,[10] direct mathematical method for calculating full-energy peak efficiency and coincidence corrections of HPGe detectors for extended sources,[11] an empirical expression for the full energy peak efficiency of an N-type high purity germanium detector,[12] and a software package using a mesh-grid method for simulating HPGe detector efficiencies.[13]

12.5.2 Energy Resolution of Ge Detectors

The energy resolution of a Ge detector is given in terms of the FWHM (Γ). The width Γ consists of the following two components:

Γ_d = width due to detector effects
Γ_e = width due to effects of electronics

Since these two components are uncorrelated, they are added in quadrature to give the total width, Γ,

$$\Gamma = \sqrt{\Gamma_d^2 + \Gamma_e^2} \tag{12.10}$$

As shown in Chapter 9, the width Γ_d is energy dependent and is given by

$$\Gamma_d = 2\sqrt{(2\ln 2)FEw} \tag{12.11}$$

where F is the Fano factor and w is the average energy needed to produce an electron-hole pair. For germanium, at the operational temperature of 77 K, $w = 2.97$ eV. Thus,

$$\Gamma_d(\text{keV}) = 0.1283\sqrt{FE(\text{keV})} \tag{12.12}$$

The width Γ_e increases when the detector capacitance increases. The detector capacitance, in turn, generally increases with detector size and may change with detector bias. Good Ge detectors have a flat capacitance-bias relationship over most of the range of bias voltage applied.

The capacitance of the detector has an effect on the energy resolution because it influences the performance of the charge-sensitive preamplifier that accepts the detector signal. The contribution of the preamplifier to the value of Γ_e increases with the input capacitance. One of the manufacturers, Canberra, reports a 0.570 eV Γ_e with zero input capacitance and a slow increase with higher values as shown in Figure 12.27. Clearly, the resolution improves if the capacitance is kept low. The other component of the input capacitance comes from items like connectors and cables. Reduction of the length of input cable and of connectors' capacitance is helpful. For the best resolution with a given system, the preamplifier should be located as close to the detector as possible.

Large Ge detectors commercially available today have capacitance as high as 30 pF, which results in a value of $\Gamma_e = 1.06$ keV (Figure 12.28). Combining the two contributions Γ_d and T_e in

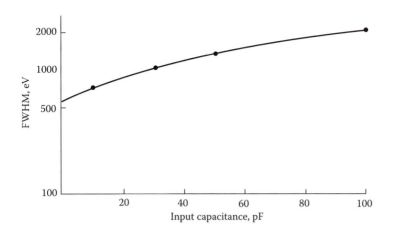

Figure 12.28 The dependence of Γ_e on input capacitance for a charge-sensitive preamplifier.

accordance with Eq. 12.10, one gets

$$\Gamma(\text{keV}) = \sqrt{(0.1283)^2 \, FE(\text{keV}) + 1.06^2} \qquad (12.13)$$

A typical value of the Fano factor for $0.1 < E < 10$ MeV is 0.16. Substitution in Eq. 12.13 gives

$$\Gamma(\text{keV}) = \sqrt{(2.63 \times 10^{-3}) \, E(\text{keV}) + 1.06^2} \qquad (12.14)$$

Equation 12.14 shows that for low energies, the resolution is determined by electronic noise. For higher energies, the energy contribution predominates. Consider two cases:

Case 1: E = 100 keV

$$\Gamma = \sqrt{(2.63 \times 10^{-3})100 + 1.06^2} = \sqrt{0.263 + 1.12} = 1.2 \text{ keV}$$

Case 2: E = 1000 keV

$$\Gamma = \sqrt{(2.63 \times 10^{-3})1000 + 1.06_2} = \sqrt{2.63 + 1.12} = 1.9 \text{ keV}$$

The energy resolution versus gamma energy is shown in Figure 12.29 for the four detectors depicted in Figures 12.22–12.25. Usually, the resolution is given in terms of the FWHM at 5.9, 122, and 1332 keV.

12.5.3 Analysis of Ge Detector Energy Spectra

Despite the superb energy resolution of Ge detectors compared to that of NaI(Tl) scintillators (see Figure 12.27), analysis of complex gamma spectra is necessary. Figure 12.30 shows a typical Ge energy spectrum. Analysis of the spectrum entails, first, assignment of energy to the peaks of the spectrum and, second, the determination of the number of counts (i.e., the area) for each peak.

The energy assignment to the peaks of the spectrum is accomplished by calibrating the detector with a source that emits gammas of known energy and intensity. As explained in Section 9.10, calibration means to determine the constants of the equation

$$E = a_1 + a_2 C + a_3 C^2 \qquad (12.15)$$

where C is the channel number (for most systems, a_3 is very small, or zero). When the energy assignment is performed, the observer should be aware of the following general features of a gamma spectrum recorded by a Ge detector:

1. For $E_\gamma < 2$ MeV, the full-energy peak is intense and almost Gaussian in shape.

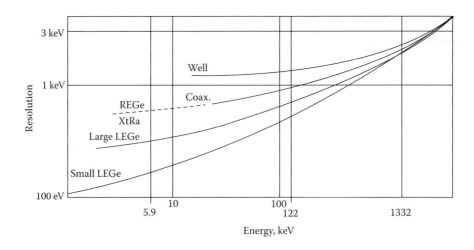

Figure 12.29 Typical energy resolution versus gamma energy for Ge detectors.

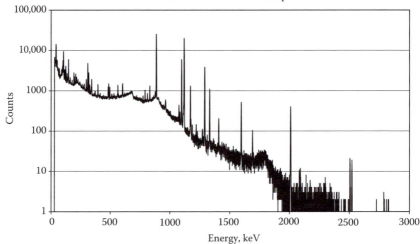

Figure 12.30 A gamma-ray energy spectrum recorded with a Ge detector. (This is the spectrum from a coal sample irradiated in a reactor.)

2. At higher energies ($E_\gamma > 2$ MeV) the double-escape peak ($E_{DE} = E_\gamma - 1.022$ MeV) becomes prominent. The single-escape peak is present too (see Figure 12.7).

3. In spectra taken with thin Ge detectors, one may see the germanium "escape" peaks. The escape peaks (EP) have energy equal to

$$E_{EP} = E_\gamma - E_k$$

where E_k is the K X-ray energy of germanium. The escape peaks are due to the loss of the energy carried away by the escaping K X-ray of germanium. This energy is equal to 9.9 keV for the K_α and 11.0 keV for the K_β X-ray of germanium. Figure 12.31 shows the ^{139}Ce X-ray spectrum with the escape peaks marked.

4. When the front surface of the detector is covered by a metal, characteristic X-rays of that metal are emitted if the incident radiation consists of photons with energy greater than the K X-ray energy of that metal. Gold, which is sometimes used, emits a K_α X-ray with energy 68 keV and five L X-rays between 9 and 13 keV.

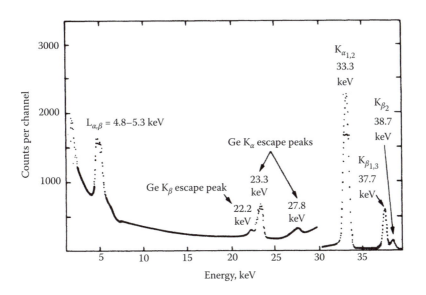

Figure 12.31 The X-ray spectrum resulting from the decay of ^{139}Ce. The Ge escape peaks are clearly seen. (From Palms, J. M. et al., *Nucl. Instrum. Meth.*, 64, 310, 1968.)

The determination of the area under a peak—that is, the absolute intensity of a particular gamma energy—is not as straightforward as the assignment of energy, because the area under a peak includes contributions from other gammas. The methods that have been developed for the determination of the area can be classified into two groups: methods that treat the data (i.e., counts per channel) directly, and methods that fit a known function to the data.

Methods that treat the data directly give the area under the peak by adding the counts from all the channels in the region of the peak and subtracting a "base background." The methods differ in the way they define the "base" and the number of channels that define the peak (a review of the methods is given in References 14 and 15). Figures 12.32 and 12.33 show graphically three of the methods.

Method 1. A straight line is used to separate the peak from the base,[14,16] and the net area under the peak (NPA) is calculated using the equation (Figure 12.32a).

$$\text{NPA} = \sum_{i=L}^{R} a_i - (a_L + a_R)\,\frac{R-L+1}{2} \tag{12.16}$$

where a_i = number of counts in channel i
L = channel number at left limit of photopeak
R = channel number at right limit of photopeak

Method 2. Here, the background (Figure 12.32b) is defined as the average count in a region equal to 3Γ (3 FWHM), extending 1.5Γ on both sides of the peak. The gross count under the peak is taken as the sum of all the counts in the channels corresponding to 3Γ. The net peak area is then given by

$$\text{NPA} = G - B_L - B_R \tag{12.17}$$

Method 3. This method, due to Quittner,[17] fits a polynomial, using the least-squares technique, to the data from $2k_L + 1$ and $2k_R + 1$ channels (Figure 12.33) on either side of the peak. The base line is constructed in such a way that at X_L and X_R (shown in Figure 12.33) it has the same magnitude (p_L, p_R) and slope (q_L, q_R) as the fitted polynomial. The net area under the peak is now given by the equation

$$\text{NPA} = \sum_{i=-W_L}^{W_R} \left(a_{x_p+i} - b_{x_p+i} \right) \tag{12.18}$$

333

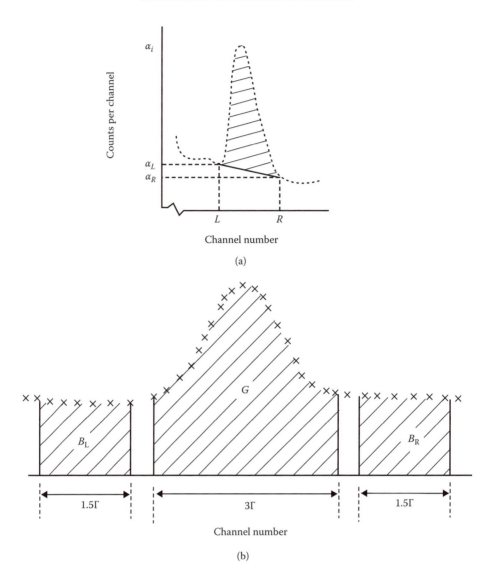

Figure 12.32 Determination of the net area under the peak, (a) a straight line is used to define the base, (b) an average background is subtracted from the gross count.

where the value of b is, in terms of a third-degree polynomial, equal to

$$b(x) = p_L + q_L(x - x_L) + \left[\frac{-q_R - 2q_L}{l_L + l_R} + \frac{3(p_R - p_L)}{(l_L + l_R)^2} \right](x - x_L)^2$$

$$+ \left[\frac{q_L + q_R}{(l_L + l_R)^2} + \frac{2(p_L - p_R)}{(l_L + l_R)^3} \right](x - x_L)^3$$

(12.19)

Quittner's method is quite accurate if the peaks are separated by about 20 channels.

Today, the most widely used methods are those that fit an analytic function to each peak. After the fit is accomplished, the analytic function is used for the calculation of quantities of interest such as the area and the centroid (position) of the peak.

The principle of obtaining the fit is simple and essentially the same for all the methods.[18–26] Let $y_i|_{i=1,N}$ be the experimental point—that is, the counts in channels $x_i|_{i=1,N}$—and $f(x, a_1, a_2, ...)$ be a

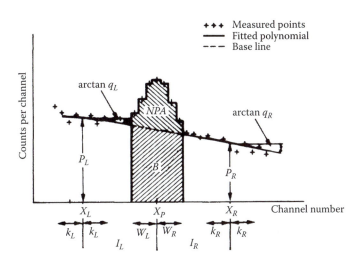

Figure 12.33 Determination of the area under the peak using the method of Quittner. (From Siffert, P. et al., *Nucl. Instrum. Meth.*, 115, 13, 1974.)

function that will represent a single peak. The parameters $a_1, a_2, \ldots$ are determined by minimizing the quantity

$$\sum_i W_i \left[y_i - f(x_1, a_1, a_2, \ldots) \right]^2$$

that is, by a least-squares technique, where the weighting factors W_i usually are the inverse of the variance of y_i. The fitting function consists of a Gaussian plus modifying functions. Three examples are

1. $f = G(x) + B + Cx = y_0 \exp\left[-(x - x_0)^2 / 2\sigma^2 \right] + B + Cx$

   Here, a Gaussian describes the peak, and the linear function $B + Cx$ describes the background.[18]

2. $f = G(x)\left[1 + a_1 (x - x_0)^{m_1} + a_2 (x - x_0)^{m_2} \right]$ (from Ref. 19)

3. $f = G(x) + B + S + D$ (from Ref. 26)

where B = linear background

   S = step function = $A\{1 - \mathrm{erf}[(x_0 - x)/\sigma\sqrt{2}]\}$

   D = tail function having an exponential form on the left of the peak and a Gaussian on the right

   In addition to the methods discussed so far, there are others that use mixed techniques.[27,28] No matter what the method is, the analysis has to be done by computer, and numerous computer codes have been developed for that purpose. Examples are the codes SAMPO,[20] HYPERMET,[28] GAMAVISION,[29] GENIE 2000,[30] and HYPERTMET_PC.[31]

   The determination of the absolute intensity of a gamma requires that the efficiency of the detector be known for the entire energy range of interest. The efficiency is determined from information provided by the energy spectrum of the calibration source. Using the known energy and intensity of the gammas emitted by the source, a table is constructed giving efficiency of the detector for the known energy peaks. The efficiency at intermediate points is obtained either by interpolation or, better yet, by fitting an analytic form to the data of the table (see Section 12.5.1).

As in the case of minimum detectable activity (Section 2.20), two types of errors are encountered when one tries to identify peaks in a complex energy spectrum.[32] Type I arises when background fluctuations are falsely identified as true peaks. Type II arises when fluctuations in the background obscure true peaks. Criteria are set in the form of confidence limits (see Section 2.20 and Reference 32) that can be used to avoid both types of errors.

12.5.4 Timing Characteristics of the Pulse

For certain measurements, like coincidence-anticoincidence counting or experiments involving accelerators, the time resolution of the signal is also important, in addition to energy resolution. For timing purposes, it is essential to have pulses with constant rise time.

No detector produces pulses with exactly the same rise time. This variation is due to the fact that electrons are produced at different points inside the detector volume, and thus traverse different distances before they reach the point of their collection. As a result, the time elapsing between production of the charge and its collection is not the same for all the carriers.

Consider a true coaxial detector, shown in Figure 12.34a (see also Figure 7.26). Since the electric field is radial, electrons and holes will follow a trajectory perpendicular to the axis of the detector. The maximum time required for collection of the charge corresponds to electron-holes being produced either at A or C. That time t is equal to $t \approx (AC)/v$, where AC is the detector thickness and v is the speed of electrons or holes. For a detector bias of about 2000 V and the size shown in Figure 12.34a, $v \approx 0.1$ mm/ns = 10^5 m/s, which gives a maximum collection time of 120 ns. The best rise time corresponds to electron-holes generated at point B (Figure 12.34a) and is equal to about 60 ns.

The pulse rise time is essentially equal to the collection time. For the detector shown in Figure 12.34a, the rise time will vary between 60 and 120 ns. For other detector geometries the variation in rise time is greater because the electrons and holes, following the electric field lines, may travel distances larger than the thickness of the detector core (Figure 12.34b). The variation in rise time for the detector of Figure 12.34b will be between 60 and 200 ns. The distribution of pulse rise times for commercial detectors is a bell-type curve, not exactly Gaussian, with a FWHM of less than 5 ns.

12.6 CdTe AND HgI$_2$ DETECTORS AS GAMMA SPECTROMETERS

A great advantage for CdTe and HgI$_2$ detectors,[33–42] compared to Ge and Si(Li) detectors, is that they can operate at room temperature (see also Section 7.5.6). At this time, they can be obtained in relatively small volumes, but they still have an intrinsic efficiency of about 75% at 100 keV because of the high atomic number of the elements involved. The energy resolution of CdTe detectors is 18% at 6 keV and 1.3% at 662 keV. The corresponding numbers for HgI$_2$ are 8% and 0.7%.[39]

The HgI$_2$ detectors are very useful for the measurement for X-rays with energy less than 10 keV. In that energy range, resolutions of 245 eV at 1.25 keV[43] and 295 eV at 5.9 keV[44] have been reported.

As with small Ge detectors, X-ray escape peaks are present in spectra taken with these CdTe and HgI$_2$ detectors. Figure 12.35 shows the spectrum of the 59.5-keV X-rays and γ-rays from ^{241}Am taken with a CdTe detector. The escape peaks due to the K$_\alpha$ X-rays of cadmium (23 keV) and tellurium (27 keV) are clearly seen.

A description of CdTe, CdZnTe, CsI, and HgI$_2$ detectors is also given in 7.5.6.

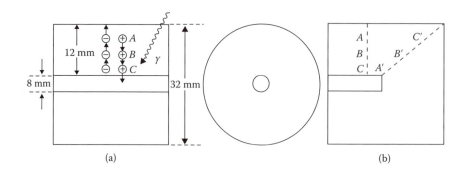

(a)   (b)

Figure 12.34 (a) In a true coaxial detector, electrons and holes travel along the direction ABC, (b) in wrap-around coaxial detectors, the carriers may travel along ABC but also along the longer path $A'B'C'$.

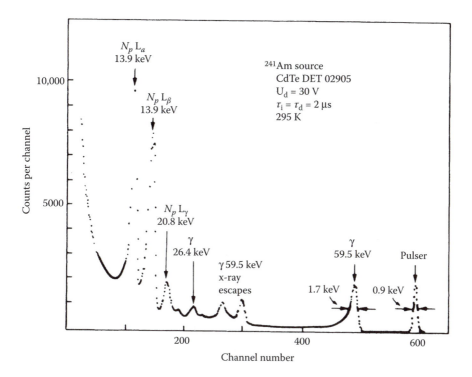

Figure 12.35 The 59.5-keV X-ray of ^{241}Am detected with a CdTe detector. (From Siffert, P., *Nucl. Instrum. Meth.*, 150, 1, 1978.)

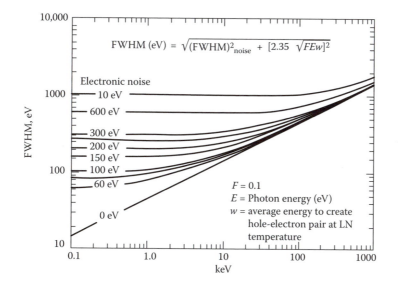

Figure 12.36 Si(Li) detector energy resolution as a function of X-ray energy. What is indicated as electronic noise is the width Γ_e. (From http://www.canberra.com.)

12.7 DETECTION OF X-RAYS WITH A Si(Li) DETECTOR

Si(Li) detectors are generally used as X-ray spectrometers for $E < 50$ keV. They need cooling and therefore require a cryostat. Their energy resolution and efficiency are better than those of a Ge detector for $E < 50$ keV. Figure 12.36 shows how the resolution changes with energy. The FWHM is again given by Eq. 12.10. Using a value of $w = 3.7$ eV per electron-hole pair for silicon at 77 K and Fano factor equal to 0.1, the width Γ becomes

$$\Gamma(\text{eV}) = \sqrt{2.05E(\text{eV}) + \Gamma_e^2} \tag{12.20}$$

The width Γ_e is indicated as "electronic noise" in Figure 12.36. Of the three types of X-ray detectors mentioned—scintillation, proportional, and semiconductor counters—the Si(Li) detector has

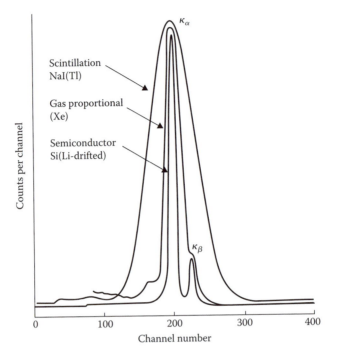

Figure 12.37 Demonstration of the superior energy resolution of Si(Li) detectors by showing the same peak recorded with a NaI(Tl) scintillator and a gas-filled proportional counter. (From Muggleton, A. H. F., *Nucl. Instrum. Meth.*, 101, 113, 1972.)

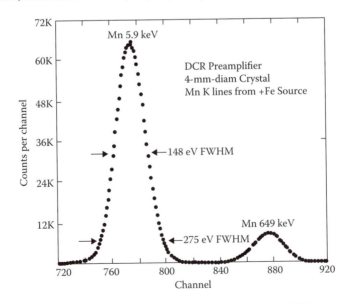

Figure 12.38 The manganese fluorescence spectrum obtained with a Si(Li) detector. (From Muggleton, A. H. F., *Nucl. Instrum. Meth.*, 101, 113, 1972.)

the best energy resolution for X-rays. This fact is demonstrated in Figure 12.37, which shows the same energy peak obtained with the three different detectors. Notice that only the Si(Li) detector can resolve K_α and K_β lines, an ability absolutely necessary for the study of fluorescent X-rays for most elements above oxygen. The manganese fluorescence spectrum obtained with a Si(Li) detector is shown in Figure 12.38.

The dependence of Si(Li) detector efficiency on the X-ray energy is shown in Figure 12.39. For $E < 3$ keV, the efficiency drops because of absorption of the incident X-rays by the beryllium detector window. For $E > 15$ keV, the efficiency falls off because of the decrease of the total linear attenuation coefficient of X-rays in silicon (Figure 12.40).

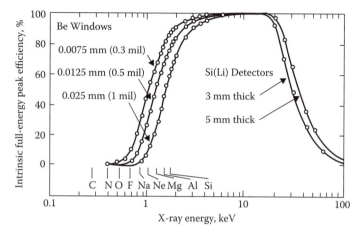

Figure 12.39 Si(Li) detector efficiency as a function of X-ray energy for different beryllium window thicknesses.

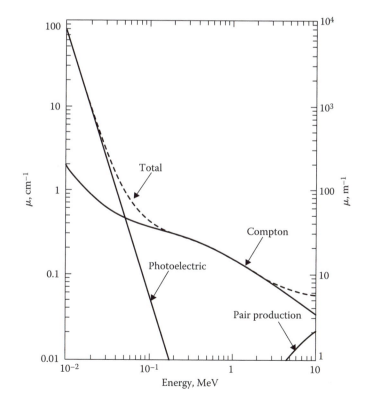

Figure 12.40 Linear photon attenuation coefficients for silicon.

Applications of X-ray systems have included semiconductor drift detectors for X- and γ-ray spectroscopy and imaging,[45] improvement in the low energy collection efficiency of Si(Li) X-ray detectors,[46] a novel high dynamic range X-ray detector for synchrotron radiation studies,[47] use of silicon drift detectors for the detection of medium-light elements in PIXE,[48] measurement of M-shell X-ray production induced by protons of 0.3–0.7 MeV on W, Au, Pb, Bi, Th, and U,[49] and design of a portable generator-based XRF instrument for nondestructive analysis at crime scenes.[50]

PROBLEMS

12.1 The Compton edge of a γ-ray peak falls at 0.95 MeV. What is the energy of the photon? What is the energy of the backscatter peak?

12.2 Sketch the energy spectrum you would expect to get from isotopes having the decay schemes shown in the figure below. Explain energy and origin of all peaks. You may assume either a NaI(Tl) or a Ge detector.

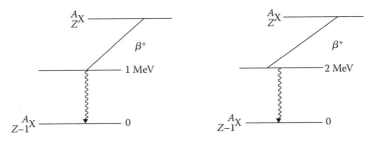

12.3 A liquid sample is contaminated with equal amounts (mass, not activity) of ^{131}I and ^{137}Cs. Sketch the energy spectra you expect to see if you use (a) a NaI(Tl) crystal with 7% energy resolution for the cesium peak, and (b) a Ge(Li) detector with energy resolution given by Eq. 12.14. Assume the same number of channels is used with both detectors. Relevant data for the two isotopes are given in the table below. Assume that the sample is placed at a distance of 0.20 m from the detectors. The NaI(Tl) is a 3 in. × 3 in. crystal. The efficiency of the Ge(Li) detector is given by Figure 12.23.

^{131}I		^{137}Cs	
E (MeV)	**Intensity (%)**	**E (MeV)**	**Intensity (%)**
0.284	5	0.662	85
0.364	82		
0.637	6.8		
0.723	1.6		
Half-life = 0.05 d		Half-life = 30 y	

12.4 An isotope emits two gammas with energies 0.8 and 1.2 MeV and intensities 30% and 100%, respectively. Assume that a Ge detector 5-mm thick is used for the measurement of this spectrum. Also assume that all the photons are normally incident upon the detector. Calculate the ratio of counts under the 0.8-MeV spectrum to counts under the 1.2-MeV spectrum.

12.5 What is the width above which the two peaks of Mn shown in Figure 12.38 cannot be resolved?

12.6 Will the peaks of Figure 12.38 be resolved with a gas-filled proportional counter, assuming the best possible resolution for that type of counter?

12.7 A ^{137}Cs gamma (0.662 MeV) is recorded with a NaI detector shows a peak with a FWHM equal to 56 keV. (a) What is the energy resolution (in %) for this peak? (b) Using the same detector, what will be the energy resolution (in %) for the 1.332 MeV gamma of ^{60}Co? (c) What will be the width of this ^{60}Co peak?

12.8 What is the efficiency of a 50-mm-long proportional counter filled with a mixture of xenon (20%) and methane at a pressure of 1 atmosphere for a parallel beam of 10-keV X-rays (geometry of Figure 12.11)?

12.9 Verify the efficiency values for 1-keV X-rays, shown in Figure 12.39 for a Si(Li) detector.

12.10 Other things being equal, what is the ratio of intrinsic efficiencies of Si(Li) and Ge detectors 3 mm thick for 50-keV X-rays?

12.11 Prove that the energy of a photon is related to its wavelength by

$$E = \frac{1.2399}{\lambda\,(\text{nm})}\,(\text{keV})$$

BIBLIOGRAPHY

Agarwal, B. K., *X-Ray Spectroscopy: An Introduction*, Springer, Berlin, 1991.

Ahmed, S. N., *Physics & Engineering of Radiation Detection*, Elsevier, Amsterdam, 2007.

Debertin, K., and Helmer, R. G., *Gamma- and X-Ray Spectrometry with Semiconductor Detectors*, North Holland, Amsterdam, 1988.

Evans, R. D., *The Atomic Nucleus*, McGraw-Hill, New York, 1972.

Gilmore, G., *Practical Gamma-Ray Spectrometry*, 2nd ed., John Wiley and Sons, New York, 2008.

Hippert, F., Geissler, E., Hodeau, J-L., Lelièvre-Berna, E., and Regnard, J-R., *Neutron and X-Ray Spectroscopy*, Springer, Berlin, 2005.

Knoll, G. F., *Radiation Detection and Measurement*, 3rd ed., John Wiley & Sons, New York, 2000.

Leroy, C., and Rancoita, P-R., *Principles of Radiation Interaction in Matter and Detection*, World Scientific, Singapore, 2004.

Vetter, K., "Recent Developments in the Fabrication and Operation of Germanium Detectors," *Ann. Rev. Nucl. Part. Sci.* **57**:363 (2007).

REFERENCES

1. http://www.ortec-online.com

2. http://www.canberra.com

3. http://www.thermo.com

4. Calibration and Usage of Germanium Detectors for Measurement of Gamma-Ray Emission of Radionuclides, ANSI N 42.14–1978 (1978).

5. Freeman, J. M., and Jenkin, J. G., *Nucl. Instrum. Meth.* **43**:269 (1966).

6. Mowatt, R. S., *Nucl. Instrum. Meth.* **70**:237 (1969).

7. McNelles, L. A., and Campbell, J. L., *Nucl. Instrum. Meth.* **109**:241 (1973).

8. Appleby, P. G., and Piliposian, G. T., *Nucl. Instrum. Meth. Phys. Res.* **B225**:423 (2004).

9. Ródenas, J., Gallardo, S., Ballester, S., Primault, V., and Ortiz. *Nucl. Instrum. Meth. Phys. Res.* **B 263**:144 (2007).

10. Montgomery, R. S., *J. Radioanal. Nucl. Chem.* **193**:71(1995).

11. Abbas, M. I., *Nucl. Instrum. Meth. Phys. Res.* **B256**:544 (2007).

12. Osae, E. K., Nvarko, J. B., Serfor-Armah, Y., and Darko, E. O., *J. Radioanal. Nucl.* **242**:617 (1999).

13. Jackman, K. R., Gritzo, R. E., and Biegalski, S. R., *J. Radioanal. Nucl. Chem.* **282**:223 (2009).

14. Baedecker, P. A., *Anal. Chem.* **43**:405 (1971).

15. Kokta, L., *Nucl. lustrum. Meth.* **112**:245 (1973).

16. Covell, D. F., *Anal. Chem.* **31**:1785 (1959).

17. Quittner, P., *Nucl. Instrum. Meth.* **76**:115 (1969).

18. Mariscotti, M. A., *Nucl. Instrum. Meth.* **50**:309 (1967).

19. Helmer, R. G., Heath, R. L., Putnam, M., and Gipson, D. H., *Nucl. Instrum. Meth.* **57**:46 (1967).

20. Routti, J. T., and Prussin, S. G., *Nucl. Instrum. Meth.* **72**:125 (1969).

21. Robinson, D. C., *Nucl. Instrum. Meth.* **78**:120 (1970).

22. Kern, J., *Nucl. Instrum. Meth.* **79**:233 (1970).

23. McNelles, L. A., and Campbell, J. L., *Nucl. Instrum. Meth.* **127**:73 (1975).

24. Jorch, H. H., and Campbell, J. L., *Nucl. Instrum. Meth.* **143**:551 (1977).

25. DeLotto, I., and Ghirardi, A., *Nucl. Instrum. Meth.* **143**:617 (1977).

26. Campbell, J. L., and Jorch, H. H., *Nucl. Instrum. Meth.* **159**:163 (1979).

27. Connelly, A. L., and Black, W. W., *Nucl. Instrum. Meth.* **82**:141 (1970).

28. Phillips, G. W., and Marlow, K. W., *Nucl. Instrum. Meth.* **137**:525 (1976).

29. http://www.ortec-online.com/pdf/a66.pdf

30. http://www.canberra.com/products/831.asp

31. Révay, Z. S., Belgya, T., and Molnár, G. L., *J. Radioanal. Nucl. Chem.* **265**:261 (2005).

32. Head, J. H., *Nucl. Instrum. Meth.* **98**:419 (1972).

33. Siffert, P., Gonidec, J. P., and Cornet, A., *Nucl. Instrum. Meth.* **115**:13 (1974).

34. Jones, L. T., and Woollam, P. B., *Nucl. Instrum. Meth.* **124**:591 (1975).

35. Siffert, P., *Nucl. Instrum. Meth.* **150**:1 (1978).

36. Dabrowski, A. J., Iwanczyk, J., Szymczyk, W. M., Kokoschinegg, P., and Stelzhammer, J., *Nucl. Instrum. Meth.* **150**:25 (1978).

37. Schieber, M., Beinglass, I., Dishon, G., Holzer, A., and Yaron, G., *Nucl. Instrum. Meth.* **150**:71 (1978).

38. Shaler, J., *Nucl. Instrum. Meth.* **150**:79 (1978).

39. Whited, R. C., and Schieber, M. M., *Nucl. Instrum. Meth.* **162**:113 (1979).

40. Markakis, J. M., *Nucl. Instrum. Meth.* **263**:499 (1988).

41. Courat, B., Fourrier, J. P., Silga, M., and Omaly, J., *Nucl. Instrum. Meth.* **269**:213 (1988).

42. McKee, B. T. A., Goetz, T., Hazlett, T., and Forkert, L., *Nucl. Instrum. Meth.* **272**:825 (1988).

43. Dabrowski, A. J., Iwanczyk, J. S., Barton, J. B., Huth, G. C., Whited, R., Ortale, C., Economou, T. E., and Turkevich, A. L., *IEEE Trans. Nucl. Sci.* **NS-28**(1):536 (1981).

44. Iwanczyk, J. S., Dabrowski, A. J., Huth, G. C., Del Duca, A., and Schnepple, W., *IEEE Trans. Nucl. Sci.* **NS-28**(1):579 (1981).

45. Fiorini, C., and Longoni, A., *Nucl. Instrum. Meth. Phys. Res.* **B266**:2173 (2008).

46. Cox, C. E., Fischer, D. A., Schwarz, W. G., and Yongwei Song, Y., *Nucl. Instrum. Meth. Phys. Res. B* **241**:436 (2005).

47. Cockerton, S., Tanner, B. K., and Derbyshire, G., *Nucl. Instrum. Meth. Phys. Res.* **B97**:561 (1995).

48. Alberti, R., Bjeoumikhov, A., Grassi, N., Guazzoni, C., Klatka, T., Longoni, A., and Quattrone, A., *Nucl. Instrum. Meth. Phys. Res.* **B266**:2296 (2008).

49. Rodríguez-Fernández, L., Miranda, J., Ruvalcaba-Sil, J. L., Segundo, E., and Oliver, A., *Nucl. Instrum. Meth. Phys. Res.* **B189**:27 (2002).

50. Schweitzer, J. S., Trombka, J. I., Floyd, S., Selavka, C., Zeosky, G., Gahn, N., Timothy McClanahan, T., and Burbine, T., *Nucl. Instrum. Meth. Phys. Res.* **B241**:816 (2005).

51. Muggleton, A. H. F., *Nucl. Instrum. Meth.* **101**:113 (1972).

52. Palms, J. M., Venugopala Rao, P., and Wood. D., *Nucl. Instrum. Meth.* **64**:310 (1968).

13

Charged-Particle Spectroscopy

13.1 INTRODUCTION

A charged particle going through any material will have interactions affecting its detection in two ways. First, the energy spectrum is distorted because of the energy loss caused by the interactions in any mass interposed between source and detector. Second, a particle entering the active detector volume will interact there at least once and will be detected, that is, the efficiency is practically 100%.

Because any energy loss outside the detector is undesirable, the task of the experimenter is to design a spectrometer with zero mass between the source and the detector. Such an ideal system cannot be built, and the only practical alternative is a spectrometer that results in such a small energy loss outside the detector that reliable corrections can be applied to the measured spectrum.

In certain measurements, the particles do not stop in the detector; but they go through it and emerge with only a fraction of their energy deposited in the detector. Then a correction to the spectrum of the exiting particles will have to be applied because of *energy straggling*, a term used to describe the statistical fluctuations of energy loss. Energy straggling should not be confused with the statistical effects that result in the finite energy resolution of the detector.

For heavy ions, a phenomenon called the *pulse height defect* (PHD) seems to have an important effect on energy calibration. Because of the PHD, the relationship between pulse height and ion energy is mass dependent. In semiconductor detectors, experiments have shown that the PHD depends on the orientation of the incident ion beam relative to the crystal planes of the detector. This phenomenon is called channeling.

To avoid unnecessary energy loss, the source of the charged particles should be prepared with special care. The heavier the ion, the more important the source thickness becomes and the more difficult the source preparation is.

This chapter discusses the subjects of energy loss and straggling, pulse height defect, energy calibration methods, and source preparation, from the point of view of their effect on spectroscopy. All the effects are not equally important for all types of particles. Based on similarity in energy loss behavior, the charged particles are divided into three groups, as given in Chapter 4:

1. Electrons and positrons

2. Alphas, protons, deuterons, tritons

3. Heavy ions ($Z > 2$, $A > 4$)

Energy straggling, which is a phenomenon common to all particles, is discussed first. Then the other effects are analyzed separately for each particle group.

13.2 ENERGY STRAGGLING

If a monoenergetic beam of charged particles traverses a material of thickness Δx, where Δx is less than the range of the particles in that medium, the beam will emerge from the material with a distribution of energies. The broadening of the beam is due to the statistical fluctuations of the energy loss processes. Simply stated, the incident particle participates in a great number of collisions as it travels the distance Δx, and loses a certain fraction of its energy in every collision. However, neither the number of collisions nor the energy lost per collision is constant, resulting in a distribution of energies called *energy straggling*.

Energy straggling plays no role in the measurement of the total energy of the charged particle. It does play a significant role, however, in transmission-type experiments where the particle emerges from a detector after depositing only a fraction of its energy in it.

Consider a monoenergetic beam of particles with kinetic energy T_0 (Figure 13.1) going through a thickness Δx that is a fraction of the particle range. The average energy $\overline{T}$ of the emerging particles is

$$\overline{T} = T_0 - \int_0^{\Delta x} \left(\frac{dE}{dx} \right) dx \tag{13.1}$$

where dE/dx is the stopping power of the medium for the incident particle (see Chapter 4). In most cases, $\overline{T} < T_p$, where T_p is the most probable energy of the particles after going through the thickness Δx.

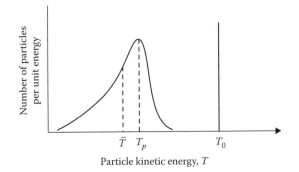

Figure 13.1 A monoenergetic beam will exhibit an energy distribution after going through a material of thickness $\Delta x < R$.

The shape of the energy distribution shown in Figure 13.1 is determined by the parameter k,

$$k = \frac{\xi}{\Delta E_{max}} \tag{13.2}$$

where ξ is roughly equal to the mean energy loss of the particle traversing the thickness Δx, and ΔE_{max} is the maximum energy transfer to an atomic electron in one collision. The expressions for ξ and ΔE_{max} are

$$\xi = 2\pi r_0^2 mc^2 \frac{Z_1^2 N Z_2}{\beta^2} \Delta x \tag{13.3}$$

$$\Delta E_{max} = 2mc^2 \frac{\beta^2}{1-\beta^2}\left[1 + \left(\frac{m}{M_1}\right)^2 + 2\frac{m}{M_1}\frac{1}{\sqrt{1-\beta^2}}\right]^{-1} \tag{13.4a}$$

All the symbols in Eqs. 13.3 and 13.4a have been defined in Section 4.3, except Z_1, the charge of the incident particle, and Z_2, the atomic number of the stopping material. For nonrelativistic particles ($\beta \ll 1$), which are much heavier than electrons, Eq. 13.4a takes the form

$$\Delta E_{max} = \frac{4mM_1}{(m+M_1)^2}T \tag{13.4b}$$

If $M \gg m$, then Eq. 13.4b takes the form $\Delta E_{max} = 4(m/M_1)T$.

For small values of $k(k \leq 0.01)$, a small number of collisions takes place in the stopping medium and the resulting distribution is asymmetric with a low-energy tail. Landau[1] first investigated this region and obtained a universal asymmetric curve. The case of intermediate k values ($0.1 < k < 10$) was first investigated by Symon[2] and later by Vavilov.[3] The Vavilov distribution was checked and was found to agree with experiment.[4] For small k, the Vavilov distribution takes the shape of the Landau result, while for large k, when the number of collisions is large, it becomes a Gaussian. Figure 13.2 shows how the distribution changes as a function of k. Many other authors have studied special cases of the energy straggling problem.[5–11]

The variance of the energy straggling distribution was first calculated* by Bohr[12] using a classical model. Bohr's result is

$$\sigma_E^2 = \overline{(\Delta T)^2} - (\overline{\Delta T})^2 = 4\pi r_0^2 (mc^2)^2 Z_1^2 Z_2 N \Delta x \tag{13.5}$$

where ΔT = energy loss in a specific case and $\overline{\Delta T}$ = average energy loss given by (Eq 13.1) $T_0 - \overline{T}$.

---

* The calculation is presented by Evans (1972) and by Segré (1965).

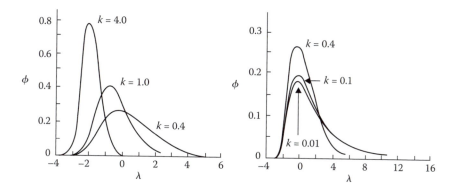

Figure 13.2 The Vavilov distribution shown for various values of the parameter k. The quantity ϕ is a measure of the probability that a particle will lose energy between T and $T + dT$ in traversing thickness Δx. The parameter $\lambda = (T - \bar{T})/\xi - 0.423 - \beta^2 - \ln k$. (From Skyrme, D. J., *Nucl. Instrum. Meth.*, 57, 61, 1967.)

The width Γ of the distribution is equal to $(2\sqrt{2\ln 2})(\sigma_E)$.

Livingston and Bethe[13] obtained a different expression by incorporating quantum-mechanical concepts into the calculation. Their result is

$$\sigma_E^2 = 4\pi r_0^2 (mc^2)^2 Z_1^2 Z \left(Z_2' + \sum_i \frac{8}{3} \frac{I_i Z_i}{\Delta E_{max}} \ln \frac{\Delta E_{max}}{I_i} \right) \Delta x \tag{13.6}$$

where Z_2' = effective atomic number of the stopping material and I_i, Z_i = ionization potential and number of electrons, respectively, in the ith atomic shell of the stopping material.

ΔE_{max} is given by Eq. 13.4a. A third expression for σ_E^2 was obtained by Titeica.[14] It is worth noting that Bohr's result (Eq. 13.5) is independent of the particle energy, while the Beth-Livingston (Eq. 13.6) and the Titeica result have a small energy dependence.

The expressions for σ_E^2 mentioned above were all obtained by taking into account electronic collisions only. Nuclear collisions (see Chapter 4) are rare, but they cause large energy losses. As a result, they do not contribute significantly to the average energy loss but they do influence the energy distribution by giving it a low-energy tail. (The energy *loss* distribution will have a high-energy tail.)

The width of the energy distribution after the beam traverses a thickness Δx consists of a partial width Γ_s due to straggling and a second one Γ_d due to the resolution and noise of the detection system. The total width Γ is obtained by adding the two partial widths in quadrature:

$$\Gamma = \sqrt{\Gamma_s^2 + \Gamma_d^2} \tag{13.7}$$

The energy straggling is measured with an experimental setup shown schematically in Figure 13.3. A source, a detector, and a movable absorber are housed in an evacuated chamber, to avoid any energy loss as the particles travel from the source to the detector. The width Γ_d is measured first by recording the particle energy spectrum with the absorber removed. Then the absorber is put into place and the measured spectrum gives the width Γ. The straggling width is, using Eq. 13.7,

$$\Gamma_s = \sqrt{\Gamma^2 - T_d^2} \tag{13.8}$$

By using absorbers of different thicknesses, the width Γ_s may be studied as a function of Δx. Measurements of this type have been performed by many people, especially with alpha particles.[15,16] For small thicknesses, the experimental results agree with theory, but for large thicknesses the theory underestimates the width. Figure 13.4 shows results for thin and

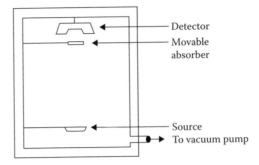

Figure 13.3 The experimental setup used in the study of energy straggling.

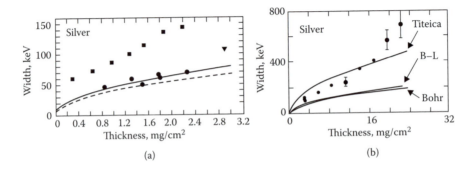

Figure 13.4 Alpha-particle energy straggling for (a) thin, and (b) thick foils of silver, [(a) Data (•) from Reference 16; (■) measurements of Sykes and Harris (Reference 72); (---) calculation from Reference 12; (—) calculation from Reference 13; (▼) measurements from Reference 15; (b) Experimental points from Reference 15; the lines are results of calculations.]

thick silver foils. It should be noted that according to the theory (Eqs. 13.5 and 13.6), the width Γ_s is proportional to $\sqrt{\Delta x}$, assuming that Z_1 does not change as the particle traverses the thickness Δx.

Energy straggling is more pronounced for electrons than for heavier particles for three reasons. First, electrons are deflected to large angles and may lose up to half of their energy in one collision. Second, large-angle scattering increases their path-length. Third, electrons radiate part of their energy as bremsstrahlung. All three effects tend to increase the fluctuations of the energy loss. Results of electron transmission and straggling measurements have been reported by many observers. A typical spectrum of straggled electrons is shown in Figure 13.5, which compares the experimental result[17] with a Monte Carlo calculation.[18]

Range straggling is a phenomenon related to energy straggling by the equation

$$\sigma_R^2 = \left(\frac{dE}{dx}\right)^{-2} \sigma_E^2 \tag{13.9}$$

where σ_R^2 is the range variance. Range straggling refers to the path length distribution of monoenergetic particles traversing the same absorber thickness (for more details see Section 22.5 of Evans, 1972). For spectroscopy measurements, only energy straggling is important.

Energy straggling still remains an important phenomenon for a wide variety of research areas including the photon self-absorption correction for thin-target-yields versus thick-target-yields in radionuclide production,[19] high-energy resolution alpha spectrometry using cryogenic detectors,[20] and design and construction of a new chamber for measuring the thickness of alpha-particle sources.[21]

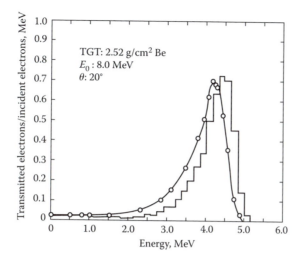

Figure 13.5 Energy spectrum of 8-MeV electrons transmitted through 2.52 g/cm² of Be and observed at 20° from the direction of the incident beam. The histogram is the result of a Monte Carlo calculation (Reference 18). The experimental points are from Reference 17.

13.3 ELECTRON SPECTROSCOPY

Under the common title of electron spectroscopy, this section discusses the most important problems of electron, positron, and beta-ray energy measurements: back-scattering, energy resolution and detector response functions, energy calibration, and source preparation.

13.3.1 Electron Backscattering

Electrons moving through a detector behave differently from heavier charged particles in two respects. First, as a result of successive collisions with atomic electrons, the incident electrons may be deflected by more than 90°, that is, they may be backscattered. Second, electrons slowing down lose part of their energy as bremsstrahlung.* In general, the effect of bremsstrahlung production on spectrum distortion is small. Backscattering in the detector, on the other hand, is important, and therefore corrections to the measured spectrum have to be applied. The effect of backscattering on electron energy spectra is discussed in this section.

Consider a monoenergetic electron beam of energy T_0 impinging normally up a detector of thickness x, where $x > R(T_0)$ (Figure 13.6) and $R(T_0)$ is the range of electrons of energy T_0 in the material of which the detector is made. Most of the incident electrons will deposit all their energy in the detector (electron A, Figure 13.6) and thus generate a pulse proportional to T_0. But some electrons (like B or C or D, Figure 13.6) are scattered out of the detector before they deposit all their energy in it. Such particles will give rise to a pulse smaller than that corresponding to energy T_0. As a result of electron backscattering, the energy spectrum of a monoenergetic source will have a full-energy peak and a low-energy tail, as shown in Figure 13.7.[22-26] The fraction of electrons recorded in the tail

1. Increases with the atomic number of the detector material

2. Changes slowly with the energy T_0

3. Increases as the incident angle of the beam deviates from the normal

An electron energy spectrum measured with a plastic scintillator is shown in Figure 13.8. It is represented extremely well by the following analytic function, which was developed by Tsoulfanidis *et al.*[23] and is shown in Figure 13.9.

$$R(E,E') = \frac{1}{2}\frac{b}{E'}\text{erfc}\left(\frac{E-E'}{\sigma\sqrt{2}}\right) + \frac{1-b}{\sigma\sqrt{2\pi}}\exp\left[-\frac{1}{2}\frac{(E-E')^2}{\sigma^2}\right] \tag{13.10}$$

---

* Every charged particle slowing down radiates part of its energy. For particles other than electrons, however, and for the energies considered here, the bremsstrahlung can be ignored (see Chapter 4).

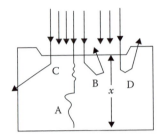

Figure 13.6 Some of the electrons incident upon the detector are backscattered and deposit only part of their energy in it.

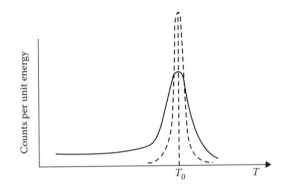

Figure 13.7 The dashed line shows the measured spectrum without backscattering in the detector. The solid line shows the same spectrum with backscattering.

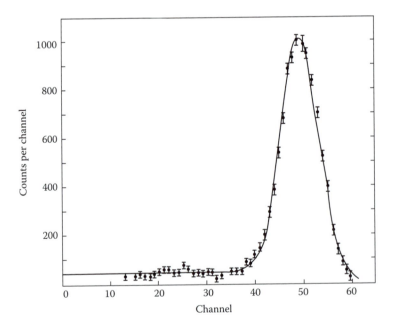

Figure 13.8 The ^{113}Sn internal conversion electron spectrum obtained with a plastic scintillator. The solid line was obtained using Eq. 13.10. (From Tsoulfanidis, N. et al., *Nucl. Instrum. Meth.*, 73, 98, 1969.)

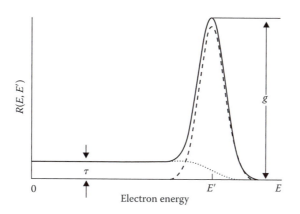

Figure 13.9   Analytical response function for monoenergetic electrons measured by a thin plastic scintillator. Shown are the backscattering tail $(\cdots)$ an the Gaussian (---).

where

$$\mathrm{erfc}(x) = 1 - \frac{2}{\sqrt{\pi}} \int_0^x \exp(-t^2)\, dt$$

and

σ = standard deviation of the Gaussian
b = fraction of electrons in the tail

The backscattering fraction b is given by

$$b = \left[1 + \sqrt{2\pi} \left(\frac{g}{\tau} - \frac{1}{2} \right) \frac{\sigma}{E'} \right]^{-1} \tag{13.11}$$

where (Figure 13.9) g is the height of the peak and τ is height of the backscattering tail.

Similar results have been obtained with a Si(Li) detector.[23,26] More references on the subject are given by Bertolini and Coche (see their Section 4.3.3). Semiempirical formulas giving the value of b as a function of Z and T have been developed by many authors[27,28] but such equations are of limited general value because the response function and the backscattering depend on the geometry of the system; for this reason, response function and backscattering should be measured for the actual experimental setup of the individual observer.

13.3.2 Energy Resolution and Response Function of Electron Detectors

The best energy resolution for electrons is obtained using silicon semiconductor detectors, with the possible exception of magnetic spectrometers. Silicon detectors may be surface barrier or Si(Li) detectors. The surface-barrier detectors operate at room temperature, while the Si(Li) detectors give best results when cooled to liquid nitrogen temperatures. The energy resolution of semiconductor detectors is determined by the electronic noise alone. It deteriorates as the area and the sensitive depth of the detector increase. For commercial detectors the full width at half maximum (FWHM) ranges from about 7 to 30 keV.

The energy resolution of scintillators, plastic scintillators in particular, is much worse. It is of the order of 8–10% at 1 MeV, which gives a FWHM of 80–100 keV. For scintillators the FWHM is roughly proportional to $\sqrt{E}$. Plastic scintillators have two advantages over semiconductor detectors: the backscattering fraction is less for scintillators because of their lower atomic number, and the timing characteristics are extremely useful for certain types of measurements. The pulse rise time is about 0.1 ns for a plastic scintillator, while for a silicon detector it is between 1 and 10 ns.

The response function of electron detectors is of the form shown in Figure 13.9. Because of the low-energy tail, if one measures a continuous spectrum (e.g., one beta spectrum or a mixture of beta spectra), the measured spectrum will be higher than the source spectrum at the low-energy end and lower at the high-energy end, as shown in Figure 13.10. Therefore, spectrum unfolding is

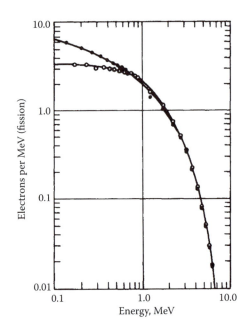

Figure 13.10 Measured (•) and unfolded (○) beta spectra from ^{235}U fission fragments. Unfolding removes the effect of backscattering in the detector; thus, it reduces the spectrum at the low-energy end. (From Tsoulfanidis, N., et al., *Nucl. Sci. Eng.*, 43, 42, 1971.)

necessary to shift back to their proper energy all the betas that were incorrectly recorded at lower energies because of backscattering. The iteration method of unfolding, described in Section 11.5.3, is suitable for beta spectra. The spectrum shown in Figure 13.10 was unfolded using that method.

13.3.3 Energy Calibration of Electron Spectrometers

The energy calibration of any spectrometer requires the use of sources of known energy and preferably of monoenergetic sources. Monoenergetic electron sources are provided by accelerators and by radioisotopes emitting internal-conversion (IC) electrons (see Chapter 3).

The advantage of the accelerators is their ability to provide a monoenergetic beam with any desired energy from zero up to the upper limit of the machine. The disadvantages are their expensive operation and the fact that the spectrometer has to be moved to the accelerator beam.

IC emitters are relatively inexpensive to obtain and very easy to handle. They have the disadvantage that they emit not only IC electrons but also gammas. Thus, when a spectrum is recorded, the result includes both IC electrons and Compton electrons created by gammas that interact in the detector. One may eliminate the Compton electrons by utilizing the X-rays that are also given off by the IC source. The X-rays are emitted in coincidence with the IC electrons, while the gammas, and therefore the Compton electrons too, are not. Thus, if the IC electrons are counted in coincidence with the X-rays, the Compton electrons will not be recorded.

IC sources emit K, L,..., electrons. The energy resolution of silicon semiconductor detectors is so good that separation of the K, L,..., electrons is possible. Figure 13.11 shows the IC electron energy spectrum of ^{207}Bi, one of the most widely used calibration sources. The excellent energy resolution of the detector distinguishes K, L, and M electrons. The K_α and K_β X-rays, which accompany the IC process, are also known.

Pure beta-emitting isotopes exist and may be used for calibration, but only after the energy spectrum is cast into a form called the *Kurie plot*. The beta spectrum is continuous and extends from zero energy up to a maximum end point kinetic energy (see Figure 13.12). Because of the shape of the spectrum, it is impossible to accurately determine the end point energy. However, from the theory of beta decay, it is known that the beta spectrum may be written as[31]

$$\frac{1}{E}\sqrt{\frac{\beta(\varepsilon)}{G(Z,E)}} = k(\varepsilon_0 - \varepsilon) \tag{13.12}$$

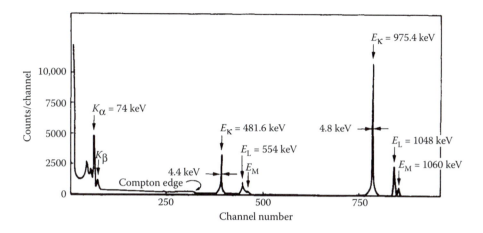

Figure 13.11 The ^{207}Bi internal conversion electron spectrum obtained with a silicon semicon-ductor detector. (From Meyer, O., and Langmann, H. J., *Nucl. Instrum. Meth.*, 39, 119, 1966.)

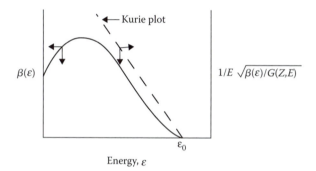

Figure 13.12 A typical beta energy spectrum (solid line) and its Kurie plot (dashed line).

where ε = beta kinetic energy in units of mc^2
$\qquad E = T + mc^2 = (\varepsilon + 1)mc^2$ = total energy
$\quad G(Z, E)$ = modified Fermi function of β decay
$\qquad\quad k$ = constant independent of energy (for allowed transitions)

If the left-hand side of Eq. 13.12 is plotted against ε, the result is a straight line that crosses the energy axis at $\varepsilon = \varepsilon_0$. The Kurie plot is a straight line for allowed beta transitions. A "forbidden" beta decay will show an upward curvature at the end.[31]

13.4 ALPHA, PROTON, DEUTERON, AND TRITON SPECTROSCOPY

Protons, deuterons, tritons, and alpha particles behave similarly as far as energy loss and strag-gling are concerned. As they travel in a medium, they are deflected very little from their direction of incidence, as a result of which backscattering is insignificant and their range is almost equal to their path length.

To avoid significant energy loss, the particles must go through as small a mass as possible when they move from the source to the detector. This is accomplished by making the source cover and the detector window as thin as possible. The entrance window of such detectors consists of a metallic layer, usually gold, with a thickness of 4×10^{-4} kg/m^2 (40 µg/cm^2) or less. The measure-ments are performed in an evacuated chamber to avoid energy loss in air.

The discussion in the rest of this section uses examples from alphas, but the points made are valid for the other particles of this group. Alphas have been studied and used much more exten-sively than the others, providing a basis for discussion.

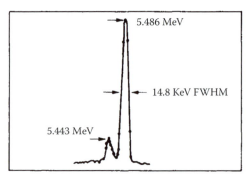

Figure 13.13 The ^{241}Am alpha spectrum obtained with a silicon surface-barrier detector.

13.4.1 Energy Resolution and Response Function of Alpha Detectors

The best energy resolution is obtained with silicon surface-barrier detectors. Most detector manufacturers quote the resolution obtained for the 5.486-MeV alphas of ^{241}Am. A typical spectrum obtained with a detector having 25 mm^2 active area and 100 µm sensitive depth is shown in Figure 13.13. The resolution deteriorates somewhat as the detector size increases. Since the response function of alpha detectors is a narrow Gaussian, there is no need to unfold a measured monoenergetic alpha spectrum.

In the past years alpha spectroscopy has been greatly studied primarily because of the naturally occurring radionuclides in the environment, nuclear fuel cycle and the potential impact that actinides have on the environment. These studies include the separation of plutonium from irradiated uranium and identified by α-spectroscopy,[32] current status of α-particle spectrometry,[33] simulation of alpha particle spectra from aerosol samples,[34] determination of ^{226}Ra in environmental and geological samples by α-spectrometry using ^{225}Ra as yield tracer,[35] evolution of chemical species during electrodeposition of uranium for alpha spectrometry,[36] high-energy resolution alpha spectrometry using cryogenic detectors,[37] micro-column solid phase extraction to determine uranium and thorium in environmental samples,[38] determination of parameters relevant to alpha spectrometry when employing source coating,[39] and a rapid method for α-spectrometric analysis of radium isotopes in natural waters using ion-selective membrane technology.[40] α-Spectroscopy is also used for fundamental measurements of the half-life of ^{246}Cm and the α-decay emission probabilities of ^{246}Cm and ^{250}Cf.[41]

13.4.2 Energy Calibration

All isotopes with $Z > 82$ emit alphas in the energy range 4 MeV $< T <$ 8 MeV, each isotope giving off more than one group of alphas. A particular isotope is selected to be used for calibration based on the energy of the alphas, the presence of other interfering radiations, and its half-life. For example, the isotope ^{241}Am is very popular because it has a 432-year half-life and its only other radiation emitted is ^{237}Np X-rays. Other isotopes frequently used are ^{210}Po, ^{226}Ra, and ^{252}Cf. Alpha sources with $T_\alpha > 8$ MeV, as well as sources of protons, deuterons, and tritons of any energy, can be provided by accelerators only.

13.4.3 Source Preparation

The main precaution taken in the preparation of an alpha source using an alpha-emitting radioisotope is to cover its front face with the thinnest possible layer of material. Commercial sources are made by sandwiching the radioisotope between two thin foils. Figure 13.14 shows a ^{241}Am source made by the Amersham Corporation.

Alpha-emitting isotopes are considered extremely hazardous when ingested, and in particular when they enter the lungs. To avoid accidental exposure, the user should always be certain that the source cover has not been damaged. The user should also be aware that all alpha sources emit a small number of neutrons, produced either through (α, n) reactions with the source-supporting material or from spontaneous fission of the radioisotope itself.

13.5 HEAVY-ION ($Z > 2$) SPECTROSCOPY

Heavy-ion spectroscopy is different from that of lighter charged particles because of the pulse-height defect (PHD), which makes the energy calibration equation mass dependent.

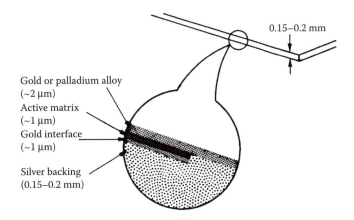

0.15–0.2 mm

Gold or palladium alloy
(~2 μm)
Active matrix
(~1 μm)
Gold interface
(~1 μm)

Silver backing
(0.15–0.2 mm)

Figure 13.14   A ²⁴¹Am source (from Amersham Corp.).

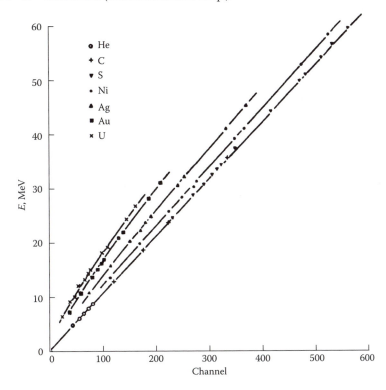

Figure 13.15   The relationship between energy and pulse height (channel) for ions with different masses. (From Wilkins, B. D., et al., *Nucl. Instrum. Meth.*, 92, 381, 1971.)

13.5.1 The Pulse-Height Defect

The measurement of particle energy with any type of detector is based on the assumption that the charge collected at the output of the detector is proportional to the energy of the incident particle. The assumption is valid if all the particle energy is lost in ionizing collisions and all the charge produced is collected, that is, no recombination takes place.

It has been known for many years that heavy ions moving in gases lose part of their energy in nonionizing collisions with nuclei. The same phenomenon, PHD, has been observed in semiconductor detectors. The PHD, which is negligible for alphas and lighter particles, is defined as the difference between the energy of a heavy ion and that of a light ion (usually an alpha particle) that generates the same pulse height in the detector. Experimental results showing this phenomenon are presented in Figure 13.15.[42] Notice that for alphas and carbon ions, the relationship between energy and pulse height is linear. Nickel and silver ions show a small PHD. Heavier ions (Au, U) show a significant PHD.

The PHD is the result of the following three contributing defects[42–47]:

1. The *nuclear defect* is due to nuclear collisions. As a result of such collisions, the moving ion imparts energy upon other nuclei. The recoiling nuclei lose their energy partly in electronic ionizing collisions and partly in nuclear nonionizing ones. The nuclear defect has been calculated[45] based on the work of Bohr[1] and of Lindhard *et al.*[48] (see also Chapter 4).

2. The *recombination defect* arises from incomplete collection of the charge produced in the detector. A heavy ion is a strongly ionizing particle. It creates a dense plasma of electron-hole pairs along its path, a plasma that reduces the electric field established by the external bias applied to the detector. The reduction of the electric field intensity hinders the drifting of the electrons and holes and thus increases the probability of recombination. The calculation of this defect is not so easy as that of the nuclear one, but an approximate calculation was performed by Wilkins *et al.*[42]

3. The *window defect* is due to energy loss in the dead layer (window) of the front surface of the detector. It can be obtained from the thickness of the window and the stopping power of the ion. The thickness of the window can be *measured by* determining the change in pulse height as a function of the incident angle (see Prob. 7.6).

The PHD for iodine and for argon ions has been measured by Moak *et al.*[49] using the *channeling effect* in silicon. Pulse-height distributions were measured by first aligning the direction of incident ions with the [110] crystal axis of the silicon surface-barrier detector and then by letting the ions impinge at an angle with respect to the same axis. In the first case, the ions moved along the channel between two planes (*channeled* ions); in the second, they did not (*unchanneled* ions). The channeled ions showed an energy resolution about three times better than that of unchanneled ones, and essentially no PHD (Figure 13.16). This result can be explained by assuming that the channeled ions traveling between atomic planes lose all their energy in ionizing collisions, all the way to the end of their track. Similar results have been obtained with ^{235}U fission fragments[43] and ^{252}Cf fission fragments.[50]

The lack of nuclear collisions for channeled ions is not the only phenomenon that affects the pulse height. It is known that the electron density is much reduced along the channel. As a result, the electronic stopping power is lower and, consequently, so is the charge density produced by the heavy ion. Thus, not only the nuclear but also the recombination defect is reduced for the channeled ions.

The PHD increases slowly with ion energy, as shown in Figure 13.16.

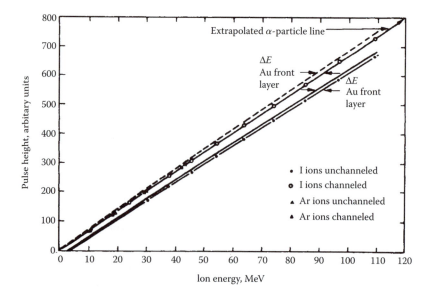

Figure 13.16 Pulse-height response of a Si surface-barrier detector for "channeled" and "unchanneled" ions. (From Moak, C. D., et al., *Rev. Sci. Instrum.*, 1131, 37, 1966.)

13.5.2 Energy Calibration: The Schmitt Method

The relationship between the pulse height h and the kinetic energy T of a heavy ion was determined by Schmitt et al.[51,52] to be of the form

$$T = (a + bM)h + c + dM \qquad (13.13)$$

where M is the mass of the ion and a, b, c, and d are constants. The calibration of the detector, that is, the determination of the constants a, b, c, and d, can be achieved in two ways. The first is an absolute calibration, and the second is a relative one.

Absolute energy calibration is performed with the help of an accelerator. One measures the pulse heights of four monoenergetic beams of ions with known mass. Substitution of the known energy, mass, and pulse height into Eq. 13.13 provides four equations with four unknowns that can be solved for the constants a, b, c, and d.

For fission-fragment measurements, a relative calibration is used. The calibration constants of Eq. 13.13 are determined in terms of two pulse heights H and L of a fission-fragment spectrum (Figure 13.17), where H and L represent pulse height corresponding to the mid-point at three-quarters maximum of the heavy or light fragment peak, respectively. The equations for the constants are

$$a = \frac{a_1}{L - H} \qquad (13.14)$$

$$b = \frac{a_2}{L - H} \qquad (13.15)$$

$$c = a_3 - aL \qquad (13.16)$$

$$d = a_4 - bL \qquad (13.17)$$

and the constants a_1 a_2, a_3, a_4 for ^{252}Cf and ^{235}U fission fragments are given in Table 13.1.

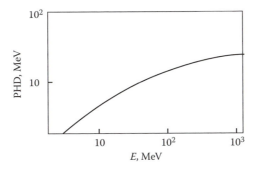

Figure 13.17 The dependence of the pulse height defect on ion energy (from Ref. 42). The numbers correspond, approximately, to uranium ions.

Table 13.1
Schmitt Calibration Constants for ^{252}Cf and ^{235}U Fission Fragments[51,55]

Constant	^{252}Cf	^{235}U
a_1	24.0203	30.9734
a_2	0.03574	0.04596
a_3	89.6083	87.8626
a_4	0.1370	0.1345

The constants a_1, a_2, a_3, and a_4 do not depend on the detector. The quality of the detector with respect to energy resolution is determined from a set of criteria developed by Schmitt and Pleasonton[52] and shown in Table 13.2. Figure 13.18 explains the symbols used (see also Ref. 54).

13.5.3 Calibration Sources

Monoenergetic heavy ions necessary for energy calibration can be provided only by accelerators. Fission fragments, which are heavy ions, cover a wide spectrum of energies (Figure 13.18).

Table 13.2
Acceptable Parameters for a ^{252}Cf Fission-Fragment Spectrum

Spectrum Parameters	Expected Values	Spectrum of Figure 13.18
N_L/N_V	2.9	2.79
N_H/N_V	2.2	2.20
N_L/N_H	1.3	1.27
$\Delta L/(L-H)$	0.36	0.37
$\Delta H/(L-H)$	0.44	0.43
$(H-HS)/(L-H)$	0.69	0.69
$(LS-D)/(L-H)$	0.48	0.50
$\Delta S/(L-H)$	2.17	2.19

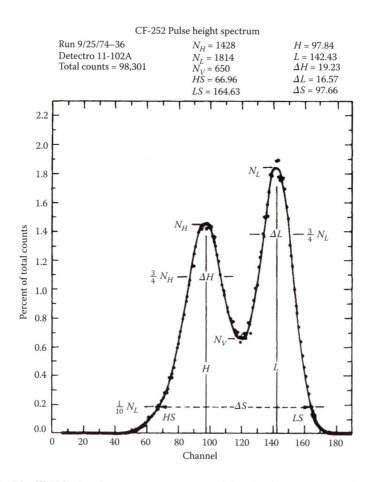

CF-252 Pulse height spectrum

Run 9/25/74–36
Detectro 11-102A
Total counts = 98,301

$N_H = 1428$
$N_L = 1814$
$N_V = 650$
$HS = 66.96$
$LS = 164.63$

$H = 97.84$
$L = 142.43$
$\Delta H = 19.23$
$\Delta L = 16.57$
$\Delta S = 97.66$

Figure 13.18 The ^{252}Cf fission-fragment spectrum used for the determination of the detector calibration constants. (From Bucher, R. G., Ph.D. thesis, University of Illinois, 1975.)

The isotope ^{252}Cf is a very convenient source of fission fragments produced by the spontaneous fission of that isotope. Uranium, plutonium, or thorium fission fragments can only be produced after fission is induced by neutrons; therefore, a reactor or some other intense neutron source is needed.

13.5.4 Fission Foil Preparation

Fission foils are prepared by applying a coat of fissile material of the desired thickness on a thin metal backing. Details of several methods of foil preparation are given in References 56–60. A technique used for the preparation of uranium foils is described here.

Enriched uranium in the form of uranium nitrate hexahydrate crystals is dissolved in ethanol until it forms a saturated solution. A small amount of collodion is added to the solution to improve its spreading characteristics. A thin metal foil—for example, nickel—that serves as the backing material is dipped into the solution and then heated in an oven in a controlled temperature environment. The heating of the foil is necessary to remove organic contaminants and to convert the uranium nitrate into uranium oxides (mostly U_3O_8). The temperature is critical because if it is too high, part of the uranium diffuses into the backing material, causing fragment energy degradation. Dipping produces a two-sided foil. If the material is applied with a paint brush, a one-sided foil is formed.

The dipping (or brush painting) and heating is repeated as many times as necessary to achieve the desired foil thickness. The thickness of the foil is determined by weighing it before and after the uranium deposition or, better yet, by counting the alphas emitted by the uranium isotopes. Most of the alphas come from ^{234}U; therefore the fraction of this isotope in the uranium must be known.

13.6 THE TIME-OF-FLIGHT SPECTROMETER

The time-of-flight (TOF) method, which is also used for the measurement of neutron energy (see Section 14.8), has been applied successfully for the determination of the mass of fission fragments and other heavy ions.

The principle of TOF is simple. A beam of ions is directed along a flight path of length L (Figure 13.19). The time t it takes the ions to travel the distance L determines their speed $V = L/t$. This information, combined with the measurement of the energy of the particle, gives the mass (nonrelativistically):

$$M = \frac{2Et^2}{L^2} \tag{13.18}$$

The errors in determining the mass come from uncertainty in energy, ΔE, in time, Δt, and in length of the flight path, ΔL. The mass resolution is then given by

$$\frac{\Delta M}{M} = \sqrt{\left(\frac{\Delta E}{E}\right)^2 + \left(\frac{2\Delta t}{t}\right)^2 + \left(\frac{2\Delta L}{L}\right)^2} \tag{13.19}$$

Usually, the system is designed in such a way that $\Delta L/L$ is negligible compared to the other two terms of Eq. 13.19. Assuming that this is the case, consider the sources of uncertainty in energy and time.

The uncertainty $\Delta E/E$ is the resolution of the detector measuring the energy of the ion. The best energy resolution that can be achieved with silicon surface-barrier detectors is about 1.5–2%. The resolution can be improved with magnetic or electrostatic analyzers (DiIorio and Wehring[60] achieved 0.3% energy resolution using an electrostatic analyzer).

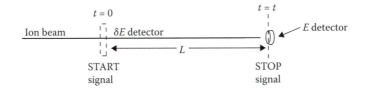

Figure 13.19 The principle of time-of-flight for the determination of the mass of heavy ions.

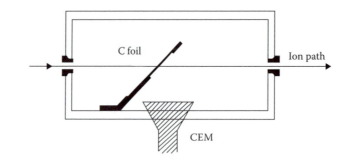

Figure 13.20 The use of a CEM as a δE detector.

The time t it takes the particle to travel the distance L is the difference between a START and a STOP signal (Figure 13.19). The STOP signal is generated by the detector, which measures the energy of the ion. This detector is usually a surface-barrier detector. The START signal is generated by a transmission counter, also called the δE detector. The ion loses a tiny fraction of its energy going through the START detector.

Several types of δE detectors have been used.[61] Examples are totally depleted surface-barrier detectors,[62,63] thin ($\sim 10^{-3}$ kg/m^2 = 100 μg/cm^2) plastic scintillators,[64] ionization chambers,[65] and secondary-electron emission detectors.[60,66–69] Secondary-electron emission detectors fall into two categories. In the first, the ions traverse a thin foil (e.g., carbon foil 10^{-4} kg/m^2 = 10 μg/cm^2 thick) and generate secondary electrons that are accelerated and focused to strike a scintillator coupled to a photomultiplier tube. In the second category belong the *channel electron multipliers* (CEM) and the *microchannel plates* (MCP).

A CEM is essentially a thin glass tube ($\sim$1 mm diameter) shaped into a spiral, with its inside surface coated with a semiconducting material that is also a good secondary electron emitter. An accelerating field is created in the tube by applying a high voltage along its length. Electrons multiply as they proceed down the tube. Figure 13.20 shows one possible arrangement for the use of a CEM.

An MCP is a glass disk perforated with a large number of small-diameter (10–100 μm) holes or channels. Each channel is a glass tube coated with a resistive secondary electron-emitting material. If a voltage is applied, each channel acts as an electron multiplier.

The state of the art of δE detector systems is such that $\Delta t < 100$ ps has been achieved and the flight path L can become long enough that the time $T \approx 100$–300 ns. Thus, the time resolution of TOF measurements is

$$\frac{\Delta t}{t} \approx \frac{100 \times 10^{-3}\,\text{ns}}{200\,\text{ns}} \approx 5 \times 10^{-4} \approx 0.05\%$$

and the mass resolution (Eq. 13.19) is essentially limited by the energy resolution. The mass measurement is actually the measurement of the mass number A ($\Delta M/M = \Delta A/A$). Since A is an integer, the lowest limit for mass resolution is $\Delta A < 1$. Assuming $\Delta A = 0.7$, Figure 13.21 gives the ion energy necessary for such resolution as a function of A and $L/\Delta t$.[70] For heavy ions, mass resolution as low as $\Delta A = 0.2$ has been reported.[63]

If the mass of the ion is known, the TOF technique can be used to determine the energy of the ion with a resolution much better than with any detector in use today. Indeed, if $\Delta M = 0$, Eq. 13.19 gives

$$\frac{\Delta E}{E} = 2\frac{\Delta t}{t} \approx \frac{200 \times 10^{-3}\,\text{ns}}{200\,\text{ns}} \approx 0.1\%$$

13.7 DETECTOR TELESCOPES (E dE/dX DETECTORS)

The TOF method discussed in the previous section measures the energy and the mass of the ion. This section presents a method that identifies the atomic number Z and the mass number A of the particle.[71]

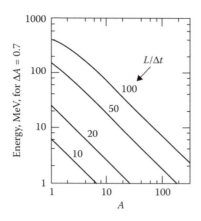

Figure 13.21 Energy of ion at which adjacent isotopes can be resolved, as a function of A, for different values of $L/\Delta t$. (From Butler, G. W., et al., *Nucl. Instrum. Meth.*, 89, 189, 1970.)

Identification of A and Z is possible by making use of a *detector telescope* consisting of a very thin detector measuring dE/dx and a thick detector that stops the particle. The geometric arrangement is similar to that shown in Figure 13.19. The particle traverses the thin detector after depositing there an energy equal to $(dE/dx)t$ (where t is the detector thickness), and stops in the "E detector." The total energy of the particle is obtained from the sum of the two detector signals. The product $E\, dE/dx$ can be written, using Eq. 4.2 or 4.33, as

$$E\frac{dE}{dx} = k_1 M Z_{ef}^2 \ln\left(k_2 \frac{E}{M}\right) \tag{13.20}$$

where Z_{ef} is the effective charge of the ion. Since the logarithmic term changes very slowly with energy, Eq. 13.20 gives a value for MZ_{ef}^2.

Another method, giving better results, is based on the fact that the range of heavy ions is given, over a limited energy range, by an equation of the form

$$R = C\frac{E^b}{MZ_{ef}^2} \tag{13.21}$$

where b is a constant. If a particle deposits energy $\delta E = (dE/dx)t$ in a detector of thickness t and then deposits energy E in the second detector, one can say that the range of the particle with energy $E + \delta E$ is t units longer than the range of the same particle with energy E. Using Eq. 13.21, one can write

$$MZ_{ef}^2 \approx \frac{\left[(E+\delta E)^b - E^b\right]}{t} \tag{13.22}$$

Thus, Eq. 13.22 provides the value of MZ_{ef}^2 since t, E, and δE are known. The constant b is also assumed to be known for the ion of interest.

Equations 13.20–13.22 were written in terms of the mass M of the particle. For nonrelativistic particles, M has a nonintegral value very close to the value of A, which in turn, is given by an integer. This is fortunate because Z also assumes integral values only, and the product MZ^2 assumes unique values for many particles. For example, for protons, deuterons, tritons, and alphas, the value of MZ^2 is 1, 2, 3, and 16, respectively.

13.8 POSITION-SENSITIVE DETECTORS

Detectors that in addition to the measurement of the energy also indicate the position of the particle have been developed for imaging devices used in biological and medical research. Biological and medical imaging devices, on the other hand, involve mostly X-rays and, in general, low-energy radiation. It is because of the latter use that position-sensitive detectors are briefly discussed here.

13.8.1 Position-Sensitive Semiconductor Detectors

Most of the position-sensitive semiconductor detectors determine the position of the incident particle by employing the method of *resistive-charge division*.[73] To illustrate the method, consider the detector in Figure 13.22. The detector is a reverse-biased p–n junction with electrodes on both front and back. The front electrode with considerable resistivity has two electrical contacts a distance L apart. The back electrode has low resistivity and provides a good electrical contact to the base material. When a particle enters the detector, electrons and holes are created that move under the influence of the electric field. If the resistivity of the front electrode is homogeneous, and charge-sensitive low-impedance amplifiers are used, the charge collected at one of the two contacts of the front electrode is proportional to the distance between the point of impact and the other contact. The total charge q_0 collected through the single contact of the back side is, of course, proportional to the energy deposited in the sensitive region of the junction. This technique of determining the position by comparing the signals from q_1 and q_0 is called the *amplitude method*.

The signal q_1 changes with time, as shown in Figure 13.23. The timescale is in units of the time constant $\tau_D = R_D C_D$, where R_D and C_D are the resistance and capacitance of the detector, respectively. Figure 13.23 shows that the rise time of the signal depends on the position. This property is the basis of a second technique for determination of position, called the *time method*. The position is now determined from the difference in arrival times of the signals from the charges q_1 and q_2 (Figure 13.22).

The detector described above is of the "continuous" type. Position-sensitive detectors of the "discrete" type have also been developed.[74] They consist of individual semiconductor elements all placed on the same base material, with each element connected to its own preamplifier-amplifier system. Two-dimensional detectors of the continuous type[75] as well as of the discrete type[76] have also been tried.

13.8.2 Multiwire Proportional Chambers

Mulitwire proportional chambers (MWPC)* have been developed for use as position-sensitive focal detectors for magnetic spectrometers.[78–87] They can provide excellent position resolution, operate with counting rates as high as 10^6 counts/s, and provide a large solid angle at the focal plane of the spectrometer.

The basic design of an MWPC is shown in Figure 13.24. A series of thin, equally spaced anode wires is positioned between two parallel plates serving as cathodes. A noble gas mixed with an organic component fills the space between the cathodes. A positive voltage is applied to the anode wires. When a particle goes through the counter, electrons and ions are created by the processes explained in Chapter 5. The electric field close to the wires is so intense that the primary electrons acquire enough energy to produce secondary ionization. Thus, an avalanche of electrons is produced that is collected within a time of 1 ns, but that leaves behind a cloud of positive ions.

It is significant that the pulse produced by the counter is not due to the motion of the electrons, but to that of the ions.[78] As the positive ions move away from the anode wires, they generate a fast-rising (~10 ns) negative pulse that gradually slows down and lasts a few microseconds. The pulse induced in the neighboring wires is positive. Thus, the active wire (the wire close to the trajectory of the particle) is distinguished from the others. The signals from the wires are processed and read by either digital or analog systems.[79]

The position resolution depends on the spacing between the wires, but it is better than the actual size of the spacing. The smallest wire spacing is about 1 mm, while position resolutions better than 100 μm have been reported.[88,89] Such space resolution has been achieved because the position is determined from the well-defined centroid of the charge distribution generated by the passage of the particle.

A variation of the detector described above is the so-called "drift chamber." The drift chamber determines the position from the time it takes the electrons produced by the incoming particle to drift to the nearest anode wire.[90] A two-dimensional MWPC has also been constructed for detection of neutrons scattered from biological samples.[91] It is a ^{3}He gas-filled counter that detects neutrons through the (n, p) reaction.

---

* Single-wire position-sensitive proportional counters operating by charge division (as described in Section 13.10.1) have also been used.[77]

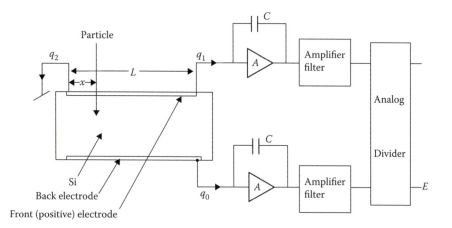

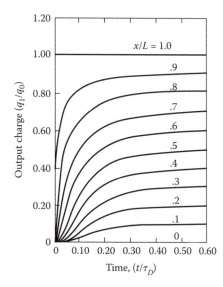

Figure 13.22 A position-sensitive semiconductor detector using resistive-charge division. (From Laegsgaard, E., *Nucl. Instrum. Meth.*, 162, 93, 1979.)

Figure 13.23 Time dependence of the position signal for different positions of incidence. (From Laegsgaard, E., *Nucl. Instrum. Meth.*, 162, 93, 1979.)

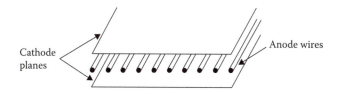

Figure 13.24 The basic design of an MWPC.

PROBLEMS

13.1 Prove that the maximum energy transfer to an electron as a result of a collision with a particle of mass M and speed $v = \beta c$ is equal to

$$\Delta E_{max} = 2mc^2 \frac{\beta^2}{1-\beta^2} \left[1 + \left(\frac{m}{M} \right)^2 + 2\frac{m}{M}\frac{1}{\sqrt{1-\beta^2}} \right]^{-1}$$

13.2 Show that ΔE_{max} of Problem 13.1 takes the form

$$\Delta E_{max} = \frac{4mM}{(m+M)^2} T$$

if $T = \frac{1}{2}Mv^2$ (nonrelativistic particle) and $m \ll M$.

13.3 Calculate the energy loss of a 6-MeV alpha particle going through an aluminum foil with thickness equal to one-fourth of the range of the alpha (remember dE/dx is not constant as the particle slows down).

13.4 A monoenergetic beam of 10-MeV alpha particles goes through a nickel foil with a thickness equal to 1/30 of the range of the alphas. What is the width of the emerging beam (in keV)? If the emerging particles are counted by a silicon detector with a resolution of 15 keV, what is the total width of the measured alpha beam (measurement performed in vacuum)?

13.5 What is the thickness of a nickel foil that will cause energy loss equal to 1/50 of the energy of a 10-MeV proton traversing it?

13.6 Show that the backscattering fraction of electrons is given by

$$b = \left[1 + \sqrt{2\pi} \left(\frac{g}{\tau} - 0.5 \right) \frac{\sigma}{E'} \right]^{-1}$$

where g, τ, and E' are defined in Figure 13.9; σ is the standard deviation of the Gaussian.

13.7 If the stopping power for a heavy ion can be represented by the equation

$$\frac{dE}{dx} = k\sqrt{E} + b$$

where k and b are constants, what is the expression for the range as a function of energy for such ions? What is the energy of the ion as a function of distance traveled?

13.8 If the expression for dE/dx given for Problem 13.7 is valid, what fraction of the initial kinetic energy of the ion is lost in the first half of its range?

13.9 A thin fission foil was prepared using natural uranium. Calculate the ratio of alpha activities due to the isotopes ^{234}U, ^{235}U, and ^{238}U. The abundance a and half-lives of the three isotopes are

$$^{234}\text{U}: a = 0.0057\% \quad T_{1/2} = 0.47 \times 10^5 \text{y}$$
$$^{235}\text{U}: a = 0.711\% \quad T_{1/2} = 7.10 \times 10^8 \text{y}$$
$$^{238}\text{U}: a = 99.283\% \quad T_{1/2} = 4.51 \times 10^9 \text{y}$$

Assume that every decay gives off an alpha; that is, neglect decay by spontaneous fission.

BIBLIOGRAPHY

Bertolini, G., and Coche, A. (eds.), *Semiconductor Detectors*, Wiley Interscience, New York, 1968.

Davidson, R. C., and Hong, Q., *Physics of Intense Charged Particle Beams in High Energy Accelerators*, Imperial College Press, London and World Scientific Publishing Company, Singapore, 2001.

Evans, R. D., *The Atomic Nucleus*, McGraw-Hill, New York, 1972.

Rohrlich, F., *Classical Charged Particles*, World Scientific Publishing Company, Singapore, 2007.

Segré, E., *Nuclei and Particles*, W. A. Benjamin, New York, 1965.

Siegbahn, K. (ed.), *Alpha, Beta and Gamma-Ray Spectroscopy*, North Holland Publishing Co., Amsterdam, 1965, vols. 1 and 2.

Soroko, Z. N., and Sukhoruchkin, S., *Nuclear States from Charged Particle Reactions*, Springer, Berlin, 2007.

REFERENCES

1. Landau, W., *J. Phys. USSR* **8**:201 (1944).

2. Symon, K., Thesis, Harvard 1958, "Experts in High Energy Particles," Prentice-Hall, Englewood Cliffs, N.J., 1952.

3. Vavilov, P. V., *JETP* **5**:749 (1957).

4. Maccabee, H. D., Raju, M. R., and Tobias, L. A., *IEEE Nucl. Sci.* **13**(3):176 (1966).

5. Blunk, O., and Leisegang, S., *Z. Physik* **128**:500 (1950).

6. Tschalar, C., *Nucl. Instrum. Meth.* **61**:141 (1968).

7. Tschalar, C., *Nucl Instrum. Meth.* **64**:237 (1968).

8. Clarke, N. M., *Nucl. Instrum. Meth.* **96**:497 (1971).

9. Ispirian, K. A., Murgarian, A. T., and Zuerev, A. M., *Nucl. Instrum. Meth.* **117**:125 (1974).

10. Ermilova, V. C., Kotenko, L. P., and Merzon, G. I., *Nucl. Instrum. Meth.* **145**:444 (1977).

11. Skyrme, D. J., *Nucl. Instrum. Meth.* **57**:61 (1967).

12. Bohr, N., *Philos. Mag.* **30**:581 (1915).

13. Livingston, M. S., and Bethe, H., *Rev. Mod. Phys.* **9**:245 (1937).

14. Titeica, S., *Bull. Soc. Roumaine Phys.* **38**:81 (1939).

15. Comfort, J. R., Decker, J. R., Lynk, E. T., Scully, M. D., and Quinton, A. R., *Phys. Rev.* **150**:249 (1966).

16. Strittmatter, R. B., and Wehring, B. W., *Nucl. Instrum. Meth.* **135**:173 (1976).

17. Lonergan, J. A., Jupiter, C. P., and Merkel, G., *J. Appl. Phys.* **41**:678 (1970).

18. Berger, M. J., and Seltzer, S. M., "ETRAN Monte Carlo Code System for Electron and Photon Transport Through Extended Media," NBS-9836 and NBS-9837 (1968).

19. Alfassi, Z. B., Persico, E., Groppi, F., and Bonardi, M. L., *Appl. Rad. Isotop.* **67**:240 (2009).

20. Leblanca, E., Coron, N., Leblancb, J., Marcillacb, P. de., Bouchard, J., and Plagnard, J., *Appl. Rad. Isotop.* **64**:1281 (2006).

21. Sánchez, A. M., Jurado, Vargas, M. J., Nuevo Sánchez, M. J. N., and Timón, A. F., *Appl. Rad. Isotop.* **66**:804 (2008).

22. Rogers, P. C., and Gordon, G. E., *Nucl. Instrum. Meth.* **37**:259 (1965).

23. Tsoulfanidis, N., Wehring, B. W., and Wyman, M. E., *Nucl. Instrum. Meth.* **73**:98 (1969).

24. Wohn, F. K., Clifford, J. R., Carlson, G. H., and Talbert, W., Jr., *Nucl. Instrum. Meth.* **101**:343 (1972).

25. Schupferling, H. M., *Nucl. Instrum. Meth.* **123**:67 (1975).

26. Dakuba, T., and Gilboy, W. B., *Nucl. Instrum. Meth.* **150**:479 (1978).

27. Tabata, T., Ito, R., and Okabe, S., *Nucl. Instrum. Meth.* **94**:509 (1971).

28. Kuzminikh, V. A., and Vorobiev, S. A., *Nucl. Instrum. Meth.* **129**:561 (1975).

29. Tsoulfanidis, N., Wehring, B. W., and Wyman, M. E., *Nucl. Sci. Eng.* **43**:42 (1971).

30. Meyer, O., and Langmann, H. J., *Nucl. Instrum. Meth.* **39**:119 (1966).

31. Segré, E., *Nuclei and Particles*, Benjamin, New York, 1965, p. 348.

32. Ramírez, G. J., Iturbe, J. L., and Solache-Ríos, M., *Appl. Radiat. Isotop.,* **47**:27 (1996).

33. Eduardo García-Toraño, E., *Appl. Radiat. Isotop.* **64**:1273 (2006).

34. Siiskonen, T., and Pöllänen, R., *Appl. Radiat. Isotop.* **60**:947 (2004).

35. Crespo, M. T., *Appl. Radiat. Isotop.* **53**:109 (2000).

36. Beesley, A. M., Crespo, M. T., Weiher, N., Tsapatsaris, N., Cózar, J. S., Esparza, H., Méndez, C. G., Hill, P., Schroeder, S. L. M., and Montero-Cabrera, M. E., *Appl. Radiat. Isotop.* **67**:1559 (2009).

37. Leblanc, E., Coron, N., Leblanc, J., De Marcillac, P., Bouchard, J., and Plagnard, J., *Appl. Radiat. Isotop.* **64**:1281 (2006).

38. Tsai, T-L., Lin, C-C., and Tieh-Chi Chu, T-C., *Appl. Radiat. Isotop.* **66**:1097 (2008).

39. Vainblat, N., Pelled, O., German, U., Haquin, Tshuva, G. A., and Alfassi, Z. B., *Appl. Radiat. Isotop.* **61**:307 (2004).

40. Purkl, S., and Anton Eisenhauer, A., *Appl. Radiat. Isotop.* **59**:245 (2003).

41. Kondev, F. G., Ahmad, I., Greene, J. P., Kellett, M. A., and Nichols, A. L. *Appl. Radiat. Isotop.* **65**:335 (2007).

42. Wilkins, B. D., Fluss, M. J., Kaufman, S. B., Cross, C. E., and Steinberg, E. P., *Nucl. Instrum. Meth.* **92**:381 (1971).

43. Sullivan, W. J., and Wehring, B. W., *Nucl. Instrum. Meth.* **116**:29 (1974).

44. Caywood, J. M., Mead, C. A., and Mayer, J. W., *Nucl. Instrum. Meth.* **79**:329 (1970).

45. Haines, E. L., and Whitehead, A. B., *Rev. Sci. Instrum.* **37**:190 (1966).

46. Finch, E. C., and Rodgers, A. L., *Nucl. Instrum. Meth.* **113**:29 (1973).

47. Finch, E. C., *Nucl. Instrum. Meth.* **113**:41 (1973).

48. Lindhard, J., Scharff, M., and Schiott, H. E., *Fys. Medd.* **33**:1 (1963).

49. Moak, C. D., Dabbs, J. W. T., and Walker, W. W., *Rev. Sci. Instrum.* **1131**:37 (1966).

50. Britt, H. C., and Wegner, H. E., *Rev. Sci. Instrum.* **34**:274 (1963).

51. Schmitt, H. W., Kiker, W. E., and Williams, C. W., *Phys. Rev.* **137**:B837 (1965).

52. Schmitt, H. W., and Pleasonton, F., *Nucl. Instrum. Meth.* **40**:204 (1966).

53. Bucher, R. G., Ph.D. thesis, University of Illinois, 1975.

54. Long, A. B., Wehring, B. W., and Wyman, M. E., *Rev. Sci. Instrum.* **1566**:39 (1968).

55. Schmitt, H. W., Gibson, W. M., Neiler, J. H., Walter, F. J., and Thomas, T. D., *Proceedings of IAEA Conf. on the Physics and Chemistry of Fission* **1**:531 (1965).

56. Van de Eijk, W., Oldenhof, W., and Zehner, W., *Nucl. Instrum. Meth.* **112**:343 (1973).

57. Lowenthal, G. C., and Wyllie, H. A., *Nucl. Instrum. Meth.* **112**:353 (1973).

58. Dobrilovic, L., and Simovic, M., *Nucl. Instrum. Meth.* **112**:359 (1973).

59. Adair, H. L., and Kuehn, P. R., *Nucl. Instrum. Meth.* **114**:327 (1974).

60. DiIorio, G. J., and Wehring, B. W., *Nucl. Instrum. Meth.* **147**:487 (1977).

61. Betts, R. R., *Nucl. Instrum. Meth.* **162**:531 (1979).

62. Pleyer, H., Kohlmeyer, B., Schneider, W. F. W., and Bock, R., *Nucl. Instrum. Meth.* **96**:363 (1971).

63. Zediman, B., Henning, W., and Kovar, D. G., *Nucl. Instrum. Meth.* **118**:361 (1974).

64. Gelbke, C. K., Hildenbrabd, K. D., and Bock, R., *Nucl. Instrum. Meth.* **95**:397 (1971).

65. Barrette, J., Braun-Munzinger, P., and Gelbke, C. K., *Nucl. Instrum. Meth.* **126**:181 (1975).

66. Dietz, E., Bass, R., Reiter, A., Friedland, V., and Hubert, B., *Nucl. Instrum. Meth.* **97**:581 (1971).

67. Schneider, W. F. W., Kohlmeyer, B., and Bock, R., *Nucl. Instrum. Meth.* **123**:93 (1975).

68. Pfeffer, W., Kohlmeyer, B., and Schneider, W. F. W., *Nucl. Instrum. Meth.* **107**:121 (1973).

69. Wiza, J. L., *Nucl. Instrum. Meth.* **162**:587 (1979).

70. Butler, G. W., Poskanzer, A. M., and Landis, D. A., *Nucl. Instrum. Meth.* **89**:189 (1970).

71. Goulding, F. S., and Harvey, B. G., *Annu. Rev. Nucl. Sci.* **25**:167 (1975).

72. Sykes, D. A., and Harris, S. J., *Nucl. Instrum. Meth.* **101**:423 (1972).

73. Laegsgaard, E., *Nucl Instrum. Meth.* **162**:93 (1979).

74.   Haase, E. L., Fawzi, M. A., Saylor, D. P., and Vellen, E., *Nucl. Instrum. Meth.* **97**:465 (1971).

75.   Kalbitzer, S., Barer, R., Melzer, W., and Stumpfi, W., *Nucl. Instrum. Meth.* **54**:323 (1967).

76.   Hofker, W. K., Ooosthock, D. P., Hoeberechts, A. M. E., Van Dantzig, X., Mulder, K., Obserski, J. E. J., Koerts, L. A., Dieperink, J. H., Kok, E., and Rumphorst, R. F., *IEEE Trans.* **NS-13**(3):208 (1966).

77.   Ford, J. L. C., Jr., *Nucl. Instrum. Meth.* **162**:277 (1979).

78.   Charpak, G., *Nature.* **270**:479 (1977).

79.   Ball, G. C., *Nucl. Instrum. Meth.* **162**:263 (1979).

80.   Allemond, R., Bourdel, J., Roudant, E., Convert, P., Ibel, K., Jacobe, J., Cotton, J. P., and Farnoux, B., *Nucl. Instrum. Meth.* **126**:29 (1975).

81.   Beardsworth, E., Fischer, J., Iwata, S., Lavine, M. J., Rudeka, V., and Thorn, C. E., *Nucl. Instrum. Meth.* **127**:29 (1975).

82.   Glussel, P., *Nucl. Instrum. Meth.* **140**:61 (1977).

83.   Kitahara, T., and Isozumi, Y., *Nucl. Instrum. Meth.* **140**:263 (1977).

84.   Fischer, J., Okuno, H., and Walenta, A. H., *Nucl. Instrum. Meth.* **151**:451 (1978).

85.   Lindgren, L. J., and Sandell, A., *Nucl. Instrum. Meth.* **219**:149 (1984).

86.   Fuzesy, R. Z., Hadley, N. J., and Robrish, P. R., *Nucl. Instrum. Meth.* **223**:40 (1984).

87.   Doll, P., Brady, F. P., Ford, T. D., Garrett, R., Krupp, H., and Klages, H. O., *Nucl. Instrum. Meth.* **270**:437 (1988).

88.   Lacy, J. L., and Lindsey, R. S., *Nucl. Instrum. Meth.* **119**:483 (1974).

89.   Bertozzi, W., Hynes, M. V., Sargent, C. P., Creswell, C., Dunn, P. C., Hirsch, A., Leitch, M., Norum, B., Rad, F. N., and Susanuma, T., *Nucl. Instrum. Meth.* **141**:457 (1977).

90.   Breskin, A., Charpak, G., Gabioud, B., Sauli, F., Trantner, N., Duinker, W., and Schultz, G., *Nucl. Instrum. Meth.* **119**:9 (1974).

91.   Alberi, J., Fischer, J., Radeka, V., Rogers, L. C., and Schoenborn, B., *Nucl. Instrum. Meth.* **127**:507 (1975).

Neutron Detection and Spectroscopy

14.1 INTRODUCTION

Since neutrons do not directly ionize atoms, they are detected "indirectly" upon producing a charged particle or a photon, which is then recorded with the help of an appropriate detector. The charged particle or the photon is the result of a neutron interaction with a nucleus. If the mechanism of the interaction is known, information about the neutron can be extracted by studying the products of the reaction. Many types of interactions are used, divided into absorptive and scattering reactions.

Absorptive reactions are (n, α), (n, p), (n, γ), or $(n, \text{fission})$. In the case of an (n, γ) reaction, the neutron may be detected through the interactions of the gamma emitted at the time of the capture, or it may be detected through the radiation emitted by the radioisotope produced after the neutron is captured. The radioisotope may emit β^- or β^+ or γ or a combination of them. By counting the activity of the isotope, information is obtained about the neutron flux that produced it. This is called the activation method. If the reaction is fission, two fission fragments are emitted; being heavy charged particles, these are detected easily.

The main scattering reaction used is neutron–proton collision, called the *proton-recoil method*. The knocked-out proton is the particle recorded.

With the exception of the proton-recoil method, which functions for fast neutrons only ($E_n > 1.0$ keV), all the other interactions can be used with neutrons of any energy. However, at every neutron energy, one method may be better than another. The best method will be selected based on the neutron energy, the purpose of the experiment (is it number or energy of neutrons measured, or both?), and the physical constraints of the measurement (e.g., inside a reactor core or outside).

This chapter discusses in detail all the neutron detection methods mentioned above, as well as the Bragg crystal spectrometer, the time-of-flight (TOF) method, compensated ion chambers, and self-powered neutron detectors (SPND). Other specialized neutron detectors, such as fission track recorders and thermoluminescent dosimeters, are described in Chapter 16.

14.2 NEUTRON DETECTION BY (n, CHARGED PARTICLE) REACTION

There are many nuclear reactions of the type (n, charged particle) used for neutron detection. In general, endothermic reactions are used for fast neutrons, and exothermic ones for thermal neutrons. The endothermic reactions will be discussed in Section 14.6.

The most useful exothermic reactions are listed in Table 14.1, along with their Q values and the value of the cross section for thermal neutrons.

The charged particles from any one of the reactions of Table 14.1 share an amount of kinetic energy equal to $Q + E_n$, where E_n is the neutron kinetic energy. The large Q values make detection of the products very easy, regardless of the value of E_n, but at the same time make measurement of the energy of slow neutrons practically impossible. The neutron energy would be measured from the pulse height, which is proportional to $Q + E_n$. However, if the pulse corresponds to energy of the order of MeV (because of the Q value), a small change in E_n will produce a variation in the pulse that is undetectable. For example, the fractional change of a 1-MeV pulse due to 1-keV change in neutron kinetic energy is $(1.001–1.000)/1.000 = 0.1\%$, which is less than the best energy resolution of alpha-particle detectors. Therefore, the measurement of neutron energy is possible only when E_n amounts to a considerable fraction of the Q value.

Table 14.1
Exothermic Reactions Used for Neutron Detection

Reaction	Charged Particles Produced	Q Value (MeV)	σ (b) for $E_n = 0.025$ eV
$^{10}_{5}\text{B}(n,\alpha)^{7}_{3}\text{Li}$	$\alpha, {}^{7}\text{Li}$	2.78	3840
$^{6}_{3}\text{Li}(n,\alpha)^{3}_{1}\text{H}$	$\alpha, {}^{3}\text{H}$	4.78	937
$^{3}_{2}\text{He}(n,p)^{3}_{1}\text{H}$	$p, {}^{3}\text{H}$	0.765	5400

14.2.1 The BF$_3$ Detector

The (n, α) reaction with $^{10}_{5}\text{B}$ is probably the most useful reaction for the detection of thermal neutrons because of the following:

1. The reaction cross section is large.

2. The energy dependence of the cross section is of the $1/v$ type.

3. $^{10}_{5}\text{B}$ is a constituent of the compound BF$_3$, which may be used as the gas of a proportional gas detector.

The BF$_3$ detector is a proportional counter filled with BF$_3$ gas, usually enriched to more than 90% in ^{10}B (about 20% of natural boron is ^{10}B; the rest is ^{11}B). The BF$_3$ detector detects the alpha and the lithium particles produced by the reaction

$$^{10}_{5}\text{B} + ^{1}_{0}n \rightarrow ^{4}_{2}\text{He} + ^{7}_{3}\text{Li} + 2.78 \text{ MeV}$$

With thermal neutrons, the ^7Li nucleus is left in an excited state about 96% of the time. In that case, the Q value of the reaction is 2.30 MeV and the ^7Li nucleus goes to the ground state by emitting a gamma with energy equal to $2.78 - 2.30 = 0.480$ MeV. This photon may also be used for the detection of the neutron.

The relationship between counting rate and neutron flux is derived as follows. Let

$n(E) \, dE$ = number of neutrons/m^3 with kinetic energy between E and $E + dE$

$\phi(E) \, dE = v(E)n(E) \, dE$ = neutron flux consisting of neutrons with kinetic* energy between E and $E + dE$

$v(E)$ = neutron speed for energy E (m/s)

E_m = upper limit of neutron energy considered

N = number of ^{10}B atoms per unit volume

V = volume of the detector

$\sigma(E) = \sigma(v)$ = cross section of the (n, α) reaction for neutron energy E, speed v

Assuming that the neutron flux is uniform over the detector volume, the reaction rate R is given by

$$R \text{ (reactions/s)} = VN \int_0^{E_m} \sigma(E)\phi(E) \, dE \tag{14.1}$$

The ^{10}B cross section has a $1/v$ dependence† over a wide range of neutron energies; that is, it can be written as

$$\sigma(E) = \sigma_0 \frac{v_0}{v(E)} = \sigma_0 \sqrt{\frac{E_0}{E}} \tag{14.2}$$

where σ_0 is the cross section at some known speed $v_0 = \sqrt{2E_0/M}$, and M is the neutron mass. If Eq. 14.2 is substituted into Eq. 14.1, the reaction rate takes the form

$$R = NV\sigma_0 v_0 n \text{ (reactions/s)} \tag{14.3}$$

where n is the total number of neutrons per unit volume, or

$$n = \int_0^{E_m} n(E) \, dE \tag{14.4}$$

Equation 14.3 shows that the reaction is proportional to the total neutron density. BF$_3$ detectors are most frequently used for the detection of thermal neutrons, for which one can calculate an

---

* In Chapter 3, the symbol T was used to denote kinetic energy. That was necessary because the discussion involved kinetic and total energy. In this chapter, E is used for the kinetic energy of the neutron.
$\dagger$ The ^{11}B cross section is quite different. Equation 14.2 represents only the ^{10}B cross section.

average neutron speed $\bar{v}$ given by

$$\bar{v} = \frac{\int_0^{E_m} v(E)n(E)\,dE}{\int_0^{E_m} n(E)\,dE} \tag{14.5}$$

and a total flux ϕ, given by

$$\phi = \int_0^{E_m} \phi(E)dE = \int_0^{E_m} v(E)n(E)\,dE \tag{14.6}$$

Under these conditions, Eq. 14.3 takes the form

$$R = NV\sigma_0 \frac{v_0}{\bar{v}}\phi \tag{14.7}$$

Eq. 14.7 shows that the reaction rate is proportional to the total neutron flux ϕ for the commonly encountered Maxwell-Boltzmann distribution of thermal neutrons, $\bar{v} = 2v_p/\sqrt{\pi}$, where v_p is the most probable neutron speed.

The derivation of Eqs. 14.3 and 14.7 was based on the assumption that the neutron flux is uniform over the volume of the detector. A measure of the flux uniformity is the value of the factor $\exp(-\Sigma_t l)$, where

Σ_t = total macroscopic cross section for the gas of the detector, averaged over all the neutron energies present

l = a characteristic dimension of the detector (usually the diameter or the length of a cylindrical detector)

If $\exp(-\Sigma_t l) \approx 1$, the flux may be taken as uniform over the detector volume.

Example 14.1 Consider a BF_3 detector with a diameter of 0.05 m (≈ 2 in.) and length 0.30 m (≈ 12 in.) filled with BF_3 gas, 96% enriched to ^{10}B, at a pressure of 1 atm and used for the detection of 0.0253 eV neutrons. Should the user take into account flux depression in the detector?

Answer The factor $\exp(-\Sigma_t l)$ should be calculated. For the worst case, consider l = length of the detector. The total macroscopic cross section is

$$\Sigma_t = N(BF_3)\left[0.96_s(^{10}B) + 0.04_s(^{11}B) + 3_s(F)\right]$$
$$\approx \frac{0.6022}{22,400}(0.96)(3840) = 0.0991\,\text{cm}^{-1} = 9.91\,\text{m}^{-1}$$

For the calculation of Σ_t, the cross sections of ^{11}B and F were neglected because they are much smaller than that of ^{10}B. The "depression factor" is

$$\exp(-\Sigma_t l) = \exp[(-9.91)(0.30)] = \exp(-2.97) = 0.05$$

If l is the diameter of the detector, then

$$\exp(-\Sigma_t l) = \exp[(-9.91)(0.05)] = \exp(-0.496) = 0.61$$

One concludes that flux depression is considerable in this case, and the flux cannot be taken as uniform over the detector volume.

If the BF_3 detector is used for the detection of a polyenergetic neutron spectrum, instead of a monoenergetic neutron source, average cross sections should be used for the calculation (Section 14.9.4 explains how average cross sections are obtained).

As a first approximation, the efficiency of a BF_3 detector is equal to

$$\varepsilon = \frac{\Sigma_a}{\Sigma_t}[1 - \exp(-\Sigma_t l)][\exp(-\Sigma_t^w t_w)] \tag{14.8}$$

where Σ_t and Σ_a are total and absorptive neutron macroscopic cross sections, respectively, for BF$_3$, and l is the dimension of the detector parallel to the direction of the neutron beam. Σ_t^w and t_w are the total macroscopic cross section and thickness, respectively, for the material of which the wall or the front window of the detector is made. Equation 14.8 was derived under the following assumptions:

1. All neutrons travel the same distance inside the detector (parallel beam).
2. Every neutron interaction in the wall or the front window of the detector removes the neutron from the beam.

Example 14.2 What is the efficiency of a BF$_3$ detector enriched to 96% in ^{10}B, 0.04 m (1.57 in) in diameter, 0.30 m (~12 in.) long, for a parallel beam of 1-eV neutrons? The BF$_3$ pressure in the detector is 53,329 Pa (40 cmHg). Consider two cases:

(a) The beam is directed parallel to the axis of the detector.
(b) The beam is perpendicular to the axis of the detector.

Assume that the wall and the window of the detector are made of aluminum and are 2 mm thick. Take the total neutron cross section for Al at 1 eV to be 1.5 b.

Answer Equation 14.8 will be used. At 1 eV for BF$_3$ enriched to 96% to ^{10}B, $\Sigma_a \approx \Sigma_t \approx \Sigma_a(^{10}$B). To find the microscopic cross section at 1 eV, use Eq. 14.2 and the value of $\sigma_a = 3840$ b

$$\sigma_a(1\text{eV}) = 3840\frac{v_0(0.025\,\text{eV})}{v(1\text{eV})} = 3840\sqrt{\frac{0.025}{1}} = 607\,\text{b}$$

$$\Sigma_a = 0.96\left(\frac{53,329}{101,325}\right)\left(\frac{0.6022}{22,400}\right)(607) = 0.0082\,\text{cm}^{-1} = 0.82\,\text{m}^{-1}$$

For aluminum,

$$\Sigma_t^w t_w = 1.5(2.7)\left(\frac{0.6022}{27}\right)(0.2) = 0.018$$

(a) If the beam is parallel to the detector axis, $l = 0.30$ m and

$$\varepsilon = [1 - \exp(\Sigma_a l)][\exp(-\Sigma_t^w t_w)] = (0.18)(0.982) = 0.214 = 21.4\%$$

(b) If the beam is perpendicular to the axis, all the neutrons do not travel the same distance inside the cylindrical detector. Assuming that the incident neutrons form a narrow beam that hits the detector at the center, l = diameter = 0.04 m. Then,

$$\varepsilon = \left\{1 - \exp\left[(-0.82)(0.04)\right]\right\}(0.982) = (0.032)(0.982) = 0.032 = 3.2\%$$

The specifications of commercial BF$_3$ detectors consist of sensitivity, dimensions, composition of the filling gas, operating voltage, and maximum operating temperature.

The sensitivity S is defined as the ratio

$$S = \frac{\text{true net counting rate}}{\text{neutron flux}} = \frac{r}{\phi} \tag{14.9}$$

and is given in terms of counts/s per neutron/(m^2 s). The parameters affecting the sensitivity can be seen by noting that

$$r = \varepsilon_p R = g - b \tag{14.10}$$

where ε_p = efficiency of the detector for detection of the charged particles produced
  b = background counting rate
  g = gross counting rate
  R = reaction rate given by Eq. 14.7

Since the charged particles are generated inside the volume of the detector, the efficiency ε_p is practically equal to 1. Also, the background rate may be made negligible because the pulses produced by the charged particles are well above the electronic noise. Thus, a proper discriminator level may be set to eliminate almost all the background. Under these conditions, $R = g$ and the equation for the sensitivity becomes

$$S = \frac{R}{\phi} = NV\sigma_0 \frac{v_0}{v} \tag{14.11}$$

Equation 14.11 indicates that for a certain neutron spectrum, the sensitivity is proportional to boron density (i.e., pressure of the BF_3 gas) and volume of the detector.

The number of boron atoms decreases with exposure, and so does the sensitivity. The decrease is expressed by a factor having the form

$$\exp\left(-\sigma_a \phi t\right)$$

where ϕt in neutrons/m^2 is the fluence to which the detector was exposed. Since the average value of σ_a for thermal neutrons reacting with ^{10}B is of the order of 10^{-25} m^2 (1000 b), the fluence necessary to cause an appreciable change in sensitivity is of the order of 10^{25} neutrons/m^2.

Typical specifications of commercial detectors are the following.

Sensitivity: 5 counts per second per n/(cm^2 s)

Dimensions: Almost any dimensions

Pressure of BF_3: From a little less than 1 to about 2 atmosphere (202 kPa). An increase in pressure requires an increase in the operating voltage.

Operating voltage: BF_3 detectors show an almost flat plateau (see Chapter 5) extending over 1000 V or more. Typical operating voltages range from 1000 to 3000 V.

Temperature: Maximum operating temperature is about 100°C.

14.2.2 Boron-Lined Detectors

Boron-lined detectors are gas-filled proportional counters that employ the same reaction as the BF_3 detector, except that the ^{10}B is coated on the walls of the detector. Since the (n, α) reactions take place in a thin layer close to the wall (Figure 14.1), only one of the two particles has a chance of entering the sensitive volume of the detector and producing a pulse; the other stops in the wall. The sensitivity increases with the thickness of the ^{10}B coating. That thickness, however, cannot

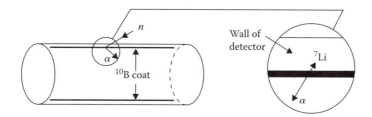

Figure 14.1 A ^{10}B-lined detector.

exceed the range of the alphas produced. The advantage of boron-lined over the BF_3 detectors is the flexibility of using a gas more appropriate than BF_3. Then the operating voltage is less and the counter is less sensitive to gamma rays. For this reason, boron-lined detectors may be used to detect neutrons in intense gamma fields.

Both BF_3 and boron-lined detectors are mainly used for the measurement of the number (not energy) of thermal neutrons.

14.2.3 ^{6}Li Detectors

Neutron detection by ^{6}Li is based on the reaction

$$^6_3\text{Li} + ^1_0n \rightarrow ^4_2\text{He} + ^3_1\text{H} + 4.78\,\text{MeV}$$

The cross section for this reaction is of the $1/v$ type up to 10 keV, with a value of 937 b at 0.025 eV. Lithium is used either as LiF or as the component of a scintillator.

A neutron spectrometer based on LiF[1] consists of a thin slice of ^{6}LiF ($\sim30 \times 10^{-5}$ kg/m^2 = 30 µg/cm^2) sandwiched between two surface-barrier silicon detectors. When neutrons strike the LiF, charged-particle pairs (^{4}He–^{3}H) are produced and are detected simultaneously by the two detectors. The pulses from the detectors are amplified and then summed to produce a single pulse, which is proportional to the energy of the neutron plus the Q value of the reaction.

There are many inorganic scintillators based on lithium. ^{6}LiI(Eu) has been used for neutron energy measurements from 1 to 14 MeV with 10% energy resolution.[2] It has good efficiency for low-energy neutrons, but activation of iodine creates some problems. The most widely used lithium scintillator was developed by Ginther and Schulman[3] and Voitovetskii et al.[4] It is a cerium-activated scintillating glass containing Li_2O. The proportion of the cerium activator affects the efficiency of luminescence. A series of measurements of many properties of commercially available glasses has been reported recently by Spowart.[5,6] Today one can buy these glasses in a large variety of thicknesses (0.5–25 mm), sizes (up to 125 mm in diameter), Li contents (up to 11%), and ^{6}Li enrichments (up to 95%). The efficiency of ^{6}Li glass as a function of neutron energy is shown in Figure 14.2.

To increase the efficiency, ^{6}Li glass scintillators with thickness about 13 mm and diameter 110 mm have been optically coupled to one or more photomultiplier tubes through light pipes.[8,9] To avoid moderation of the incident neutrons, the light pipe should not contain hydrogenous material. One problem with such thick scintillators is considerable scattering of the incident neutrons. The scattered neutrons add an exponential tail to the primary neutron signal, a tail that should be included in the time resolution function of the instrument in TOF measurements.

A different type of arrangement, using a ^{6}LiI(Eu) scintillator to detect the neutrons after they are moderated, is the *Bonner* ball (or sphere).[10,11] Using the Bonner ball, neutrons are detected by

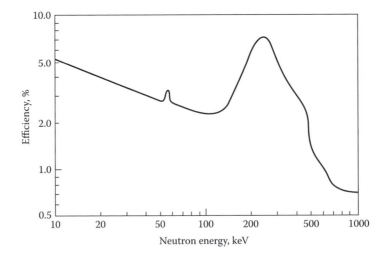

Figure 14.2 The efficiency of ^{6}Li glass as a function of neutron energy, with glass thickness 12.7 mm. (From Hill, N. W., et al., 147, ORNL-4743, 1972.)

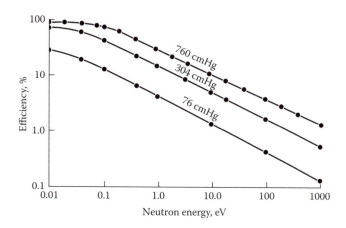

Figure 14.3 Efficiency of ^{3}He-filled proportional counter as a function of neutron energy.

a ^{6}LiI(Eu) scintillator, a BF$_3$ tube, or an ^{3}He detector placed at the center of polyethylene spheres with sizes ranging from 51 to 305 mm (2–12 in.) in diameter. Using the Bonner sphere, the neutron energy is determined on the basis of the difference in moderating efficiency for the spheres of different sizes. Because the Bonner sphere is primarily used as a neutron dosimeter, it is discussed in more detail in Chapter 16.

14.2.4 ^{3}He Detectors

Neutron detection by ^{3}He is based on the reaction

$$^3_2\text{He} + ^1_0 n \rightarrow ^1_1\text{H} + ^3_1\text{H} + 765\,\text{keV}$$

The cross section for this reaction is quite high for thermal neutrons (5400 b at 0.025 eV) and varies as $1/v$ from 0.001 to 0.04 eV.

One type of helium neutron spectrometer consists of two surface-barrier silicon detectors facing each other, with the space between them filled with helium at a pressure of a few atmospheres (this is similar to the ^{6}LiF spectrometer described in Section 14.2.3).

Proportional counters filled with ^{3}He are widely used, especially in TOF measurements (see Section 14.8). The efficiency of the detector can be increased by increasing the pressure. Figure 14.3 shows the efficiency of a ^{3}He detector as a function of neutron energy for three different pressures. One of the problems of ^{3}He detectors is the *wall effect*. If the reaction takes place close to the wall of the detector, there is a high probability that only a fraction of the charged-particle energy will be deposited in the detector. As a result, smaller size pulses are produced which do not come under the main peak. There are two ways to overcome this effect. One is to use a large-diameter detector, in which case the fraction of reactions occurring near the wall is smaller. The other is to increase the stopping power of the gas. Increase in stopping power is achieved either by increasing the pressure of the gas or by adding a small fraction of another gas, such as krypton, with a higher stopping power. Increase in stopping power is accompanied, however, by an increase in gamma sensitivity and a decrease in pulse rise time. Fast rise time is important for TOF measurements.

The pressure of the gas and the operating voltage are higher in ^{3}He than in BF$_3$ detectors. The pressure of the ^{3}He is usually between 404 and 1010 kPa (4–10 atmosphere), and the operating voltage is 3000–5000 V.

14.3 FISSION CHAMBERS

Fission chambers are gas-filled detectors that detect the fragments produced by fission. The fission fragments, being massive charged particles with $Z \approx +20e$ and kinetic energy 60–100 MeV, have a short range even in a gas. They produce such an intense ionization that gas multiplication is not necessary. Thus, fission chambers operate in the ionization region.

In the most common type of fission chamber, the interior surface of the detector is coated with a fissile isotope (Figure 14.4). When fission takes place, one of the fission fragments (denoted as FF$_1$

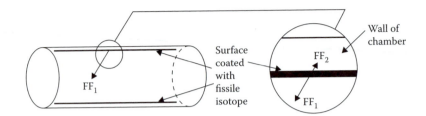

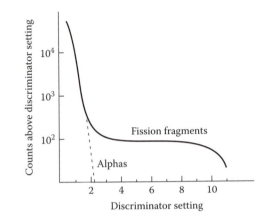

Figure 14.4 A fission chamber.

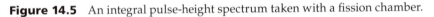

Figure 14.5 An integral pulse-height spectrum taken with a fission chamber.

in Figure 14.4) is emitted toward the center of the detector and is detected. The other (FF$_2$) stops in the fissile deposit or the wall of the detector. The counting rate of a fission chamber is proportional to the fission rate, which in turn, is proportional to the neutron flux. The relationship among these three quantities is similar to the equations given for the BF$_3$ detector.

For relative measurements, the thickness of the fissile material coating is not very critical except that it should be less than the range of the fission fragments. For absolute measurement, however—that is, measurements for which every fission should be detected—at least one fission fragment from each fission should produce a recorded pulse. To achieve this, the thickness of the coating must be limited so that fission fragments being produced anywhere in the layer of the fissile material generate a pulse larger than that of alphas, betas, or gammas, which are always present.* Pulses from gammas present a problem only when the chamber is used in an extremely intense gamma field. Pulses from alphas, however, are always present and should be discriminated from those of fission fragments. Fortunately, the difference between the ionization produced by alphas and that produced by fission fragments is so large that such discrimination is easy. Figure 14.5 shows a typical integral pulse-height spectrum. Notice that the alpha pulses start at such a low level that a discriminator level can be set to cut them off.

Fission chambers may be used for detection of either fast or thermal neutrons. If the chamber wall is coated with ^{235}U, essentially only thermal neutrons are detected because the ^{235}U fission cross section for thermal neutrons is about 500 times higher than that of fast neutrons. If the chamber is coated with ^{238}U or ^{232}Th, only fast neutrons with kinetic energy greater than 1 MeV are detected because the fission cross section of these isotopes has a threshold at about that energy.

Fission chambers are used for detection of the number and not the energy of neutrons. They can be used, however, for differentiation of thermal and fast-neutron flux by using a combination of ^{235}U- and ^{238}U-coated chambers (see also Section 14.6).

---

* All fissile isotopes emit alpha particles. Betas and gammas may be emitted either by the fissile isotope or by fission fragments.

The sensitivity of a fission chamber decreases with exposure because of the depletion of the fissile isotope (the same phenomenon as depletion of boron atoms—see Section 14.2.1). The decrease in sensitivity may be halted, to a certain extent, if the chamber wall is coated with a mixture of fertile and fissile materials. One such combination is 90% ^{234}U and 10% ^{235}U. The ^{235}U is partially replenished with new atoms produced by neutron capture in ^{234}U. A ^{235}U–^{238}U combination will have a similar effect, thanks to ^{239}Pu produced as a result of neutron capture in ^{238}U.

Fission chambers are used extensively for both out-of-core and in-core measurements of neutron flux in nuclear flux in nuclear reactors. In out-of-core situations, they monitor the neutron population during the early stages of power ascension when the neutron flux level is very low. For in-core measurements, fission chambers are used for flux mapping (and consequently, determination of the core power distribution). They are manufactured as long thin cylindrical probes that can be driven in and out of the core with the reactor in power. Typical commercial fission chambers for in-core use have diameters of about 1.5 mm (0.06 in.), use uranium enriched to at least 90% in ^{235}U as the sensitive material, and can be used to measure neutron fluxes up to 10^{18} neutrons/(m$^2 \cdot$ s) (10^{14} neutrons/[cm^2 s]).

Another method of measuring fission rates is by using fission track detectors, as discussed in Section 16.9.3.

14.4 NEUTRON DETECTION BY FOIL ACTIVATION

14.4.1 Basic Equations

Neutron detection by foil activation is based on the creation of a radioisotope by neutron capture, and subsequent counting of the radiation emitted by that radioisotope. Foil activation is important not only for neutron flux measurements but also for neutron activation analysis, which is the subject of Chapter 15. This section presents the basic equations involved.

Consider a target being irradiated in a neutron flux $\phi(E)$, where

$\sigma_i(E)$ = neutron absorption cross section of isotope A_i at neutron energy E
λ_{i+1} = decay constant of isotope with atomic mass number $A_i + 1(A_{i+1})$
$\sigma_{i+1}(E)$ = neutron absorption cross section of isotope A_{i+1} at neutron energy E
$N_i(t)$ = number of atoms of nuclide with atomic mass number A_i, present at time t
m = mass of target (normally this is the mass of the element whose isotope A_i captures the neutron)
α_i = weight fraction in the sample of isotope A_i

As a result of neutron absorption, the following processes take place:

1. Target atoms of atomic mass number A_i are destroyed.

2. Atoms with atomic mass number A_{i+1} are produced.

3. Atoms of type A_{i+1} decay.

4. Atoms of type A_{i+1} may be destroyed by absorbing a neutron.

For the target isotope $\left(^{A_i}_{Z_i}X\right)$, the reaction involved is

$$^{A_i}_{Z_i}X + n \rightarrow {}^{A_{i+1}}_{Z_i}X$$

The destruction of these atoms proceeds according to the equation

$$-\frac{dN_i(t)}{dt} = N_i(t)\int_0^\infty dE\sigma_i(E)\phi(E) \tag{14.12}$$

In Eq. 14.12 and all others in this section, it is assumed that the presence of the target does not disturb the flux; that is, the foil does not cause depression of the flux. Corrections that take into account foil self-absorption can be found in Chapter 11 in Beckurts and Wirtz and in Reference 12. The integral over energy in Eq. 14.12 is usually expressed as

$$\int_0^\infty dE\sigma_i(E)\phi(E) = \bar{\sigma}_i\int_0^\infty \phi(E)dE = \bar{\sigma}_i\phi = \sigma_i\phi \tag{14.13}$$

That is, an average cross section is used, even though the overbar that indicates averaging is normally dropped. From now on, Eq. 14.13 will be used without the overbar, but the reader should keep in mind that σ is an average over the neutron energy spectrum.

The solution of Eq. 14.12 is, using Eq. 14.13,

$$N_i(t) = N_i(0)e^{-\sigma_i \phi t} \tag{14.14}$$

where

$$N_i(0) = \frac{a_i m N_A}{A_i} = \text{number of atoms of isotope } A_i \text{ at } t = 0$$

The net production of the A_{i+1} isotope is expressed by

$$\frac{dN_{i+1}(t)}{dt} = \text{production-destruction-decay}$$

or

$$\frac{dN_{i+1}(t)}{dt} = N_i(t)\sigma_i\phi - N_{i+1}(t)\sigma_{i+1}\phi - \lambda_{i+1}N_{i+1}(t) \tag{14.15}$$

With initial condition $N_{i+1}(t) = 0$, the solution of Eq. 14.15 is

$$N_{i+1}(t) = \frac{\sigma_i N_i(0)\phi}{\lambda_{i+1} + \sigma_{i+1}\psi - \sigma_i \psi}\left\{\exp(-\sigma_i\phi t) - \exp\left[-(\lambda_{i+1} + \sigma_{i+1}\phi)t\right]\right\} \tag{14.16}$$

The activity of this target, $A_{i+1}(t)$ is, after irradiation for time t,

$$A_{i+1}(t) = N_{i+1}(t)\lambda_{i+1} = \frac{\sigma_i N_i(0)\phi}{1 + (\sigma_{i+1} - \sigma_i)\phi/\lambda_{i+1}}\left\{\exp(-\sigma_i\phi t) - \exp\left[-(\lambda_{i+1} + \sigma_{i+1}\phi)t\right]\right\} \tag{14.17}$$

Equation 14.17 refers to the most general case. In practice, targets are selected in such a way that

1. The fraction of target nuclei destroyed is negligible, that is, $\sigma_i\phi t \ll 1$.
2. The radioisotope produced has a neutron absorption cross section such that $\lambda_{i+1} \gg \sigma_{i+1}\phi$.

If conditions (1) and (2) are met, Eq. 14.17 takes the form

$$A_{i+1}(t) = \sigma_i N_i(0)\phi\left[1 - \exp(-\lambda_{i+1}t)\right] \tag{14.18}$$

which is the more familiar form of the activity or activation equation.

If one plots activity as a function of irradiation time, the result is Figure 14.6. Two regions are observed:

1. For irradiation times that are short compared to the half-life of the radioisotope produced, the activity increases linearly with time. Indeed, if $\lambda_{i+1}t \ll 1$, then $e^{-\lambda_{i+1}t} \approx 1 - \lambda_{i+1}t$

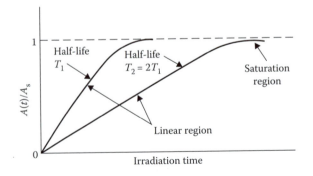

Figure 14.6 Activity versus irradiation time (shown as a fraction of saturation activity).

Table 14.2
Fraction of Saturation Activity Produced as a Function of Irradiation Time

Irradiation Time (in Half-Lives)	$A_{i+1}(t)/A_s$
4	0.937
5	0.969
6	0.984
7	0.992
8	0.996

and

$$A_{i+1}(t) \approx \sigma_i N_i(0)\phi\lambda_{i+1}t = \sigma_i N_i(0)\phi\frac{\ln 2}{T_{i+1}}t$$

where T_{i+1} is the half-life of the isotope produced.

2. For irradiation times many times longer than the half-life of the radioisotope, the activity reaches a saturation value (A_s). Theoretically, the saturation activity

$$A_s = \sigma_i N_i(0)\phi \tag{14.19}$$

is reached for $t = \infty$. In practice, the activity produced is taken as equal to A_s for $t \approx 6$–7 half-lives. Table 14.2 gives the fraction of saturation activity produced for several irradiation times.

Example 14.3 The isotope ^{197}Au is irradiated in a thermal neutron flux of 10^{18} neutrons/(m² · s). The cross section for neutron capture is 99 b, and the half-life of the radioactive ^{198}Au produced is 2.7 days: (a) How long does the sample have to be irradiated for 0.1% of the target atoms to be destroyed? (b) What is the irradiation time necessary to produce 95% of saturation activity? and (c) If the mass of the sample is 4×10^{-6} kg, what is the irradiation time necessary to produce 7.4×10^4 Bq (2 μCi) of activity?

Answer

(a) Using Eq. 14.14,

$$\frac{N(t)}{N(0)} = 0.999 = e^{-\sigma\phi t} \text{ or } t = \frac{1}{\sigma\phi}\ln\frac{1}{0.999}$$

$$t = \frac{1}{(99\times10^{-28})10^{18}}\ln\frac{1}{0.999} = 1.01\times10^5 \text{ s} = 28\text{ h}$$

(b) Using Eq. 14.18, the irradiation time t should be such that $1 - \exp(-\lambda t) = 0.95$ or

$$t = \frac{T}{\ln 2}\ln\left(\frac{1}{1-0.95}\right) = 11.67 \text{ days}$$

(c) Using Eq. 14.18,

$$A(t) = \sigma N(0)\phi(1-e^{-\lambda t}) = \sigma m\frac{N_A}{A}\phi(1-e^{\lambda t})$$

It is useful to evaluate A_s first, because if A_s is less than the activity desired, it is impossible to obtain such activity under the conditions given.

The saturation activity is

$$A_s = \sigma m\frac{N_A}{A}\phi = (99\times10^{-28})(4\times10^{-6})\frac{0.6022\times10^{24}}{197\times10^{-3}}\times10^{18}$$

$$= 1.21\times10^{11} \text{ Bq}$$

In this example, A_s is greater than $A(t)$ and the required irradiation time t is

$$t = \frac{T}{\ln 2}\ln\left[1-\frac{A(t)}{A_s}\right] = -\frac{2.7 \text{ days}}{\ln 2}\ln\left(1-\frac{7.4\times10^4}{1.21\times10^{11}}\right) = 0.21\text{ s}$$

14.4.2 Determination of the Neutron Flux by Counting the Foil Activity

As shown in Eq. 14.18, the activity of the irradiated foil is proportional to the neutron flux. Determination of the flux requires measurement of the activity, a task accomplished as follows.

Let the irradiation time be t_0. In practice, counting of the foil starts some time after irradiation stops, and it is customary to consider the end of irradiation as time $t = 0$ (Figure 14.7). At time t after irradiation stops, the activity is, using Eq. 14.18,

$$A_{i+1}(t) = N_i(0)\sigma_i\phi\left[1-\exp(-\lambda_{i+1}t_0)\right]e^{-\lambda_{i+1}t} \tag{14.20}$$

If the sample is counted between t_1 and t_2, the number of disintegrations in that period is

$$D(t_1,t_2) = \int_{t_1}^{t_2} A_{i+1}(t)dt = \frac{N_i(0)\sigma_i\phi}{\lambda_{i+1}}\left[1-\exp(-\lambda_{i+1}t_0)\right]\left(e^{-\lambda_{i+1}t_1} - e^{-\lambda_{i+1}t_2}\right) \tag{14.21}$$

Assuming that one counts particles with energy E_k for which e_k is the probability of emission per decay, and the counting system is such that

ε_k = the efficiency of the detection of particles with energy E_k
Ω = solid angle
B = background counts recorded in time $T = t_2 - t_1$

then the gross counts recorded, G_k, will be

$$G_k = \varepsilon_k e_k F\Omega \frac{N_i(0)\sigma_i\phi}{\lambda_{i+1}}\left[1-\exp(-\lambda_{i+1}t_0)\right]\left(e^{-\lambda_{i+1}t_1} - e^{-\lambda_{i+1}t_2}\right) + B \tag{14.22}$$

The factor F in Eq. 14.22 takes into account any other corrections (i.e., backscattering, foil self-absorption) that may be necessary (see Section 8.3). If dead-time correction is necessary, it should be applied to G_k.

The flux ϕ is determined from Eq. 14.22 if all the other factors are known. There are two types of factors in Eq. 14.22:

1. Factors that depend on the sample $\left[N_i(0),\sigma_i,\lambda_{i+1},e_k\right]$, which are assumed to be known with negligible error

2. Factors that depend on the counting system (ε, F, Ω, B), which are the main sources of error

To determine the flux distribution only, not the absolute value of the flux, foils are placed at known positions x_j and are irradiated for a time t_0. The foils are then counted using the same detector. At any point x_j, the flux may be written as

$$\phi(x_j) = \frac{L\left[G_k(x_j)-B_j\right]}{m_j C_j} \tag{14.23}$$

where the subscript j indicates position of the foil and m_j = mass of foil at position j.

$$C_j = \left[\exp(-\lambda_{i+1}t_1)-\exp(-\lambda_{i+1}t_2)\right]_j$$

$$L = \lambda_{i+1}\left(\varepsilon_k e_k F\Omega \frac{N_A}{A_i}\left[1-\exp(-\lambda_{i+1}t_0)\right]\right)^{-1}$$

(L includes all the factors that are common to all the foils.)

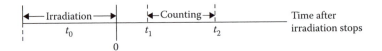

Figure 14.7 Timescale for counting an irradiated sample. Time $t = 0$ coincides with the end of the irradiation period.

The title of this section includes the word *foil* because the sample to be irradiated is used in the form of a thin foil of the order of 1 mm thick or less. The mass of the foil is only a few milligrams. Small thin foils are used because of the following considerations:

1. A thick sample will absorb so many neutrons that the radiation field will be perturbed and the measurement will not give the correct flux.

2. A thick sample will cause a depression of the flux in its interior. In such a case, correction factors will have to be applied to all the equations of this section that contain the flux ϕ.

3. If the radioisotope emits β particles, increased thickness will not necessarily increase the counting rate, because only particles emitted close to the surface within a thickness less than the range will leave the target and have a chance to be recorded.

4. There is no purpose in producing more activity than is necessary.

Foil activation may be used for detection of the number of either fast or thermal neutrons. The use of foils for fast-neutron energy measurements is discussed in Section 14.6. Foil activation is not used generally for measurement of the energy of thermal neutrons.

14.5 MEASUREMENT OF A NEUTRON ENERGY SPECTRUM BY PROTON RECOIL

Detection of neutrons by proton recoil is based on collisions of neutrons with protons and subsequent detection of the moving proton. Since neutrons and protons have approximately the same mass, a neutron may, in one collision, transfer all its kinetic energy to the proton. However, there is a possibility that the struck proton may have any energy between zero and the maximum possible, as a result of which the relationship between a neutron energy spectrum and a pulse-height distribution of the struck protons is not simple. It is the objective of this section to derive a general expression for this relationship. The sections that follow show its application for specific detectors.

Consider the case of a neutron with kinetic energy E_n colliding with a proton at rest (Figure 14.8). To calculate the proton kinetic energy after the collision, one must apply the equations of conservation of energy and linear momentum (Eqs. 3.81–3.83) using $Q = 0$ and $M_n = M_p$. The result for E_p, the proton kinetic energy as a function of the recoil angle θ, is

$$E_p = E_n \cos^2 \theta \tag{14.24}$$

In a neutron–proton collision, the maximum value of angle θ is 90°, and the minimum 0°; therefore, the limits of the proton energy are $0 \le E_p \le E_n$. For neutron energies up to about 14 MeV, the $(n-p)$ collision is isotropic in the center-of-mass system; as a consequence, there is an equal probability for the proton to have any energy between zero and E_n in the laboratory system. That is, if $p(E)\,dE$ is the probability that the proton energy is between E and $E + dE$, after the collision, then

$$p(E)dE = \frac{dE}{E_n} \tag{14.25}$$

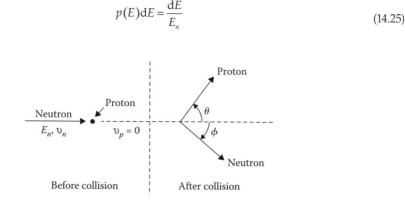

Figure 14.8   Neutron–proton collision kinematics.

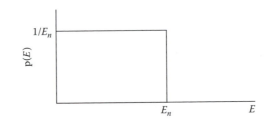

Figure 14.9 The proton energy distribution after a (n, p) collision that is isotropic in the center of mass system of the two particles.

The function $p(E)$ is shown in Figure 14.9. What is important for the observer is not $p(E)$ but the proton pulse-height distribution produced by the detector. The relationship between the pulse-height distribution and the neutron spectrum is derived as follows. Let

$\phi(E_n)dE_n$ = neutron energy spectrum = flux of neutrons with energy between E_n and $E_n + dE_n$

$N(E_p) dE_p$ = proton recoil energy spectrum = number of protons produced (by collisions with neutrons) with energy between E_p and $E_p + dE_p$

$R(E, E_p) dE$ = response function of the detector = probability that a proton of energy E_p will be recorded as having energy between E and $E + dE$ (defined before in Section 11.5)

$M(E) dE$ = measured spectrum = number of protons measured with energy between E and $E + dE$

The measured spectrum $M(E)$ is the pulse-height distribution in energy scale. The response function $R(E, E_p)$ takes into account the finite energy resolution of the detector and the relationship between energy deposition and pulse height.

Assuming isotropic scattering in the center-of-mass system, the proton energy spectrum is

$$N(E_p)\, dE_p = N_H T \int_0^{E_{max}} \sigma(E_n)\phi(E_n)dE_n \frac{dE_p}{E_n} H(E_n - E_p) \tag{14.26}$$

where N_H = number of hydrogen atoms exposed to the neutron beam

T = time of measurement of the recoil protons

$H(E_n - E_p)$ = step function; $H(E_n - E_p) = 1 | E_n \geq E_p$, zero otherwise

$\sigma(E_n)$ = elastic scattering cross section for (n, p) collisions

The measured energy spectrum is then given by

$$M(E)dE = \int_0^{E_{max}} dE\, R(E, E_p) N(E_p) dE_p \tag{14.27}$$

In Eqs. 14.26 and 14.27, the energy E_{max} is the upper limit of the neutron energy spectrum. Equation 14.27 may be rewritten in the form

$$M(E) = \int_0^{E_{max}} dE_n k(E, E_n) \phi(E_n) \tag{14.28}$$

where

$$k(E, E_n) = \int_0^{E_{max}} dE_p R(E, E_p) N_H T \frac{\sigma(E_n)}{E_n} H(E_n - E_p) \tag{14.29}$$

Equation 14.28 has the form of the folding integral (see also Section 11.5), while Eq. 14.29 gives the "composite" response function for the proton recoil spectrometer.

Example 14.4 As a first application of Eq. 14.28, consider the case of a monoenergetic neutron spectrum and a detector with a Gaussian response function. What is the measured spectrum?

Answer   Substituting the Gaussian response function

$$R(E,E_p) = \frac{1}{\sqrt{2\pi}\sigma}\exp\left[-\frac{(E-E_p)^2}{2\sigma^2}\right]$$

into Eq. 14.29 and performing the integration, assuming $E/\sigma \gg 1$, one obtains*

$$k(E,E_n) = \frac{N_H T \sigma(E_n)}{2E_n}\left[1 + \mathrm{erf}\left(\frac{E_n - E}{\sqrt{2}\sigma}\right)\right]$$

where

$$\mathrm{erf}\left(\frac{x}{\sqrt{2}}\right) = \sqrt{\frac{2}{\pi}}\int_0^x e^{-t^2/2}\,dt$$

Substituting the value of $k(E, E_n)$ and the monoenergetic flux $\phi(E_n) = S\delta(E_n - E_0)$ into Eq. 14.28 and performing the integration, one obtains

$$M(E) = S\frac{N_H T \sigma(E_0)}{2E_0}\left[1 + \mathrm{erf}\left(\frac{E_0 - E}{\sqrt{2}\sigma}\right)\right] \qquad (14.30)$$

The function $M(E)$ given by Eq. 14.30 is shown in Figure 14.10. It is essentially the same function as that shown in Figure 14.9, except for the rounding off at the upper energy limit caused by the Gaussian detector response.

The task of neutron spectroscopy is to obtain the neutron energy spectrum $\phi(E)$, which means to unfold Eq. 14.28. Two general methods used to unfold this equation are discussed next.

14.5.1 Differentiation Unfolding of Proton Recoil Spectra
If $R(E, E_p) = \delta(E - E_p)$, then the response function of the proton recoil spectrometer is (using Eq. 14.29)

$$k(E,E_n) = N_H T \frac{\sigma(E_n)}{E_n}$$

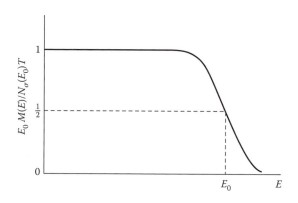

Figure 14.10   The measured monoenergetic neutron spectrum obtained with a detector having a Gaussian response ($S = 1$).

---

* There are two σ's involved here: $\sigma(E)$ is the cross section at energy E; σ without an argument is the standard deviation of the Gaussian.

and Eq. 14.28 takes the form

$$M(E) = N_H T \int_E^{E_{max}} dE_n \frac{\sigma(E_n)}{E_n} \phi(E_n) \tag{14.31}$$

The lower limit of the integral is set equal to E because at any energy E, only neutrons with energy $E_n > E$ can contribute to $M(E)$. Equation 14.31 may be solved by differentiation to give

$$\phi(E) = \frac{E}{N_H T \sigma(E)} \left| \frac{dM(E)}{dE} \right| \tag{14.32}$$

The evaluation of the derivative in Eq. 14.32 is performed by numerical techniques, since it is not the spectrum $M(E)$ that is measured but its "binned" equivalent,

$$\int_{E_i}^{E_{i+1}} M(E) dE = M_i \tag{14.33}$$

where M_i is the number of counts in channel i of the spectrometer. Several investigators[13–17] applied least-squares fit techniques to numerically perform the differentiation of Eq. 14.32. Usually, each M_i is assigned the energy corresponding to the midpoint of the channel and an M th-order polynomial is least-squares fit to that point and the preceding and following N points [this is an M th-order, $(2N + 1)$-point fit]. The derivative of the polynomial at the mid-point is used as the derivative $[dM(E)/dE]_i$. This method has the disadvantage that it slightly hardens the unfolded spectrum.

An improved differentiation technique[18] consists of first smoothing the true spectrum $\phi(E)$ to obtain a "smoothed" true spectrum $\phi_s(E)$, given by

$$\phi_s(E) = \int_0^\infty G(E, E') \phi(E') dE' \tag{14.34}$$

where $G(E, E')$ is a smoothing function normalized to 1. Substituting Eq. 14.32 into Eq. 14.34, one obtains[18]

$$\phi_s(E) \simeq \sum_{i=1}^{NC} \frac{1}{NT} \left[\frac{E_{i+1}}{\sigma(E_{i+1})} G(E, E_{i+1}) - \frac{E_i}{\sigma(E_i)} G(E, E_i) \right] M_i \tag{14.35}$$

where NC is the number of channels, M_i is given by Eq. 14.33, and the assumption is made that the quantity $[E'/\sigma(E')]G(E, E')$ approaches zero at both limits of integration of Eq. 14.34. Results obtained with Eq. 14.35 and a Gaussian smoothing function show no spectrum hardening.

14.5.2 Proportional Counters Used as Fast-Neutron Spectrometers

Proportional counters filled with hydrogen or methane are used for the measurement of neutron spectra in the energy range 1 keV $< E_n <$ 2 MeV. Neither hydrogen nor methane are equally useful over the full energy range. Hydrogen-filled detectors are used for $E_n < 100$ keV. For higher neutron energy, greater stopping power is needed, and for this reason, methane is used instead of hydrogen. Methane-filled detectors do not give good results for $E_n < 100$ keV because of spectrum distortion from carbon recoils.

The efficiency of a proportional counter, like that of any other gas-filled detector, depends on its size, the composition and pressure of the gas, and the energy of the incident neutrons. Knowledge of the neutron cross section for interactions with hydrogen and carbon over the energy range of interest is necessary for efficiency calculations. The hydrogen cross section is known to better than 1% for neutron energies between 0.2 and 22 MeV. The carbon cross section is less accurately known for $E_n > 2$ MeV. In addition to elastic scattering, other carbon reactions, such as (n, n'), (n, α), and (n, p), are important for $E_n > 4.8$ MeV and should be included in the response function of the detector. Table 14.3 gives hydrogen and carbon cross sections for $0.2 < E_n < 20$ MeV.

The relationship between the neutron spectrum and the measured pulse-height distribution is given by Eq. 14.28. The response function $k(E, E_n)$ (Eq. 14.29) may be measured or calculated. In either case, the following effects have to be taken into account in obtaining $k(E, E_n)$[14–17,19,20]:

Table 14.3
Total Hydrogen and Carbon Cross Sections for $0.2 < E_n < 20$ MeV[21]

Neutron Energy (MeV)	$\sigma_t(^1\text{H})$ (b)	$\sigma_t(^{12}\text{C})$ (b)
0.2015	9.523	4.260
0.3220	7.719	3.857
0.4265	6.712	3.583
0.6170	5.532	3.174
0.9035	4.515	2.720
1.205	3.902	2.345
1.613	3.287	1.965
2.234	2.737	1.606
3.329	2.183	1.760
4.236	1.827	1.943
4.919	1.616	1.198
6.017	1.422	1.094
7.038	1.246	1.023
8.029	1.137	1.466
10.00	0.940	1.140
11.98	0.801	1.259
14.01	0.693	1.360
16.06	0.605	1.431
17.81	0.541	1.448
19.91	0.483	1.450
21.81	0.442	1.445

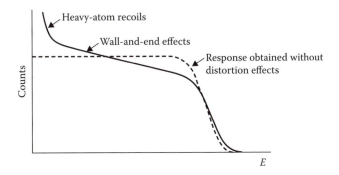

Figure 14.11 Distortion of the response function because of wall-and-end effects and heavy-atom recoils.

Wall-and-end effects. Tracks of protons generated close to the wall or close to the ends of the detector have a high probability for incomplete energy deposition and collection of ionization. Proton-recoil tracks close to the wall are truncated by collisions with the wall material before the struck proton deposits all its energy in the detector. Protons being produced close to the end of the detector and moving toward either end enter into a region of lower electric field where there is little or no gas multiplication and where there is, consequently, incomplete collection of the charge produced.

As a result of wall-and-end effects, lower energy pulses are generated that tend to increase the response function toward the lower energies (Figure 14.11). Corrections for this effect have been calculated,[17,22] but unfortunately each calculation applies only to a specific geometry.

The magnitude of wall-and-end effects increases as the size of the detector decreases. It also increases as the neutron energy increases. In fact, for neutrons in the MeV range, the distortion of the spectrum due to these effects becomes so significant that it sets the upper energy limit (~2 MeV) for the use of a proportional counter as a spectrometer.

Electric field distortion. The gas multiplication in a proportional counter depends on the intensity of the electric field. Close to the ends of a cylindrical detector, the strength of the electric field becomes gradually less intense than in most of the detector volume. This effect produces lower pulses from proton recoils at the ends of the detector. Detectors with large length-to-diameter ratio are less affected by this problem. Theoretical corrections of this effect have been developed and successfully applied.[17]

Effect of carbon recoils. Neutrons detected by methane-filled detectors collide not only with hydrogen nuclei but also with carbon atoms. The ionization produced by carbon recoils is indistinguishable from that produced by protons. However, carbon recoils produce pulses that are smaller than those from protons because of differences in both kinematics and ionization ability. The maximum fraction of neutron energy that can be imparted to a carbon nucleus in one collision is 0.28 (vs. 1 for a hydrogen nucleus), and the relative ionization efficiency of a carbon to a proton recoil is about 0.5.[20] Thus, the effect of carbon recoils is to add pulses at the low-energy region of the response function (Figure 14.11). Carbon-recoil effects are so significant for $E_n < 100$ keV that methane-filled detectors are not usable below this energy.

Variation of energy needed to produce one ion pair. The measured spectrum may differ from the proton-recoil spectrum because of nonlinearity in the relationship between proton energy and ionization produced. That relationship is expressed by the quantity $w(E)$, defined as the energy needed by a proton of energy E to produce one electron-ion pair. To be able to use a detector as a spectrometer, the value of $w(E)$ should be accurately known for the gas of the detector for all energies below the maximum neutron energy measured. Experiments have shown that the value of $w(E)$ is essentially constant for hydrogen for neutron energies about ~20 keV. Below that energy, $w(E)$ changes slightly with energy.[17,20] For methane, $w(E)$ seems to be essentially constant between 100 keV and 1 MeV.[20]

Gamma-ray discrimination. Proportional counters used in a mixed neutron-gamma field detect both types of radiation. Discrimination of γ-ray pulses has been accomplished by utilizing the fact that the Compton electrons produced by the gammas have longer range than proton recoils. The *time-of-rise method*, which is now almost universally used, takes advantage of the faster rise time of the proton pulse relative to that of the electrons. Proton range is so much shorter than electron range that all the ions produced by the proton arrive at the anode at about the same time and generate a pulse with a fast rise time. On the other hand, ions produced by electrons along their path arrive at the anode over a period of time and generate a pulse with a slower rise time. Thus, using appropriate electronics, the pulses from gammas can be rejected.

Finite resolution of proton detector. The resolution of a proportional counter for monoenergetic protons is derived from two factors. One is a statistical broadening that depends on the number of ion pairs produced. The other is a "mechanical" broadening due to imperfections in the design of the detector and impurities in the filling gas. At an energy of 615 keV, the energy resolution is of the order of 4%, but it deteriorates to about 60% at 1 keV.

Response functions of proportional counters have been measured and calculated by several people. Verbinski and Giovannini[20] gave a critical study of response functions of gas-filled detectors as well as a comparative study of the different codes used to unfold their spectra. Figures 14.12 and 14.13 show measured and calculated response functions for methane- and hydrogen-filled proportional counters.

Coarse calibration of proportional counters is achieved by using ^{3}He and N_2 as additives in the gas of the detector and employing the reactions

$$^{14}N(n,p)^{14}C, \quad Q = 626 \text{ keV}$$

$$^3He(n,p)^3H, \quad Q = 765 \text{ keV}$$

Fine calibration is obtained by placing the detector inside neutron filters made of aluminum, NaCl, and Teflon.[23] The filters generate dips in the unfolded spectrum, which coincide with the energies of cross-section resonances of the corresponding isotope. Fine calibration is achieved when the energies of the dips of the unfolded spectrum coincide with the energies of the resonances.

14.5.3 Organic Scintillators Used as Fast-Neutron Spectrometers

Organic scintillators have proven to be excellent fast-neutron detectors because they have high and known efficiency, good energy resolution, and low sensitivity to gammas. The high efficiency is due to their hydrogen content (1.1 hydrogen atoms per carbon atom, density about 10^3 kg/m^3 = 1 g/cm^3), the relatively high hydrogen cross section (2.5 b for 2.5-MeV neutrons), and the ability to make and use

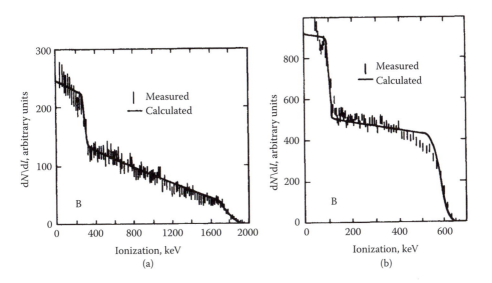

Figure 14.12 Measured and calculated response functions for a methane-filled proportional counter (1.5 in. diameter, 3.5 atm pressure), (a) At 75°C, 1772 keV. (b) At 75°C, 592 keV (dN/dI is the proton ionization spectrum). (From Verbinski, V. V., and Giovannini, R., *Nucl. Instrum. Meth.*, 114, 205, 1974.)

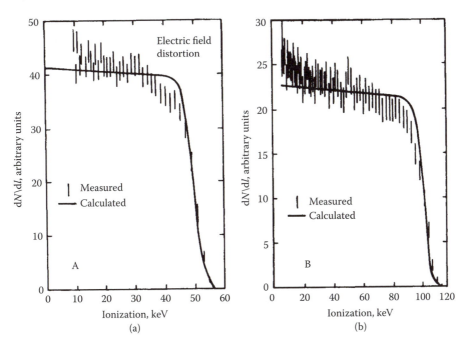

Figure 14.13 Measured and calculated response functions for a cylindrical hydrogen-filled proportional counter (1.5 in. diameter, 2.6 atm pressure, H_2 + 5% N_2). (a) Electric field distortion, 45°C, 50 keV, (b) 45°C, 100 keV (dN/dI is the proton ionization spectrum). (From Verbinski, V. V., and Giovannini, R., *Nucl. Instrum. Meth.*, 114, 205, 1974.)

them in large sizes. Organic scintillators are the main detectors used for neutron spectroscopy from ~10 to 200 MeV. An excellent review of organic scintillator properties is given in Reference 24.

Stilbene scintillators were used as early as 1957. Stilbene as a crystal is very sensitive to mechanical and thermal shock and shows an anisotropic response to neutrons—that is, neutrons incident from different directions, with respect to the crystal lattice, produce different light output. Liquid organic scintillators have none of these problems; in addition, they have higher H/C ratio, and light production from carbon recoils relatively lower than in stilbene. For all these reasons, liquid organic scintillators are almost exclusively used for detecting fast neutrons.

The NE series* of organic scintillators has been studied in detail and used extensively,[20,21,24,25] in particular NE 213. The NE 213 scintillator, which is most commonly used, consists of xylene, activators, the organic compound POPOP (as a wavelength shifter), and naphthalene, which is added to improve light emission. The density of NE 213 is about 870 kg/m³ (0.87 g/cm³), and its composition is taken to be CH_{121}.

As the size of an organic scintillator increases, the efficiency increases, the energy resolution deteriorates, and the background increases. The optimum size for MeV neutrons seems to be a scintillator with a volume 10^{-4} m³ (100 cm³), that is, a cylinder 50 mm in diameter and 50 mm tall. The efficiency of the NE 213 scintillator has been determined by Verbinski et $al.$[21] using a combination of measurements and Monte Carlo calculations for 20 neutron energies between 0.2 and 22 MeV.

The response of an organic scintillator to monoenergetic neutrons depends on effects similar to those discussed in the previous section for proportional counters, with the exception of electric field distortions. The most important cause of a response different from the ideal rectangular distribution shown in Figure 14.9 is the nonlinear relation between the energy of the proton and the amount of light produced by the scintillation process. For organic scintillators, the light production by protons and heavier ions is essentially proportional to the $\frac{3}{2}$ power of the energy deposited[21] in the energy range $0.3 < E < 4$ MeV, and linear for lower energies.[26] The light production by electrons varies almost linearly with energy[26] (Figure 14.14).

Response functions for the NE 213 organic scintillator were first obtained by Verbinski et $al.$[21] These authors measured the NE 213 response for 20 energies between 0.2 and 22 MeV and then normalized the spectra to Monte Carlo calculations. Ingersoll and Wehring,[27] and Johnson et $al.$[29] using an interpolation scheme, expanded these data into an 81-column matrix, and used it successfully to unfold neutron spectra up to 20 MeV. Figure 14.15 shows typical response functions for monoenergetic neutrons up to 8.12 MeV. Figure 14.16 shows a pulse-height distribution and an unfolded spectrum. Two other unfolding methods can be found in References 30 and 31.

Neutron-gamma discrimination is essential for satisfactory performance of an organic scintillator as a neutron spectrometer. Fortunately, rejection of gamma pulses can be achieved by electronic means. The method is called pulse-shape discrimination (PSD) and is based on the difference in scintillator response to gamma- and neutron-associated events. The electrons, which are produced by gammas, cause scintillations at a rate faster than that due to protons produced by neutrons. Thus, the electron pulses, which are associated with photon interactions, have a faster rise time than the proton pulses associated with neutrons. There are many PSD circuits. All of them generate a pulse with amplitude dependent upon the fast and slow components of the

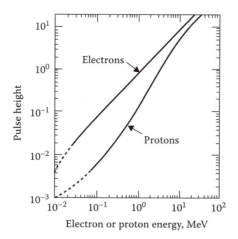

Figure 14.14 Light response of NE 110 (plastic) and NE 213 scintillator as a function of electron and proton energy. (From Harvey, J. A., and Hill, N. W., *Nucl. Instrum. Meth.*, 162, 507, 1979.)

---

* Manufactured by Nuclear Enterprises, Winnipeg, Ontario, Canada.

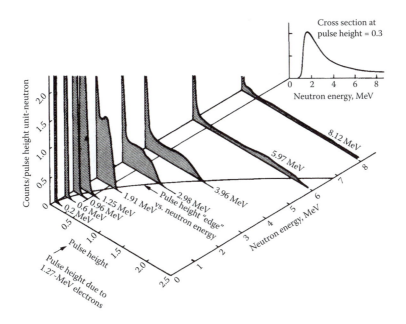

Figure 14.15 Response functions of NE 213 spectrometer. The curves represent the quantity Eq. 14.37. (From Burrus, W. R., and Verbinski, V. V., *Nucl. Instrum. Meth.*, 67, 181, 1969.)

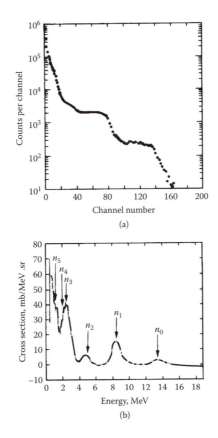

Figure 14.16 Pulse-height distribution and unfolded neutron spectrum obtained with a 46-mm × 46.5-mm diameter NE 213 organic scintillator. (From Burrus, W. R., and Verbinski, V. V., *Nucl. Instrum. Meth.*, 67, 181, 1969.)

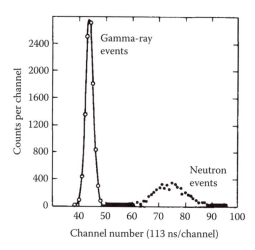

Figure 14.17   Rise time spectrum of gamma-ray and neutron events recorded by a NE 213 organic scintillator. (From Johnson, R. H., et al., *Nucl. Instrum. Meth.*, 145, 337, 1977.)

scintillation. The PSD circuit used by Burrus and Verbinski[32] produces a large positive pulse for neutrons and a small positive or large negative pulse for gammas.

Johnson et al.[29] used a time-to-amplitude converter to generate a signal proportional to the rise time of the pulses produced in the scintillator (Figure 14.17).

Organic scintillators can be used in TOF measurements because they have shown timing resolution less than 1 ns[24] (see Section 14.8).

14.6 DETECTION OF FAST NEUTRONS USING THRESHOLD ACTIVATION REACTIONS

Detection of fast neutrons by threshold activation reactions (or threshold detectors) is based on the existence of an energy threshold for certain reactions of neutrons with nuclei. Thus, if one activates a foil made of such nuclei, the activity of the foil will give a measure of the neutron flux above the threshold. Consider, for example, the (n, α) and $(n, 2n)$ cross sections of ^{27}Al an ^{46}Ti shown in Figure 14.18. If Al and Ti foils are irradiated, the activity produced (activity of ^{24}Na and ^{45}Ti) will be a measure of the neutron flux above ~5 MeV and ~13 MeV, respectively.

The main advantages of this technique, over the use of other spectrometers, are as follows:

1. The foils have a small volume and a low cross section; therefore, they do not disturb the neutron field.

2. The foils are almost insensitive to gammas.

3. Their small size makes the location of foils possible in places where no other spectrometer would fit.

4. The counting equipment does not have to be carried to the radiation area.

As shown in Section 14.4, the saturation activity A_s of a foil is given by

$$A_s = N\int_{E_{th}}^{\infty} \sigma(E)\phi(E)\,dE \tag{14.36}$$

where N, $\sigma(E)$, and $\phi(E)$ have been defined in Section 14.4 and E_{th} is the energy threshold for the cross section $\sigma(E)$. Table 14.4 gives a partial list of the many reactions one can use to cover a given neutron energy range. In general, reactions are selected according to the energy range of the neutron spectrum and the counting equipment available. There are, however, criteria that make certain reactions and certain foils more desirable than others:

1. The cross section for the reaction should be well known as a function of energy.

2. The type, energy, and relative intensity of the radiations of the product of the reaction should be well known.

3. The half-life of the radionuclide produced should be well known and should be at least several minutes long.

4. The foil material should be available in high purity, to avoid interference reactions caused by impurities.

To determine the neutron flux as a function of energy by the threshold reaction technique, one irradiates n foils and obtains n equations for the saturation activity per target nucleus,*

$$A_i = N \int_{E_{th,i}}^{\infty} \sigma_i(E)\phi(E)dE \mid i = 1, n \tag{14.37}$$

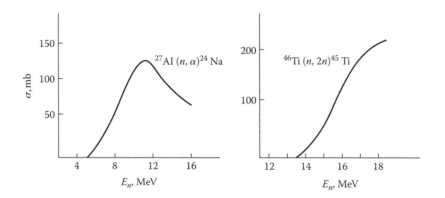

Figure 14.18 The (n, α) and $(n, 2n)$ cross sections for ^{27}Al and ^{46}Ti.

Table 14.4
A List of Threshold Reactions

Material	Reaction	Threshold (MeV)	Half-Life of Product
Teflon	^{19}F$(n, 2n)^{18}$F	11.6	109.7 min
Li	^{7}Li$(n, \alpha n')^{3}$H	3.8	–
Mg	^{24}Mg$(n, p)^{24}$Na	6	15 h
Al	^{27}Al$(n, \alpha)^{24}$Na	4.9	15 h
Al	^{27}Al$(n, p)^{27}$Mg	3.8	9.45 min
Ti	^{46}Ti$(n, p)^{46}$Sc	5.5	83.3 d
Ti	^{47}Ti$(n, p)^{47}$Sc	2.1	3.41 d
Ti	^{48}Ti$(n, p)^{48}$Sc	6.8	43.7 h
Fe	^{56}Fe$(n, p)^{56}$Mn	4.9	2.58 h
Fe	^{54}Fe$(n, p)^{54}$Mn	2.2	312.5 d
Co	^{59}Co$(n, \alpha)^{56}$Mn	5.2	2.58 h
Ni	"Ni$(n, 2n)^{57}$Ni	13	36 h
Ni	^{58}Ni$(n, p)^{58}$Co	2.9	71.3 d
Cu	^{63}Cu$(n, 2n)^{62}$Cu	11.9	9.8 min
Cu	^{63}Cu$(n, \alpha)^{60}$Co	6.1	5.27 y
Zn	^{64}Zn$(n, p)^{64}$Cu	2	12.7 h
I	^{127}I$(n, 2n)^{126}$I	9.3	13 d
Au	^{197}Au$(n, 2n)^{196}$Au	8.6	6.17 d
Np	^{237}Np$(n, $ fission$)^{140}$ Ba[a]	0.5	12.8 d
U	^{238}U$(n, $ fission$)^{140}$ Ba[a]	1.45	12.8 d

[a] Various fission products are available for counting, for example, ^{95}Zn, ^{103}Ru, ^{140}La. ^{140}Ba is shown as typical.

---

* If saturation activity cannot be obtained because of long half-life, an appropriate time-correction factor is used.

Note that the activity A_i is not the total activity of the foil but only the activity due to the reaction associated with the cross section $\sigma_i(E)$. For example, if one irradiates an aluminum foil, the total activity will be the result of the (n, α) and (n, p) reactions listed in Table 14.4 and the (n, γ) reaction that will also occur. If the user intends to examine the (n, p) reaction, the activity that should be used in Eq. 14.41 is that of ^{27}Mg. Activity due to ^{24}Na [from the (n, α) reaction] and ^{28}Al [from the (n, γ) reaction] should be disregarded.

It is advantageous, but not necessary, to choose reactions that result in the same type of radiation being emitted by all the foils used. Then the same counting equipment can be used with all the foils. The most common choice is gammas, and the detector is a Ge spectrometer.

Example 14.5 Consider two foils made of materials with neutron absorption cross sections as shown below:

The foils were exposed to a fast-neutron flux for 2 h. The half-life of the radioisotope produced by the first foil is 10 min, and by the second is 5 h.

(a) Write the activation equations and sketch activity produced versus irradiation time for both foils.
(b) What information about the neutron spectrum can one obtain from this measurement?

Answer The activation equation is

$$A_i(t) = N_i\sigma_i\phi(1-e^{-\lambda_i t}) = A_{sat,i}(1-e^{\lambda_i t})$$

The saturation activity is an integral over the neutron spectrum (Eq. 14.40):

$$A_{sat1} = N_1\int_{0.5}^{\infty}\sigma_1(E)\phi(E)dE, \quad A_{sat2} = N_2\int_{1.75}^{\infty}\sigma_2(E)\phi(E)dE$$

Since the cross sections are constants,

$$A_{sat1} = N_1\sigma_1\int_{0.5}^{\infty}\phi(E)dE, \quad A_{sat2} = N_2\sigma_2\int_{1.75}^{\infty}\phi(E)dE$$

These expressions show that (a) from foil 1, one can get the number of neutrons above 0.5 MeV, and (b) from foil 2, one can get the number of neutrons above 1.75 MeV. Subtracting the two activities gives information about the number of neutrons between 0.5 and 1.75 MeV.

The calculation of $\phi(E)$ based on Eq. 14.41 is another case of unfolding. Usually the flux is expressed in terms of a number of energy groups G. If $G < n$, unfolding of Eq. 14.41 is a simple case of least-squares fit. Unfortunately, in most cases of practical interest, $G > n$, and the only way to obtain $\phi(E)$ is to assume a certain a priori form for it and then try to improve upon this initial guess. The result depends on the choice of the input spectrum, the set of threshold reactions chosen, the errors of the measured activities, and the uncertainties of the cross sections involved. The several unfolding codes that are used differ mainly in the choice of the input spectrum. Descriptions of four such codes, SAND-II,[33] SPECTRA,[34] relative deviation minimization method (RDMM),[35] and LSL-M2[36] have been published.

More recent evaluations of unfolding codes have appeared. These include a neutron spectrum unfolding,[37] estimating the uncertainty in unfolding,[38] comparing different unfolding codes,[39] measurement and unfolding neutron spectra using Bonner spheres,[40] and unfolding neutron energy spectra from foil activation detector measurements with the gold algorithm.[41]

14.7 NEUTRON ENERGY MEASUREMENT WITH A CRYSTAL SPECTROMETER

The measurement of neutron energy with a crystal spectrometer is based on the Bragg diffraction principle.

A neutron with kinetic energy E has a de Broglie wavelength equal to

$$\lambda = \left(\frac{h}{p}\right) = \frac{0.028602}{\sqrt{E(eV)}} \, nm = \frac{0.28602}{\sqrt{E(eV)}} \tag{14.38}$$

where h = Planck's constant and $p = Mv = \sqrt{2ME}$ = linear momentum of the neutron.

Neutrons with wavelength λ incident upon a crystal with interplanar distance d are scattered by the atoms of the crystal. As a result of constructive interference, a diffracted neutron beam appears at an angle θ satisfying the Bragg condition (Figure 14.19),

$$n\lambda = 2d \sin \theta \qquad (14.39)$$

where n is the order of reflection. If the incident beam is polyenergetic, the neutron detector set at an angle θ will detect neutrons of wavelength λ satisfying Eq. 14.43, that is, neutrons having kinetic energy E related to λ by

$$E(\text{eV}) = \frac{8.191 \times 10^{-4}}{\lambda^2 (\text{nm})^2} \qquad (14.40)$$

Neutron crystal spectrometers are used either to analyze a polyenergetic neutron source or, more frequently, to provide a source of monoenergetic neutrons. Considering again Figure 14.19, even though the incident beam may consist of neutrons of many energies, the neutrons diffracted at an angle θ constitute a monoenergetic neutron beam of energy given by Eq. 14.40.

Monoenergetic neutron sources at the energy range provided by crystal spectrometers are necessary for the study of low-energy neutron cross sections with resonances. Consider, as an example, the total neutron cross section of iridium shown in Figure 14.20. To be able to measure the resonances of this cross section, one needs neutron energy resolution less than 0.1 eV, resolution that can be achieved only with crystal spectrometers[42,43] or TOF measurements (see Section 14.8).

The resolving power of a neutron crystal spectrometer is given (based on Eqs. 14.38 and 14.39) by

$$\frac{E}{\Delta E} = \frac{\lambda}{2\Delta\lambda} = \frac{\tan\theta}{2\Delta\theta} = \frac{nk}{4(\Delta\theta)d\sqrt{E}}\left(1 - \frac{n^2k^2}{4d^2E}\right)^{-1/2} \qquad (14.41)$$

where $k = 0.028602$ nm $(\text{cV})^{1/2}$, the constant of Eq. 14.42. This function, as well as its inverse (which is the energy resolution*), is shown in Figure 14.21, assuming $d = 0.2$ nm, $\Delta\theta = 0.3°$, and $n = 1$. One can improve the resolution, as in the use of X-ray spectrometers, by decreasing $\Delta\theta$ and choosing a crystal with short interatomic distances. Using a beryllium crystal with $d = 0.0732$ nm and $\Delta\theta = 7.8$ min, a resolution of 2% at 1 eV has been achieved.[43]

The energy range over which the crystal spectrometer can be used is determined from the Bragg condition (Eq. 14.43) and the requirement that $0 < \sin \theta < 1$. Using Eqs. 14.42 and 14.43, one obtains for first-order reflection

$$\sin\theta_{\min} < \frac{0.028602}{2\sqrt{E(\text{eV})}\, d(\text{nm})} < 1 \qquad (14.42)$$

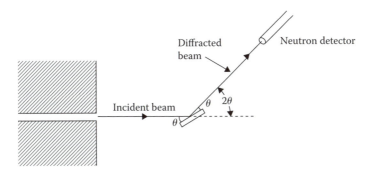

Figure 14.19 The arrangement of a neutron diffraction spectrometer.

---

* Better resolution would be obtained with higher order reflections ($n > 1$), but unfortunately, the intensity is much lower for $n > 1$.

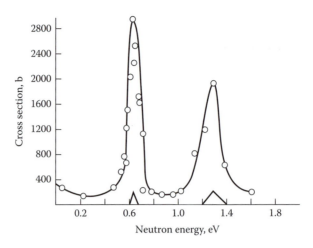

Figure 14.20 The total neutron cross section of iridium. (From Sawyer, R. B., et al., *Phys. Rev.*, 72, 109, 1947.)

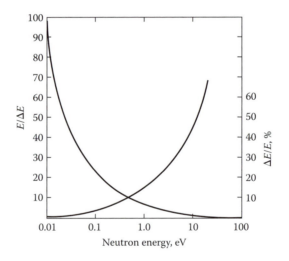

Figure 14.21 Resolving power and energy resolution of a neutron crystal spectrometer ($n = 1$, $d = 0.2$ nm, $\Delta\theta = 0.3°$).

which shows that the energy range is a function of the crystal (interplanar distance d) and the minimum observable angle θ_{min}. If one assumes $d = 0.2$ nm (LiF crystal) and $\theta_{min} = 0.5°$, the energy range becomes $0.005 < E < 67$ eV. Both energy limits increase if a crystal with smaller interplanar distance d is used. In practice, the upper limit is determined by the energy resolution that is acceptable for the experiment. As Figure 14.21 shows, the resolution deteriorates rather rapidly as energy increases. Neutron crystal spectrometers are generally used for $E < 100$ eV. Crystals that have been used include LiF, calcite, mica, beryllium, and copper.

14.8 THE TIME-OF-FLIGHT (TOF) METHOD

The TOF method determines the neutron energy with a resolution that is better than with any other detector. The principle of neutron TOF is the same as for heavy ions (see Section 13.6). As was pointed out in Section 13.6, by using the TOF technique, the particle energy can be measured extremely accurately if the mass of the particle is known. The mass of the neutron is known (to within 3 keV), and energy resolution as good as 0.1% has been achieved.

In a TOF measurement, one determines the speed of the neutron v from the time t it takes to travel a flight path of length L. The kinetic energy of the neutron is given by

$$E = Mc^2 \left(\frac{1}{\sqrt{1-\beta^2}} - 1 \right) = Mc^2 \left(\frac{1}{\sqrt{1-L^2/c^2t^2}} - 1 \right) \tag{14.43}$$

where $Mc^2 = 939.552$ MeV = rest mass energy of the neutron.

The nonrelativistic equation is the familiar one,

$$E_{NR} = \frac{1}{2}Mv^2 = \frac{1}{2}M\frac{L^2}{t^2} \tag{14.44}$$

Which equation should be used depends on the energy range measured and the resolution of the experiment. At 1 MeV, the nonrelativistic equation, Eq. 14.48, introduces an error of 0.16%.

The energy resolution is, using Eqs. 14.43 and 14.44,

Relativistic:

$$\frac{\Delta E}{E} = \frac{E + Mc^2}{E} \frac{\beta^2}{1-\beta^2} \sqrt{\left(\frac{\Delta L}{L}\right)^2 + \left(\frac{\Delta t}{t}\right)^2} \tag{14.45}$$

Nonrelativistic:

$$\frac{\Delta E}{E} = 2\sqrt{\left(\frac{\Delta L}{L}\right)^2 + \left(\frac{\Delta t}{t}\right)^2} \tag{14.46}$$

In neutron TOF experiments, the neutron source is a burst of neutrons generated either by a velocity selector (chopper) or by an ion beam, as explained later in this section. The TOF t is the difference between the time of production of the neutron burst and the time of neutron detection.

The uncertainty Δt consists of three parts:

1. Δt_s is the uncertainty in the time of neutron emission; it is essentially equal to the width of the neutron burst and ranges from a few hundred nanoseconds to less than a hundred picoseconds.

2. Δt_d is the uncertainty in the time of neutron detection; it depends on the pulse rise time, since it is the pulse rise time that signals the time of detection. Neutron detectors used today have a pulse rise time equal to 5 ns or less.[44]

3. Δt_m is the uncertainty in neutron slowing-down time if the source is surrounded by a moderator.

The uncertainty ΔL is due to the finite thicknesses of the neutron-producing target and the neutron detector. The uncertainty in the measurement of L itself can be made negligible. The longer the flight path is, the smaller the uncertainty $\Delta L/L$ becomes. As the length L increases, however, the intensity of the source should increase, and by a greater factor, so that the counting rate in the detector stays the same.

It is customary to use the quantity t/L as a *figure of merit* for TOF experiments. From Eqs. 14.47 and 14.48, one obtains

$$\left(\frac{t}{L}\right)_{Rel} = \frac{1}{c}\left[1 - \left(\frac{Mc^2}{E + Mc^2}\right)^2\right]^{-1/2} \tag{14.47}$$

$$\left(\frac{t}{L}\right)_{NR} = \frac{1}{c}\sqrt{\frac{Mc^2}{2E}} \tag{14.48}$$

Table 14.5
Typical Values of *t/L* for Several Neutron Energies

E (eV)	t/L (µs/m)	E (MeV)	t/L (ns/m)
0.01	722	0.1	228
0.1	228.5	1	72.3
1	72.2	2	51.2
10	22.8	5	32.4
100	7.2	10	23
1000	2.3	20	16

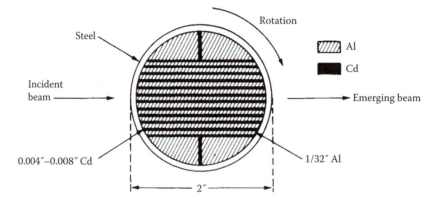

Figure 14.22 The Fermi chopper.

Table 14.5 gives typical t/L values.

The requirements for slow-neutron TOF experiments fall in the µs/m range, and those of fast neutrons in the ns/m range. Because of this large difference in timing requirements, it is impossible to span the whole neutron energy range (eV to MeV) with the same TOF spectrometer. The change of resolution with neutron energy is the same for TOF and crystal spectrometers. In both systems, the energy spread ΔE changes, essentially, as $E^{3/2}$ (compare Eqs. 14.45 and 14.50):

$$\Delta E = 2E\sqrt{\left(\frac{\Delta L}{L}\right)^2 + \left(\frac{\Delta t}{t}\right)^2} \approx 2E\frac{\Delta t}{t} = 2E\frac{L}{t}\frac{\Delta t}{L} \approx Ev \approx E^{3/2}$$

14.8.1 The Neutron Velocity Selector (Neutron Chopper)

The first velocity selector was designed by Fermi and his coworkers in the 1940s and is now known as the Fermi chopper.[45] The Fermi chopper consisted of a multiple sandwich of aluminum and cadmium foils that fit tightly into a steel cylinder about 38 mm (1.5 in) in diameter (Figure 14.22). The cylinder was rotated at speeds of up to 15,000 r/min, thus allowing only bursts of neutrons to go through the aluminum channels. Based on the geometry of Figure 14.22, no neutrons from a parallel beam would go through the channel when the chopper was more than $\Delta\theta/2$ degrees from its fully open position (Figure 14.23), where $\Delta\theta =$ (width of channel)/(radius of cylinder). The spinning cylinder was viewed with two photocells, one giving a direct measure of the rotation and the other sending to the neutron detector a signal used for the measurement of the TOF of the transmitted neutrons.

Fermi's chopper was a "slow" chopper, the word slow referring to the speed of the neutrons, and was used for neutrons up to 1 eV. Fast choppers have also been developed for use with neutron energy up to the keV range,[46–48] with the rotating cylinders of the chopper having different design, depending on the requirements of the measurement.

The most important characteristic of any chopper is the width of the neutron burst. In all choppers, the shape of the pulse is essentially triangular with the base of the triangle being inversely

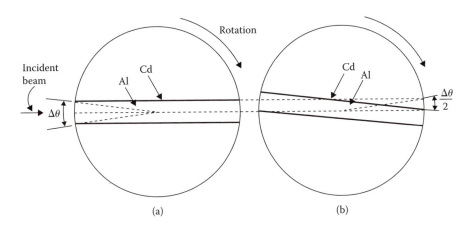

Figure 14.23 (a) The channel is fully open, (b) the chopper has rotated by $\Delta\theta/2$, and the channel is closed.

proportional to the rotating speed of the shutter. The shape of the pulse changes slightly with the neutron speed and the shape of the channel. The width of the channel, which also affects the pulse, is a compromise between acceptable time resolution and adequate counting rate. Using choppers, neutron bursts with widths as low as 0.5 μs have been achieved.[44] The time resolution due to such a width is adequate for energies up to 10 keV. At higher energies, ion beams from accelerators are used to provide the neutron burst.

14.8.2 Pulsed-Ion Beams

Narrow and intense bursts of neutrons for TOF experiments are obtained by using ion beams. The ions are accelerated, strike a target, and produce neutrons through a (charged particle, n)-type reaction. Examples of such reactions are

$$^{2}\mathrm{H}(d,n),\, ^{3}\mathrm{H}(p,n),\, ^{7}\mathrm{Li}(p,n),\, ^{9}\mathrm{Be}(p,n),\, ^{9}\mathrm{Be}(\alpha,n),\, ^{12}\mathrm{C}(d,n),\text{ and}$$
$$^{13}\mathrm{C}(\alpha,n)$$

Neutrons produced by these reactions are in the MeV range.

The first accelerator to be used for neutron production was the cyclotron.[49] Since that time, other types of accelerators have been utilized in TOF experiments[50–53] such as electron linear accelerators; the Los Alamos Meson Physics Facility (LAMPF), which accelerates protons to 800 MeV[54,55]; and the ORELA facility at Oak Ridge, Tenn.[56] The width of neutron bursts produced by accelerators can be lower than 100 ps.[57,58] If the burst becomes as narrow as ~50 ps, the resolution is limited by the time response of the neutron detector.

14.9 COMPENSATED ION CHAMBERS

Neutron detectors located close to a reactor core are subjected to both neutron and gamma bombardment. Although a neutron detector—for example, a $^{10}\mathrm{B}$ detector—is mainly sensitive to neutrons, it responds to gammas too. At low reactor power, when the neutron flux is small, the neutron signal is overshadowed by a signal due to gammas emitted from fission products that had been accumulated from earlier reactor operation. To eliminate the effect of the gammas, a *compensated ion chamber* is used.

Compensated ion chambers operate in such a way that the gamma signal is subtracted from the total ($n + \gamma$) signal and the output is proportional to the neutron signal only. Figure 14.24 shows the basic principle of a compensated ion chamber. The detector consists of two compartments. One, coated with boron, is sensitive to both neutron and gammas and produces a signal proportional to the total radiation field. The other is sensitive to gammas only and produces a signal proportional to γ radiation only. As Figure 14.24 shows, the circuitry is such that the γ signal is subtracted from the ($\gamma + n$) signal, thus giving a signal proportional to the neutron field only. The signal, in the form of a current, is measured by a picoammeter.

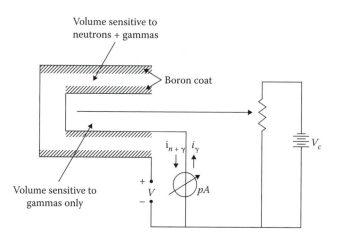

Figure 14.24 A compensated ion chamber.

Correct compensation is achieved when the signal is zero in a pure gamma field. This is accomplished by using the proper combination of volumes for the two compartments or by changing the voltages or by a combination of voltage and volume change. Typical compensation voltages (V_c) are of the order of –25 V; the positive voltage V is of the order of + 800 V. Without compensation, a detector of this type has a useful range from 2×10^8 to 2×10^4 neutrons/(m$^2 \cdot$ s). With compensation, the useful range is extended downward by about two orders of magnitude. The sensitivity of compensated ion chambers is of the order of 10^{-18} A/[neutrons/(m$^2 \cdot$ s)].

14.10 SELF-POWERED NEUTRON DETECTORS (SPND)

Self-powered detectors, as their name implies, operate without an externally applied voltage. The incident radiation (neutrons or gammas or both) generates a signal in the form of a current proportional to the bombarding flux. The detectors are usually constructed in coaxial configuration (Figure 14.25). The central conductor is called the *emitter* and is the material responsible for the generation of the signal. The outer conductor, called the *collector*, is separated from the emitter by an insulator. The collector is made of inconel alloy, and has the form of a metallic sheath encasing the insulator and the emitter.

The principle of signal generation in a self-powered detector is simple. As a result of bombardment by radiation, the emitter releases electrons (betas) that escape to the insulator and leave the emitter positively charged. If the emitter is connected to the collector through a resistor (Figure 14.25), current flows, which when measured, gives an estimate of the incident flux. Note that this is not an emitter-collector system: any beta particle escaping from the emitter contributes to the current, regardless of whether or not it reaches the collector.

Because self-powered detectors have been developed for use inside the core of power reactors, they are designed to have small size (a few millimeters in diameter), to be able to operate for rather long periods of time (years) in the intense radiation field of the reactor core without appreciable deterioration in performance, and finally, to operate without an external power supply.

The performance of a self-powered detector is given in terms of its sensitivity S, defined by

$$S(t) = \frac{\Delta I(t)}{\Delta \phi} \tag{14.49}$$

where $I(t)$ = detector current after exposure to the flux ϕ for time t and ϕ = neutron flux.

Thus, the sensitivity represents the change in detector current per unit change in the flux.

Many elements have been considered as emitters for self-powered detectors.[60–63] The ideal emitter should be such that the detector has

1. High sensitivity

2. Low burnup rate

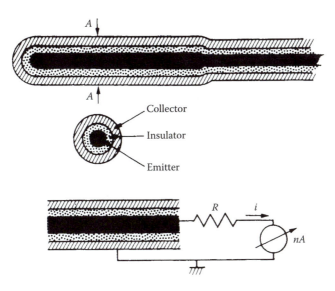

Figure 14.25 Configuration of a self-powered detector.

3. Prompt response

4. Sensitivity to neutrons only

The material properties that determine these characteristics are discussed in Sections 14.10.1 and 14.10.2, after the equations for the detector current and sensitivity are derived.

The properties of the insulator are also important. The insulator must have a resistance of about 10^{12} ohms at room temperature and 10^9 ohms at reactor operating temperature. The two insulators commonly used are magnesium oxide (MgO) and aluminum oxide (Al_2O_3). Experiments have shown[64] that the resistance of MgO decreases with exposure to radiation, while that of Al_2O_3 does not change. For this reason, Al_2O_3 is gradually replacing MgO as an insulator for self-powered detectors.

The self-powered neutron detectors are divided into those with delayed response and those with prompt response. The characteristics of these types of self-powered detectors are presented in Sections 14.10.1 and 14.10.2.

14.10.1 SPNDs with Delayed Response

Rhodium, vanadium, cobalt, and molybdenum have been used as emitters for SPNDs. Since rhodium SPNDs are the main in-core instruments for the determination of power distribution in pressurized-water reactors (PWR), they are discussed first and in greater detail than the others.

The signal of rhodium SPNDs is produced as a result of activation of the emitter (^{103}Rh) by the incident neutrons, and subsequent decay of the isotope ^{104}Rh that is produced. The decay scheme of ^{104}Rh is shown in Figure 14.26. An isomeric state with a 4.4-min half-life is produced with a 12-b cross section. The ground state of ^{104}Rh has a 43-s half-life and is formed with a cross section of 138 b (with thermal neutrons). It decays to ^{104}Pd with a maximum β^- energy of 2.5 MeV. This decay, which takes place 98.5% of the time, is primarily responsible for the signal of the rhodium detector. The isomeric state with the 4.4-min half-life contributes very little to the signal, but is responsible for a residual current after reactor shutdown.

To identify the factors that improve sensitivity of the detector and lengthen its life, one should look at the processes responsible for the generation of the detector signal. This is done below, and equations for current and sensitivity are derived for any emitter material.

Consider an emitter with an average neutron absorption cross section σ exposed to a total neutron flux ϕ, and upon absorption of a neutron becoming radioactive with a half-life T [or decay constant $\lambda = (\ln 2)/T$]. The number of radioactive atoms $N(t)$ present after exposure for time t is (see Eq. 14.16)

$$N(t) = \frac{s\sigma_e \phi N_0}{\lambda - \sigma_a \phi} \left(e^{-s\sigma_a \phi t} - e^{-\lambda t} \right)$$

(14.50)

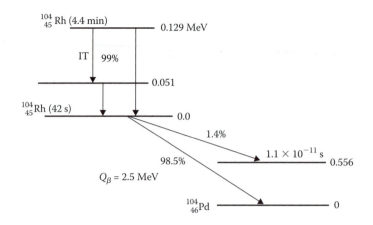

Figure 14.26 The decay scheme of ^{104}Rh.

where N_0 = number of emitter atoms at $t = 0$
σ_a = absorption cross section of emitter
σ_e = cross section that leads to the state that contributes to the signal
s = self-shielding factor ($s < 1$)

The self-shielding factor s corrects for the fact that the target (emitter) is thick, as a result of which the flux in the emitter is depressed. Thus, interior atoms are "shielded" from exposure to the full flux by the atoms close to the surface. The shielding factor is less than 1 and decreases as the diameter of the emitter increases (see Figure 14.27). If every decay of the radioisotope releases a particle with charge q, the current at time t is equal to

$$I(t) = kqN(t)\lambda = kq\frac{s\sigma_e\phi N_0}{1-\sigma_a\phi/\lambda}\left(e^{-s\sigma_a\phi t} - e^{-\lambda t}\right) \tag{14.51}$$

where k is a constant that takes into account such effects as self-absorption of betas in the emitter or loss of betas in the insulator (see Figure 14.27). The factor $\sigma_a\phi/\lambda$ in the denominator can be neglected because, for all emitters of interest, $\sigma_a\phi/\lambda \ll 1$. The exponential factors of Eq. 14.51 have the following meaning.

The factor $\exp(-\lambda t)$ gives the response of the detectors. If the flux undergoes a step increase, as shown in Figure 14.28, the signal will rise exponentially to its saturated value. If the flux goes down suddenly, the signal will decay again exponentially. The speed of response is determined by the half-life of the isotope involved. Rhodium, with a half-life of 42 s, reaches saturation after about 5 min. Vanadium (^{52}V), with a half-life equal to 3.76 min, reaches saturation after about 25 min.

The factor $\exp(-s\sigma_a\phi t)$ gives the burnup rate of the emitter. It is the factor that determines the lifetime of the detector, because, as seen below, the decrease in sensitivity with time is essentially given by this same factor.

Assuming saturation, the sensitivity of the detector is given (using Eqs. 14.49 and 14.51) by

$$S(t) = \frac{\Delta I(t)}{\Delta\phi} = ks\sigma_e N_0 q\left(1-s\sigma_e\phi t\right)e^{-s\sigma_a\phi t}\,\text{A}\,/\,[\text{neutrons}/(\text{m}^2\cdot\text{s})] \tag{14.52}$$

If the emitter diameter is D, its length is L, its density is ρ, and its atomic weight is A, then the number of atoms is $N_0 = (\rho\pi D^2/4)(LN_A/A)$, where N_A is Avogadro's number.

Substituting into Eq. 14.52, one obtains an equation for the sensitivity per unit length:

$$\frac{S(t)}{L} = ksq\sigma_e\rho\frac{\pi D^2}{4}\frac{N_A}{A}\left(1-s\sigma_a\phi t\right)e^{-s\sigma_a\phi t}(\text{A}/\text{m})/[\text{neutrons}/(\text{m}^2\cdot\text{s})] \tag{14.53}$$

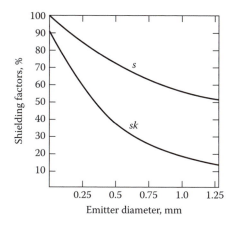

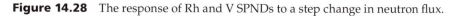

Figure 14.27 The self-shielding factors for rhodium detectors with 10-mil (0.254-mm) thick MgO insulator. (From Hawer, J. M., and Beckerley, J. G., TID-25952-P1, vol. 1, 1973.)

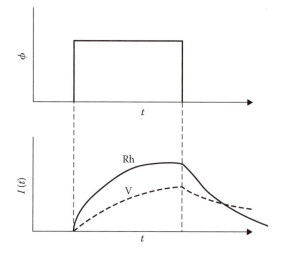

Figure 14.28 The response of Rh and V SPNDs to a step change in neutron flux.

Example 14.6 What is the sensitivity of Rh detectors 0.5 mm in diameter, per unit detector length under saturation conditions, for a new detector?

Answer For rhodium, $\sigma_e = 139$ b, $\sigma_a = 150$ b, $\rho = 12.4 \times 10^3$ kg/m^3, $A = 103$, and $q = 1.602 \times 10^{-19}$ C. At the beginning of life ($t = 0$),

$$\frac{S(0)}{L} = ks\,(1.602 \times 10^{-19}\,\mathrm{C})\,(139 \times 10^{-28}\,\mathrm{m}^2)\,(12.4 \times 10^3\,\mathrm{kg/m}^3)$$

$$\times \pi \frac{(5 \times 10^{-4})^2\,\mathrm{m}^2}{4}\left(\frac{6.022 \times 10^{26}}{103}\,\mathrm{atm/kg}\right)$$

$$= 3.17 \times 10^{-23}\,ks\,(\mathrm{A/m})/\left[\mathrm{neutrons}/(\mathrm{m}^2 \cdot \mathrm{s})\right]$$

Typical values of ks are about 0.4.[68] Since the flux in a large power reactor (1000 MWe) is about 10^{17} neutrons/(m$^2 \cdot$ s)[10^{13} neutrons/(cm$^2 \cdot$ s)] at full power, and the typical detector has a length of about 0.10 m, the expected current is of the order of

$$(0.8)\,(0.8 \times 10^{-23})\,(\mathrm{A/m})/[\mathrm{neutrons}/(\mathrm{m}^2 \cdot \mathrm{s})][10^{17}\,\mathrm{neutrons}/(\mathrm{m}^2 \cdot \mathrm{s})](0.1\,\mathrm{m})$$

$$= 1.92 \times 10^{-7}\,\mathrm{A} \approx 200\,\mathrm{nA}$$

Equation 14.53 shows that to achieve high sensitivity, one should select an emitter with high cross section σ_e and large diameter D. The diameter affects the sensitivity through the factor D^2 and through the shielding factors k and s, which decrease* as the diameter increases. The net result is that the sensitivity changes roughly as the first power of the diameter.[68]

A high cross section increases the sensitivity but also increases the rate at which the sensitivity decreases with time. Indeed, from Eq. 14.64, the ratio of the sensitivity after exposure for time t to its value at time $t = 0$ is

$$\frac{S(t)}{S(0)} = (1 - s\sigma_a \phi t)\, e^{-s\sigma_a \phi t} \tag{14.54}$$

Table 14.6 gives the characteristics of several self-powered detectors. Rhodium detectors have the best sensitivity but also the largest burnup rate. Their change of sensitivity with time is important and necessitates a correction before the signal is used for the determination of power. The correction is not trivial because of changes in self-absorption effects in the emitter, and may introduce errors unless the detector is calibrated properly.

Despite the drawback of large burnup, rhodium SPNDs are used extensively in nuclear power plants, especially in PWRs for the determination of power distribution, fuel burnup, and other information related to the performance of the core. The detectors are inserted into a certain number of "instrumented" fuel assemblies through guide tubes. Every instrumented assembly has seven equally spaced SPNDs (a background detector and a thermocouple are also included in the package; see Figure 14.29) for the measurement of the flux at seven axial locations. The outputs of the detectors, corrected for background, are transmitted to the plant computer, where after appropriate corrections are applied, the power, fuel burnup, plutonium production, etc., are calculated. Every PWR has a least 50 instrumented assemblies, which means that the flux is monitored at more than 350 locations.

14.10.2 SPNDs with Prompt Response

Neutron-sensitive self-powered detectors with prompt response operate on a different principle than rhodium and vanadium SPNDs. The emitter, in this case, absorbs a neutron and emits gammas at the time of capture. It is these capture gammas that are responsible for the signal, and since they are only emitted at the time of the neutron capture, the detector response is instantaneous. Cobalt (^{59}Co) and molybdenum (^{95}Mo) are two elements seriously considered as emitters. The subsequent discussion is based on cobalt, but the processes involved in the signal generation are the same for molybdenum.

Consider, then, cobalt SPND (Figure 14.30). Most of the capture gammas traverse the emitter, the insulator, and the collector without an interaction. Those that do interact produce electrons through the photoelectric or Compton reactions. As these fast electrons travel, they produce an outward flow of charge that generates a current. Because relatively few gammas interact, the sensitivity of a cobalt detector is lower than that of either rhodium or vanadium detectors (see Table 14.6).

Table 14.6
Characteristics of Self-Powered Detectors with 0.5-mm Emitter Diameter[59,62]

Emitter	Sensitivity (A/m)/[2 × 10^{17} Neutrons/(m$^2 \cdot$ s)]	Burnup^a/Year (%) for $\phi = 2 \times 10^{17}$ Neutrons/(m$^2 \cdot$ s)	Response
Rh	2.4×10^{-6}	5	Delayed
V	1.5×10^{-7}	0.3	Delayed
Co	3.4×10^{-8}	2.3	Prompt
Mo	1.7×10^{-8}	0.9	Prompt
Pt	2.6×10^{-7}	0.2	Prompt

^{a} Burnup $\approx 1 - \exp(\sigma_a \phi t)$.

---

* The factor s is not constant over prolonged exposure. It tends to increase as the emitter burnup continues because a smaller number of emitter atoms is left for self-shielding.[61]

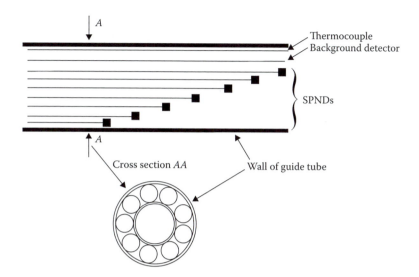

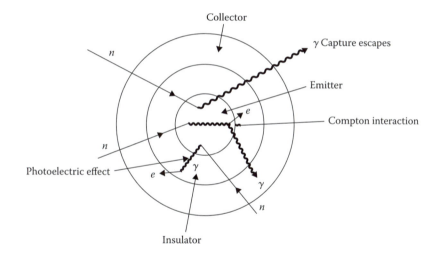

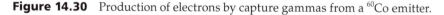

Figure 14.29 Each detector tube contains seven SPNDs, one background detector, and one thermocouple.

Figure 14.30 Production of electrons by capture gammas from a ^{60}Co emitter.

The Co detector has one undesirable characteristic due to the product of the neutron capture. That product is ^{60}Co, a β^- emitter with a 5.3-year half-life. The betas from ^{60}Co produce a background signal that builds up with exposure. A way to suppress this background, using platinum shields, has been reported by Goldstein and Todt.[65]

More recent applications of SPND include the performance of hafnium and gadolinium self powered neutron detectors in the TREAT reactor,[66] characterization of a hybrid self-powered neutron detector under neutron irradiation,[67] a novel solid state self-powered neutron detector that expands upon the basic concept of coating a p–n junction solar cell with a neutron detection layer that typically employs either ^{6}Li or ^{10}B,[68] development of an Inconel self powered neutron detector for in-core reactor monitoring,[69] and the development of a bismuth self-powered detector.[70]

14.11 CONCLUDING REMARKS

Neutron detection is, in general, more complicated and more difficult than detection of either charged particles or photons for two reasons. First, neutrons have no charge and can only be detected indirectly through photons or charged particles that they generate. Second, the neutron

Table 14.7
Summary of Neutron Detectors and Their Range of Application

Neutron Energy	Measurement of Number of Neutrons Only	Measurement of Energy and Number of Neutrons
$0 < E < 1$ keV	BF$_3$, boron-coated, SPND, TLD,[a] ^{6}Li, ^{3}He Foil activation	Crystal spectrometer Time-of-flight
	Fission track detectors[a]	
1 keV $< E \leq 2$ MeV	BF$_3$, ^{6}Li (both with low efficiency)	Proton recoil (proportional counters, organic scintillators)
	Foil activation, SPND	Time-of-flight
	Bonner ball[a]	Threshold reactions
$E \gtrsim 1$ MeV	Foil activation	Organic scintillators Threshold reactions Time-of-flight

[a] See Chapter 16.

energy range spans at least 10 decades (10^{-3} eV $< E < 10^8$ eV), over which the type and cross sections for neutron reactions change drastically. Table 14.7 gives a summary of all the methods for neutron detection and spectrometry.

PROBLEMS

14.1 Prove that for thermal neutrons, the kinetic energies of the alpha particle and the lithium in the ^{10}B$(n, \alpha)^7$Li reaction are given by

$$T_\alpha = \frac{M_{Li}}{M_\alpha + M_{Li}} Q, \qquad T_{Li} = \frac{M_\alpha}{M_\alpha + M_{Li}} Q$$

14.2 If the neutron energy is 1 MeV, what is the maximum energy of the proton in the ^{3}He$(n, p)^3$H reaction?

14.3 Obtain the efficiency curve as a function of neutron energy for a 1-in-diameter proportional counter filled with ^{3}He at 10 atm. Assume $1/v$ cross section from 0.01 to 1000 eV. The neutron beam is perpendicular to the counter axis. Compare your result with that of Figure 14.3.

14.4 Show that the sensitivity of a neutron detector (BF$_3$ or boron-lined or fission chamber) decreases with time as $\exp(-\sigma_a \phi t)$.

14.5 What is the maximum thickness of ^{235}U coating inside a fission chamber if it is required that a 60-MeV fission fragment lose no more than 10% of its energy as it goes through the uranium deposit? For the fission fragment, assume $Z = 45$, $A = 100$.

14.6 How long should one irradiate an ^{115}In foil (100 mm^2 area, 1 mm thick) in a thermal neutron flux of 10^{14} neutrons/(m$^2 \cdot$ s) to obtain 1 mCi (3.7×10^7 Bq) of activity? ($\sigma = 194$ b, $\rho = 7.3 \times 10^3$ kg/m^3, $T_{1/2} = 54$ min.)

14.7 An aluminum foil is left in a reactor for 15 s in a flux of 10^{16} neutrons/(m$^2 \cdot$ s). What is the activity produced? ($\sigma = 0.23$ b, $T_{1/2} = 2.3$ min, $m = 10^{-6}$ kg.)

14.8 The betas from the Al foil of Problem 14.7 were counted in a 2π detection system with $\varepsilon = 0.95$. Counting started 1 min after the end of the irradiation and stopped 2 min later. If the background is 20 counts/min, how many counts will the sealer record?

14.9 A 1mg Eu foil was irradiated in a thermal neutron flux of 2.5×10^{13} n/cm^2.s for 10 days. What are the activities present at the end of the irradiation? Give your result in Bq and Ci. Eu has two natural isotopes: ^{151}Eu, abundance 48%, ^{153}Eu, abundance 52%. Cross sections for thermal neutrons: ^{151}Eu, $\sigma = 5900$b, ^{152}Eu, $\sigma = 5000$b, $t_{1/2} = 12.7$ y, ^{153}Eu, $\sigma = 320$b, ^{154}Eu, $\sigma = 1400$b, $t_{1/2} = 16$ y.

14.10 A 1-gram foil of Mn is exposed to a thermal neutron flux. (a) How long should the foil be irradiated in order to produce 90% of the maximum possible activity? (b) What is the neutron flux that will produce maximum activity of 10 mCi? (c) What activity will be produced if the foil is irradiated for 1.5 h (neutron flux that from part b)? For Mn: $\sigma\,(^{55}Mn)= 14b$, $t_{1/2}\,(^{56}Mn)=$ 2.6 h.

14.11 Calculate the irradiation time needed to produce ^{198}Au in such a quantity that the gross counting rate is 1000 counts/min using a counter with $\Omega = 10^{-2}$, $\varepsilon = 0.90$, and $F = 1$. The background is 100 counts/min. It takes 10 min to get the sample from the reactor, place it under the counter, and start counting, $[\phi = 10^{16}$ neutrons/$(m^2 \cdot s)$, $t_{1/2} = 2.7$ days, $\sigma = 99$ b, $m(^{197}Au) = 10^{-9}\,kg]$

14.12 In a light-water reactor, how long will it take for the initial amount of ^{235}U to be reduced by 50%? ($\phi = 10^{17}$ neutrons/$(m^2 \cdot s)$, $\sigma_a = 670$ b)

14.13 Prove Eq. 14.24. Also show that the neutron and the proton directions after collision are 90° apart ($\theta + \phi = 90°$, Figure 14.8).

14.14 Calculate the measured neutron spectrum obtained by the proton recoil method if the detector response is a δ-function and the source spectrum is the "square" function shown in the figure below. Assume $\sigma(n, p)$ is constant for the range $E_1 \le E \le E_2$.

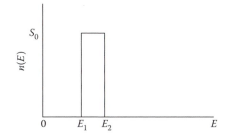

14.15 Assuming that the threshold-reaction cross sections are ideal step functions, as shown in this figure, indicate how the neutron energy spectrum could be obtained. There are N such cross sections with thresholds at $E_i|_{i=1,N}$ and $E_{i+1} - E_i = \Delta E = $ constant (see Figure below).

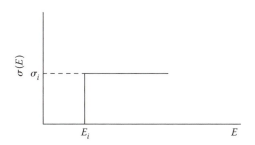

14.16 Prove Eq. 14.38.

14.17 Prove Eq. 14.45 and show that it takes the form of Eq. 14.46 in a nonrelativistic region.

14.18 A neutron TOF experiment will be designed for the measurement of 1-MeV neutrons, with the requirement that the energy resolution is 0.1%. What should the length of the flight path be if $\Delta t = 1$ ns and $\Delta L/L$ is negligible?

14.19 The original Fermi chopper consisted of a cylinder 1.5 in. in diameter with a maximum rotational speed of 15,000 r/min. The open channels consisted of aluminum sheets 1/32 in. thick.
(a) Calculate the angle during which the channel is fully open.
(b) Assuming the maximum rotational speed, what is the minimum neutron speed necessary for a neutron to make it through the channel?

14.20 What is the burnup rate per month of an SPND using ^{235}U as the emitter and being exposed to a thermal flux of 2×10^{17} neutrons/$(m^2 \cdot s)$?

14.21 How long will it take for the sensitivity of a rhodium SPND to decrease to 50% of its initial value? Assume a thermal flux of 10^{18} neutrons/(m^2 · s).

14.22 Using semiquantitative arguments, show that the sensitivity of a Co detector increases as D^m, where D is the emitter diameter and $2 < m < 3$.

BIBLIOGRAPHY

"A Review of Radiation Energy Spectra Unfolding," Proceedings of a Seminar-Workshop, ORNL/RSIC-40 (1976).

Beckurts, K. H., and Wirtz, K., *Neutron Physics*, Springer-Verlag, Berlin, 1964.

Fleischer, R. L., Price, P. B., and Walker, R. M., *Nuclear Tracks in Solids*, University of California Press, Berkeley, CA, 1975.

Furrer, A., Mesot, J., and Strässle, T., *Neutron Scattering in Condensed Matter Physics (Neutron Techniques and Applications (Series on Neutron Techniques and Applications)*, World Scientific Publishing Company, 2009.

https://www/canberra.com

https://www.ortec-online.com

Knoll, G., *Radiation Detection and Measurement*, 3rd ed., Wiley, New York, 2000.

REFERENCES

1. Bishop, G. B., *Nucl. Instrum. Meth.* **62**:247 (1968).

2. Murray, R. B., *Nucl. Instrum. Meth.* **2**:237 (1958).

3. Ginther, R. J., and Schulman, J. H., *IRE Trans. Nucl. Sci.* **N.S-5**(3):92 (1958).

4. Voitovetskii, V. K., Tolmcheva, N. S., and Arsaev, M. I., *Atomn. Energ.* **6**:321 (1959).

5. Spowart, A. R., *Nucl. Instrum. Meth.* **135**:441 (1975).

6. Spowart, A. R., *Nucl. Instrum. Meth.* **140**:19 (1977).

7. Hill, N. W., Harvey, J. A., Slaughter, G. G., and St. James, A., 147, **ORNL-4743** (1972).

8. Haacke, L. C., Hewett, J. S., and McNeill, K. S., *Nucl. Instrum. Meth.* **144**:507 (1977).

9. Benjamin, R. W., Ahlfeld, C. E., Harvey, J. A., and Hill, N. W., *Nucl. Sci. Eng.* **55**:440 (1974).

10. Bramblett, R. L., Ewing, R. L., and Bonner, T. W., *Nucl. Instrum. Meth.* **9**:1 (1960).

11. Bonner, T. W., *Nucl. Phys.* **23**:116 (1961).

12. Hanna, G. C., *Nucl. Sci. Eng.* **15**:325 (1963).

13. Brunfelter, B., Kockum, J., and Zetterstrom, H. O., *Nucl. Instrum. Meth.* **40**:84 (1966).

14. Bennett, E. F., *Nucl. Sci. Eng.* **27**:16 (1967).

15. Bennett, E. F., *Nucl. Sci. Eng.* **27**:28 (1967).

16. Bennett, E. F., Gold, R., and Olson, I. K., ANL-7394 (1968).

17. Bennett, E. F., and Yule, T. J., ANL-7763 (1971).

18. Johnson, R. H., ORNL/RSIC-40, p. 41 (1976).

19. Bennett, E. F., and Yule, T. J., *Nucl. Sci. Eng.* **46**:236 (1971).

20. Verbinski, V. V., and Giovannini, R., *Nucl. Instrum. Meth.* **114**:205 (1974).

21. Verbinski, V. V., Burrus, W. R., Love, T. A., Zobel, W., Hill, N. W., and Textor, R., *Nucl. Instrum. Meth.* **65**:8 (1968).

22. Snidow, N. L., and Warren, H. D., *Nucl. Instrum. Meth.* **51**:109 (1967).

23. Vehar, D. W., Clikeman, F. M., and Johnson, R. H., *ANS Trans.* **33**:697 (1979).

24. Brooks, F. D., *Nucl. Instrum. Meth.* **162**:477 (1979).

25. Nordell, B., *Nucl. Instrum. Meth.* **224**:547 (1984).

26. Furuta, Y., Kinbara, S., and Kaieda, K., *Nucl. Instrum. Meth.* **84**:269 (1970).

27. Ingersoll, D. T., and Wehring, B. W., *Nucl. Instrum. Meth.* **147**:551 (1977).

28. Harvey, J. A., and Hill, N. W., *Nucl. Instrum. Meth.* **162**:507 (1979).

29. Johnson, R. H., Ingersoll, D. T., Wehring, B. W., and Doming, J. J., *Nucl. Instrum. Meth.* **145**:337   (1977).

30. Numata, S., Abe, T., and Shin, K., *Nucl. Instrum. Meth.* **269**:261 (1988).

31. Shin, K., and Hamamoto, M., *Nucl. Instrum. Meth.* **269**:272 (1988).

32. Burrus, W. R., and Verbinski, V. V., *Nucl Instrum. Meth.* **67**:181 (1969).

33. McElroy, W. N., Berg, S., and Crockett, T., AFWL-TR-67–41, vols. I-IV (1967).

34. Greer, C. R., and Walker, J. V., SC-DC-66–1512 (1966).

35. DiCola, G., and Rota, A., *Nucl. Sci. Eng.* **23**:344 (1965).

36. "LSL-M2 Least-Squares Logarithmic Adjustment of Neutron Spectra," PSR-233, RSIC/ ORNL (July 1987).

37. Poyarkov, V. A., Sadovnikova, T. S., Chikai, J., and Shudar, Sh., *Atomic Energy* **67**:708(1989).

38. Grigor'ev, E. I., Troshin, V. S., and Yaryna, V. P., *Measurement Techniques,* **45**:102 (2002).

39. Koohi-Fayegh, R., Green, S., and Scott, M. C., *Nucl. Instrum. Meth. Phys. Res. A* **460**:391 (2001).

40. Esposito, A., and Nandy, M., *Rad. Protect. Dosim.* **110**:555 (2004).

41. Seghour, A., and Seghour, F. Z., *Nucl. Instrum. Meth. Phys. Res. A* **457**:617 (2001).

42. Sawyer, R. B., Wollan, E. O., Bernstein, S., and Peterson, K. C., *Phys. Rev.* **72**:109 (1947).

43. Borst, L. B., and Sailor, V. L., *Rev. Sci. Instrum.* **24**:141 (1953).

44. Firk, F. W. K., *Nucl. Instrum. Meth.* **162**:539 (1979).

45. Fermi, E., and Marshall, L., *Phys. Rev.* **72**:193 (1947).

46. Selove, W., *Rev. Sci. Instrum.* **23**:350 (1952).

47. Seidl, F. G. P., Hughes, D. J., Palevsky, H., Levin, J. S., Kato, W. Y., and Sjostrand, N. G., *Phys. Rev.* **95**:476 (1954).

48. Bollinger, L. M., Cote, R. E., and Thomas, G. E., *Proceedings of the 2nd Intern. Conference on Peaceful Uses of Atomic Energy* **14**:239 (1958).

49. Alvarez, L. W., *Phys. Rev.* **54**:609 (1938).

50. Bareford, C. F., and Kelliker, M. G., *Phillips Technol. Rev.* **15**:1 (1953).

51. Firk, F. W. K., and Bowey, E. M., "Comptes Rendus du Congres Intern, de Physique Nucleaire," P. Gugenberger (ed.), C.N.R.S., Paris, 1964.

52. Wasson, O. A., and Draper, J. E., *Nucl. Phys.* **73**:499 (1965).

53. Harvey, J. A., "Proc. Int. Conf. on Interactions of Neutrons with Nuclei," E. Sheldon (ed.), US ERDA Tech Info. Center, Oak Ridge, Tenn., p. 144.

54. Cokinos, D. M., and Van Tuyle, G. J., *ANS Trans.* **69**:423 (1993).

55. Bowman, C. D., Venneri, F., and Waters, L. S., *ANS Trans.* **69**:425 (1993).

56. Dickens, J. K., *ANS Trans.* **69**:424 (1993).

57. Norris, N. J., and Hanst, R. K., *Edgerton, Germeshausen and Grier, Prog. Rep. EGG*, 1183–2142 (1967).

58. Laszewski, R. M., Holt, R. J., and Jackson, H. E., *Phys. Rev. Lett.* **38**:813 (1977).

59. Hilborn, J. W., *Nucleonics* **22**(2):69 (1964).

60. Joslin, C. W., *Nucl. Eng. Intern.* (May 1972).

61. Kroon, J. C., Smith, F. M., and Taylor, R. I., *ANS Trans.* **23**:459 (1976).

62. Balcar, E., Bock, H., and Hahn, F., *Nucl. Instrum. Meth.* **153**:429 (1978).

63. Bozarth, D. P., and Warren, H. D., *ANS Trans.* **23**:517 (1976).

64. Hawer, J. M., and Beckerley, J. G., TID-25952-P1, vol. 1 (1973).

65. Goldstein, N. R., and Todt, W. H., *IEEE Trans.* **NS-26**:1 (1979).

66. Imel, G. R., and Hart, P. R., *Nucl. Instrum. Meth. Phys. Res. B* **111**:325 (1996).

67. Nakamichi, M., Nagao, Y., Yamamura, C., Nakazawa, M., Kawamura, H., *Fusion Engin. Design* **51–52**:837 (2000).

68. LiCausi, N., Dingley, J., Yaron, D., Lu, J-Q., and Bhat, I. B., Hard X-Ray, Gamma-Ray, and Neutron Detector Physics X. edited by Burger, A., Franks, L. A. and James, R. B. Proceedings of the SPIE, 7079:707908 (2008).

69. Alex, M., and Ghodgaonkar, M. D., *Nucl. Instrum. Meth. Phys. Res. A* **574**:127 (2007).

70. Alex, M., Prasad, K. R., and Kataria, S. K., *Nucl. Instrum. Meth. Phys. Res. A* **523**:163 (2004).

15

Activation Analysis and Related Techniques

15.1 INTRODUCTION

Activation analysis achieves a qualitative and quantitative analysis of major, minor and trace elements of an unknown sample by irradiating the sample and thus producing radioactive nuclides usually from stable isotopes. In nuclear spent fuel unstable transuranic isotopes can also undergo neutron activation to also produce further unstable isotopes. For example ^{239}Pu with a half-life of 24,100 years can be activated with neutrons to give ^{240}Pu with a half-life of 6560 years.

The radioactive nuclides can then be identified from properties of the radiations they emit:

1. Type of radiation (photons, charged particles, neutrons)

2. Energy of radiation

3. Intensity of radiation

4. Half-life

The basic principle of activation analysis is not new. It was applied for the first time in 1936 by Hevesy and Levi,[1] who determined the amount of dysprosium in an yttrium sample. The dysprosium in the sample became radioactive when bombarded with neutrons from a Ra-Be source. Two years later, Seaborg and Livingood[2] determined the gallium content in an iron sample by bombarding it with deuterons. The sensitivity of the method increased considerably with the availability of high neutron fluxes from nuclear reactors. Although charged particles, high energy gamma rays, fast neutron (MeV) energies from isotopic sources, 14-MeV neutrons from deuterium–tritium (d,t) reactions and 2.5-MeV neutrons from deuterium–deuterium (d,d) reactions may be used as the bombarding particles, thermal neutrons are, by far, the particles most frequently utilized for the irradiation of the sample.

Activation analysis has become, because of its extremely high sensitivity, an indispensable tool in a wide variety of fields in science and engineering[3] industry,[4,5] minerals exploration,[6] medicine,[7] environmental monitoring,[8–10] and forensic science.[11,12] The purpose of this chapter is not to present all the aspects, details, and applications of this field, but to discuss the major steps that comprise the method, the interpretation of the results, the errors and sensitivity of the method, and certain representative applications. The techniques of activation analysis has also become of great interest in the nuclear fuel cycle especially for artificial transmutation of waste (ATW). This technique employs thermal and fast neutrons to transmute long-lived radioactive isotopes to short-lived ones in spent fuel after chemical separation of specific radionuclides.

The reader will find many more details and an extensive list of applications in the bibliography and the references given at the end of the chapter. The International Atomic Energy Agency (http://www.iaea.org) has a multitude of activation analysis technical documents of which many are freely available to be downloaded.

The activation analysis method consists of the following major steps, to be discussed next:

1. Selection of the optimum nuclear reaction

2. Preparation of the sample for irradiation

3. Irradiation of the sample

4. Counting of the irradiated sample

5. Analysis of the counting results.

15.2 SELECTION OF THE OPTIMUM NUCLEAR REACTION

The optimum nuclear reaction is chosen with the following considerations in mind:

1. Production of large activity should occur within a reasonable irradiation time.

2. The radioisotope produced should have a reasonable half-life ($T > $ min).

3. The type and energy of the radiation emitted by the radioisotope should not present great counting difficulties.

4. A minimum number of interfering reactions should be involved.

If the sample is completely unknown, one starts with neutron irradiation because neutrons are absorbed by almost all isotopes. If the composition of the sample is known, then the best reaction for the identification of the isotope of interest should be chosen. Sometimes there is more than one reaction available for the same isotope. For example, aluminum bombarded with fast neutrons may be detected by three different reactions:

$$^{27}Al(n, p)^{27}Mg$$

$$^{27}Al(n, \alpha)^{24}Na$$

$$^{27}Al(n, \gamma)^{28}Al$$

The "optimum nuclear reaction" depends not only on the isotope and the bombarding particles but also on the composition of the sample that is analyzed. For example, the $^{27}Al(n, \gamma)^{28}Al$ reaction may be the best for detection of aluminum in a certain sample. However, if the sample contains silicon in addition to aluminum, the reaction $^{28}Si(n\ p)^{28}Al$ also produces ^{28}Al, and thus causes an interference to the measurement. If silicon is present, it may be better to use the reaction $^{27}Al(n, p)^{27}Mg$ or $^{27}Al(n, \alpha)^{24}Na$. More details about interfering reactions are given in Section 15.9.

The most commonly used neutron reaction is the (n, γ) reaction, which takes place with almost all isotopes (although with different probability) and has no threshold. In general, the (n, γ) cross section is higher for thermal than for fast neutrons. Other neutron interactions are $(n\ \alpha)$, (n, p), and $(n, 2n)$ reactions; except for a few exothermic (n, α) reactions, the others have a threshold; therefore, they can occur with fast neutrons only. Table 15.1 lists neutron reactions for the identification of several elements. Details for many more elements and reactions can be found in the bibliography of this chapter.

Charged-particle reactions are also used in activation analysis. Their disadvantage over neutron reactions is that charged-particle reactions are mostly endothermic, that is, they have a threshold. Table 15.2 gives several examples of such reactions.

Photon activation complements neutron and charged-particle activation. Photons are better than neutrons in certain cases. For example, photons are preferred if the product of the neutron activation is an isotope that has a very short half-life or emits only low-energy betas or low-energy X-rays. The cross sections for photonuclear reactions are generally smaller than those for neutrons and charged particles. Table 15.3 gives several photonuclear reactions that are used in activation analysis.

15.3 PREPARATION OF THE SAMPLE FOR IRRADIATION

A sample should be prepared properly and then placed in a container before it is irradiated. The person who prepares the sample should be extremely careful not to contaminate it. Activation analysis is so sensitive that it can determine traces of elements undetectable by chemical methods. If the sample is left on a table for a certain period of time, it collects dust that acts as a contaminant. Touch by hand may transfer enough salt to cause the irradiated sample to show the presence of sodium and chlorine. To avoid contamination, samples should be handled in dry boxes or in clean rooms. The person who prepares the sample should use clean instruments (knife, file, tweezers, etc.) and also wear clean plastic gloves.

Solid samples should have their surfaces cleaned with a suitable cleaning fluid to remove any surface contamination. The weight of the sample should be determined after cleaning it. For maximum accuracy, the weight is determined again after irradiation and counting are completed.

Liquids and powders cannot be cleaned, so they are handled in clean containers, avoiding contamination from the container wall. For liquid samples, care should be exercised to avoid loss of fluid when the fluid is transferred in and out of the container. In the case of powder (or pulverized) samples, the observer should be certain that a truly representative sample has been prepared. This is especially important if the main sample under analysis is not homogeneous.

Table 15.1
Typical Neutron Activation Reactions

Element	Symbol	Reaction	Threshold Energy (MeV)	Half-Life of Product	Main Radiation Emitted and Its Energy (MeV)
Aluminum	Al	$^{27}Al(n, \gamma)^{28}Al$	(–)ᵃ	2.3 min	β^- (2.85), γ (1.78)
		$^{27}Al(n, p)^{27}Mg$	1.9	9.46 min	β^- (1.75), γ (0.84, 1.013)
		$^{27}Al(n, \alpha)^{24}Na$	3.27	15 h	β^- (1.389), γ (1.369, 2.754)
Arsenic	As	$^{75}As(n, \gamma)^{76}As$	(–)	26.4 h	β^- (2.97), γ (0.559)
		$^{75}As(n, \alpha)^{72}Ga$	(–)	14.1 h	β^- (3.15), γ (0.835)
		$^{75}As(n, 2n)^{74}As$	8.27	17.9 d	β^- (3.15), γ (0.835)
Cadmium	Cd	$^{110}Cd(n, \gamma)^{111m}Cd$	(–)	48.6 min	e^-, γ (0.247)
		$^{110}Cd(n, p)^{110m}Ag$	2.12	235 d	γ (0.658)
Calcium	Ca	$^{48}Ca(n, \gamma)^{49}Ca$	(–)	8.8 min	γ (3.07), β (1.95)
Chlorine	Cl	$^{37}Cl(n, \gamma)^{38}Cl$	(–)	37.2 min	β^- (4.91), γ (1.6, 2.17)
		$^{37}Cl(n, p)^{37}S$	3.6	5 min	β^-, γ (3.09)
Copper	Cu	$^{63}Cu(n, 2n)^{62}Cu$	11.01	9.76 min	γ (0.511)
Fluorine	F	$^{19}F(n, \alpha)^{16}N$	1.57	7.15 s	γ (6.13)
Gold	Au	$^{197}Au(n, \gamma)^{198}Au$	(–)	2.7 d	β^- (0.962), γ (0.412)
		$^{197}Au(n, 2n)^{196}Au$	7.36	6.18 d	β^-, γ (0.356)
Iodine	I	$^{127}I(n, \gamma)^{128}I$	(–)	25 min	β^- (2.12), γ (0.441)
Iron	Fe	$^{58}Fe(n, \gamma)^{59}Fe$	(–)	45.5 d	β^-, γ (1.095, 1.292)
		$^{56}Fe(n, p)^{56}Mn$	2.98	2.57 d	β^-, γ (0.847)
Lead	Pb	$^{208}Pb(n, 2n)^{207m}Pb$	7.45	0.885 s	γ (0.570)
Mercury	Hg	$^{200}Hg(n, 2n)^{199m}Hg$	8.11	43 min	γ (0.158)
		$^{196}Hg(n, \gamma)^{197}Hg$	(–)	65 h	γ (0.077)
		$^{202}Hg(n, \gamma)^{203}Hg$	(–)	46.9 d	γ (0.279)
Nickel	Ni	$^{58}N(n, 2n)^{57}Ni$	12.09	36 h	β^+, γ (0.511, 1.37)
Nitrogen	N	$^{14}N(n, 2n)^{13}N$	11.31	10 min	γ (0.511)
Oxygen	O	$^{16}O(n, p)^{16}N$	10.2	7.1s	γ (6.13)
Phosphorus	P	$^{31}P(n, \alpha)^{28}Al$	2	2.3 min	β^-, γ (1.78)
Potassium	K	$^{39}K(n, 2n)^{38}K$	13.41	7.7 min	γ (0.511, 2.17)
Silicon	Si	$^{28}Si(n, p)^{28}Al$	3.99	2.3 min	γ (1.78)
Silver	Ag	$^{109}Ag(n, \gamma)^{110}Ag$	(–)	24 s	γ (0.66)

ᵃ (–) = No threshold.

Table 15.2
Charged-Particle Reactions

Element	Symbol	Reaction	Threshold Energy (MeV)	Half-Life of Product	Main Radiation Emitted and Its Energy (MeV)
Boron	B	$^{10}B(\alpha, n)^{13}N$	–	10 min	β^+, γ (0.511)
		$^{10}B(p, \gamma)^{11}C$	0.4	20.4 min	β^+, γ (0.511)
		$^{11}B(p, n)^{11}C$	2.76		
Carbon	C	$^{12}C(p, n)^{12}N$	18.12	11 ms	
Nitrogen	N	$^{14}N(p, \alpha)^{11}C$	2.88	20.4 min	
Oxygen	O	$^{16}O(\alpha, d)^{18}F$	35	109.8 min	
Sodium	Na	$^{23}Na(\alpha, n)^{26m}Al$	4.62	6.7 s	β^+, γ (0.511)
Aluminum	Al	$^{27}Al(\alpha, n)^{30}P$	3.38	2.55 min	β^+, γ (0.511)
Copper	Cu	$^{63}Cu(\alpha, n)^{66}Ga$	7.69	9.45 h	
Phosphorus	P	$^{31}P(d, p)^{32}P$	–	14.3 d	β^-, (E_{max} = 1.17)
Iron	Fe	$^{54}Fe(d, n)^{55}Co$	–	18.2 h	β^+, γ (0.511, 0.93)

Table 15.3
Photonuclear Reactions

Element	Symbol	Reaction	Threshold Energy (MeV)	Half-Life of Product	Main Radiation Emitted and Its Energy (MeV)
Carbon	C	$^{12}C(\gamma, n)^{11}C$	18.7	20.4 min	β^+, γ (0.511)
Fluorine	F	$^{19}F(\gamma, n)^{18}F$	10.5	109.8 min	β^+, γ (0.511)
Nitrogen	N	$^{14}N(\gamma, n)^{13}N$	10.5	10 min	β^+, γ (0.511)
Oxygen	O	$^{16}O(\gamma, n)^{15}O$	15.7	2.03 min	β^+, γ (0.511)
Copper	Cu	$^{63}Cu(\gamma, n)^{62}Cu$	10.8	9.7 min	β^+, γ (0.511)
		$^{65}Cu(\gamma, n)^{64}Cu$	9.9	12.8 h	β^+, γ (0.511)
Silver	Ag	$^{107}Ag(\gamma, n)^{106}Ag$	9.4	24 min	β^+, γ (0.511)
Sodium	Na	$^{23}Na(\gamma, n)^{22}Na$	12.4	2.62 y	γ (0.511)
Lead	Pb	$^{204}Pb(\gamma, n)^{203}Pb$	8.4	52.1 h	γ (0.279)

The packaging material or container should

1. Have high radiation and thermal resistance (i.e., it should not decompose, melt, or evaporate in the irradiation environment)

2. Have low content of elements that become radioactive

3. Be inexpensive and easy to handle

Materials that are used most frequently are polyethylene, silica, and aluminum foil. Polyethylene satisfies requirements 2 and 3 listed above, but it has low resistance to radiation and temperature. It becomes brittle after exposure to a fluence of 10^{21}–10^{22} neutrons/m². Polyethylene tubes of different diameters are routinely used in radiation laboratories. The tubes can be easily cleaned and sealed. Silica containers are not as useful as polyethylene because they are not as pure, they become radioactive, and sealing is more complicated. Aluminum foil is useful for packaging solids, but it becomes radioactive through $^{27}Al(n, \gamma)$ ^{28}Al and $^{27}Al(n, \alpha)$ ^{24}Na reactions. The second reaction is more troublesome than the first because the half-life of ^{24}Na is 15 h, whereas the half-life of ^{28}Al is only 2.24 min.

15.4 SOURCES OF RADIATION

Intensities, energies, and special characteristics of the various radiation sources are briefly discussed in this section.

15.4.1 Sources of Neutrons

Neutron sources include nuclear reactors, accelerators, and isotopic sources. Nuclear reactors are, by far, the most frequently used irradiation facilities. They provide high fluxes [upper limit 10^{18} neutrons/(m² · s)] of mostly thermal neutrons ($E < 1$ eV). Fast neutrons in the keV range are also available, but at lower flux levels.

When short-lived isotopes are involved, a higher activity is produced by irradiating the sample in a reactor that can be pulsed (see Lenihan et al., 1972). Such a reactor producing a high flux of about 10^{20} neutrons/(m² · s) for a short period of time (milliseconds) is the TRIGA reactor, marketed by General Atomics, San Diego.

Accelerators produce fast neutrons as products of charged-particle reactions. The most popular device is the so-called neutron generator, which operates on the reaction

$$_{1}^{2}H + _{1}^{3}H \rightarrow _{0}^{1}n + _{2}^{4}He + 17.586 \text{ MeV}$$

The cross section for this exothermic reaction peaks at a deuteron kinetic energy of about 120 keV with a value of about 5 b. The neutrons produced have an energy of about 14 MeV. (The neutron kinetic energy changes slightly with the direction of neutron emission.) The maximum neutron flux provided by a neutron generator is of the order of 10^{12} neutrons/s.

Neutrons with an average energy of about 2.5 MeV are produced by the (d, d) reaction

$$\,_1^2\mathrm{H} + \,_1^2\mathrm{H} \;\rightarrow\; \,_0^1 n + \,_2^3\mathrm{H} + 3.266 \text{ MeV}$$

The cross section for this reaction peaks at about 2-MeV bombarding deuteron energy with a value of about 100 mb. At acceleration voltages normally used in neutron generators (~150 kV), the cross section is about 30 mb. The (d,d) reaction offers neutron fluxes of the order of 10^9 neutrons/($m^2 \cdot$ s). It is important to note that both the (d,t) and the (d,d) reactions produce essentially monoenergetic neutrons.

Isotopic neutron sources are based on (α, n) and (γ, n) reactions, and on spontaneous fission (^{252}Cf). They all produce fast neutrons. The (α, n) and (γ, n) sources produce the neutrons through the reactions

$$\,_2^4\mathrm{He} + \,_4^9\mathrm{Be} \;\rightarrow\; \,_0^1 n + \,_6^{12}\mathrm{C}$$
$$\gamma + \,_4^9\mathrm{Be} \;\rightarrow\; \,_0^1 n + \,_4^8\mathrm{Be}$$

The isotope ^{252}Cf is the only spontaneous fission (SF) source of neutrons commercially available. It provides fission spectrum neutrons with an average of 2.3 MeV. The characteristics of isotopic neutron sources are given in Table 15.4.

15.4.2 Sources of Charged Particles

Apart from certain α-emitting radioisotopes, accelerators are the only practical sources of charged particles. They can provide almost any charged particle or ion for bombardment of the target for a wide range of energies. The particles most commonly used as projectiles are protons, deuterons, alphas, tritons, and ^{3}He nuclei. The beam current of the accelerator is related to particles per second hitting the target by the equations (see Ricci, 1972).

$$I(\text{particles/s}) = 6.2 \times 10^{15}\, \frac{i\,(\text{mA})}{z} \tag{15.1}$$

where i = the beam current in mA

  z = the charge of the accelerated particle in units of the electronic charge. Knowing the number of particles per second hitting the target and the cross section for the reaction, one can calculate the reaction rate.

Table 15.4
Isotopic Neutron Sources

Source	Reaction	Yield	Average Neutron Energy (MeV)	Half-Life of Isotope Involved
^{226}Ra–Be	(α, n)	7×10^5 neutrons/s·g (Ra)	~4	1600 y
^{210}Po–Be	(α, n)	1×10^{10} neutrons/s·g (Po)	~4	138.4 d
^{239}Pu–Be	(α, n)	1.0×10^5 neutrons/s·g (Pu)	~4	24,131 y
^{241}Am–Be	(α, n)	7×10^6 neutrons/s·g (Am)	~4	432 y
^{124}Sb–Be	(γ, n)	1.0×10^{10} neutrons/s·g	0.024	60.2 d
^{242}Cm–Be	(α, n)	1×10^{10} neutrons/s·g (Cm)	4	162.8 d
^{252}Cf	(SF)	2×10^{12} neutrons/s·g	2.3	2.646 y

Source: From Lenihan, M. M. A., et al., *Advances in Activation Analysis*, vol. 2, Academic, New York, 1972.

15.4.3 Sources of Photons

The sources of photons include radioisotopes, nuclear reactions, and bremsstrahlung radiation. There are many radioisotopes that emit gamma rays. The most useful ones are as follows:

1. ^{24}Na with a 15-h half-life emitting γ's with energy 1.37 and 2.75 MeV.

2. ^{60}Co with 5.3-year half-life emitting γ's with energy 1.17 an 1.33 MeV.

3. ^{124}Sb with a 60-d half-life emitting γ's with energy 1.71, 2.1, and 2.3 MeV.

4. ^{22}Na with a 2.6-y half-life emitting a gamma with energy 1.275 MeV (^{22}Na being a positron emitter is also a source of 0.511-MeV gamma rays).

Photons of extremely high energy may be produced by nuclear reactions. Examples are:

$$^{3}H(p, \gamma)^{4}He, \quad E_{\gamma} = 19.8 \text{ MeV}$$

$$^{7}Li(p, \gamma)^{8}Be, \quad E_{\gamma} = 14.8 \text{ and } 17.6 \text{ MeV}$$

$$^{11}B(p, \gamma)^{12}C, \quad E_{\gamma} = 11.7 \text{ and } 16.1 \text{ MeV}$$

Unfortunately, the gamma fluxes generated by these reactions are very small, relative to neutron fluxes produced by reactors.

Bremsstrahlung is produced with the help of electron accelerators. The electrons are accelerated to a certain energy and then are allowed to hit a solid target. The radiation produced has a continuous energy spectrum, extending from zero energy up to the maximum electron kinetic energy. Large photon fluxes are produced, and may be used for activation of rather large samples.

15.5 IRRADIATION OF THE SAMPLE

Depending on the selected reaction, irradiation of the sample may take place in a reactor, in an accelerator, or with an isotopic source. After the selection of an irradiation facility, the next step is a decision about the irradiation time. If the sample contains known isotopes at approximately known amounts, it is easy to estimate the proper irradiation time. If, on the other hand, the sample is completely unknown, one irradiates the sample for an arbitrary time, checks some of the isotopes present (from the emitted radiations), and then irradiates the sample again for a time that will provide enough activity for proper isotope identification with the desired accuracy.

Since neutrons are, by far, more frequently used for activation analysis than other particles, neutrons will be assumed to be the projectiles for the equations discussed next. However, it should be noted that the same equations apply when some other radiation is used as the bombarding particle.

The equation that gives the activity produced after irradiating the sample for time t_0 is (for derivation, see Section 14.4)

$$A(t_0) = a_i m \frac{N_A}{A_i} \sigma_i \phi (1 - e^{-\lambda_{i+1} t_0}) \tag{15.2}$$

where m = mass of the element of interest in the sample
$\quad$ a_i = weight fraction (abundance) of isotope with atomic mass A_i (A_i is an isotope of the element with mass m, the element of interest)
$\quad$ λ_{i+1} = decay constant of the radioisotope produced σ_i = cross section for the reaction that makes the isotope with atomic weight A_i radioactive
$\quad$ ϕ = particle flux [particles/(m$^2 \cdot$ s)]
$\quad$ N_A = Avogadro's number

Equation 15.2 is valid if (see Section 14.4)

1. The number of target nuclei stays essentially constant, that is, $\sigma_i \phi t \ll 1$.

2. The radioisotope produced has such a small reaction cross section that $\lambda_{i+1} \gg \phi \sigma_{i+1}$.

3. The flux is uniform throughout the target.

If the half-life of the radioisotope is much shorter than t_0 ($t_0 \geq 6\tau_{1/2}$), saturation activity (A_{sat}) is obtained, given by

$$A_{\text{sat}} = a_i m \frac{N_A}{A_i} \sigma_i \phi \qquad (15.3)$$

Equation 15.3 indicates that, for a particular isotope, the activity increases by irradiating a larger mass m in a higher flux ϕ.

The size of the sample (mass m) is dictated by four factors:

1. The maximum activity that can be safely handled under the conditions of the laboratory (i.e., shielding of source and detector, existence of remote control, automated remote handling of samples).

2. The size of the sample holder.

3. The self-absorption of the radiation emitted by the sample. This is particularly important if the radiation detected is betas or soft X-rays.

4. The size of the detector. Little, if anything, is gained by using a sample much larger than the detector size.

The flux ϕ is determined by the limitations of the irradiation facility. At the present time, the maximum flux is about 10^{18} neutrons/(m² · s) (thermal neutron flux).

15.6 COUNTING OF THE SAMPLE

After irradiation is completed, the sample is counted using an appropriate system. The qualitative and quantitative determination of an isotope is based on the analysis of the energy spectrum of the radiations emitted by the radioisotope of interest. Sometimes it may be necessary to use information about the half-life of the isotope(s). In such a case, counting may have to be repeated several times at specified time intervals.

The counting system depends on the radiation detected. Modern activation analysis systems depend on the detection of gamma rays and X-rays and very seldom on detection of other particles. For this reason, the discussion in the rest of this chapter is based on the assumption that the irradiated sample emits photons.

A basic counting system for activation analysis consists of a detector [Ge or Si(Li)], electronics (i.e., preamplifier, amplifier), and a multichannel analyzer (MCA). Modern MCAs do much more than record the data. They are minicomputers or are connected to computers that store and analyze the recorded data. Examples are the ADCAM the architectures offered by ORTEC and CANNBERA (see chapter bibliography).

15.7 ANALYSIS OF THE RESULTS

The analysis of an activation analysis spectrum is based on the procedures described in Section 12.5.3. It is performed either by the MCA itself, if that instrument has such capability, or by a digital computer. Several computer codes have been written for that purpose.[13-16] Two of the most popular commercially peak fitting software are GENIE 2000 from CANBERRA[17] and GammaVision from ORTEC.[18]

Activation analysis may be qualitative or quantitative. In a qualitative measurement, only identification of the element is involved. This is accomplished, as shown in Section 12.5.3 from the energies and intensities of the peaks of the spectrum. In a quantitative measurement, on the other hand, in addition to identification, the amount of element in the sample is also determined. To illustrate how the mass is determined and what the errors and sensitivity of the method are, consider the energy spectrum of Figure 15.1 as an example.

Assume that the mass of an element in the sample will be determined from the full-energy peak at E_k. Using the notation of Section 14.4, the mass m is given by (see Eq. 14.22)

$$m = \frac{P_k A_i \lambda_{i+1}}{\varepsilon(E_k) e_k a_i N_A \sigma_i \phi (1 - e^{-\lambda_{i+1} t_0})(e^{-\lambda_{i+1} t_1} - e^{-\lambda_{i+1} t_2})} \qquad (15.4)$$

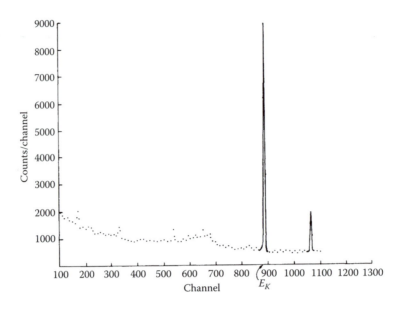

Figure 15.1 A typical Ge energy spectrum. The element will be identified using the main peak at channel 888.

where P_k = net number of counts under the peak (determined by one of the methods described in Section 12.5.3)

  $\varepsilon(E_k)^*$ = absolute full-energy peak detector efficiency at energy E_k

  e_k = probability that a photon of energy E_k is emitted per decay of the isotope (also known as intensity of this gamma)

  $t_2 - t_1$ = counting time = T

The error in the value of m depends on the errors of the quantities that comprise Eq. 15.4, such as P_k, λ, ε, σ_i, and ϕ. In the most general case, the standard derivation σ_m is

$$\sigma_m = \sqrt{\left(\frac{\partial m}{\partial P_k}\right)^2 \sigma_{P_k}^2 + \left(\frac{\partial m}{\partial \phi}\right)^2 \sigma_\phi^2 + \left(\frac{\partial m}{\partial t}\right)^2 \sigma_t^2 + \cdots} \tag{15.5}$$

In practice, certain errors are always negligible when compared to others. The quantities A_i, λ, e_k, and σ_i are known very accurately for most isotopes. Also, the flux ϕ and the efficiency ε can be determined with a known but small error. The error in the times t_1 and t_2 can be negligible. Thus, the major contribution to the error of m comes from the error of P_k, that is, the error of the area under the peak. Assuming that σ_{P_k} is the only important error, the standard error of m is

$$\sigma_m = \left(\frac{\partial m}{\partial P_k}\right)\sigma_{P_k} \tag{15.6}$$

or the relative error (using Eq. 15.4) is

$$\frac{\sigma_m}{m} = \frac{\sigma_{P_k}}{P_k} \tag{15.7}$$

---

* In activation analysis, the efficiency is determined in such a way as to include the solid angle Ω and the other correction factors $F(E_k)$ discussed in Chapter 8.

Thus, the relative error of m is equal to the relative error of P_k, in this case.

It should be emphasized that only one well-identified peak, and not the whole spectrum, is needed for quantitative determination of an element in the unknown sample. The other peaks, if used, should give results consistent with the one chosen for the analysis, and they should also be utilized as additional check points to remove any doubts in the identification of the unknown. For example, if the k peak leads to the identification of isotope X, other peaks in the spectrum should agree, in energy and intensity, with additional gammas emitted by isotope X.

In practice, the objective of the measurement is often to identify the mass of a particular trace element in the sample. Then the unknown mass is determined in a simpler way by irradiating, along with the unknown, a standard sample with a known mass of the trace element and counting both samples with the same counter. If m_s is the mass of the standard and m_x the mass of the unknown, using Eq. 15.4, one obtains

$$m_x = m_s \frac{(P_k)_x (e^{-\lambda_i+1 t_1} - e^{-\lambda_i+1 t_2})_s}{(P_k)_s (e^{-\lambda_i+1 t_1} - e^{-\lambda_i+1 t_2})_x} \tag{15.8}$$

where the times t_1 and t_2 are different for the standard and the unknown, but in both cases, $t_2 - t_1 = T$. Use of Eq. 15.8 constitutes a relative method, in contrast to the use of Eq. 15.4, which represents an absolute method. In the last years there has a lot successful usage of the k-zero method, which takes into account all aspects of the absolute method.[19,20]

15.8 SENSITIVITY OF ACTIVATION ANALYSIS

Sensitivity of the activation analysis method for a particular element refers to the minimum mass of that element that can be reliably detected. The minimum detectable mass is determined from Eq. 15.4 by assuming the most favorable conditions for the measurement and by setting an upper limit for the acceptable error of the result. The process is similar to the determination of the minimum detectable activity discussed in Section 2.20.

Assuming that the observer is willing to accept a maximum error σ_m such that

$$\sigma_m \le fm \tag{15.9}$$

($f < 1$) and that the only error in the determination of the mass comes from the error in the number of counts under the peak, a limiting counting rate can be defined as follows. If one defines a net counting rate r_k as

$$r_k = \frac{G_k}{T} - b_k \tag{15.10}$$

where*

$$G_k = P_k + b_k T = r_k T + b_k T \tag{15.11}$$

then, using Eqs. 2.101, 15.7, and 15.11, one obtains for the minimum acceptable counting rate r_k,

$$r_k \ge \frac{1 + \sqrt{1 + 4f^2 T (b_k + T\sigma_b^2)}}{2f^2 T} \tag{15.12}$$

Example 15.1   What is the minimum mass of gold that can be detected by neutron activation analysis under the conditions listed below?

$$\phi = 10^{16} \text{ neutrons/(m}^2\text{·s)} \quad \varepsilon = 0{,}30$$
$$t_0 = 2 \text{ h,} \quad t_1 = 5 \text{ min}$$

---

* Note that both G_k and b_k refer to gross counts and background of the peak k, and not the whole spectrum.

$$f = 0,50, \quad t_2 = 125 \text{ min}$$

$$b = 20 \pm 0.2 \text{ counts/min}$$

For gold, $A = 197$, $a = 1$, $e_k = 1$, $\sigma = 99$ b, and $T_{1/2} = 2.7$ days.

Answer   Using Eq. 15.12, the minimum acceptable counting rate is

$$r_k = \frac{1 + \sqrt{1 + 4(0.5)^2 20(120) + 4(0.5)^2(120)^2(0.2)^2}}{2(0.5)^2 120} = 0.93 \text{ counts/min}$$

Using Eqs. 15.10 and 15.11,

$$P_k = r_k T = 0.93(120) = 111.12 \text{ counts}$$

Since the half-life of ^{198}Au is 2.7 days,

$$\lambda = \frac{\ln 2}{2.7 \text{ d}} = 2.971 \times 10^{-6} \text{ s}^{-1}$$

and the exponential factors in Eq. 154 become

$$\exp(-\lambda_{r+1} t_0) = \exp[(-2.971 \times 10^{-6}) 2 (3600)] = 0.979$$
$$\exp(-\lambda_{r+1} t_1) = \exp[(-2.971 \times 10^{-6}) 300] = 0.999$$
$$\exp(-\lambda_{r+1} t_2) = \exp[(-2.971 \times 10^{-6}) 125(60)] = 0.978$$

Equation 15.4 gives

$$m = \frac{111.12 \, (197 \times 10^{-3}) \, (2.971 \times 10^{-6}) \text{ s}^{-1}}{0.30 \, (6.022 \times 10^{23}) \, (99 \times 10^{-28} \text{m}^2) \, (10^{16} \text{ n/m}^2 \text{ s}) \, (1 - 0.979) \, (0.999 - 0.978)}$$
$$= 8.24 \times 10^{-15} \text{ kg} = 8.24 \text{ pg}$$

Example 15.2   What is the absolute minimum mass of an element that can be detected under the most favorable conditions?

Answer   The absolute minimum mass will be determined if one assumes

Efficiency 100% ($\varepsilon = 1$)

Intensity 100% ($e_k = 1$)

Saturation activity $(1 - e^{-\lambda t_0} = 1)$

Maximum thermal neutron flux [$\sim 10^{18}$ neutrons/($m^2 \cdot$ s)]

Then Eq. 15.4 takes the form

$$m = \frac{P_k A_i \lambda_{i+1}}{a_i N_A \sigma_i \phi (e^{-\lambda_{i+1} t_1} - e^{-\lambda_{i+1} t_2})}$$ (15.13)

Factors that may further affect the result given by Eq. 15.13 depend on the background of the counting system and the maximum acceptable error (both background and acceptable error affect the minimum acceptable value of P_k).

15.9 INTERFERENCE REACTIONS

One source of error in activation analysis is interference reactions. These are reactions that produce the same isotope as the one being counted, through bombardment of a different isotope in the sample. As an example, assume that a sample is analyzed for magnesium by using fast-neutron activation. The reaction of interest is $^{24}Mg(n, p)^{24}Na$. Therefore, the activity of ^{24}Na will be recorded, and from that the amount of ^{24}Mg can be determined. If the sample contains ^{23}Na and ^{27}Al, two other reactions may take place which also lead to ^{24}Na. They are

$$^{23}Na(n, \gamma)^{24}Na$$

$$^{27}Al(n, \alpha)^{24}Na$$

If this is the case and the investigator does not consider these last two reactions, the mass of ^{24}Mg will be determined to be higher than it is.

Interference reactions are discussed in detail in many activation analysis books (see Rakovic, 1970; Nargolwalla and Przybylowicz, 1973). A few representative examples are given below:

$$^{68}Zn(n, \gamma)^{69m}Zn \quad \text{and} \quad ^{69}Ga(n, p)^{69m}Zn \quad \text{and} \quad ^{72}Ge(n, \alpha)^{69m}Zn$$

$$^{32}S(n, p)^{32}P \quad \text{and} \quad ^{31}P(n, \gamma)^{32}P \quad \text{and} \quad ^{35}Cl(n, \alpha)^{32}P$$

$$^{17}O(n, \alpha)^{14}C \quad \text{and} \quad ^{13}C(n, \gamma)^{14}C \quad \text{and} \quad ^{14}N(n, p)^{14}C$$

$$^{98}Tc(n, \gamma)^{99m}Tc \quad \text{and} \quad ^{98}Mo(n, \gamma)^{99}Mo^{\beta^-} \to ^{99m}Tc$$

$$^{31}P(n, \gamma)^{32}P \quad \text{and} \quad ^{30}Si(n, \gamma)^{31}Si^{\beta^-} \to ^{31}P(n, \gamma)^{32}P$$

$$^{55}Mn(n, \gamma)^{56}Mn \quad \text{and} \quad ^{54}Cr(n, \gamma)^{55}Cr^{\beta^-} \to ^{55}Mn(n, \gamma)^{56}Mn$$

$$^{19}F(n, 2n)^{18}F \quad \text{and} \quad ^{17}O(p, \gamma)^{18}F \quad \text{and} \quad ^{18}O(p, n)^{18}F$$

In the last two reactions, the proton is produced by the incident fast neutrons interacting with the target nuclei.

15.10 ADVANTAGES AND DISADVANTAGES OF THE ACTIVATION ANALYSIS METHOD

One of the greatest advantages of activation analysis is its ability to detect most of the isotopes with an extremely high sensitivity. Other advantages are that the method

1. Is nondestructive (in most cases)

2. Needs a sample with a very small mass

3. Can detect more than one element at a time

4. Identifies different isotopes of the same element

5. Provides results rapidly

6. Is not affected by the chemical form of the element of interest.

The greatest disadvantage of the method is that it requires expensive equipment, and the analysis of the results is not trivial. Also, activation analysis does not provide information about the chemical compound in which the nuclide of interest belongs.

Overall, activation analysis is a very powerful technique, as demonstrated by its wide use in so many different fields—that is, chemistry, biology, medicine, forensic medicine, industry, archaeology, and environmental research. For details regarding these special applications, the reader is referred to the references and to the bibliography of this chapter, in particular, to the books by Nargolwalla and Przybylowicz (1973) and by Rakovic (1970).

15.11 PROMPT GAMMA ACTIVATION ANALYSIS

Prompt gamma activation analysis, commonly known as PGAA or PGNAA, is a technique which has gained a lot of acceptance as a complimentary method to traditional neutron activation analysis. The technique employs a neutron beam which irradiates a sample and the gamma-rays from the compound nucleus are detected. The target nucleus ^AZ absorbs a neutron n and forms a compound nucleus ^{A+1}Z in an excited state. A prompt gamma ray (γ) is emitted when the compound nucleus de-excites as shown in Eq. 15.14.

$$n + {}^AZ \to {}^{A+1}Z^* \to {}^{A+1}Z + \gamma \tag{15.14}$$

PGAA is ideally suited for stable isotopes that have very good cross-sections and have no ^{A+1}Z radioactive products that can be measured or these products have poor gamma-ray characteristics to be detected (e.g., weak branching ratio). Elements such as H, B, and others are excellent candidates to be determined by PGAA.[21-23] Typically, a lot of shielding is needed to prevent neutron damage of the detector.[24] One should keep in mind that emission of the prompt gammas stops when the neutron beam is turned off as opposed to emission of decay gammas, emitted by radioactive isotopes with a certain half-life, that continues after the neutron beam is off.

15.12 NEUTRON DEPTH PROFILE

Neutron depth profile known as NDP, is a near surface analysis technique based on the absorption of neutrons in a target that emits charged particles which are then detected by a silicon barrier detector. The concentrations of certain elements can then be determined. Since NDP relies on the emission of charged particles, there are only a few reactions that can benefit from this technique. These include among others $^3H(n,p)^3H$, $^6Li(n,\alpha)$ ^3H, $^{10}B(n,\alpha)^7Li$, and $^{14}N(n,p)^{14}C$.

Most of the work involves characterization of samples in materials science and engineering.[25-27]

15.13 NEUTRON RADIOGRAPHY

Neutron radiography uses the unique interaction probabilities of neutrons to create images of the internal features of an object. This imaging technique is nondestructive; however, samples receive significant radiation exposure. X-ray and neutrons operate in complimentary fashion to create radiographs. X-rays are best attenuated best by high atomic Z materials. Thus X-rays pass through the water in a human body easily but they are attenuated once they encounter bones. Conversely, neutrons move through high Z material easily and but they attenuated considerably by low Z materials, such as water.

Neutron radiography can also be used to create images of hidden samples due to the fact that a neutron beam can pass through a significant amount of this type of material without being completely attenuated.

Samples are placed in front of a phosphor scintillation screen. Neutrons pass through the sample and strike the scintillation screen. Neutrons ionize the phosphorus in the screen, which cause it to produce flashes of light. The flashes of light are recorded by a camera and converted to numbers in a matrix.[28] A typical configuration of a neutron radiography system is shown in Figure 15.2 while a radiograph of a combination lock is seen in Figure 15.3.

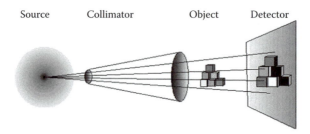

Figure 15.2 A typical configuration of a neutron radiography system.

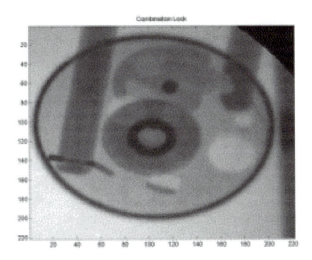

Figure 15.3 Radiograph of a combination lock.

Some typical applications include high-speed neutron radiography for monitoring the water absorption by capillarity in porous materials,[29] nuclear materials identification system for fissile material transfers,[30] flow visualization of refrigerant in a self-vibration heat pipe by neutron radiography,[31] and visualization of oil behavior in a small 4-cycle engine with electrical motoring by neutron radiography.[32]

PROBLEMS

15.1 For a target containing 1.00 mg of manganese in a nuclear reactor of flux 1.0×10^{13} neutrons/cm^2*s, calculate the activity of ^{56}Mn ($t_{1/2} = 2.6$ h) formed in 5.2 h. What would be the activity after 520 h irradiation? The isotopic abundance of ^{56}Mn is 100%, and $s = 13.4$ barns.

15.2 ^{127}I(n, γ)^{128}I is 6.3 barns. The half-life of ^{128}I is 25 min. If 12.7 g pure ^{127}I is placed in a reactor in which the neutron flux is 2.0×10^5 neutrons/cm^2*s and is bombarded for 25 min, what is the activity of ^{128}I in the sample 100 min after bombardment has stopped?

15.3 How long must 5.0 g of the pure nuclide ^{31}P be irradiated with neutron flux 2.0×10^5 neutrons/cm^2 s to achieve an activity of ^{32}P of 1.00×10^3 dis/min 100 h after the bombardment has topped ($t_{1/2} = 14.3$ days)?

15.4 Traces of manganese are suspected in an unknown sample that has been irradiated for 30 min in a flux of 10^{16} neutrons/(m$^2 \cdot$ s). Counting started 5 min after the irradiation ended. The 0.8-MeV gamma of ^{56}Mn was detected by a counter with a 4% efficiency ($F.\varepsilon.\Omega$). The sample gave 500 counts in 5 min, while the background was 30 ± 1 counts/min. Based on this information, calculate the mass of manganese in the sample and the standard error of this measurement. For ^{55}Mn, $\sigma = 14$ b. For ^{56}Mn, $T_{1/2} = 2.58$ h.

15.5 What should the minimum activity of a sample be if it is required that in the worst case $\sigma_r = 0.8r$, and the sample is counted in a system for which $\varepsilon = 0.50$, and the background is 100 ± 5 counts/min. The sample can only be counted for 1 min.

15.6 In a neutron activation analysis experiment, a 10%-efficient Ge(Li) detector with 12 ± 0.1 counts/min background is used for the measurement of 0.6-MeV γ's emitted by the sample. If the counting time is 5 min.
  (a) What is the minimum acceptable counting rate if the maximum acceptable error is 60%?
  (b) What is the minimum mass that can be detected if the isotope under investigation has $A = 75$, $\sigma = 0.21$ b, $t_{1/2}(^{76}X) = 20$ h, irradiation time = 8 h, and $\phi = 10^{16}$ neutrons/(m$^2 \cdot$ s)?

15.7 Prove Eq. 14.12.

15.8 What is the absolute minimum mass of phosphorus that can be detected using the reaction (n, α) under irradiation with 14-MeV neutrons? Assume $\phi = 10^{12}$ neutrons/(m$^2 \cdot$ s), $\sigma = 0.150$ b, counting system background $= 15 \pm 0.5$ counts/min, counting time $= 2$ min, and maximum acceptable error is 30%.

BIBLIOGRAPHY

Alfassi, Z. B., *Chemical Analysis by Nuclear Methods*, John Wiley & Sons, 1994.

Alfassi, Z. B., *Activation Analysis*, Vol I, CRC Press, 1990.

Alfassi, Z. B., *Activation Analysis*, Vol II, CRC Press, 1990.

Alfassi, Z. B., *Instrumental Multi-Element Chemical Analysis,* Springer, 1998.

Alfassi, Z. B., *Elemental Analysis by Particle Accelerators*, CRC Press, 1991.

Bowen, H. J. M., and Gibbons, D., *Radioactivation Analysis*, Oxford University Press, London, 1963.

Coomber, D. I. (ed.), *Radiochemical Methods in Analysis*, Plenum Press, New York, 1975.

De Soete, D., *Neutron Activation Analysis,* Wiley-Interscience, 1972.

Heydorn, K., *Neutron Activation Analysis for Clinical Trace Element Research*, CRC Press, 1983.

http://ww.canberra.com

http://www.ortec-online.com

International Atomic Energy Agency (http://www.iaea.org).

Lederer, C. M., and Shirley, V. S. (eds.), *Table of Isotopes*, 7th ed., Wiley, New York, 1978.

Lenihan, M. M. A., Thomson, S. J., and Guinn, V. P., *Advances in Activation Analysis*, Academic, New York, 1972, vol. 2.

Lutz, G. J., Boreni, R. J., Maddock, R. S., and Wing, J. (eds.), "NBS Technical Note 467 Activation Analysis: A Bibliography Through 1971," National Bureau of Standards, 1972.

Lyon, W. S. (ed.), *Guide to Activation Analysis*, Van Nostrand, Princeton, New Jersey, 1964.

Molnar, G., *Handbook of Prompt Gamma Activation Analysis: With Neutron Beams,* Springer, 2004.

Muecke, G. K., *Short Course in Neutron Activation Analysis in the Geosciences*, Mineralogical Association of Canada, 1980.

Nargolwalla, S. S., and Przybylowicz, E. P., *Activation Analysis with Neutron Generators*, Wiley, New York, 1973.

Overwater, R. M. W., *The Physics of Big Sample Instrumental Neutron Activation Analysis (Delft Studies in Integrated Water Management , No 5)*, Delft University Press, 1994.

Parry, S., *Activation Spectrometry in Activation Analysis, Volume 19 In Chemical Analysis*, Wiley-Interscience, 1991.

Radiochemical Methods of Analysis, Proceedings of a Symposium on Radiochemical Methods of Analysis, IAEA, Salzburg, Austria, 1964.

Rakovic, M., *Activation Analysis* (English translation), CRC Press, Cleveland, Ohio, 1970.

Ricci, E., Charged Particle Activation Analysis, in J. M. A. Lenihan, S. J. Thomson, and V. P. Guinn (eds.), *Advances in Activation Analysis*, Vol. 2, Academic, London, 1972.

REFERENCES

1. Hevesy, G., and Levi, H., *K. Danske Vidensk. Sels. Mat.-fys. Medd.* **14**:5 (1936).

2. Seaborg, G. T., and Livingood, J. J., *J. Am. Chem. Soc.* **60**:1784 (1938).

3. The 11th International Conference on Modern Trends in Activation Analysis, *J. Radioanal. Nucl. Chem.* **271** (2007).

4. Khrbish, Y. S., Abugassa, I. O., Benfaid, N., and Bashir, A. A., *J. Radioanal. Nucl. Chem.* **271**:63 (2007).

5. Hancock, R. G. V., Pidruczny, A. E., and Johnson, J. M., *J. Radioanal. Nucl. Chem.* **271**:203 (2007).

6. Tsipenyuk, Y. M., *J. Radioanl. Nucl. Chem.* **271**:107 (2007).

7. Moon, J. H., Kang, S. H., Chung, Y. S., and Lee, O. K., *J. Radioanal. Nucl. Chem.* **271**:155 (2007).

8. Miyamoto, Y., Saito, Y., Magara, M., Sakurai, S., and Usuda, S., *J. Radioanal. Nucl. Chem.* **271**:83 (2007).

9. Menezes, M. A. B., Maia, E. C. P., Albinati, C. C. B., and Amaral1, A. M., *J. Radioanal. Nucl. Chem.* **271**:119 (2007).

10. Freitas, M. C., and Pacheco, A. M. G., *J. Radioanal. Nucl. Chem.* **271**:185 (2007).

11. Hofstetter, K. J., Beals, D. M., Halverson, J., Villa-Aleman, E., and Hayes, D. W., *J. Radioanal. Nucl. Chem.* **248**:683 (2001).

12. Giordani, L., Rizzio, E., and Brandone, A., *J. Radioanal. Nucl. Chem.* **263**:739 (2005).

13. Beeley, P. A., Heimlich, M. S., Edward, J. B., Bennet, L. G. I., and Page, J. A., *J. Radioanal. Nucl. Chem.* **169**:453 (1993).

14. Aleklett', K., Lilienzin, J-O., and Loveland, W., *J. Radioanal. Nucl. Chem.* **193**:187 (1995).

15. Aarnio, P. A., Nikkinen, M. T., and Routti, J. T., *J. Radioanal. Nucl. Chem.* **193**:179 (1995).

16. Révay, Z. S., Belgya, T., and Molnár, G. L., *J. Radioanal. Nucl. Chem.* **265**:261 (2005).

17. http;//ww.canberra.com

18. http://www.ortec-online.com

19. Abugassa, I. O., Khrbish, Y. S., Abugassa, S. O., Ben Faid, N., Bashir, A. T., and Sarmani, S., *J. Radioanal. Nucl. Chem.* **271**:27 (2007).

20. De Corte, F., *J. Radioanal. Nucl. Chem.* **271**:37 (2007) (and references therein).

21. Naqvi, A., Nagadi, M. M., and Al-Amoudi, O. S. B., *J. Radioanal. Nucl. Chem.* **271**:151 (2007).

22. Alvarez, E., Biegalski, S. R., and Landsberger, S., *Nucl. Instrum. Meth. Phys. Res. B* **262**:333 (2007).

23. Harrison, R. K., and Landsberger, S., *Nucl. Instrum. Meth. Phys. Res. B* **267**:513 (2009).

24. Révay, Z. S., Harrison, R. K., Alvarez, E., Biegalski, S. R., and Landsberger, S., *Nucl. Instrum. Meth. Phys. Res. A* **577**:611 (2007).

25. Whitney, S. M., Downing, R. G., Biegalski, S., and O'Kelly, D. S., *J. Radioanal. Nucl. Chem.* **276**:257 (2008).

26. Çetiner, S. M., Ünlü, K., and Downing, D. G., *J. Radioanal. Nucl. Chem.* **276**:623 (2008).

27. Çetiner, S. M., Ünlü, K., and Downing, D. G., *J. Radioanal. Nucl. Chem.* **271**:275 (2007).

28. http://www.me.utexas.edu/~nuclear/index.php/netl/netl-experiment-facilities.

29. Cnudde, V., Dierick, M., Vlassenbroeck, J., Masschaele, B., Lehmann, E., Jacobs, P., and Van Hoorebeke, L., *Nucl. Instrum. Meth. Phys. Res. B* **266**:155 (2008).

30. Hausladen, P. A., Bingham, P. R., Neal, J. S., Mullens, J. A., and Mihalczo, J. T., *Nucl. Instrum. Meth. Phys. Res. B* **261**:387 (2007).

31. Sugimoto, K., Kamata, Y., Yoshida, T., Asano, H., Murakawa, H., Takenaka, N., and Mochiki, K., *Nucl. Instrum. Meth. Phys. Res. A* **605**:200 (2009).

32. Nakamura, M., Sugimoto, K., Asano, H., Murakawa, H., Takenaka, N., and Mochiki, K., *Nucl. Instrum. Meth. Phys. Res. A* **605**:204 (2009).

16

Health Physics Fundamentals

16.1 INTRODUCTION

Health physics is the discipline that consists of all the activities related to the protection of individuals and the general public from potentially harmful effects of ionizing radiation. Ionizing radiation comes from two sources:

1. Natural or background radiation which is radiation emitted by radioisotopes that exist on or inside the earth, in the air we breathe, in the water we drink, in the food we eat, and in our bodies, as well as radiation incident upon the earth from outer space (cosmic rays). Humans have been exposed to this natural radiation for as long as they have existed on this planet.

2. Man-made radiation which is radiation emitted by radioisotopes that have been produced by nuclear reactors and accelerators, radiation produced by machines used in medical installations for diagnostics and/or therapy (e.g., X-ray machines, accelerators, positron emission tomography for imaging of organs etc.).

Health physics is concerned with protection of people from radiation. Since the background radiation has been, is, and will always be on our planet at about the same level everywhere, there is not much a health physicist can do to protect individuals or populations from background radiation. Hence, health physics is concerned with protection of people from natural and technologically produced radiation.

A health physicist performs many tasks. He or she, most importantly,

1. Is responsible for the detection and measurement of radiation in areas of work and in the environment

2. Is responsible for the proper operation and calibration of detection instruments

3. Is responsible to design and implement shielding for radiation protection

4. Inspects at regular intervals the facilities where radiation sources are used

5. Enforces federal and state regulations dealing with proper handling of radiation sources and establishment of acceptable levels of radiation fields at places of work

6. Keeps records of exposure for all individuals under his or her jurisdiction

7. Knows how to clean areas that have been contaminated with radioactive materials

8. Acts as the liaison representative between the regulatory agencies and his or her organization

Although the term health physics was coined after 1940, and a Health Physics Society was established in 1955, the concern about the harmful effects of radiation had been born much earlier—but probably not early enough. The first recorded radiation damage case occurred in 1896, only a year after the discovery of X-rays, yet the first limits concerning X-ray exposure were set in the 1920s. Today, both national and international groups exist that act as advisory bodies* to the appropriate regulatory agencies.

Since improperly handled radiation may produce deleterious effects to humans, it is important that individuals who use radiation sources learn the fundamentals of dosimetry, definition of dose units, biological effects of radiation, standards for radiation protection, and operation of health physics instruments. This chapter briefly discusses all these items. If more detailed treatment of these topics is needed, consult the bibliography and references given at the end of the chapter.

Although the SI units for radiation are the accepted norms, we have included the rem and mrem units as well, both for historical purposes and the fact that in the United States these older terms are still widely used in academic institutions and the nuclear industry.

---

* International Commission on Radiological Units and Measurements (ICRU); International Commission on Radiological Protection (ICRP); in the United States, the National Council on Radiation Protection and Measurements (NCRP).

16.2 UNITS OF EXPOSURE AND ABSORBED DOSE

Protection of individuals against radiation necessitates the completion of two tasks:

1. Development of safe radiation exposure limits including modeling of radiation interactions with biological systems

2. Construction of instruments that measure the intensity of radiation

Neither of these tasks can be accomplished without the means of quantitative description of radiation, that is, without defining radiation units.

The radiation effect is measured in terms of exposure or dose. Exposure is defined as charge released per unit mass of air. Dose is defined as energy absorbed per unit mass of material. The first radiation unit to be defined was the *roentgen* (symbol R):

1 R = exposure due to X-rays or gamma-rays of such intensity that the electrons produced by this radiation in 1 cm³ of dry air, at standard temperature and pressure, generate along their tracks electron–ion pairs carrying a total charge of 1 electrosurgical unit (esu) of either sign

The SI unit of exposure is defined as 1 C/kg air, without any new name proposed for it. Numerically,

$$1 \text{ R} = 2.58 \times 10^{-4} \text{ C/kg air}$$

The roentgen suffers from two limitations:

1. It was defined in terms of electromagnetic radiation only.

2. It was defined in terms of air only.

Radiation protection may involve other types of radiation, and media other than air. For this reason, another unit was defined called the *radiation absorbed dose* or *rad*, defined as

$$1 \text{ rad} = 100 \text{ erg/g}$$

The SI unit of absorbed dose is the *Gray* (Gy), defined as

$$1 \text{ Gy} = 1 \text{ J/kg} = 100 \text{ rad}$$

The rad (or the Gy) has a simple definition and is a unit independent of both type of radiation and material. But the measurement of absorbed dose in terms of rad (or Gy) is neither simple nor straightforward, because it is very difficult to measure energy deposited in a certain mass of tissue. Fortunately, one can bypass this difficulty by measuring energy deposited in air, which is proportional to the exposure, and then relate it to the absorbed dose.

The measurement of exposure is achieved by using ionization chambers, and the result is given in roentgens. Based on the definitions of the roentgen, the following relationship can be established between roentgens and rads.

$$1 \text{ R} = \frac{1 \text{ esu}}{1.293 \times 10^{-3} \text{ g}} (2.082 \times 10^{9}) \text{ ion pairs/esu}$$
$$\times (34 \text{ ev/pair}) 1.602 \times 10^{-12} \text{ ergs/eV}$$
$$= 88 \text{ ergs/g} = 0.88 \text{ rad} = 8.8 \text{ mGy}$$

If D is the absorbed dose in air, and X is the exposure in air, the relationship between the two is

$$D = 0.88X \tag{16.1}$$

For media other than air, the relationship is obtained as follows. The absorbed dose rate in material i is (in terms of energy deposited per unit mass per unit time)

$$\dot{D}_i = \varphi \left[\text{part.}/(\text{cm}^2) \right] \mu_{a,i} (\text{m}^2/\text{kg}) E (\text{J/part.}/\text{cm}^2) \tag{16.2}$$

The absorbed dose rate in air is

$$\dot{D}_{air} = \varphi \left[\text{part.}/(\text{cm}^2) \right] \mu_{a,air} (\text{m}^2/\text{kg}) E(\text{J}/\text{part.}) \tag{16.3}$$

The ratio of Eqs. 16.2 to 16.3 gives

$$\dot{D}_{a,i} = \frac{\mu_{a,i}}{\mu_{a,air}} = \dot{D}_{a,i} = \frac{\mu_{a,i}}{\mu_{a,air}} (0.88) X_{air} \tag{16.4}$$

Equations 16.1 and 16.4 express the fact that the measurement of absorbed dose* is a two-step process:

1. Exposure (or exposure rate) is measured.

2. Absorbed dose (or dose rate) is calculated from the measured exposure using Eq. 16.4.

In practice, the instruments that measure radiation dose are, usually, properly calibrated to read rad or Gy.

16.3 THE RELATIVE BIOLOGICAL EFFECTIVENESS—THE DOSE EQUIVALENT

The units of absorbed dose defined in the previous section are quite adequate for the quantitative assessment of the effects of radiation to inanimate objects, like irradiated transistors or reactor fuel. For protection of people, however, the important thing is not the measurement of energy deposited—that is, the absorbed dose—but the biological effects due to radiation exposure. Unfortunately, biological effects and absorbed dose do not always have one-to-one correspondence, and for this reason a new unit had to be defined: a unit that takes into account the biological effects of radiation.

The ideal unit for the measurement of biological effect should be such that a given dose, measured in that unit, produces a certain biological effect regardless of the *type and energy of radiation* and also regardless of the *biological effect* considered. Unfortunately, such a unit cannot be established because of the different modes by which radiation deposits energy in tissue, the intricate way by which the energy deposition is related to a given biological effect, and the complexity of biological organisms. An ideal unit may not exist, but some unit that "equalizes" biological effects had to be defined.

The first step toward that task was the introduction of a factor called the relative biological effectiveness (RBE), defined as

$$\text{RBE}_i = \frac{\left[\begin{array}{c} \text{absorbed dose from X-ray or gamma radiation (200–300 keV)} \\ \text{producing a certain biological effect} \end{array} \right]}{\left[\begin{array}{c} \text{absorbed dose from radiation type } i \\ \text{producing the same biological effect} \end{array} \right]} \tag{16.5}$$

In understanding the meaning of RBE, note the following:

1. RBE is defined in terms of photons; therefore, it follows that RBE = 1 for electromagnetic radiation. Also, although the definition of RBE specifies the energy of the photons to be 200–300 keV, RBE is taken as equal to 1 for photons of all energies.

2. A given type of radiation does not have a single RBE, because RBE values depend on the energy of the radiation, the cell, the biological effect being studied, the total dose, dose rate, and other factors.

3. It is a well-known fact that the biological damage increases as the energy deposited per unit distance, the *linear energy transfer* (LET), increases. Thus, heavier particles (alphas, heavy ions, fission fragments) are, for the same absorbed dose, more biologically damaging than photons, electrons, and positrons.

---

* Equations 16.2 and 16.3 give dose rate, not dose; the meaning of Eq. 16.4, however, is the same if one uses either dose or dose rate.

In 1963, the International Commission on Radiological Units and Measurements (ICRU) proposed the replacement of RBE by a new factor named the *quality factor* (QF). Here is an excerpt from their recommendation.

In radiation protection it is necessary to provide a factor that denotes the modification of the effectiveness of a given absorbed dose by LET (Linear Energy Transfer). Unlike RBE, which is always experimentally determined, this factor *must* be *assigned* on the basis of a number of considerations and it is recommended that it be termed the *quality factor* (QF). Provisions for other factors are also made. Thus a *distribution factor* (DF) may be used to express the modification of biological effect due to nonuniform distribution of internally deposited radionuclides. The product of absorbed dose and modifying factors is termed the dose equivalent, (*H*).

In 1973 the ICRU[1] recommended dropping the "F" from QF, a suggestion that has now become practice. In 1977 the ICRP[2] recommended that the dose equivalent (*H*) at a point in tissue be written as

$$H = NQD \tag{16.6}$$

where Q = quality factor
D = absorbed dose
N = product of all the modifying factors. The suggested value of N is 1.

RBE is now used only in radiobiology, whereas Q is used in radiation protection. A detailed discussion of similarities and differences between the two factors is given in Reference 3. For the radiations and energy ranges considered in this book, RBE and Q are practically the same, and from this point on, only the factor Q will be mentioned. Table 16.1 gives Q values for various radiations commonly encountered.

When the unit of absorbed dose is multiplied by the corresponding Q value, the unit of dose equivalent (*H*) is obtained. The *H* units are

$$1 \text{ rem} = Q \times 1 \text{ rad}$$

and the SI unit

$$1 \text{ Sievert (Sv)} = Q \times 1 \text{ Gy}$$

Thus

$$1 \text{ Sv} = 100 \text{ rem}$$

Because it is only the dose equivalent that equalizes biological effects from different types and energy of radiation, *only Sv (or rem) should be added*, not Gy (or rad); Gy or rad may be added if the dose is due to two or more sources of the same radiation type.*

Table 16.1
Quality Factors for Several Types of Radiation

Radiation Type	Q	Radiation Type	Q
γ-rays	1	Neutrons	
X-rays	1	Thermal	3
Beta particles	1	0.005 MeV	2
Electrons	1	0.02 MeV	5
Positrons	1	0.10 MeV	7.5
Protons ($E < 14$ MeV)	10	0.50 MeV	11
Alpha particles		1.00 MeV	11
($E < 10$ MeV)	20	5.0 MeV	8
Recoil nuclei ($A > 4$)	20	10 MeV	6.5

Sources: From *1977 Recommendations of the International Commission on Radiological Protection* (ICRP Publication 26), Pergamon, 1977; and Standards for Protection against Radiation, *Code of Federal Regulations*, Title 10, Part 20, December 1992, Chapter 1.

---

* The ICRP approved in 1990 (ICRP 60; 1991) a change from "dose equivalent" to "equivalent dose" and instead of using the quality factors "Q" to use "tissue weighting factors." As of 2010, the US NRC has not implimented these changes yet.

Example 16.1   At the open beam port of a research reactor, the absorbed dose rate consists of 10 mrad/h due to gammas, 10 mrad/h due to fast neutrons, and 6 mrad/h due to thermal neutrons. What is the total dose a person will receive by standing in front of the beam for 5 s?

Answer   Calculate the dose equivalent H, as shown below:

From	Abs. Dose Rate (mrad/h)	Q	$\dot{H}$ (mrem/h)	$\dot{H}$ (mSv/h)
Gammas	10	1	10	0.1
Fast neutrons	10	10	100	1.0
Thermal neutrons	6	2	12	0.12
Total			122	1.22

The total dose received by the individual is

$$H = 122 \ \text{mrem/h}\left(\frac{5}{3600}\right) = 0.17\,\text{mrem} = 1.7 \times 10^{-6}\ \text{Sv}$$

16.4 DOSIMETRY FOR RADIATION EXTERNAL TO THE BODY

The general dosimetry problem is defined as follows: Given the intensity of the radiation field at a certain point in space, calculate the dose rate received by an individual standing at that point. The radiation field, outside the body, is assumed to be known in terms of the type, energy, and number of particles involved. The calculation that follows disregards the possible perturbation of the field from the presence of the human body. The calculation is different for charged particles, photons, and neutrons.

16.4.1 Dose due to Charged Particles

Consider a point in space where it is known that the charged-particle radiation field is given by

$\phi(E)\ dE$ = charged particles per m² s with kinetic energy between E and $E + dE$

A person exposed to this field will receive a radiation dose because of energy deposited by these charged particles. The dose equivalent rate is given by

$$\dot{H} = \int_E dE\varphi(E)\left[\text{part.}/(\text{m}^2\text{s})\right]\left(\frac{dE}{dx}\right)(\text{MeV}/m)\,\frac{Q(E)}{\rho}\,(\text{kg}/\text{m}^3) \qquad (16.7)$$

where dE/dx = stopping power of tissue for particles of energy E
$\qquad\quad \rho$ = density of tissue
$\qquad\ Q(E)$ = quality factor for particles of energy E

The units of Eq. 16.7 are MeV/(kg s). To obtain the result in Sv/s, one needs to transform MeV to J (1 MeV = 1.602 × 10⁻¹³ J).

Most of the time in practice, the radiation field is computed not as an analytic function $\phi(E)$ but as an energy group distribution, where

$$\phi_g = \int_{E_g}^{E_{g-1}} \phi(E)\,dE = \text{number of particles per m}^2\text{ s with energy between } Eg \text{ and } Eg\ 1\ (Eg < Eg\ 1)$$

Using the multigroup structure, Eq. 16.7 takes the form

$$\dot{H} = \sum_{g=1}^{G} \varphi_g \left(\frac{dE}{dx}\right)_g \frac{Q_g}{\rho}\ (\text{Sv}/\text{s}) \qquad (16.8)$$

429

where G is the total number of energy groups and Q_g is the average quality factor for group g. In principle, Eqs. 16.7 and 16.8 are valid for any charged-particle flux that hits a human body from the outside. In practice, for the particles and energies considered here, these equations are useful for electron and beta beams only, since alphas with $E < 10$ MeV do not penetrate the human skin. The dose from external beams of betas will be confined to a depth in tissue equal to the range of these particles.

The division between electrons and betas is necessary (although beta particles are electrons) because an electron beam consists of monoenergetic electrons; a beam of beta particles consists of electrons emitted by the beta decay of a nucleus. Therefore, as explained in Chapter 3, the beta particles have an energy spectrum with a maximum energy E_{max} and an average energy $1/3E_{max}$.[*]

To calculate the dose rate from an electron or a beta beam, one can use Eq. 16.7 or Eq. 16.8 with $Q(E) = 1$. In practice, the actual calculation is shortened by using tables that provide flux-to-dose-rate conversion factors (Table 16.2). In terms of the flux-to-dose-rate conversion factors, the dose rate is written as

$$\dot{H} = \int_E C(E)\varphi(E)\, \mathrm{d}(E) \tag{16.9}$$

or, in terms of energy groups,

$$\dot{H} = \sum_{g=1}^{G} C_g \varphi_g \tag{16.10}$$

with the group conversion factor defined by

$$C_g = \frac{1}{E_{g-1} - E_g} \int_{E_g}^{E_{g-1}} C(E)\mathrm{d}E \tag{16.11}$$

Table 16.2
Flux-to-Dose-Rate Conversion Factors for Electrons and Betas

	Electrons			Betas	
E_{max} (MeV)	(Sv/s)/ [Particles/ (m² s)]	(mrem/h)/ [Particles/ (cm² s)]	E (MeV)$_{max}$	(Sv/s)/ [Particles/ (m² s)]	(mrem/h)/ [Particles/ (cm² s)]
0.02	2.104 – 13[a]	0.758	0.2	1.160 – 13	0.417
0.100	0.649 – 13	0.234	0.4	0.772 – 13	0.278
0.200	0.440 – 13	0.158	0.6	0.578 – 13	0.208
0.300	0.369 – 13	0.133	0.8	0.433 – 13	0.156
0.400	0.337 – 13	0.121	1	0.386 – 13	0.119
0.600	0.309 – 13	0.111	1.5	0.330 – 13	0.119
0.800	0.297 – 13	0.107	2	0.317 – 13	0.114
1	0.293 – 13	0.105	3	0.303 – 13	0.109
2	0.297 – 13	0.107			
3	0.303 – 13	0.109			
5	0.320 – 13	0.115			
7	0.324 – 13	0.118			
10	0.342 – 13	0.123			

Source: From *Engineering Compendium on Radiation Shielding*, vol. 1: *Shielding Fundamentals and Methods*, Springer-Verlag, 1968.
[a] Read as 2.104×10^{-13}.

---

[*] A more accurate equation for average beta energy is given in Reference 6.

16.4.2 Dose due to Photons

The dose rate due to a beam of photons is calculated based on an equation similar to Eq. 16.7:

$$\dot{H} = \int_0^\infty dE\,\varphi(E)\,E\mu_a(E)\,(Sv/s)$$ (16.12)

where $\mu_a(E)$ (m²/kg) = energy absorption coefficient in tissue for a photon of energy E.

Notice the two main differences between Eq. 16.7 and Eq. 16.12. For photons, $Q(E) = 1$ and dE/dx is replaced by the product $E\mu_a(E)$. As with charged particles, the analytic form of $\phi(E)$ and $\mu_a(E)$ is seldom known. Instead, one has to work with a set of energy groups, and Eq. 16.12 takes the form

$$\dot{H} = \sum_g \phi_g\, E_g\, \mu_a^{tiss}(E_g)\,(Sv/s)$$ (16.13)

where ϕ_g and E_g have the same meaning as before (see 16.4.1).

As with charged particles, tables have been developed that provide a flux-to-dose-rate conversion factor as a function of photon energy (Table 16.3 and Figure 16.1). The units of the conversion factor given in Table 16.3 are (Sv/s)/(gamma/m²·s). To convert to (rem/h)/(gamma/cm²·s) multiply by (100) (3600)(10⁴) which equals 3.6×10^9. Using the conversion factor $C(E)$ Eqs. 16.12 and 16.13 take the form

$$\dot{H} = \int_0^\infty dE\,\varphi(E)\,C(E)$$ (16.14)

$$\dot{H} = \sum_g \varphi_g\, C_g$$ (16.15)

Table 16.3
Gamma-Ray Flux-to-Dose-Rate Conversion Factors

Photon Energy (MeV)	(Sv/s)/[Particles/(m² s)]
0.01	2.20 – 18
0.015	5.70 – 18
0.02	9.12 – 18
0.03	1.38 – 17
0.05	1.80 – 17
0.08	2.20 – 17
0.10	2.37 – 17
0.15	4.36 – 17
0.20	6.02 – 17
0.30	9.49 – 17
0.40	1.30 – 16
0.50	1.64 – 16
0.60	1.98 – 16
0.80	2.64 – 16
1.0	3.27 – 16
1.5	4.68 – 16
2.0	5.93 – 16
3.0	8.19 – 16
4.0	1.02 – 15
5.0	1.21 – 15
6.0	1.40 – 15
8.0	1.78 – 15
10.0	2.16 – 15

Source: Annals of the ICRP, Data for Use in Protection of the International Commission on Radiological Protection, Publication, Vol. 17, No 2/3 Permagon Press, 1987.

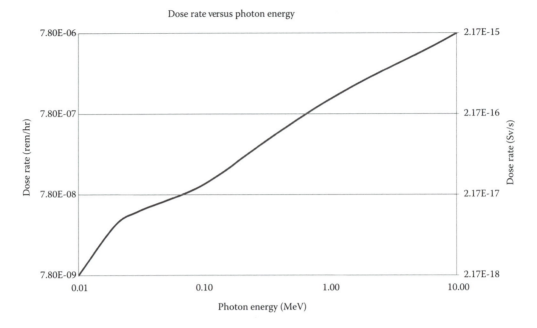

Dose rate versus photon energy

Figure 16.1 Photon flux-to-dose factors from 0.01 to 10 MeV. (From Neutron and Gamma-Ray Flux-to-Dose Rate Factors, American National Standard, ANSI/ANS-6.1.1, 1991.)

Example 16.2 What is the dose rate at 1 m away from 3.7×10^{10} Bq (1 Ci) of ^{137}Cs, if (a) the attenuating medium is water or (b) the attenuating medium is air? ^{137}Cs emits a 0.662-MeV gamma 0.85% of the time.

Answer In both cases the dose rate is calculated by using Eq. 16.15 with a dose-to-flux conversion factor obtained from Table 16.3 for 0.662-MeV gammas.

Using linear interpolation, that factor is 2.185×10^{-16} (Sv/s)/[γ/(m² · s)].

The flux at r meters from the source is given by $BSe^{-\mu r}/4\pi r^2$,

where B = buildup factor (from Appendix E)
　　　S = source strength
　　　μ = total attenuation coefficient for 0.662-MeV gammas in air or water

From Appendix E, the buildup factor is

$$B = 1 + a\,\mu r\,[\exp(b\,\mu r)]$$

where (using linear interpolation) $a = 1.96$ and $b = 0.054$.

(a) For water, the value of the total attenuation coefficient is (from Appendix D, using linear interpolation)

$$\mu = 0.00861\,\mathrm{m^2/kg} = 0.0861\ \mathrm{cm^2/g}$$

The number of mean free paths (mfp) in water is

$$\mu r = (0.00861\,\mathrm{m^2/kg})\,(10^3\ \mathrm{kg/m^3})\,(1\,\mathrm{m}) = 8.61\,\mathrm{mfp}$$

Thus,

$$B\,(\text{water}) = 1 + (1.96)\,(8.61)\,[\exp(8.61 \times 0.054)] = 27.86$$

The dose rate in water is

$$\dot{H} = \dot{D} = (27.86)\,(3.7 \times 10^{10})\,(\times\,0.85)\left[\frac{e^{-8.61}}{4\pi\,(1^2)}\gamma/m^2\cdot s\right]$$

$$\times(2.2\times10^{-16}\,(Sv/s)/[\gamma/m^2\cdot s])$$

$$= 3.07\times10^{-9}\,Sv/s = 1.01\,mrem/h$$

(b) For air, the value of the total attenuation coefficient is (from Appendix D, using linear interpolation)

$$\mu = 0.0082\,m^2/kg = 0.082\,cm^2/g$$

$$\mu r = (0.0082\,m^2/kg)\,(1.29\,kg/m^3)\,(1m) = 0.01\,mfp$$

The buildup factor is

$$B\,(air) = 1 + (1.96)(0.01)[exp(0.01 \times 0.054)] = 1.02$$

The dose rate in air is

$$\dot{H} = \dot{D} = (1.02)\,(3.7\times10^{10})\,(\times\,0.85)\left[\frac{e^{-0.01}}{4\pi\,(1^2)}\gamma/m^2\cdot s\right]$$

$$\times(2.2\times10^{-16}\,(Sv/s)/[\gamma/m^2\cdot s])$$

$$= 0.57\times10^{-9}\,Sv/s = 0.20\,rem/h$$

As pointed out in Section 4.8.6, where buildup factors are defined and discussed in detail, the value of the buildup factor is significantly greater than 1 if the distance in mfps is significantly greater than 1. In Example 16.2 the distance in water is 8.61 mfp and B = 27.86, while in air the distance is 0.01 mfp and B = 1.02.

The buildup factor constants given in Appendix E apply to a point isotropic source in an infinite medium (Example 16.2 is such a case). The same constants can be used, however, in other geometries if no better values are available.

For example, one could use the constants given in Appendix E to calculate the dose rate from a point isotropic source in a semi-infinite medium or from a point isotropic source located behind a slab shield. In such cases, the use of the constants from Appendix E results in an overestimate of the buildup factor. Buildup factors for many different geometries are given in Reference 5.

16.4.3 Dose due to Neutrons

Neutrons hitting the human body deliver energy to it through elastic and inelastic collisions with nuclei, and through secondary radiation emitted by the radioisotopes produced after neutrons are captured.

If an individual is exposed to fast neutrons, most of the energy transfer takes place through elastic collisions with hydrogen (~90%) and, to a lesser extent, through collisions with oxygen and carbon nuclei (the average neutron energy loss per collision with hydrogen [proton] is 50% of the incident neutron energy; the corresponding fractions for carbon and oxygen are 14% and 11%). These "recoil" nuclei are charged particles, which lose their energy as they move and slow down in tissue. This is true for neutron energies down to about 20 keV. When the neutron energy reaches or becomes lower than a few keV, the importance of elastic collisions decreases, and the reaction $^{14}N(n,\,p)^{14}C$ produces more significant effects. As discussed in Chapter 14, this is an exothermic reaction producing protons with kinetic energy of 584 keV. Radioactive ^{14}C is also produced, emitting betas with a maximum energy of 156 keV. The biological damage comes mainly from the protons, not from the betas of ^{14}C.

Thermal neutrons are absorbed in the body mainly through the reaction $^{1}H(n,\,\gamma)^{2}H$, which results in the emission of a 2.2-MeV gamma. A reaction of secondary importance is $^{23}Na(n,\,\gamma)^{24}Na$.

The isotope ^{24}Na has a 15-h half-life and emits two energetic gammas with energy 1.37 and 2.75 MeV. Thus, when thermal neutrons are absorbed, damage is caused by the energetic gammas that are produced as a result of the neutron capture.

The general equation for the dose rate has the form

$$H(r, E) = \sum_{i=1}^{M} \phi(r,E) \left[\sum_{s}^{i}(E) \frac{2A_i E}{(A_i + 1)^2} + \sum_{\gamma}^{i} f_{\gamma}^{i} E_{\gamma}^{i} + \sum_{q}^{i} Q_i + \cdots \right] Q(E) \, (\text{Sv/s}) \tag{16.16}$$

where $\phi(r, E)$ = neutron flux $[n/(m^2 \, s)]$ at point r, of neutrons with energy E
$\sum^{i}(E)$ = macroscopic cross sections, for neutrons of energy E for elastic scattering, capture, charged-particle-producing reactions, etc., for isotope i
M = total number of isotopes present
f_{γ} = fraction of gamma energy deposited at the capture site
Q_i = the Q value of the charged-particle reaction; all Q_i are assumed to be deposited at site of the reaction
$Q(E)$ = quality factor for neutrons of energy E

Equation 16.16 neglects inelastic scattering, which is negligible for neutrons in tissue for the energies considered here. If neutrons of many energies are present, the calculation should be repeated for all energies, and the results summed to give the total dose rate. Flux-to-dose-rate conversion factors have been developed for neutrons as well (Table 16.4, Figures 16.2 and 16.3). Using the conversion factor, Eq. 16.16 takes the form

$$\dot{H}(r,E) = \varphi(r,E)C(E) \tag{16.17}$$

If the neutron spectrum is known in terms of energy groups, Eq. 16.17 becomes

$$\dot{H}(r) = \sum_{g} C_g \varphi_g(r) \tag{16.18}$$

Although Eq. 16.16 is not normally used for everyday dose calculations, it is instructive to present it so that the reader may comprehend the various contributors to the neutron dose.

Table 16.4
Neutron Flux-to-Dose-Rate Conversion Factors

Neutron Energy (MeV)	(Sv/s)/[Neutrons/(m² s)]	(rem/h)/[Neutrons/(cm² s)]
2.5 – 08[a]	1.02 – 15	3.67 – 06
1.0 – 07	1.02 – 15	3.67 – 06
1.0 – 06	1.23 – 15	4.44 – 06
1.0 – 05	1.23 – 15	4.44 – 06
1.0 – 04	1.19 – 15	4.28 – 06
1.0 – 03	1.02 – 15	3.67 – 06
1.0 – 02	9.89 – 16	3.56 – 06
1.0 – 01	5.89 – 15	2.12 – 05
5.0 – 01	2.56 – 14	9.23 – 05
1.0	3.69 – 14	1.33 – 04
2.5	3.44 – 14	1.24 – 04
5.0	4.33 – 14	1.56 – 04
7.0	4.17 – 14	1.50 – 04
10.0	4.17 – 14	1.50 – 04
14.0	5.89 – 14	2.12 – 04
20.0	6.25 – 14	2.25 – 04

Source: From Standards for Protection against Radiation, *Code of Federal Regulations*, Title 10, Part 20, December 1992, Chapter 1.
[a] Read as 2.5×10^{-8}.

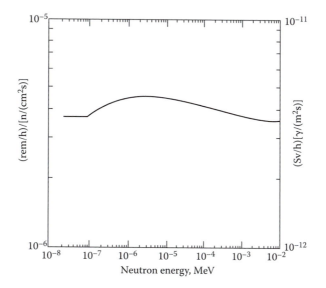

Figure 16.2 Neutron flux-to-dose-rate factors for energies 10^{-8}–10^{-2} MeV. (From Neutron and Gamma-Ray Flux-to-Dose Rate Factors, American National Standard, ANSI/ANS-6.1.1, 1991.)

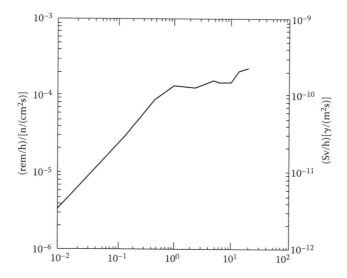

Figure 16.3 Neutron flux-to-dose-rate factors for energies 10^{-2} to 20 MeV. (From Neutron and Gamma-Ray Flux-to-Dose Rate Factors, American National Standard, ANSI/ANS-6.1.1, 1991.)

Example 16.3 At the open beam port of a research reactor, the neutron flux at a certain power level consists of 1.6×10^8 neutrons/(m² · s) with energy 100 keV, and 3.5×10^9 neutrons/(m² · s) with an average energy of 0.025 eV. What is the total dose rate at that point?

Answer Using Table 16.4, the dose rate is

$$\dot{H} = (1.6 \times 10^8)\,(5.89 \times 10^{-15}) + (3.5 \times 10^8)\,(1.02 \times 10^{-15})$$
$$= 1.30 \times 10^{-6}\,\text{Sv/s} = 0.467\ \text{rem/H}$$

435

16.5 DOSIMETRY FOR RADIATION INSIDE THE BODY

16.5.1 Dose from a Source of Charged Particles Inside the Body

If the charged particles (e, p, α) are created or deposited inside the body, the calculation of the dose is easier because the range of the particles considered (energy less than 10 MeV) is millimeters or less and all the energy is deposited in a very small volume. (In the case of electrons, a fraction of the energy escapes as bremsstrahlung, but it represents a small correction; neglecting bremsstrahlung, one obtains a conservative answer.) The dose rate equivalent is given in this case by

$$H = \frac{\int_0^\infty ES(E)Q(E)dE}{(\text{mass in which the particle energy was deposited})} \quad (\text{Sv/s}) \qquad (16.19)$$

where $S(E)\, dE$ = number of particles emitted per second (activity) with energy between E and $E + dE$.

If the particle spectrum is known in multigroup form, Eq. 16.19 becomes

$$H = \sum_g E_g S_g Q_g / \text{mass} \qquad (16.20)$$

with

$$S_g = \int_{E_g} dES(E)$$

If the charged-particle source is localized, that is, it can be considered a point isotropic source, the mass in the denominator of Eq. 16.20 is equal to

$$(\text{Mass where energy was deposited}) = \frac{4}{3}\pi R^3 \rho$$

where R = range of charged particle in tissue
ρ = density of tissue

If the source is deposited in an organ, for example, liver, thyroid, or spleen, then the mass in the denominator of Eq. 16.20 is the mass of that organ and the result of this calculation is the average dose rate for this organ. By using the mass of the organ, the tacit assumption is made that all the energy emitted by the radioactive source is absorbed in that volume. It is a conservative estimate, since some particles will be borne very close to the surface of the organ and escape from it after depositing only part of their energy there.

Example 16.4   What is the dose rate from 1 pCi of an alpha source emitting 6-MeV alphas in tissue?

Answer   The range of this alpha particle in tissue 4.7×10^{-5} m. Thus,

$$\dot{H} = \frac{(3.7 \times 10^{-2}\,\alpha/s)(6\,\text{MeV}/\alpha)(1.602 \times 10^{-13}\,\text{J/MeV})(20)}{(4/3)\pi(4.7 \times 10^{-5})^3(10^3\,\text{kg/m}^3)}$$

$$= 1.64 \times 10^{-3}\,\text{Sv/s} = 589\,\text{rem/h}$$

This is an extremely large dose rate, the result of the energy being deposited in a very small volume.

Example 16.5   What is the dose rate due to the alphas of Example 16.4 if it is known that the source is uniformly distributed in the lungs?

Answer   In this case, the mass affected is that of the lungs, which is (for a 70-kg person) about 1 kg. The

$$H = \frac{(3.7 \times 10^{-2})(6)(1.602 \times 10^{-13}\,\text{J})(20)}{(1)}$$

$$= 7.11 \times 10^{-13}\,\text{Sv/s} = 2.56 \times 10^{-7}\,\text{rem/h}$$

Example 16.6   What is the dose rate to the thyroid gland due to the betas emitted by 1 mCi of ^{131}I?

Answer   The isotope ^{131}I emits two betas, one with $E_{mas}^{(1)} = 0.608$ MeV, 85% of the time, and a second with $E_{mas}^{(2)} = 0.315$ MeV, 15% of the time. The range of these betas in tissue is about 2 mm and 0.9 mm, respectively.

The thyroid gland has a mass of about 0.025 kg (i.e., a volume of about 25 cm³); therefore, all the beta energy will be deposited in it.

The dose rate is obtained using Eq. 16.20:

$$\dot{H} = \frac{(3.7 \times 10^7)\,[0.85(0.608/3) + 0.15(0.315/3)](1.62 \times 10^{-13})}{(25 \times 10^{-3})}$$

$$= 4.46 \times 10^{-5} \text{ Sv/s} = 16.0 \text{ rem/h}$$

16.5.2 Dose from a Photon Source Inside the Body

Since photons have, essentially, an infinite range, the previous calculation for charged particles does not apply. A source of photons located anywhere in the body will deliver some dose to all the other parts of that body. The calculation of the dose rate proceeds as follows.

Consider an internal organ containing a uniform concentration of a radioisotope emitting a gamma with energy E at the rate of $S_v [\gamma/(\text{s m}^3)]$ inside the volume V_s, called the source volume (Figure 16.4). The dose rate received by another organ with volume V_T, called the target, is given by the expression

$$\dot{H} = \frac{\int_{V_s} \int_{V_t} (S_V dV_s / 4\pi r^2)\, e^{-\mu r} \left[\gamma/(\text{m}^2\, \text{s})\right] E(\text{MeV}/\gamma)\mu_a^{\text{tiss}}(\text{m}^2/\text{kg}) B(\mu r)\rho(\text{kg/m}^3) dV_T}{\rho\,(\text{kg/m}^3) V_T}$$

$$\times 1.602 \times 10^{-13}\, \text{J/MeV} \tag{16.21}$$

where μ = total linear attenuation coefficient in tissue for gammas of energy E

$\quad \mu_a^{\text{tiss}}$ = mass energy absorption coefficient in tissue for gammas of energy E

$B(\mu r) = B(\mu r, E)$ = buildup factor for gammas of energy E

$\quad \rho$ = density of tissue

Or, pulling out of the integral the quantities that are constant in space, Eq. 16.21 becomes

$$\dot{D} = \dot{H} = \frac{S_V}{4\pi} E \mu_a^{\text{tiss}} g \tag{16.22}$$

where the quantity

$$g = \frac{1}{V_T} \int_{V_T} dV_T \int_{V_S} dV_S \frac{e^{-\mu r}}{r^2} B(\mu r, E) \tag{16.23}$$

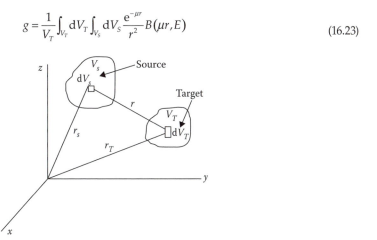

Figure 16.4   The geometry used in the calculation of the dose to a target organ (V_T) from a radioisotope uniformly distributed in another volume (V_s).

Table 16.5
Average Geometry Factors for Cylindrical Organs Containing a Uniformly Distributed Photon Source

Cylinder Height (cm)	Radius of Cylinder (cm)							
	3	5	10	15	20	25	30	35
2	17.5	22.1	30.3	34.0	36.2	37.5	38.6	39.3
5	22.3	31.8	47.7	56.4	61.6	65.2	67.9	70.5
10	25.1	38.1	61.3	76.1	86.5	93.4	98.4	103
20	25.7	40.5	68.9	89.8	105	117	126	133
30	25.9	41.0	71.3	94.6	112	126	137	146
40	25.9	41.3	72.4	96.5	116	131	143	153
60	26.0	41.6	73.0	97.8	118	134	148	159
80	26.0	41.6	73.3	98.4	119	135	150	161
100	26.0	41.6	73.3	98.5	119	136	150	162

Source: Hine, G. J., and Brownell, G., *Radiation Dosimetry*, Academic, New York, 1956.

is called the geometry factor. Note that the factor g has dimensions of length. Values of g have been calculated and tabulated (see Cember, 2008). The usefulness of g stems from the fact that g values can be calculated for a relatively small number of cases and then, by interpolation, other geometry factor values may be computed and used. Once g is known, the dose rate to an organ can be calculated from Eq. 16.22, and such calculation will have an uncertainty mainly from the value of g. There is a certain similarity in the use of g and the use of buildup factors. Buildup factors are also tabulated for a limited number of cases, and additional values are obtained by interpolation.

One common case utilizing the concept of the geometry factor is the calculation of the dose rate in an organ from a radioisotope deposited in that organ. For example, what is the dose rate to the thyroid from radioactive iodine given to a patient? If the organ is further assumed to be spherical, Eq. 16.23 takes the simple form

$$g = \int_0^R 4\pi r^2 dr \frac{e^{-\mu r}}{r^2} = \frac{4\pi}{\mu}(1 - e^{-\mu R}) \tag{16.24}$$

Using this value of g with Eq. 16.22 gives the dose rate at the center of the sphere. To obtain the average dose rate, an average value of g should be used. For a sphere the average value of the geometry factor is (see Cember, 2008)

$$\bar{g} = 0.75 \, (g)_{center} \tag{16.25}$$

Average geometry factors for cylindrical bodies are given in Table 16.5. An example of using geometry factors is given in Section 16.6.

The equations given above for the dose rate from photons are valid for monoenergetic sources. If the deposited radioisotope emits many discrete gammas or if a multigroup energy spectrum of the source is provided, the calculation should be repeated for all gammas (or groups) and the results added to obtain the total dose rate.

16.6 INTERNAL DOSE TIME DEPENDENCE—BIOLOGICAL HALF-LIFE

Radioisotopes may enter the body by inhalation, drinking, eating, injection, or through broken skin (wound). If the radiation source is inside the body, the exposure is internal and more damaging. No attenuation is provided by skin or clothes, and the person cannot walk away from the source. The exposure continues until the radioisotope decays completely or is excreted by the body.

A radioisotope is rejected by the body at a rate that depends upon the chemical properties of the element. All isotopes of the same element are rejected at the same rate, whether they are stable or not. For most radioisotopes, the rate of rejection is proportional to the amount of the isotope in the body. This leads to an exponential elimination law as a result of the combination of decay and rejection.*

---

* Other rejection laws have been proposed (References 8 and 9).

Let

> $N(t)$ = number of radioactive atoms at time t
> λ_R = radiological decay constant
> λ_B = biological decay constant
> = probability of rejection (by the body) per atom per unit time

The rate of change of $N(t)$ is $dN(t)/dt = -\lambda_R N(t) - \lambda_B N(t)$, with solution

$$N(t) = N(0)_e^{-(\lambda_R + \lambda_B)t} = N(0)e^{-\lambda_e t} \tag{16.26}$$

where $\lambda_e = \lambda_R + \lambda_B$ = effective decay constant.

A *biological half-life* is defined in terms of λ_B:

$$T_B = \frac{\ln 2}{\lambda_B} \tag{16.27}$$

and an *effective half-life* is defined by

$$T_e = \frac{T_B T_R}{T_B + T_R} = \frac{\ln 2}{\lambda_e} \tag{16.28}$$

The biological excretion rate of an element from the human body is not necessarily the same for the whole body and for a particular organ. In fact, in most cases, the biological elimination rates are different for different organs and for the body as the whole. For example, the biological half-life of iodine is 138 days for rejection from the thyroid, 7 days for the kidneys, 14 days for the bones, and 138 days for the whole body. For this reason, a table of biological and effective half-lives ought to include the organ of reference. Table 16.6 gives radiological, biological, and effective half-lives for certain common isotopes. The reader should remember that the biological half-life is the same for all isotopes of the same element, but the effective half-life is not.

If $T_B \gg T_R$, the decay removes the material much faster than the body rejects it. An example of such a case is ^{131}I, with T_B = 138 days (thyroid), T_R = 8 days, and T_e = 7.6 days. If $T_B \ll T_R$, the biological elimination is mainly responsible for the removal of the isotope. An example of such a case is tritium, with T_B = 12 days, T_R = 12 years, and T_e = 12 days.

As a result of the combined radioactive and biological elimination of a radioisotope from the whole body or from an organ, the dose rate to the body or the organ is not constant over time. Consider an amount of a certain radioisotope that delivers a dose rate equal to $H(0)$ at the time the radioisotope entered the body. If the effective half life of the isotope is T_e, the total dose delivered over a period of time T, taken as 50 years for regulatory limits, is

$$H_T = \int_0^T \dot{H}(0)e^{\lambda_e t}\, dt = \frac{\dot{H}}{\lambda_e}(1 - e^{\lambda_e T}) = \frac{\dot{H}(0)}{\ln 2}T_e(1 - e^{\ln 2(T/T_e)}) \tag{16.29}$$

If $T \gg T_e$, then

$$\dot{H}_T = \frac{\dot{H}(0)}{\ln 2}T_e \tag{16.30}$$

Example 16.7   What is the total dose received by an individual who drank, accidentally, 10^{-7} kg of ^{3}H$_2$O?

Answer   Assuming that the ^{3}H$_2$O is uniformly distributed, the dose rate at the time of the accident ($t = 0$) is given by Eq. 16.20. Tritium is a beta emitter with E_{max} = 18.6 keV, T_R = 12 years, and T_B = 12 days. At $t = 0$, the source strength (i.e., the activity) is

$$S = N\lambda = (10^{-7}\text{kg})\,(2\text{ atm/molecule})\frac{6.022 \times 10^{23}\text{ molecules/mol}}{22 \times 10^{-3}\text{ kg/mol}}$$

$$\times \left[\frac{\ln 2}{12\,(3.15 \times 10)^7}\right] = 1.00 \times 10^{10}\text{Bq} = 271\text{ mCi}$$

Considering an average-size person (70 kg), the dose rate at $t = 0$ is

$$H = \frac{\left[10^{10}(0.0186/3)\, \text{MeV/s}\right](1.602 \times 10^{-13}\, \text{J/MeV})(1)}{70\, \text{kg}}$$

$$= 1.42 \times 10^{-4}\, \text{Sv/s} = 51\, \text{mrem/h}$$

Table 16.6
Radiological, Biological, and Effective Half-Lives of Certain Common Isotopes

Isotope	Organ of Reference	Radiological Half-Life	Biological Half-Life	Effective Half-Life
^{3}H	Total body	12.3 y	12 d	12 d
^{14}C	Total body	5700 y	10 d	10 d
	Fat		12 d	12 d
	Bone		40 d	40 d
^{32}P	Total body	14.3 d	257 d	13.5 d
	liver		18 d	8 d
	Bone		1115 d	14.1 d
	Brain		257 d	13.5 d
^{40}K	Total body	1.28×10^9 y	58 d	58 d
^{55}Fe	Total body	1100 d	800 d	463 d
	Spleen		600 d	388 d
	Lungs		3200 d	819 d
	Liver		554 d	368 d
	Bone		1680 d	665 d
^{57}Fe	Total body	45.1 d	800 d	42.7 d
^{99m}Tc	Total body	0.25 d	1 d	0.2 d
	Kidneys		20 d	0.25 d
	Lungs		5 d	0.24 d
	Skin		10 d	0.24 d
	Liver		30 d	0.25 d
	Bone		25 d	0.25 d
^{129}I	Total body	1.726×10^7 y	138 d	138 d
	Thyroid		138 d	138 d
	Kidneys		7 d	7 d
	Liver		7 d	7 d
	Spleen		7 d	7 d
	Testes		7 d	7 d
	Bone		14 d	14 d
^{131}I	Thyroid	8 d	138 d	7.6 d
^{235}U	Total body	7.12×10^8 y	100 d	100 d
	Kidneys		15 d	15 d
	Bone		300 d	300 d
^{238}U	Total body	4.66×10^9 y	100 d	100 d
^{239}Pu	Total body	24,000 y	175 y	175 y
	Liver		82 y	82 y
	Kidneys		87.7 y	87.7 y

Source: Report of the ICRP Committee II on Permissible Dose for Internal Radiation (1959). *Health Phys.*, 3, 146, June 1960.

The total dose is obtained by using Eq. 16.30 (with $T = 50$ y):

$$H_T = \frac{1.42 \times 10^{-7} \text{Sv/s}}{\ln 2} (12 \text{ days}) (86,400 \text{ s/day})$$
$$= 0.215 \text{ sv} = 21.2 \text{ rem}$$

Example 16.8 A patient was given 10 μCi of ^{131}I in an attempt to kill a thyroid tumor. Assuming that all the iodine is concentrated in the thyroid, calculate (a) the dose rate to the patient at the time of the injection and (b) the total dose received by this patient. Iodine emits 0.364-MeV gammas 82% of the time and 0.606-MeV betas 92% of the time. For ^{131}I, $T_R = 8$ days, $T_B = 138$ days; mass of the thyroid is 0.020 kg. The radius of the thyroid, taken as a sphere, is $R = 16.8$ mm.

Answer a) The dose rate from the betas will be obtained with the assumption that all the beta energy is deposited in the thyroid. Thus,

$$\dot{D} = \dot{H} = \frac{\text{Energy/s}}{\text{mass}} = \frac{0.92(3.7 \times 10^5)(0.606/3)(1.602 \times 10^{-13}) \text{ J/s}}{20 \times 10^{-3} \text{ kg}}$$
$$= 5.51 \times 10^{-7} \text{ Sv/s} = 0.198 \text{ rem/h}$$

The dose rate from gammas is given by Eq. 16.22, and g is calculated using Eqs. 16.24 and 16.25.

b) For 0.364-MeV gammas in tissue, $\mu = \mu_{tot} = 0.0101$ m^2/kg=0.101 cm^{-1}, $\mu_a = 0.00325$ m^2/kg. The volumetric gamma source strength needed for Eq. 16.22 is

$$S_V = (3.7E5)/V = 3.7E5/(20 \times 10^{-6}) = 1.85 \times 10^{10} \text{ Bq/m}^3$$

$$\bar{g} = 0.75 \frac{4\pi}{\mu} (1 - e^{-\mu r}) = 0.75 \times 19.3 = 14.5 \text{ cm} = 0.145 \text{ m}$$

$$\dot{H} = \frac{0.82(1.85 \times 10^{10} \gamma)/\text{m}^3 \text{ s})}{4\pi} 0.364 \text{MeV} (1.602 \times 10^{-13} \text{ J/MeV})$$
$$= (0.00325 \text{ m}^2/\text{kg}) 0.145 \text{ m} = 3.32 \times 10^{-8} \text{ Sv/s} = 11.9 \text{ mrem/h}$$

The total dose from betas and gammas = 5.51×10^{-7} Sv/s + 3.32×10^{-8} Sv/s + 5.84×10^{-7} Sv/s or 210 mrem/h, (during 50 years after the intake) is given by Eq. 16.30, since

$$T_e = \frac{T_R T_B}{T_R + T_B} = \frac{8 \times 138}{8 + 135} \text{ days} = 7.77 \text{ days} \ll T(=50 \text{ y})$$

$$H = \frac{\dot{H}(0)}{\ln 2} T_E = \left[\frac{5.84 \times 10^{-7} \text{ Sv/s}}{\ln 2}\right] 7.77 \text{ days} (86,400 \text{ s/day})$$
$$= 0.566 \text{ Sv} = 56.6 \text{ rem}$$

16.7 BIOLOGICAL EFFECTS OF RADIATION

The study of the biological effects of radiation is a very complex and difficult task for two main reasons:

1. The human body is a very complicated entity with many organs of different sizes, functions, and sensitivities.

2. Pertinent experiments are practically impossible with humans. The existing human data on the biological effects of radiation come from accidents, through extrapolation from animal studies, and from experiments *in vitro*.

How and why does radiation produce damage to biological material? To answer the question, one should consider the constituents and the metabolism of the human body. In terms of compounds, about 61% of the human body is water. Other compounds are proteins, nucleic acids, fats, and enzymes. In terms of chemical elemental composition, the human body is, by weight, about 10% H, 18% C, 3% N, 65% O, 1.5% Ca, 1% P, and other elements that contribute less than 1% each. To understand the basics of the metabolism, one needs to consider how the basic unit of every organism, which is the cell, functions.

16.7.1 Basic Description of the Human Cell

The cell, the basic unit of every living organism, consists of a semipermeable membrane enclosing an aqueous suspension of a liquid substance called the cytoplasm. The cell exchanges material with the rest of the organism through the membrane. A typical cell size is about 10^{-5} m (size of a typical atom is 10^{-10} m). At the center of the cell, there is another region called the nucleus, also enclosed by a semipermeable membrane. The nucleus is the most important part of the cell because it controls cell activities. Nucleic acids and chromosomes are the cell's most significant contents.

The two nucleic acids found in the nucleus of a cell are ribonucleic acid (RNA) and deoxyribonucleic acid (DNA). The RNA controls the synthesis of proteins. The DNA contains the genetic code of the species. The structure of the DNA has been determined to be a double helix, or staircase with the stairsteps consisting of paired molecules of four bases: adenine (A), guanine (G), cytosine (C), and thymine (T). It is the combination of these four compounds, A, G, C, and T, that makes the genes (a gene is a segment of DNA) that contain the instructions for the metabolism of the cell. The DNA molecules have a molecular weight of about 10^9. They are usually coiled inside the cell, but when extended like a string, the width of the double-stranded helix of the DNA is about 2 nm.

The chromosomes are threadlike assemblies that are extremely important because they contain the genes that transmit the hereditary information. Every species has a definite number of chromosomes. The human species has 23 pairs, one chromosome of each pair being contributed by each parent. Every cell has 23 pairs of chromosomes with the exception of the egg and the sperm, which have 23 chromosomes each. When fertilization occurs, the first cell of the new organism contains 23 pairs of chromosomes, equally contributed by each parent.

The relationship between a chromosome, DNA and gene is presented in Figure 16.5.

Cells multiply by a dividing process called mitosis. Just before mitosis is to take place, each chromosome of the cell splits in two. Thus, each of the two new cells has exactly the same number of chromosomes as the parent cell. There are some human cells that do not divide, such as the blood cells and the nerve cells. The blood cells are regenerated by the blood-forming organs, primarily by the bone marrow. The nerve cells, when destroyed, are not supplied again.

Radiation may damage the cell when it delivers extra energy to it because that energy may be used to destroy parts or functions of the cell. For example, as a result of irradiation, chromosomes or DNA molecules may break. The break may occur either by direct collision with an incoming fast particle (e.g., fast neutron) or as the result of chemical activity initiated by the radiation. It has been determined experimentally that the energy imparted by the radiation may be used to break chemical bonds and create free radicals, which are always chemically active and which may

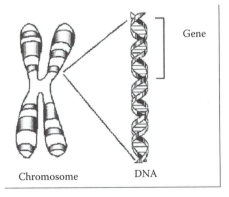

Gene

Chromosome   DNA

Figure 16.5 Relationship between chromosome, gene and DNA (http://virginia.edu/medicine/interdis/huntdisease/DNA.gif).

produce new chemical compounds unhealthy for the organism. For example, a water molecule may break into two radicals that, in turn, may form hydrogen peroxide (H_2O_2)*:

$$H_2O^\bullet \rightarrow HO^\bullet + H$$
$$HO^\bullet + HO^\bullet \rightarrow H_2O_2$$

A damaged cell may react in different ways. It may recover, or die, or grow out of control if the radiation has damaged the RNA and DNA molecules that provide the instructions to feed and divide the cell. Obviously, the net result of the damage to the organism depends on many factors, such as the number and the type of cells destroyed. Another effect of irradiation may be damage to the DNA of the germ cells, the sperm and the egg, that carry the genetic code of the new organism. This type of damage (called genetic; see Section 16.7.2) will appear in the offspring of the irradiated cell or organism.

16.7.2 Stochastic and Nonstochastic Effects

As explained in the previous section, radiation imparts energy to the cell, which may trigger mechanisms that result in biological damage. This "damage," which starts at the microscopic (cell) level, may, in some cases, manifest itself as a macroscopic observable biological effect.

The biological effects of radiation are divided into different categories, depending upon the objective of the discussion. Examples are somatic (effects appearing on the individual being irradiated), genetic (appearing in the offspring of the irradiated person), short-term effects, long-term effects, and so on. The division to be used here is stochastic and nonstochastic effects because it is this characterization that leads to a better understanding of the dose–effect relationship.

Examples of nonstochastic (or deterministic) effects are erythema, nausea, loss of hair, cataracts, sterility, and so on. Stochastic (or probabilistic) effects are cancer and genetic defects (birth defects; see Table 16.7). Genetic effects are abnormalities that may appear in the offspring of persons exposed to radiation, one or many generations after the exposure.

An important difference between stochastic and nonstochastic effects is that nonstochastic effects have a threshold; stochastic effects do not. The "threshold" is a minimum radiation dose that has to be received in a relatively short time period for the effect to appear (Figure 16.6). A dose below the threshold will not produce a nonstochastic effect. A dose above the threshold will definitely cause the effect. The threshold line in Figure 16.6a is shaded to emphasize the point that the threshold dose is not a single one but a range of doses that depends on the effect considered and on the individual receiving the dose. Different effects have different threshold doses. For example, the threshold dose for erythema is much less than that for death. For stochastic effects, it is believed today that there is no threshold. All one can say is that there is a probability that the effect *may* appear (some time later, probably years) after any amount of radiation exposure above zero. It is also accepted today that the probability that the effect will appear increases with dose received. There is no scientific proof that a threshold *does not* exist. Also there is no concrete

Table 16.7
Biological Effects of Radiation

	Stochastic (Probabilistic)	Nonstochastic (Deterministic)
Somatic	Cancer	Erythema
		Loss of Hair
		Nausea
		Sterility
		Cataracts
		Fever
		Death
		etc.
Genetic	Birth defects	

---

* Chemical poisoning by H_2O_2 shows many of the radiation sickness symptoms.

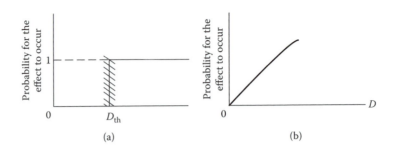

Figure 16.6 (a) The probability for a nonstochastic effect to occur versus dose D. (b) The probability for a stochastic effect to occur versus dose D.

scientific proof that all radiation effects are detrimental. However, in the absence of proof that a threshold does exist and that radiation may, at certain dose levels, be beneficial, the conservative approach is taken, which is no threshold and any radiation dose is damaging. Notice, however, the word *may*. It is not certain that the effect will appear; all one can say is that there is a probability that it may happen.

An example of stochastic versus nonstochastic effects can be made using alcohol. If a person drinks 20 glasses of wine in a short period of time, it is certain that the individual will get drunk. Drunkenness is a nonstochastic effect caused by alcohol (the "threshold" is not 20 glasses for everybody; it depends on the individual, on the rate of wine consumption, etc.). Examples of stochastic effects caused by alcohol are cirrhosis of the liver and birth defects to a child whose mother was drinking during pregnancy. One glass of wine, just once in a lifetime, or one glass per day may cause cirrhosis or produce a child with birth defects; on the other hand, it may cause neither. Regulatory limits are based on the assumption that no threshold exists for stochastic effects.

Nonstochastic effects appear after relatively high doses in the Sv (rem) range. The first measurable effect of a whole-body irradiation appears after a dose of 0.25–0.50 Sv (25–50 rem). The individual exposed to this dose will feel nothing, and clinical tests will not show any symptoms of illness or injury. Depending on the person, a clinical test may show changes in the blood. A dose of 4–5 Sv (400–500 rem) is indicated as LD-50, meaning that it is lethal to about 50% of the persons so exposed (death will occur in months). A dose of 10 Sv (1000 rem) or more to the whole body is considered lethal (death will occur in days), no matter what treatment may be applied. In the range of about 1–10 Sv (100–1000 rem), symptoms that may appear are nausea, vomiting, fever, diarrhea, loss of hair, inability of the body to fight infection, and so on. These symptoms have been observed in victims of accidents and in patients undergoing radiation treatment.

The incidence of stochastic effects can only be treated in a probabilistic manner. Consider cancer first. The estimate of nonfatal radiation induced cancer has been previously been reported[9] and updated in 2006.[10] In a lifetime, ~45 male persons out of 100 and ~37 female persons will be diagnosed with cancer from all causes.

From a dose of 0.10 Gy (10 rad), above BG, 1 person in 100 is expected to be diagnosed with cancer attributed to that 0.10 Gy. In a lifetime, ~22 male persons out of 100 and ~18 female persons are expected to die of cancer.

From a dose of 0.10 Gy (10 rad), above BG, less than 1 person in 100 is expected to die of cancer attributed to the 0.10 Gy. The BG level (including Radon) is about 250–300 mrem per year. Over a lifetime (80 y), the total dose received from BG is ~0.20–0.24 Sv (20–24 rem). These estimates are shown in Table 16.8.

Genetic effects are those related to the transmission of harmful hereditary information from one generation to the next. It is known today that the carriers of the hereditary code are the genes, which are parts of DNA molecules and are contained in the chromosomes. The gene is an extremely stable entity. Its structure is transmitted from generation to generation without any changes, which means that it transmits identical information from generation to generation. But sometimes a gene may change and become a *mutation*. The mutated gene may be transmitted through many generations without any further change, or it may change again to its original form or to a new third form. It is generally believed by geneticists that most mutations are harmful; therefore, conditions that increase the rate of mutations should be avoided.

Table 16.8
Probability of Cancer from Exposure to Radiation with a Population Sample Size Assumed 100,000 and Age Distribution Similar to That of the US population, the Risk Shown Is the Result from a Single Dose of 0.10 Gy (10 rad)

	All Solid Cancers: For Males	All Solid Cancers: For Females
Excess cases (including non fatal ones)	970 (490–1920)[a]	1410 (740–2690)
Number of *cases* in the absence of exposure (BG-yes)	45,500	36,900
Excess deaths	480 (240–980)	740 (370–1500)
Number of *deaths* in the absence of exposure (BG-yes)	22,100	17,500

[a] The numbers in parenthesis represent the 95% error BG is background.

The current incidence of human genetic disorders is 107,000 per 10^6 births, or about 0.11 per birth. These effects constitute the so-called spontaneous mutation rate. The genetic risk from radiation is expressed in terms of the ratio

$$\frac{m_r}{m_s} = \frac{\text{(radiation-induced mutations)}}{\text{(spontaneous mutaions)}}$$

This ratio is equal to (2–0.4)/Sv [(0.02–0.004)/rem]. One quantity that is always reported along with this risk is the "doubling dose," that is, the dose that if inflicted to a population over many generations will eventually result in doubling the rate of spontaneous mutations. From the ratio given above, the doubling dose is 0.5–2.5 Sv (50–250 rem).

Is radiation the only agent that causes mutations? Definitely not. Known mutagenic agents include certain chemicals, certain drugs, elevated temperature, and ionizing radiation. It is quite possible that many other mutagenic substances or environments may exist but are still unknown. Humans have been exposed to ionizing radiation since first appearing on this planet. The level of this background radiation is not constant at every point on the surface of the earth, but at sea level it is about 1.5–3 mSv/y (150–300 mrem/y). Every individual receives this exposure every year of his life. There is no doubt that genetic effects have been caused as a result of this exposure. Yet it should be pointed out that (a) there is no proof that radiation causes *only* detrimental genetic effects, and (b) despite the continuous exposure during thousands of years, there is no evidence of genetic deterioration of the human race.

16.8 RADIATION PROTECTION GUIDES AND EXPOSURE LIMITS

All regulations relevant to the protection of humans can be found in Title 10, Chapter 1, Part 20 *Code of Federal Regulations* (10CFR20). The 10CFR20 that is in force today,[4] which became effective January 1, 1994, is based on the recommendations of the International Commission on Radiological Protection (ICRP),[2] published in 1977 as ICRP Publication 26. The general principles upon which the new 10CFR20 radiation protection guides have been established are as follows:

1. No person should be exposed to any man-made radiation unless some benefit is derived from the exposure.

2. Radiation exposure limits are set at such levels that nonstochastic biological effects do not occur.

3. Radiation exposure limits are set at such levels that stochastic effects are minimized and become acceptable in view of the benefits derived from the exposure.

4. In every activity that may involve radiation exposure, it is not enough to keep exposure limits below the maximum allowed. Instead, every effort should be made to keep the exposure as low as reasonably achievable (ALARA).

The ALARA principle is strictly enforced by the U.S. Nuclear Regulatory Commission (NRC).

Since nonstochastic effects have a threshold, all that is needed to satisfy requirement 2 is to set the exposure limits below that threshold. For nonstochastic effects, the maximum allowed dose is set at 0.5 Sv (50 rem) for any tissue, except for the lens of the eye, for which the limit is set at 0.15 Sv (15 rem). For stochastic effects the limits are set at an acceptable level of risk. Ideally, the limit should be zero, since any exposure is supposed to increase the probability for stochastic effects to occur. Obviously, a zero limit is not practical. For stochastic effects the 10CFR20 sets the limiting exposure on the basis that the risk should be equal regardless of whether the whole body is irradiated uniformly or different tissues receive different doses. Recognizing the fact that different tissues have different sensitivities and, therefore, the proportionality constant between dose and effect is not the same for all tissues, the limit is expressed in terms of the "effective dose equivalent" (H_E), defined as

$$H_E = \sum_T w_T H_T \tag{16.31}$$

where H_T is the dose equivalent to tissue or organ T and w_T is a weighting factor for tissue T. The values of w_T are presented in Table 16.9.

The interpretation of the weighting factors is as follows. Consider the factor $w_T = 0.25$ for the gonads. This means that irradiation of the gonads alone would present about one-fourth the risk for stochastic effects expected to appear after uniform irradiation of the whole body at the same dose level. The risk per Sv due to irradiation of the gonads is derived from

$$(\text{Risk/Sv-gonads}) = (\text{Risk/Sv-whole body})w_T$$
$$= 1.65 \times 10^{-2}(0.25) = 4.1 \times 10^{-3}$$

(or 1 in 250). Table 16.9 presents various maximum exposure limits. For comparison, limits of the old 10CFR20, the 1994 10CFR20 and the ICRP in 2007[11] are given in Table 16.10.

For the dose from radioisotopes inhaled or ingested by the body, several other doses have been defined, as follows:

Committed dose equivalent ($H_{50,\,T}$): It is the dose equivalent to organs or tissues that will be received from an intake of radioactive material by an individual, during a 50-year period following the intake.

Committed effective dose equivalent ($H_{E,\,50}$): It is the sum, over all relevant tissues or organs, of the product of the factor w_T times the corresponding committed dose equivalent:

$$H_{E,\,50} = \sum_T w_T H_{50,T} \tag{16.32}$$

Deep dose equivalent: It applies to external whole body exposure and is the dose equivalent at a depth of tissue of 1 cm (1000 mg/cm^2).

Table 16.9
Organ Dose Weighting Factors

Organ or Tissue	w_T	Risk Coefficient (Sv)	Probability (Sv)
Gonads	0.25	4.14 × 10⁻³	1 in 250
Breast	0.15	2.5 × 10⁻³	1 in 400
Red bone marrow	0.12	2 × 10⁻³	1 in 500
Lung	0.12	2 × 10⁻³	1 in 500
Thyroid	0.03	5 × 10⁻⁴	1 in 2000
Bone surface	0.03	5 × 10⁻⁴	1 in 2000
Remainder[a]	0.30	5 × 10⁻³	1 in 200
Total (whole body)	1.0	1.65 × 10⁻²	1 in 60

[a] The remainder is 0.06 for each of five remaining organs, excluding the skin and the lens of the eye, which receive the highest doses.

Table 16.10
Various Maximum Exposure Limits[a]

	Old 10CFR20	10CFR20 Effective January 1, 1994	ICRP 103 (2007)
Whole body	5 rem/yr	50 mSv/yr (stochastic) (5 rem/yr)	20 mSv/yr, averaged over 5 yr, less than
		500 mSv/yr (nonstochastic) (50 rem/yr)	50 mSv in any single year
Lens of eye	1.25/quarter, 5.0 rem/yr	15 mSv/yr (15 rem)	15 mSv/yr
Extremities	18.75 rem/quarter, 75 rem/yr	500 mSv/yr (50 rem/yr)	500 mSv/yr
Thyroid-skin	7.5 rem/quarter, 30 rem/yr	50 mSv/yr (stochastic) (5 rem/yr)	50–500 mSv
		500 mSv/yr (nonstochastic) (50 rem/yr)	
Dose to minors Public	0.5 rem/yr	10% of adult limit 1 mSv/yr (100 mrem)	10% of adult limit 1 mSv/yr
Dose to fetus	—	5 mSv (500 mrem)	1 mSv (100 mrem)

[a] The NRC limit for the general public is 0.1 mSv (100 mrem) and is 5 mSv (500 mrem) for a fetus (see Nuclear Regulatory Commission).

Table 16.11
Sources of Background Radiation in the United States

Source	Gonads	Lung	Bone Surface	Bone Marrow	GI Tract
Cosmic radiation[a]	28.0	28	28	28	28
Cosmogenic nuclides	0.7	0.7	0.8	0.7	0.7
External terrestrial	26	26	26	26	26
Inhaled nuclides	—	100–450	—	—	—
Radionuclides in body[b]	27	24	60	24	24
Total (rounded)	80	180–350	115	80	80

All values are in units of mrem/yr.
[a] The cosmic-ray component is given at sea level; it increases with altitude.
[b] Radionuclides in the body are, primarily, ^{14}C and ^{40}K.

Total effective dose equivalent (TEDE): It is the sum of the deep dose equivalent (for external exposure) and the committed effective dose equivalent (for internal exposure)

$$\text{TEDE} = H_d + H_{E,50} \tag{16.33}$$

The annual limit for radiation workers is the more limiting of the following two: (a) TEDE being equal to 50 mSv (5 rem); or (b) the sum of the deep dose equivalent and the committed dose equivalent to any individual organ or tissue other than the lens of the eye being equal to 0.5 Sv (50 rem).

Annual limit of intake (ALI): the amount of a radioactive material taken into the body of an adult worker in one year, by inhalation or ingestion, that would result in an effective committed dose equivalent of 0.05 Sv (5 rem) or a committed dose equivalent of 0.5 Sv (50 rem) to any single tissue or organ.

Derived air concentration (DAC): the concentration for a given radioisotope is that concentration in air that, if breathed by an adult for a working year of 2000 hours under conditions of light activity (inhalation rate 2.0E^4 mL/min), would result in total intake of 1 ALI. Values of ALI and DAC for several radioisotopes are given in Appendix *B* of 10CFR20.[4]

For radon (Rn) and its daughters the radiation limits are given in terms of the working level (WL) and working level month (WLM), where 1 WL is the amount, in 1 *l* of air, of any combination of Rn and its daughters that results in the release of 1.3 × 10^5 MeV of alpha-particle energy. This number is approximately the energy released by the short-lived daughters in equilibrium with 100 pCi and Rn. One WLM is equal to exposure to 1 WL for 170 h (170 = 2000/12). As an example, if a worker is exposed to 1 WL for 50 h, the exposure is (50/170) × 1 = 0.294 WLM.

In addition to exposure from man-made radiation, humans are exposed to natural radiation. The components of the natural or background radiation are shown in Table 16.11.

16.9 HEALTH PHYSICS INSTRUMENTS

A health physics instrument is a device that can provide information about the dose rate or dose at the location where the instrument is placed. Health physics instruments are detectors like those discussed in Chapters 5–7. They have to satisfy some unique requirements, however, because their purpose is to measure dose equivalent, which is the absorbed dose in tissue times a quality factor. Radiation detectors provide a signal that, in general, depends on the energy deposited in the material of which the detector is made; that material does not necessarily have the same response as tissue to the radiation field being investigated. Even if the detector material responds to the radiation exactly like tissue, the problem still exists of getting dose equivalent from absorbed dose (which the detector signal provides). Thus, the dose measurement involves three steps:

1. Measurement of energy deposition in the detector (a quantity proportional to D in the detector material).

2. Determination of D_{tissue} by comparing the response to the incident radiation of tissue versus the response of the material of which the detector is made.

3. Computation of dose equivalent H, from D_{tissue} by incorporating the appropriate quality factor.

In practice, the instruments are properly calibrated to read directly Sv (or rem), or Gy (or rad). For some neutron detection instruments, the neutron flux is recorded. Then the dose equivalent is obtained after multiplying the flux by the conversion factor given in Table 16.4. Since different detectors do not have the same efficiency or sensitivity for all types of radiation and for all energies, there is no single instrument that can be used for all particles (α, β, γ, n) and all energies.

Health physics instruments are divided into two general groups according to the way they are used:

1. Survey instruments—portable and nonportable

2. Personnel monitoring instruments (dosimeters)

The rest of this section discusses the most commonly used devices. Survey instruments are described in Section 16.9.1. Dosimeters are described in Sections 16.9.2–16.9.8.

16.9.1 Survey Instruments

The portable survey instruments are detectors like those described in detail in Chapters 5–7. The most commonly used are GM counters and the so-called "cutie pie" meter. Many commercial GM counters offer a fixed thin window that will allow betas and photons to traverse it and be counted, and one or more movable windows that will stop the betas and allow only the gammas to enter the counter. In a mixed β–γ field, such an instrument will provide information for β and γ separately. The cutie pie is an ionization counter that can be used to detect X-rays, alphas, and some high-energy betas.

The nonportable instruments are set at fixed locations to detect the radiation field; they are usually equipped with an alarm that will provide an audio and/or visual signal when the field intensity exceeds a preset limit. Examples are continuous air monitors and personnel monitors (e.g., hand, foot, and whole body). A list of several health physics instruments and their characteristics is given in Table 16.12.

16.9.2 Thermoluminescent Dosimeters

Thermoluminescent dosimeters (TLDs) are based on the property of thermoluminescence, which can be understood if one refers to the electronic energy-band diagram of crystals (see also Chapter 7). When ionizing radiation bombards a crystal, the energy given to the electrons may bring about several results (Figure 16.7). The electron may acquire enough energy to move from the valence to the conduction band, in which case the event is called ionization. Or the electron acquires enough energy to move to an excited state (to the exciton band) and form an exciton. An exciton, consisting of an electron and a hole bound electrostatically, can migrate through the crystal. Electrons, holes, and excitons may be caught in many "traps" that exist in the solid. Traps are formed in a variety of ways. Foreign atoms (impurities), interstitial atoms, dislocations, vacancies, and imperfections may act as traps. The trapped carriers remain in place for long periods of time if the temperature of the crystal stays constant or decreases. If the temperature is raised, however, the probability of escape increases. As electrons and holes are freed and return to the ground

Table 16.12
A List of Common Health Physics Instruments

Instrument	Detector	Radiation Detected	Range (Nominal)	Remarks
Portable				
Film badge	Photographic emulsions	$\gamma, \beta, n_f, n_{th}$	0.1–1000 rad of mixed radiations	Measurement of integrated dose
Pocket chamber (direct reading)	Ionization chamber (air)	γ	5–200 mR; available with higher ranges	Visual check on gamma exposure
Personnel radiation monitor	G-M tube	γ, X-ray, high-level β	Maximum audible warning at ~0.5 R/h	Audible warning of radiation field
Cutie-pie survey meter	Ionization chamber (air)	γ, X-ray, high-energy β	5–10,000 mrad/h	Dose-rate meter for γ and X-rays (0.008–2 MeV within 10%)
PIC-6	Proportional gas-filled counter	γ	1 mR/h to 1000 R/h	Measures γ dose rate
PAC-4G	Proportional gas-filled counter	α		
Rem ball	Bonner sphere with BF_3 or LiI as the detector	n		Measures neutron dose equivalent
Nonportable				
Continuous β-γ particulate air monitor	GM tube (shielded)	β, γ	0–5000 cpm	Light and audible alarms for preset levels
Continuous α particulate air monitor	ZnS(Ag) for α	α, β	0–1000 cpm	Light and audible alarms for preset levels
Hand and foot monitor	Halogen-quenched GM tube	β, γ	Low-level	Simultaneous detection of β and γ contamination of hands and shoes

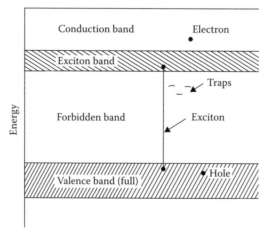

Figure 16.7 Energy-band diagram of a crystal.

state, they emit light (Figure 16.8). The emission of this light is called thermoluminescence and is the property upon which the operation of TLDs is based.

A TLD is essentially a piece of a thermoluminescent material, exposed to the radiation being measured. After irradiation stops, the TLD is heated under controlled conditions (Figure 16.9), and the light intensity is measured either as a function of temperature or as a function of the time during which the temperature is raised. The result of such a measurement is a graph called the *glow curve* (Figure 16.10). Glow curves have more than one peak, corresponding to traps at various energy levels. The amplitudes of the peaks are proportional to the number of carriers trapped in the corresponding energy traps. The absorbed dose may be measured either from the total light emitted by the glow curve or from the height of one or more peaks of the glow curve. The TLD is annealed—that is, it returns to its original condition—and is ready to be used again after being heated long enough that all the traps have been emptied; then it is left to cool down to room temperature. Measurement of the light from the glow curve, and subsequent annealing, are performed by instruments generally called *readers*, which are available commercially.

There are many thermoluminescent materials, but those useful for dosimetry should have the following characteristics:

1. Retention of trapped carriers for long periods of time at temperatures encountered during the exposure

2. Large amount of light output

3. Linear response over a large dose range

4. Perfect annealing to enable repetitive use

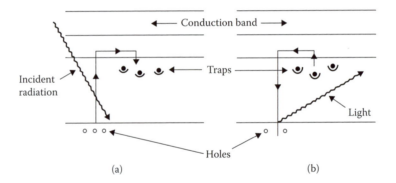

(a)   (b)

Figure 16.8 (a) As a result of irradiation, some carriers fall into traps, (b) upon heating, the carriers are given enough energy to escape from the traps and return to the valence band, with the emission of light.

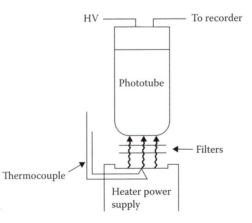

Figure 16.9 A setup used to read a TLD.

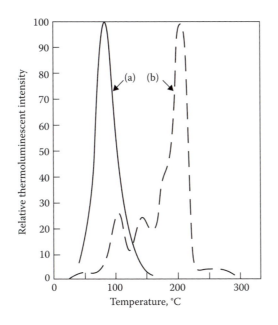

Figure 16.10 Typical thermoluminescent glow curves, (a) glow curve of $CaSO_4$: Mn heated at 6°C/min. (b) LiF (TLD-100) exposed to 10^4 R and heated at 20° C/min. (From Attix, F. H., and Roesch, W. C. (eds.), *Radiation Dosimetry*, II, Academic Press, New York, 1966.)

Table 16.13
Properties of Certain TLDs

Material	Radiation to Which It Responds	Z_{eff}	Range
$CaSO_4$ (Mn)	γ	15.3	μR-10^3 R
CaF_2 (Natural)	γ	16.3	mR-10^3 R
CaF_2 (Mn)	γ, n (thermal, low response)	16.3	mR-10^5 R
LiF (TLD-100)[a]	γ, n (thermal) β	8.2	mR-10^4 R
LiF (TLD-600)	γ, n (thermal), p	8.2	mR-10^5 R
LiF (TLD-700)	γ, α	8.2	mR-10^5 R

Source: Attix, F. H., and Roesch, W. C. (eds.), *Radiation Dosimetry*, II, Academic Press, New York, 1966.
[a] TLD-100, TLD-600, TLD-700 are products of the Harshaw Chemical Company

Materials commercially available that satisfy most of these requirements are $CaSO_4$: Mn, CaF_2 (natural), CaF_2: Mn, $Li_2B_4O_7$: Mn, and LiF. $CaSO_4$ and CaF_2 are used for gammas only. Other materials that have been studied are $CaSO_4$: Dy, BeO, and Al_2O_3. A commercial dosimetry package, based on a combination of $CaSO_4$ and $Li_2B_4O_7$, and known as Panasonic UD-802 or UD-854 is used in nuclear power plants. The main features of some of these materials are given in Table 16.13. Various uses are described in References 13–19. The book by Shani (1991) gives many details for the construction, use, and performance characteristics of all TLD materials studied.

The three LiF TLDs listed in Table 16.13 have found wide use for measurements in mixed neutron-gamma fields. The TLD-100 containing natural lithium (92.6% ^{7}Li, 7.4% ^{6}Li) responds to gamma and thermal neutrons. Thermal neutrons are detected through the (n, α) reaction with ^{6}Li, which has a cross section equal to 950 b for thermal energies. If a TLD-100 is exposed to about 3×10^{11} neutrons/m^2 (thermal), its light output is equivalent to that from 1 R of gamma radiation. The TLD-600, containing lithium enriched to 95.62% in ^{6}Li, is extremely sensitive to thermal neutrons and also to gammas. The TLD-700, containing 99.993% ^{7}Li, is sensitive to gammas only, because the neutron cross section for ^{7}Li is very small (about 0.033 b for thermal neutrons). In a mixed neutron-gamma field, one can achieve γ–n discrimination by exposing to the radiation a TLD-100 plus a TLD-700, or a TLD-600 plus a TLD-700. The difference in response between the two dosimeters of either pair gives the dose due to the neutrons only. A very sensitive LiF TLD has been reported by Nakajima *et al.*[19] It consists of LiF with three dopants: Mg, Cu, and P. Its sensitivity is supposed to be 23 times that of TLD-600. More recently there has been a significant amount of research exploring and characterizing various materials for TLD development. These materials include Al_2O_3:C,[20] microwave plasma assisted chemical vapor deposition diamond,[21] CaF_2:Mn,[22] CaF_2:Tm,[23] DyF_3,[24] and LiF:Mg,Cu, LiF:Mg,Cu,P, and LiF:Mg,Cu,Si.[25]

16.9.3 Optically Stimulated Luminescence Dosimetry

Although not as widespread as TLDs, optically simulated luminescence (OSL) technology is gaining in popularity in the fields of personal and environmental monitoring, space dosimetry, medical dosimetry, and so on. In certain crystalline structures such as feldspars, quartz and aluminum oxide, that are subjected to radiation, electron-hole pairs are created where the electrons are in the conduction bands and holes are in the valence bands. However, sometimes electrons that cannot completely jump the gap between valence and conduction bands are trapped in between these two. The use of a laser may enable these electrons to make the jump to the conduction band, while emitting light which can be detected with a photomultiplier tube. The main advantage of OSL is its ability to give instant readings that can be repeated while the TLD badge takes 20 or 30 s for a one time reading A comprehensive book on OSL dosimetry has published (see Boetter-Jensen *et al.*, 2003). Some recent applications of OSL have included low dose environmental dosimetry,[26] radiation measurements in space exploration,[27] and medical dosimetry.[28] One interesting application has to do with archaeological and sediment dating,[29,30]

16.9.4 The Bonner Sphere (the Rem Ball)

The Bonner sphere,[31] named after one of the first people to study its features and use it, is a neutron detector. It consists of a polyethylene sphere, at the center of which a neutron detector is placed (LiI scintillator or BF_3 or ^{3}He detector). With any one of these materials, the neutrons are detected through the reactions

$$^6Li(n, \alpha)^3H \quad ^{10}B(n, \alpha)^7Li \quad ^3He(n, p)^3H$$

The polyethylene serves as the moderating material. The Bonner sphere has been found to be very useful for neutron dose measurements because the response of a 0.25- to 0.30-m (10–12 in.) diameter sphere has an energy dependence very close to the dose equivalent delivered by neutrons (Figures 16.11 and 16.12). The line indicated as RPG in these two figures is the dose rate H per unit neutron flux. The similarity between detector response and dose equivalent H is just a coincidence. This coincidence is utilized, in practice, for the determination of neutron dose rate H in a neutron field of unknown energy. Note from Figure 16.11 that the sensitivity (efficiency) of the Bonner sphere is high for high-energy neutrons (with high Q value) and lower for low-energy neutrons (low Q value). For this reason, the number of counts recorded by the sphere placed in an unknown spectrum $\phi(E)$ automatically includes a weighting factor for all energies.

The match between response and neutron dose rate is not perfect, as Figure 16.12 clearly shows, particularly for energy less than 100 keV. If the unknown spectrum contains many neutrons in the energy range of discrepancy between the two curves, then the dose rate will be overestimated for the following reason. Consider $E = 10$ keV. The calculated response is about 1.7; the RPG is about 0.3. The inverse of RPG is about 3.3, which means that the detector will record 3.3 when the response ought to be close 1.7; hence, dose rate is overestimated. Another point to mention is the underresponse at $E > 10$ MeV (Figure 16.11); luckily, this energy is not important for neutron fields encountered at nuclear power plants.

Another advantage of the Bonner sphere, in addition to its convenient response function, is its complete insensitivity to gammas. This is the result of relying for neutron detection on charged-particle reactions with high Q value, thus making possible the complete rejection of pulses due to gammas with the use of a proper discriminator level.

A critical review of Bonner sphere spectrometers has been published,[32] while some applications have included measurement of neutron spectra in a medical facility[33] and spent fuel casks[34] and unfolding of neutron spectra.[35,36]

In the nuclear industry the Bonner sphere is known as the rem ball or rem meter.

16.9.5 The Neutron Bubble Detector

The neutron bubble detector (trade name BD-100R) is a reusable, passive integrating dosimeter that allows instant, visible detection of neutron dose. The bubble detector consists of a glass

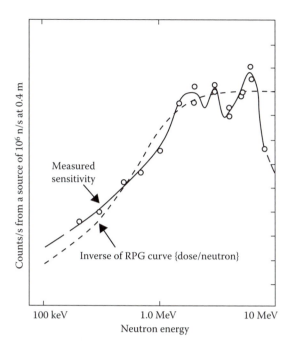

Figure 16.11 Sensitivity of a 10-in. (0.254 m) diameter Bonner sphere surrounding a 4 mm × 4 mm LiI scintillator. For comparison, the inverse of the response function RPG is also shown. The RPG gives dose rate, H, per unit neutron flux. (From Hankins, D. E., LA-2717, 1962.)

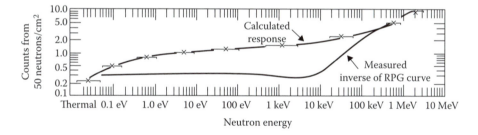

Figure 16.12 Calculated sensitivity of a 10-in. (0.254 m) diameter Bonner sphere compared with the measured response. (From Hankins, D. E., LA-2717, 1962.)

tube filled with thousands of superheated liquid drops in a stabilizing matrix. When exposed to neutrons, these droplets vaporize, forming visible permanent bubbles in an elastic polymer. The total number of bubbles formed is proportional to the neutron dose equivalent H. The bubbles can be counted manually or by a machine. Figure 16.13 shows the response of the bubble detector as a function of neutron energy.

The bubble dosimeter is reusable; it is insensitive to gammas (the formation of the bubbles is based on the stopping power of the recoil nuclei produced by collisions with neutrons); it responds to a neutron energy range from 200 keV to about 14 MeV. The dose range extend from less than 1 mrem to 1 rem; its useful life in 3 months, if recycled; its shelf life is 1 year. Some recent applications include a position sensitive bubble detectors for three dimensional spectrometry and dosimetry,[36] for application in the measurement of jet aircrew exposure to natural background radiation,[37] and for of a superheated drop as a neutron spectrometer.[38]

16.9.6 The Pocket Ionization Dosimeter

One of the most popular personal radiation monitoring instrument and still used worldwide is the pocket ionization dosimeter or sometimes known as a quartz fiber dosimeter. Its

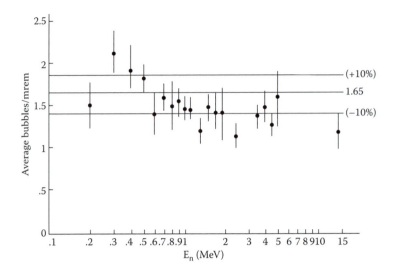

Figure 16.13 The response of the neutron bubble dosimeter as a function of neutron energy (http://www.medical.siemens.com).

attractiveness is primarily due to its reliability, instant readout and inexpensive cost. It is essentially a quartz glass fiber that is embedded in an ionization chamber which is air-filled. In the chamber one end is a metal electrode attached to a terminal for recharging. At the other end is the quartz fiber. As ionizing radiation passes through the instrument, it ionizes the air, creating electrons and positively and charged ions and reducing the charge on the electrode. The quartz fiber then moves toward the electrode which can be seen by a small microscope. The amount of movement is proportional to the amount of radiation dose. While these dosimeters are mainly used to monitor gamma radiation a neutron pocket dosimeter based on the ^{6}Li$(n, \alpha)^3$H reaction has also been tested.[39]

16.9.7 The Electronic Personal Dosimeter

The electronic personal dosimeter (EPD) was developed by England's National Radiological Protection Board. The EPD, having the size and weight of a small pocket pager, uses the latest in integrated circuitry technology. It is a solid-state device based on silicon diodes. Complete details of the design are proprietary.

The objective of the EPD project was to produce a dosimeter that is accurate over a wide energy and dose range, rugged, and can communicate with a computer for data storage and subsequent analysis. All indications are that the EPD satisfies these requirements. It is small and rugged and the energy range is from 20 keV to 7 MeV for gammas and 250 keV to 1.5 MeV (average energy) for betas. In terms of dose, the dosimeter can display doses from 1 μSv to 1 Sv (0.1 mrem to 100 rem) and dose rates from 1 μSv/h to 10 Sv/h and is equipped with audible and visual alarms that can be set only by authorized persons.

The EPD uses infrared links to interface with a computer, thus providing the data to authorized persons who can read the dose received, add up previous exposures, and establish dose-alarm threshold settings. A custom lithium battery that lasts for a year powers the EPD. Each unit will be issued to a radiation worker for a full year.

Recent applications of electronic dosimeters include gamma-ray and fast neutron responses,[42] calculations for personal electronic dosimetry purpose, response[43] to ^{60}Co and abnormal responses of electronic pocket dosimeters caused by high frequency electromagnetic fields emitted from digital cellular telephones.[44]

16.9.8 Foil Activation Used for Neutron Dosimetry

Neutron dosimetry by foil activation is not used so much to record doses received by personnel as it is to record doses to materials, instruments, or other components that may suffer radiation damage as a result of neutron bombardment. The principle of this method was presented in Sections 14.4 and 14.6. A target, in the form of a thin small foil, is exposed to the neutron field and becomes radioactive. The relationship between activity and neutron flux is

$$A(t) = N\sigma\phi(1 - e^{-\lambda t})$$ (16.34)

where N = number of targets
    σ = neutron absorption cross section
    ϕ = neutron flux
    λ = decay constant of the radioisotope produced

After irradiation, the activity $A(t)$ is counted and the flux is determined from Eq. 16.34. Depending on the foil used (reaction involved), information about the neutron energy spectrum may also be obtained. Information about the neutron spectrum $\phi(E)$ is necessary for the determination of the neutron dose equivalent H.

16.10 PROPER USE OF RADIATION

Since radiation may be hazardous, it is important that individuals who handle ionizing radiation follow certain rules to avoid accidents. The official rules to be followed by all persons licensed to handle radioactive materials have been studied and proposed by such bodies as the ICRP and the National Research Council (NRC), which is an arm of the National Academy of Sciences. The proposed standards are adopted and enforced by federal agencies such as the U.S. NRC and the Environment Protection Agency (EPA). The NRC and EPA standards for protection against radiation are contained in *Code of Federal Regulations*.[45] The exposure limits, based on these guidelines, were discussed in Section 16.8.

To protect personnel, areas where radiation sources are used are marked with certain signs. The definitions of "radiation areas" and the corresponding signs are as follows.[4]

"Restricted area" means any area to which access is controlled by the licensee for purposes of protection of individuals from exposure to radiation and radioactive materials. "Restricted area" shall not include any areas used as residential quarters, although a separate room or rooms in a residential building may be set apart as a restricted area.

"Radiation area" means an area, accessible to individuals, in which radiation levels could result in an individual's receiving a dose equivalent in excess of 0.050 mSv (5 mrem) in 1 h at 0.30 m from the radiation source or from any surface that the radiation penetrates.

"High radiation area" means an area, accessible to individuals, in which radiation levels could result in an individual's receiving a dose equivalent in excess of 1 mSv (100 mrem) in 1 h at 0.30 m from the radiation source or from any surface that the radiation penetrates.

Radiation areas should be marked with the radiation symbol shown in Figure 16.14 and with cautionary signs. If necessary, the radiation area should be roped off; or if it is a room, should be locked to keep people out.

People who work in radiation areas or use radioisotopes should keep in mind the following simple principle:

Radiation exposure should be avoided, if at all possible. If exposure is necessary, the risk from the exposure should be balanced against the expected benefit. The exposure is justified if the benefit outweighs the risk.

If exposure is justified, the employer and the employee should obey the ALARA principle (Section 16.8). Satisfying ALARA means not just keeping exposure below the maximum allowed limits; it means taking all possible "reasonably achievable" measures to minimize the dose received in any task that has to be performed. The best assurance against violating ALARA is constant education and training of the radiation workers. Given below are some commonsense rules that have proved helpful in reducing exposure.

1. Try to avoid internal exposure. Substances enter the body by mouth (eating, drinking), by breathing, through wounds, and by injection. Therefore, in places where radioactive materials are handled, do not eat, do not drink, and cover all wounds. If the air is contaminated, wear a mask. Hands should be washed after the operation is over especially if no protective gloves were used.

2. Stay close to the source of radiation for as short a time interval as possible.

3. Use protective covers, if this is the suggestion of the health physicist.

4. Place the source behind a shield or in a proper container.

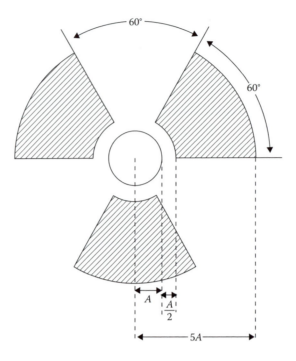

Figure 16.14 The standard radiation symbol with dimensions as shown has a yellow background, with the hatched area being magenta or purple.

5. If practical, wait for the radiation to decay to a safer level before handling it. The exponential decay law is a helpful ally.

6. Stay as far away from the source as practical. The flux of a point isotropic source decreases as $1/r^2$ (r = distance away from the source).

The shielding medium that should be used is not the same for all types of radiation. Here are simple suggestions for the three types of radiation considered in this book.

Charged particles. Charged particles have a definite range. Therefore, to stop them completely, a shielding material with thickness at least equal to the range (in that material) of the most penetrating particle should be placed between the source and the worker. A few millimeters of metal will definitely stop all charged particles emitted by radioisotopic sources. Some bremsstrahlung may get through, though.

Gammas. A beam of gammas going through a material of thickness t is attenuated by a factor $\exp(-\mu t)$, where μ is the total linear attenuation coefficient of the gamma in that medium. The higher the value of μ is, the smaller the thickness t that reduces the intensity of the beam by a desired factor. In theory, the beam cannot be attenuated to zero level. In practice, the attenuation is considered complete if the radiation level equals the background.

The attenuation coefficient μ increases with the atomic number of the material. The most useful practical element for gamma shielding is lead ($Z = 82$). Lead is relatively inexpensive, and it is easy to melt it and make shields with it having the desired shape, size, and thickness.

Neutrons. Shielding against neutrons is more difficult than shielding either against charged particles or photons. If the source emits fast neutrons, the first step is to provide a material that will thermalize the neutrons. Such materials are water, wax, or paraffin.

Thermalized neutrons are easily absorbed by many isotopes. Examples are ^{115}In, ^{113}Cd, and ^{10}B. Of these, the most practical to use is boron. It can be used in powder form or be dissolved in water or liquid wax. A very simple but effective shield for a source of fast neutrons is 0.15–0.30 m of wax or paraffin to which boron has been added. The thickness of this borated material may change, depending on the strength of the source and the amount of boron added. Cadmium is very useful in sheet form. A cadmium sheet 3–6 mm [(1/8) – (1/4 in.)] will stop a thermal neutron beam almost completely. Finally, shields have been manufactured that are flexible, like rubber, yet are excellent neutron attenuators.

PROBLEMS

16.1 What is the total dose received by an individual standing in front of the open beam port of a research reactor for 10 s under the radiation levels listed in the figure below?

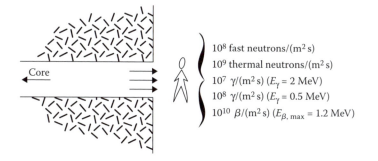

10^8 fast neutrons/(m² s)

10^9 thermal neutrons/(m² s)

10^7 γ/(m² s) ($E_γ$ = 2 MeV)

10^8 γ/(m² s) ($E_γ$ = 0.5 MeV)

10^{10} β/(m² s) ($E_{β, max}$ = 1.2 MeV)

16.2 What is the dose rate per curie of ²⁴Na if it is shielded by 0.025 m of lead as shown below? ²⁴Na emits a 1.37-MeV gamma and a 2.75-MeV gamma 100% of the time, betas with E_{max} = 4·17 MeV 0.003% of the time, and betas with E_{max} = 1.389 MeV 100% of the time.

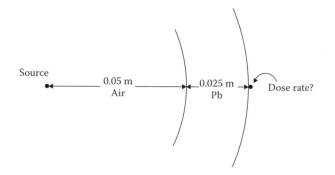

16.3 As a result of carelessness, a worker inhaled 1 μg of ²⁴¹Am. Considering alpha particles only and assuming that the americium is spread uniformly in the bones, (a) what is the dose rate at the time of the accident and (b) what is the total dose to that individual? For ²⁴¹Am, T_R = 433 years and T_B = 200 years, mass of bones = 10 kg.

16.4 What is the dose rate 0.30 m away from 1 Ci of ⁶⁰Co if (a) the attenuating medium is air, and (b) if the source is shielded by 0.01 m of aluminum?

16.5 What is the thickness of a lead container that will result in a dose rate at its surface of 2.5 mrem/h = 2.5 × 10⁻⁵ Sv/h, if it is used to store 1 Ci of ¹²⁴Sb? ¹²⁴Sb emits the following gammas:

Energy (MeV)	Intensity (%)	Energy (MeV)	Intensity (%)	Energy (MeV)	Intensity (%)
0.603	97	0.967	2.4	1.37	5
0.644	7	1.048	2.4	1.45	2
0.720	14	1.31	3	1.692	50
				2.089	7

16.6 What is the dose rate due to the ¹²⁴Sb of Problem 16.5 at a distance equal to the thickness of the container, if the attenuating medium is air?

16.7 A 1-Ci sample of ⁶⁰Co is stored behind a concrete and lead shield as shown below. What is the dose rate at point *P*?

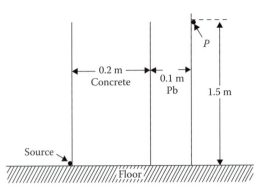

16.8 What is the dose rate at 1' away from a ^{56}Mn foil with activity 1 mCi? Consider only the dominant gamma emitted.

16.9 The isotope ^{99m}Tc is used *in vivo* for diagnostic purposes in humans. It emits X-rays with energy 140 keV and betas with $E_{max} = 0.119$ MeV. If a person is injected with 1 μCi of this isotope, what is the total dose to the brain, assuming that all of the isotope is uniformly distributed there? (Mass of the brain = 1.5 kg.)

16.10 If all the water in the human body were suddenly changed to $T_2O(T = {}^3H)$, what would be the total dose to that individual? Assume 61% of the body is water.

16.11 If 1 pCi of ^{239}Pu is inhaled by breathing and gets into the lungs, what is going to be the total dose to that individual? ($T_R = 24{,}000$ years, $T_B = 200$ years, mass of lungs = 1 kg)

16.12 Calculate the total dose rate at the center of a spherical submarine submerged in contaminated water with activity 1 Ci/m^3 of ^{137}Cs. The submarine is made of steel 0.025 m thick. Its radius is 1.5 m and it is filled with air at 1 atm.

16.13 What is the annual dose to a person due to the ^{40}K found in every human body? The isotopic abundance of ^{40}K in potassium is 0.0119%. The human body contains 1.7×10^{-3} kg of potassium per kg. ^{40}K emits the following radiations:

Particle	Energy (MeV)	Intensity (%)
β^-	1.31	89
β^+	0.49	10^{-3}
γ	1.46	11
γ	0.511	2×10^{-3}

16.14 A "hot particle" (a tiny speck of radioactive material) was lodged on the palm of a radiation worker, from where it was removed after 10 min of scrubbing. What is the estimated total dose received by the tissue exposed to this radiation if it was determined that the radioactivity came from a beta emitter with $E_{max} = 1.7$ MeV, half-life equal to 2 days, and initial estimated activity equal to 1.3 mCi?

16.15 Show that the doubling dose for radiation-induced mutations is between 0.5 and 2.5 Sv (50–250 rem) if the probability to induce a mutation by irradiation is 2–0.4 per Sv.

BIBLIOGRAPHY

Attix, F. H., *Introduction to Radiological Physics and Radiation Dosimetry*, Wiley, 1986.

Attix, F. H., and Tochilin, E., (eds.), *Radiation Dosimetry I, II, III*, Academic, New York, 1969.

Boetter-Jensen, L., McKeever, S. W. S., and Wintle, A. G., *Optically Stimulated Luminescence Dosimetry*, Elsevier Science, 2003.

Cember, H., and Johnson, T., *Introduction to Health Physics*, 4th ed., McGraw-Hill, 2008.

Fitzgerald, J. J., Brownell, G. L., and Mahoney, F. J., *Mathematical Theory of Radiation Dosimetry*, Gordon and Breach, 1967.

Forshier, S., *Essentials of Radiation, Biology and Protection*, 2nd ed., Delmar Cengage Learning, 2008

Hine, G. J., and Brownell, G., *Radiation Dosimetry*, Academic, New York, 1956.

Martin, A., and Harbison, S. A., *An Introduction to Radiation Protection*, A Hodder Arnold Publication, 2006.

Miller, K. L., *Handbook of Management of Radiation Protection Programs*, 2nd ed., CRC Press, Boca Raton, FL, 1992.

Morgan, K. Z., and Turner, J. E., (eds.), *Principles of Radiation Protection*, Wiley, 1967.

Nuclear Regulatory Commission; http://www.nrc.gov/reading-rm/doc-collections/efr/part020/

Prasad, K. N., *Handbook of Radiobiology*, CRC Press, Boca Raton, FL, 1984.

Seeram, E., *Rad Tech's Guide to Radiation Protection*, Wiley-Blackwell, 2001.

Shani, G., *Radiation Dosimetry Instrumentation and Methods*, CRC Press, Boca Raton, FL, 1991.

Shapiro, J., *Radiation Protection: A Guide for Scientists, Regulators and Physicians*, Harvard University Press, 2002.

Stabin, M.G., *Radiation Protection and Dosimetry: An Introduction to Health Physics*, Springer, 2007.

Turner, J. E., *Atoms, Radiation, and Radiation Protection*, 3rd ed., Wiley-VCH, 2007.

REFERENCES

1. Dose Equivalent ICRU Report 19 (suppl.), Washington, D.C. (1973).

2. *1977 Recommendations of the International Commission on Radiological Protection* (ICRP Publication 26), Pergamon, 1977.

3. Report of the RBE Committee to ICRU, *Health Phys.* **9**:357 (1963).

4. Standards for Protection against Radiation, *Code of Federal Regulations*, Title 10, Part 20, Dec. 1992, Chapter 1.

5. *Engineering Compendium on Radiation Shielding*, vol. 1: *Shielding Fundamentals and Methods*, Springer-Verlag, 1968.

6. Stamatelatos, M. G., and England, T. R., *Nuclear Sci. Eng.* **63**:204 (1977).

7. Neutron and Gamma-ray Flux-to-Dose Rate Factors, American National Standard, ANSI/ANS-6.1.1- 1991.

8. Report of the ICRP Committee II on Permissible Dose for Internal Radiation (1959). *Health Phys.* **3**:146 (June 1960).

9. Health Effects of Exposure to Low Levels of Ionizing Radiation (the BEIR-V report), National Academy Press, Washington, D.C., 1990.

10. Health Risks from Low Exposure to Low Levels of Ionizing Radiation (the BEIR-VI Pahse 2 Report), National Academies Press, 2006.

11. *Recommendations of the International Commission on Radiological Protection* (ICRP Publication 103), Pergamon, 2006.

12. Dudley, R. A., in F. H. Attix and W. C. Roesch (eds.), *Radiation Dosimetry*, II, Academic Press, New York, 1966.

13. Charalambous, S., and Petridou, C, *Nucl. Instrum. Meth.* **137**:441 (1976).

14. Simons, G. G., and Emmons, L. L., *Nucl. Instrum. Meth.* **160**:79 (1979).

15. Ohno, A., and Matsuura, S., *Nucl. Tech. Meth.* **47**:485 (1980).

16. Bacci, C., Calicchia, A., Pugliani, L., Salvadori, P., and Furetta, C., *Health Phys.* **38**:21 (1980).

17. Vohra, K. G., Bhatt, R. C., Chandra, B., Pradhan, A. S., Lakshmanan, A. R., and Shastry, S. S., *Health Phys.* **38**:113 (1980).

18. Liu, N. H., Gilliam, J. D., and Anderson, D. W., *Health Phys.* **38**:359 (1980).

19. Nakajima, T., Murayama, Y., Matsuzawa, T., and Kayano, A., *Nucl. Instrum. Meth.* **157**:155 (1978).

20. Edmund, J. M., Andersen, C. E., and Greilich, S., *Nucl. Instrum. Meth. Phy. Res. B* **262**:261 (2007).

21. Gastélum, S., Cruz-Zaragoza, E., Chernov, V., Meléndrez, R., Pedroza-Montero, M., and Barboza-Flores, M. *Nucl. Instrum. Meth. Phys. Res. B* **260**:592(2007).

22. Yazici, A. N., Bedir, M., and Sibel Sokucu, A., *Nucl. Instrum. Meth. Phy. Res. B* **259**:955 (2007).

23. Massillon, G., Gamboa-deBuen, I., Buenfil, A. E., Monroy-Rodrıguez, M. A., and Brandan, M. E., *Nucl. Instrum. Meth. Phys. Res. B* **266**:772 (2008).

24. Soliman, C., *Nucl. Instrum. Meth. Phys. Res. B* **267**:2423 (2009).

25. Yang, B., Wang, L., Townsend, P. D., and Gao, H., *Nucl. Instrum. Meth. Phys. Res. B* **266**:2581 (2008).

26. Zacharias, N., Stuhec, M., Knezevic, Z., Fountoukidis, E., Michael, C. T., and Bassiakos, Y., *Nucl. Instrum. Meth. Phys. Res. A* **580**:698 (2007).

27. Zhou, D., Semones, E., Gaza, R., Johnson, S., Zapp, N., and Weyland, M., *Nucl. Instrum. Meth. Phys. Res. A* **580**:1283 (2007).

28. Andersen, C. E., Edmund, J. M., Medin, J., Grusell, E., Jain, M., and Mattsson, S., *Nucl. Instrum. Meth. Phys. Res. A* **580**:466 (2007).

29. Afouxenidis, D., Stefanaki, E. C., Polymeris, G. S., Sakalis, A., Tsirliganis, N. C., and Kitis, G., *Nucl. Instrum. Meth. Phys. Res. A* **580**:705 (2007).

30. Jain, M., Andersen, C. E., Hajdas, W., Edmund, J. M., and Bøtter-Jensen, L., *Nucl. Instrum. Meth. Phys. Res. A* **580**:652 (2007).

31. Bonner, T. W., *Nucl. Phys.* **23**:116 (1961).

32. Thomas, D. J., and Alevra, A. V., *Nucl. Instrum. Meth. Phys. Res. A* **476**:12 (2002).

33. Thomas, D. J., Bardell, A. G., and Macaulay, E. M., *Nucl. Instrum. Meth. Phys. Res. A* **476**:31 (2002).

34. Rimpler, A., *Nucl. Instrum. Meth. Phys. Res. A* **476**:468 (2002).

35. Braga, C. C., and Dias, M. S., *Nucl. Instrum. Meth. Phys. Res. A* **476**:252 (2002).

36. Bedogni, R., Domingo, C., Esposito, A., and Fernández, F., *Nucl. Instrum. Meth. Phys. Res. A* **580**:1301 (2007).

37. D'Errico, F., Nath, R., Holland, S. K., Lamba, S., Patz, S., and Rivard, M. J., *Nucl. Instrum. Meth. Phys. Res. A* **476**:113 (2002).

38. Hankins, D. E., LA-2717 (1962) http://library.lanl.gov/cgi-bin/getfile20020991.pdf.

39. Marshall, T. O., Bartlett, D. T., Burgess, P. H., Cranston, D. J., Higginbottom, D. J., and Sutton, K. W., Occupational Radiation Exposure, BNES Paper 44, London (1991), p. 129.

40. Tume, P., Lewis, B. J., Bennett, L. G. I., and Cousins, T., *Nucl. Instrum. Meth. Phys. Res. A* **406**:153 (1998).

41. Das, M., Chatterjee, B. K., Roy, B., and Roy, S. C., *Nucl. Instrum. Meth. Phys. Res. A* **452**:273 (2000).

42. Aroua, A., and Hofert, M., *Nucl. Instrum. Meth. Phys. Res. A* **372**:318 (1996).

43. Jung, M., Teissier, C., and Siffert, P., *Nucl. Instrum. Meth. Phys. Res. A* **458**:527 (2001).

44. Shizuhiko, D., and Nishizawa, K., *Health Phys.*, **89**:224 (2005).

45. Standards for Protection against Radiation, *Code of Federal Regulations*, Title 40, Part 190–192, December 1992.

17

Applications of Radiation Detection

17.1 INTRODUCTION

The advances in miniaturization, detector efficiency, and newer materials have greatly improved radiation monitoring. More recently (after 2001), the rise of terrorism, homeland security, and concerns about the security of nuclear materials have stimulated the advances in sophisticated radiation detection systems. Many of these systems include the use of large array of detectors for smuggling interdiction, neutron interrogation methods for special nuclear materials and various γ-ray detectors for low level radiation monitoring. Equally important has been the advances of detection systems in nuclear medicine both for diagnosis and therapy.

This chapter cannot adequately cover all the modern applications of radiation monitoring and detection in great detail. What is presented is a sampling of the most common utilizations of radiation detection technology not previously presented in the first two editions. All the basic fundamental understanding of these techniques has been covered in the previous chapters.

17.2 HEALTH PHYSICS WITHIN NUCLEAR POWER PLANTS AND RADIOLOGICAL FACILITIES

17.2.1 Active Personal Dosimeters

Typically about a thousand persons are employed in a nuclear power plant, as reactor operators, engineers, technicians, security guards, administrators, and so on. Exposure of workers to radiation occurs, primarily, at the time of refueling, when the highly radioactive spent fuel is taken out of the core and moved to an on-site facility and new fuel is placed in the reactor core. Typically, for U.S. nuclear plants, refueling takes place every eighteen months to two years. Exposure of radiation can also happen when workers enter areas for maintenance. Following U.S. Nuclear Regulatory Commission (and other similar international regulatory bodies) and the ALARA principle, strict and accurate monitoring of the workers is required. Electronic dosimeters (see Chapter 16) have been in existence for many years,[1-5] however, online or active personal dosimeters are now more common. These devices give instant readings of radiation fields. It has also been reported the development of an on-line hybrid electronic radiation dosimeter for a mobile robot[6] inside a nuclear power plant.

17.2.2 Continuous Air Monitors and Continuous Air Particulate Monitors

For many years continuous air monitors (CAM) or continuous air particulate monitors (CAPM) have been employed in nuclear power plants and nuclear facilities to detect alpha, beta, and gamma radiation. The basic principle relies on a pump that draws air to a filter medium which is placed above a detector. Typically, GM counters, NaI detectors or plastic scintillators are used for gross beta-gamma counting. In some circumstances either a high purity germanium detector or alpha detector is used for more detailed radionuclide identification.

Other applications include the use of a continuous air monitor in a plutonium facility,[7] determination of gaseous radionuclide forms in the stack air of nuclear power plants,[8] the effectiveness of the alpha–beta-pseudo-coincidence-difference methods in air-borne alpha-activity monitoring,[9] gaseous standards preparation with the radionuclide ^{41}Ar for stack monitors calibration and verification in nuclear facilities,[10] and performance evaluation of the sampling head and annular kinetic impactor in the Savannah River site alpha continuous air monitor.[11] The measurement of the size of plutonium particles from internal sputtering into air can be performed using size-segregated filters[12]; performance testing of continuous air monitors for alpha-emitting radionuclides has also been done.[13]

17.2.3 Area Monitors and Environmental Monitoring

Besides radiation monitoring within nuclear power plants and radiological facilities, site perimeter and environmental monitoring are legally mandated. The detection systems often called radiation area monitors are placed on-site and at the perimeter of nuclear power plants. Typically, GM counters, Na(I) detectors, and ^{129}I particulate detection systems are used which are then remotely controlled. For environmental monitoring usually done at least once a year, biological and geological samples are collected and then analyzed for alpha, beta and gamma radiation. Samples may include soil vegetation, water, local dairy milk, and aquatic life. Often because of very low levels of radiation, radiochemistry procedures to concentrate the sample need to be implemented before the standard techniques as outlined in previous chapters can be employed.

Other related research has included the development of radiation area monitors for accelerator applications,[14] daily variations of indoor air-ion and radon concentrations,[15] and Latin American and Caribbean intercomparison of radiation protection area monitoring instruments.[16]

17.2.4 Foot and Hand Surface Contamination Monitors

Workers who actively handle radioactive materials are usually required to check themselves before leaving the premises, on foot and hand surface contamination monitors. These large devices typically employ proportional counters for beta detection and gas flow proportional counters for alpha and beta detection. A more complete description of these devices is available from industrial manufacturers.[17,18]

17.2.5 Whole Body Counters

Whole body counters are primarily used for quick gamma radiation monitoring of people leaving nuclear power plants and radiological facilities as well as more thorough nonevasive examinations of individuals working in radiation environments. For quick scans, usually a NaI is employed. However, for more detailed determination of potential inhaled or ingested gamma emitting radionuclides, a single or multitude of germanium detectors are used. These counters are calibrated using the BOttle MAnikin ABsober phantom (BOMAB)[19] for various geometrical positions including, standing, sitting and lying down. First developed in 1949, the phantom consists of ten polyethylene bottles, either cylinders or elliptical cylinders that represent the head, neck chest, abdomen, thighs, calves, and arms. Each section is filled with a radioactive solution, in water, that has the amount of radioactivity proportional to the volume of each section. This simulates a homogeneous distribution of material throughout the body.[20] Some recent investigations include the calculation of a size correction factor for a whole-body counter,[21] depth of γ-ray interaction positron emission tomography system,[22] and a MCNP-based calibration method and a voxel phantom for in vivo monitoring of ^{241}Am in skull.[23]

17.3 PORTAL MONITORS AND PASSIVE DETECTION

Portal monitors are essentially very large detection systems that are employed to detect radiation or special nuclear materials (SMN) entering or leaving places of security concern. These monitors incorporate an array of gamma scintillators and/or ^{3}He neutron detectors for passive detection of gamma emitting radionuclides or special nuclear materials such as enriched uranium or plutonium. . There are essentially two types of portal systems: one for pedestrians and one for vehicles. There are several factors that need to be taken into account to maximize the probability of detection. In γ-ray identification shielding of the material either by design (as in smuggling) or just by the presence of other nearby objects is the main obstacle. However, for isotopic content of SNM, the chemical composition and size of the SNM particles, ambient background, electronic noise, distance between the detectors, type, number or size of radiation detectors and passage speed all can affect performance of the portal monitors. False positive γ-ray identification is another problem usually associated with the presence of naturally occurring radioactive materials.

Recent research in this area has included sensitivity tests and risk evaluation for steelworks portal systems,[24] a methodology for improving throughput using portal monitors,[25] sensitivity studies,[26] test sources for portal monitors used for homeland security,[27] response of radiation portal monitors to medical radionuclides at border crossings,[28] and development of international standards for instrumentation used for detection of illicit trafficking of radioactive material.[29] More detailed specifications can be found elsewhere.[17,18]

Responding to airline hijackings in the 1970s the first use of radiation technology at airports was and still is X-ray imaging. While X-ray imaging is suitable to reveal objects of concern in luggage or suitcases such as guns or other weapon-like devices, it is not employed for the scanning of passengers. Instead a recent method entitled X-ray backscatter is used to ascertain not only the concealment of weapons but also of unusual objects, packages or liquid containers on the passenger. The principle of X-ray backscatter is that the imaging system detects the radiation which scatters back from the target. This technique has been pioneered by American Science and Engineering[30] and currently several systems have been deployed at many airports. The dose accumulated per scan is 0.005 millirem (0.00005 mSv),[31] which is orders of magnitude lower than any background radiation of around 300 mrem or 3mSv per year. This technique of interrogation has also been applied to cargo and vehicle inspection.[30]

17.4 INTERACTIVE RADIATION DETECTION SYSTEMS

Since the 1990s interrogation techniques have been used to assay uranium and plutonium in waste containers. These methods has been further developed as part of nonproliferation measures and various homeland security programs. One common technique is the online use of neutrons to interact with the material itself and then detect the emitted prompt capture γ-rays or even neutrons. Typically 14 MeV fast neutrons produced from (D,T) reactions give rise to reactions of the type $A(n, n'\gamma)A$. Various types of research have been done including the use of the 14 MeV neutrons in the detection of explosives,[32] chemical composition identification using fast neutrons,[33] comparison of neutron and high-energy X-ray dual-beam radiography for air cargo inspection,[34] fast neutron and γ-ray interrogation of air cargo containers,[35] a review of neutron based nonintrusive inspection technique,[36] measurement of 14 MeV neutron-induced prompt γ-ray spectra from 15 elements found in cargo containers,[37] and the role of neutron based inspection techniques.[38] There are also portable neutron interrogation systems,[39,40] militarily fielded thermal neutron activation sensor for landmine detection,[41] photofission in uranium by nuclear reaction γ-rays,[42] active detection of small quantities of shielded highly-enriched uranium,[43] and a portable neutron generator for the associated particle technique.[44]

17.5 UNMANNED AERIAL VEHICLES FOR RADIATION DETECTION

As a result of the 1986 Chernobyl reactor accident the use of unmanned aerial vehicles (UAVs) for radiation detection has gained significant prominence. UAVs can also be deployed for nuclear surveillance in the case of a successful terrorist attack. The main advantages of UAVs are their ability to fly close to the source of radiation as well as to fly at high altitudes which would be very important in characterizing a radioactive plume. Typically UAVs carry GM, NaI or CZT detectors which do not need cooling unlike those made from germanium. The usual standard electronics for data acquisition and a GPS system is incorporated onto a UAV. In 1990 a mobile survey of environmental gamma radiation and fall-out levels in Finland was performed to check contamination from the Chernobyl accident.[45] Related research conducted has included radiation surveillance using an UAV[46,47] and design of an air sampler for a UAV.[48]

17.6 COINCIDENCE AND ANTICOINCIDENCE DETECTION SYSTEMS

Coincidence–anticoincidence techniques have been presented in Section 10.8. Since the appearance of the second edition of the book, significant changes and new applications of this technique have appeared. These new techniques/applications are discussed briefly in this section. The use of sophisticated coincidence and anticoincidence techniques to lower backgrounds and thus increase analytical sensitivities for radiation detection have been in existence for many years. Nuclear physicists have used these methods primarily to deduce complex decay schemes and assign spin numbers to the various nuclear levels. In the 1970s coincidence and anticoincidence methods (primarily Compton suppression) began to appear in very low level environmental γ-ray detection measurements. Two review articles appeared in the 1990s on these methods.[49,50] In coincidence techniques specific signals of alpha or beta particles, or γ-rays from radioactive isotopes are electronically gated (within a short time sequence usually ~100 ns) with two detectors to reduce the background levels. Recent work has included $\beta-\gamma$ coincidence for radioxenon detection,[51] a neutron induced prompt γ-ray spectroscopy system using a ^{252}Cf neutron source for quantitative analysis of aqueous samples,[52] analysis of ^{26}Al in meteorite samples by coincidence γ-ray spectrometry,[53] application of neutron well coincidence counting for plutonium determination in mixed oxide fuel fabrication plant,[54] and coincidence and anticoincidence measurements in prompt gamma neutron activation analysis with pulsed cold neutron beams.[55]

In Compton suppression methods γ-rays that scatter from the primary detector into another detector usually a NaI or BGO (bismuth germanate), are suppressed. This effect leads to a reduction of the background up to a factor or ten or more. This technique has been effectively used in a wide range of neutron activation analysis and fission product identification research. Recent topics have included assessment of a Compton-event suppression γ-spectrometer for the detection of fission products at trace levels,[56] low-level gamma-spectrometry using beta coincidence and Compton suppression,[57] analysis for trace elements in some cereals and vegetables using short-lived nuclides,[58] the determination of uranium in food samples by Compton suppression epithermal neutron activation analysis,[59] and an evaluation of Compton suppression neutron activation analysis for determination of trace elements in geological samples.[60]

17.7 NUCLEAR MEDICINE

Radiation measurements and detection systems in nuclear medicine cover a wide spectrum. A good review of these detectors has previously appeared.[61] Production of radiopharmaceuticals requires stringent quality control practices which includes determining the purity of the radionuclide. This is especially true for isotopes that are produced in the uranium fission process in which there are a host of other radionuclides simultaneously produced. Before administering a radiopharmaceutical to the patient, information on the expected dose is needed, which includes both the purity and activity. Purity is regulated by identifying any other photons which may be in the radiopharmaceutical, which is normally done with a germanium or NA(I) detector. A dose calibrator is a very common detector used in nuclear medicine. It is typically an ionization chamber that is calibrated for activity measurement for 0. 37 MBq (10 μCi) to 0.37 − 0.74 GBq (1–2 Ci). For instance, a NaI crystal is used as a thyroid uptake probe. This device monitor is placed near the thyroid to monitor the uptake of radioactive ^{131}I or ^{123}I for either diagnosis or therapy. Lymphoscintigraphy is the imaging of pathways of lymphatic flow and lymph nodes after injection of a radiopharmaceutical that is absorbed by the lymphatics. A small gamma intraoperative detector such as CdTe or CdZnTe probe is used as a guide to find the radiolabeled lymph nodes. NaI or Ge well counters are used in the determination of activity concentrations of *in vitro* blood or urine samples for radiotracer bioassays.

Ideally radiopharmaceuticals used in imaging should be single-photon-emitting or positron-emitting radionuclides. Nuclear imaging methods can be acquired in two distinct ways. The first way is to administer a radiopharmaceutical to a specific organ or area of the body and then take pictures with a detector sensitive to the specific radiation used. Such pictures include Single Photon Emission Computed Tomography (SPECT) images. Typically NaI detectors are used; however attempts have been made to replace these scintillators with semiconductor detectors with better spectrometric characteristics to improve contrast and quantitative measurements such as CdTe or CdZnTe.[62] Transmission images are produced by placing the subject in front of a radiation source and detecting the transmission radiation with a digital a camera. Information about the distribution of radiation-attenuating or radiation-absorbing matter in the subject is thus collected. A typical X-ray examination would fall under this category. In the case of SPECT an array of detectors is used so a series of pictures can be reconstructed with sophisticated algorithms to give a 3D tomographic image.

Further research includes detector technology challenges for nuclear medicine and positron emission tomography,[63] applications of semiconductor detectors to nuclear medicine,[64] suitability of nuclear medicine gamma cameras as gamma spectrometers in the event of a radiological emergency,[65] a nuclear medicine γ-ray detector based on germanium strip detector technology,[66] detector design issues for compact nuclear emission cameras dedicated to breast imaging,[67] photodetectors for nuclear medical imaging,[68] and lanthanum scintillation crystals for γ-ray imaging.[69]

17.8 DETECTION OF NUCLEAR MATERIALS/NONPROLIFERATION ISSUES

The nonproliferation of nuclear materials and their acquisition by unauthorized governments or groups or individuals is a complicated geopolitical issue that requires technological solutions including the implementation of sophisticated equipment for detection thus deterring trafficking in SNM or radiological dispersive devices. There is currently a variety of detection capabilities employed for nuclear forensics. Some of the research includes a compact μ-XRF spectrometer for SNM,[70] age-dating of highly enriched uranium by gamma-spectrometry,[71] basic characterization of highly enriched uranium by gamma-spectrometry,[72] verification of the ^{239}Pu content, isotopic composition and age of plutonium in Pu–Be neutron sources by gamma-spectrometry,[73] detection of previous neutron irradiation and reprocessing of uranium materials for nuclear forensic purposes,[74] revealing smuggled nuclear material covered by a legitimate radioisotope shipment using CdTe-based γ-ray spectrometry,[75] determination of plutonium origin by nuclear forensics,[76] documentation of a model action plan to deter illicit nuclear trafficking,[77] and environmental monitoring as an important tool for safeguards of nuclear material and nuclear forensics.[78]

REFERENCES

1. Kawano, T., and Ebihara, H. G. B., *Health Phys.* **65**:313 (1993).

2. Hirning, C. R., and Yuen, P. S., *Health Phys.* **69**:46 (1995).

3. Wielunskia, M., Wahla, W., EL-Faramawy, N., Rühma, W., Luszik-Bhadra M., and Roos, H., *Rad. Meas.* **43**:1063 (2008).

4. Bolognese-Milsztajn, T., Ginjaume, M., Luszik-Bhadra, M., Vanhavere, F., Wahl, W., and Weeks, A., *Rad. Prot. Dosim.* **112**:141 (2004).

5. Clairand, I., Struelens, L., Bordy, J.-M., Daures, J., Debroas, J., Denozières, M., Donadille, L., Gouriou, J., Itié, C., Vaz, P., and d'Errico F., *Rad. Protec. Dosim.* **129**: 340 (2008).

6. Lee, N. H., Cho, J. W., Kim, S. H., and Youk, G. U., *Intelligent Techniques and Soft Computing in Nuclear science and Engineering Proceedings of The 4th International FLINS Conference*, Bruges, Belgium, 28–30 August 2000, pp. 478–487.

7. Whicker, J. J., Rodgers, J. C., Fairchild, C. I., Scripsick, R. C., and Lopez, R., *Health Phys.* **72**:734 (1997).

8. Tecl, J., and Svetlik, I., *Appl. Rad. Isotop.* **67**: 950 (2009).

9. Shevchenko, S. V., Saltykov, L. S., and Slusarenko, L. I., *Appl. Rad. Isotop.* **64**:55 (2006).

10. Kovar, P., and Pavel Dryak, P., *Appl. Rad. Isotop.* **66**:796 (2008).

11. Chen, B. T., Hoover, M. D., Newton, G. J., Montano, S. J., and Gregory, D. S., *Aerosol. Sci. Tech.* **31**:24 (1999).

12. Cheng, Y-S., Holmes, T. D., Timothy, George, G., and Marlow, W. H., *Nucl. Instrum. Meth. Phys. Res. B* **234**:219 (2005).

13. Hoover, M. D., and Newton, G. J., *Rad. Prot Dosim.* **79**:499 (1998).

14. Krueger, F., and Larson, J., *Nucl. Instrum. Meth. Phys. Res. A* **495**:20 (2002).

15. Kolarz, P. M., Filipovic, D. M., and Marinkovic, B. P., *Appl. Rad. and Isotop.* **67**:2062 (2009).

16. Ramos, M. M. O., Da Cunha, P. G., and Suárez, R. C., *Rad. Prot. Dosim.* **102**:315 (2002).

17. http://www.canberra.com.

18. http://www.thermo.com.

19. Bush, F. *Br. J. Radiol.* **22**:96 (1949)

20. http://en.wikipedia.org/wiki/Bomab

21. Carinou, E., Koukouliou, V., Budayova, M., Potiriadis, C., and Kamenopoulou, V., *Nucl. Instrum. Meth. Phys. Res. A* **580**:197(2007).

22. Ohi, J., and Tonami, H., *Nucl. Instrum. Meth. Phys. Res. A* **571**:223(2007).

23. Moraleda, M., Gómez-Ros, J. M., López, M. A., Navarro, T., and Navarro, J. F., *Nucl. Instrum. Meth. Phys. Res. A* **526**:551(2004).

24. Campi, F., and Porta, A. A., *Rad. Meas.* **39**:161(2005).

25. Kramer, G. H., Capello, K., Hauck, B., Moodie, G., DiNardo, A., Burns, L., Chiang, A., Marro, L., and Brown, J. T., *Rad. Prot. Dosim.* **134**:152 (2009).

26. Kramer, G. H., Capello, K., Hauck, B. M., and Brown, J. T., *Radiat. Prot. Dosim.* **127**:249 (2007).

27. Lucas, L., Pibida, L., Unterweger, M., and Karam, L., *Radiat. Prot. Dosim.* **113**:108 (2005).

28. Kouzes, R. T., and Siciliano, E. R., *Rad. Meas.* **41**:499 (2006).

29. Voytchev, M., Chiaro, P., and Radev, R., *Rad. Meas.* **44**:1 (2009).

30. http://www.as-e.com/

31. http://www.as-e.com/pdf/Health-Physics-Screening-Individuals.pdf

32. Sudac, D., Blagus, S., Matika, D., Kollar, R., Grivicic, T., and Valkovic, V., "The Use of the 14 MeV Neutrons in the Explosive Detection," *Proceedings from the International Conference on Requirements and Technologies for the Detection, Removal and Neutralization of Landmines and UXO, Vol.2,* Brussels, Belgium, September 15–18, pp. 749–754 (2003).

33. Sudac D., Blagus, S., and Valkovic, V., *App. Rad. Isotop.* **61**:73 (2004).

34. Liu, Y., Sowerby, B. D., and Tickner, J. R., *App. Rad. Isotop.* **66**:463 (2008).

35. Eberhardt, J. E., Liu, Y., Rainey, S., Roach, G. J., Stevens, R. J., Sowerby, B. D., and Tickner, J.R., *International Workshop on Fast Neutron Detectors, Cape Town, 3–6 April 2006, Proceedings of Science (FNDA2006) 092* (2006).

36. Gozani, T., *Conference on Technology for Preventing Terrorism,* Hoover Institution National Security Forum, Stanford University, 12–13 March 2002, p. 12.

37. Perota, B., Carascoa, C., Bernarda, S., Mariania, A., Szabob, J.-L., Sannieb, G., Valkovic, V., *et al., App. Rad. Isotop.* **66**:421 (2008).

38. Gozani, T., *Nucl. Instr. and Meth. Phys. Res. B* **213**:460 (2004).

39. Nebbia, G., Pesente, S., Lunardon, M., Viesti, G., LeTourneur, P., Heuveline, F., Mangeard, M., and Tcheng, C., *Nucl. Instr. Meth. Phys, Res. A* **533**:475 (2004).

40. http://www.ortec-online.com

41. Clifforda, E. T. H., McFeeb, J. E., Inga, H., Andrewsa, H. R., Tennanta, D., Harpera, E., and Faustb, A. A., *Nucl. Instrum. Meth. Phys. Res. A* **579**:418 (2007).

42. Morse, D. H., Antolak, A. J., and Doyle, B. L., *Nucl. Instrum. Meth. Phys. Res. B* **261**:378 (2007).

43. Kerr, P., Rowland, M., Dietrich, D., Stoeffl, W., Wheeler, B., Nakae, L., Howard, D., Hagmann, C., Newby, J., and Porter, P., *Nucl. Instrum. Meth. Phys. Res. B* **261**:347 (2007).

44. Chichester, D. L., Lemchak, M., and Simpson, J. D., *Nucl. Instrum. Meth. Phys. Res. B* **241**:753 (2005).

45. Arvela, H., Markkanen, M., and Lemmela, H., *Rad. Prot. Dosim.* **32**:177 (1990).

46.   Pöllänen, R., Toivonen, H., Peräjärvi, K., Karhunen, T., Ilander, T., Lehtinen, J., Rintala, K., Katajainen, T., Niemelä, J., and Juusela, M., *App. Rad. Isotop.* **67**:340 (2009).

47.   Kurvinena, K., Smolandera, P., Pöllänen, R., Kuukankorpia, S., Kettunenc, M., and Lyytinend, J., *J. Environ. Rad.* **81**:1 (2005).

48.   Peräjärvi, K., Lehtinen, J., Pöllänen, R., and Toivonen, H., *Rad. Prot. Dosim.* **132**:328 (2008).

49.   Landsberger, S., and Peshev, S., *J. Radioanal. Nucl. Chem.* **202**:203 (1996).

50.   Landsberger, S., *J. Radioanal. Nucl. Chem.* **179**:67 (1994).

51.   Ward, R. M., Biegalski, S. R., Haas, D. A., and Hennig, W., *J. Radioanal. Nucl. Chem.* **282**:693 (2009).

52.   Park, Y. J., Song, B. C., Chowdhury, M. I., and Jee, K. Y., *J. Radioanal. Nucl. Chem.* **260**:585 (2004).

53.   Povinec, P. P., Sýkora, I., Porubčan, V., and Ješkovský, M., *J. Radioanal. Nucl. Chem.* **282**:805 (2009).

54.   Kumar, P., and Ramakumar, K. L., *J. Radioanal. Nucl. Chem.* **277**:419 (2008).

55.   Zeisler, R., Lamaze, G. P., and Chen-Mayer, H. H., *J. Radioanal. Nucl. Chem.* **248**:35 (2001).

56.   Peerani, P., Carbol, P., Hrnecek, E., and Maria Betti, M., *Nucl. Instrum. Meth. Phys. Res. A* **482**:42 (2002).

57.   Grigorescu, E. L., De Felice, P., Razdolescu, A.-C., and Luca, A., *App. Rad. and Isotop.* **61**:191 (2004).

58.   Nyarko, B. J. B., Akaho, E. H. K., Fletcher, J. J., and Chatt, A., *App. Rad. Isotop.* **66**:1067 (2008).

59.   Landsberger, S., and Kapsimalis, R., *App. Rad. Isotop.* **67**:2097 (2009).

60.   Landsberger, S., and Kapsimalis, R., *App. Rad. Isotop.* **67**:2104 (2009).

61.   Ranger, N. T., *RadioGraphics* **19**:481(1999).

62.   Scheiber, C., *Nucl. Instrum. Meth. Phys. Res. A* **448**:513 (2000).

63.   Marsden, P. K., *Nucl. Instrum. Meth. Phys. Res. A* **513**:1 (2003).

64.   Barber, H. B., *Nucl. Instrum. Meth. Phys. Res. A* **436**:102 (1999).

65.   Engdahl, J. C., and Bharwani, K., *Nucl. Instrum. Meth. Phys. Res. A* **553**:569 (2005).

66.   Hall, C. J., Helsby, W. I., Lewis, R. A., Nolan, P., and Boston, A., *Nucl. Instrum. Meth. Phys. Res. A* **513**:47 (2003).

67.   Levin, C. S., *Nucl. Instrum. Meth. Phys. Res. A* **497**:60 (2003).

68.   Moses, W. W., *Nucl. Instrum. Meth. Phys. Res. A* **610**:11 (2009).

69.   Pani, R., Bennati, P., Betti, M., Cinti, M. N., Pellegrini, R., Mattioli, M., Cencelli, V. O., *et al.*, *Nucl. Instrum. Phys. Res. A* **567**:294 (2006).

70. Vittiglio, G., Bichlmeier, S., Klinger, P., Heckel, J., Fuzhong, W., Vincze, L., Janssens, K., *et al., Nucl. Instrum. Meth. Phys. Res. B* **213**:693 (2004).

71. Nguyen, C. T., *Nucl. Instrum. Meth. Phys. Res. B* **229**:103 (2005).

72. Nguyen, C. T., and Zsigrai, J., *Nucl. Instrum. Meth. Phys. Res. B* **246**:417(2006).

73. Nguyen, C. T., *Nucl. Instrum. Meth. Phys. Res. B* **251**:227 (2006).

74. Varga, Z., and Surányi, G., *App. Rad. Isotop.* **67**:516 (2009).

75. Lakosi, L., Tam, N. C., Zsigrai, J., and Sáfár, J., *App. Rad. Isotop.* **58**:263 (2003).

76. Wallenius, M., Peerani, P., and Koch, L., *J. Radioanal. Nucl. Chem.* **246**:317 (2000).

77. Smith, D. K., Kristo, M. J., Niemeyer, S., and Dudder, G. B., *J. Radioanal. Nucl. Chem.* **276**:415 (2008).

78. Buchmann, J. H., Sarkis, J. E. S., Kakazu, M. H., and Rodrigues, C., *J. Radioanal. Nucl. Chem.* **270**:291 (2006).

Appendix A

Useful Constants and Conversion Factors

Table A1
Useful Constants

Constant	Symbol or Definition	Value 0.6022169
Avogadro's number	N_A	0.6022045×10^{24} at/mol
Elementary charge	e	$1.6021917 \times 10^{-19}$ C = 4.803250×10^{-10} esu
Atomic mass unit	u	1.66040×10^{-27} kg = 931.481 MeV
Atomic mass unit	$\frac{1}{12}$ of mass $^{12}_{6}$C	
Electron rest mass	m	9.109558×10^{-31} kg = 0.511 MeV
Proton rest mass	M_p	1.672622×10^{-27} kg = 938.258 MeV
Neutron rest mass	M_n	1.674928×10^{-27} kg = 939.552 MeV
Planck constant	h	6.626196×10^{-34} J·s
Boltzmann constant	k	1.380622×10^{-23} J/K
Standard atmosphere		101,325 Pa = 14.696 lb/in.^{2}
Fine-structure constant	$\alpha = e^2/(h \cdot c)$	1/137.14
Classical electron radius	$R_0 = e^2/(m \cdot c^2)$	2.818042×10^{-15} m
Bohr radius	$\hbar^2/(m \cdot e^2)$	0.529177×10^{-10} m
Compton wavelength	$h/(m \cdot c)$	2.424631×10^{-12} m

Table A2
Conversion Table

To Convert	Multiply by	To Obtain (symbol)
MeV	1.602×10^{-13}	Joules (J)
Pounds (lb)	0.4536	Kilograms (kg)
Inches	2.54×10^{-2}	Meters (m)
Lb/in^2	6.8946×10^3	Pascal (Pa)
Btu/h	0.29307	Watts (W)
Pounds force	4.4482	Newtons (N)
Flux [particles/(cm^2 s)]	10^4	Flux (particles/m$^2 \cdot$ s)
Density (g/cm^3)	10^3	Density (kg/m^3)
μ (cm^2/g)	10^{-1}	μ (m^2/kg)
μ (cm^{-1})	100	μ (m^{-1})
Σ (cm^{-1})	100	Σ (m^{-1})
Range (g/cm^2)	10	Range (kg/m^2)
Curies	3.7×10^{10}	Becquerels (Bq)
Rad	10^{-2}	Grays (Gy)
Rem	10^{-2}	Sieverts (Sv)

Table A3
Prefix and SI Symbols of Multiplication Factors

10^{12} tera (T)	10^{-6} micro (μ)
10^9 giga (G)	10^{-9} nano (n)
10^6 mega (M)	10^{-12} pico (p)
10^3 kilo (k)	10^{-15} femto (f)
10^{-3} milli (m)	10^{-18} atto (a)

Appendix B

Atomic Masses and Other Properties of Isotopes[a]

Name	Symbol	Z	A	Isotopic Mass (u)[b]	Natural Abundance (%)	Density of Element (in 10^3 kg/m^3)	σ_a(b) for (n, γ) (0.0253-eV neutrons)
Aluminum	Al	13	27	26.98153	100	2.7[c]	0.235
Antimony	Sb	51	121	120.9038	57.25	6.62	5.9
			123	122.9041	42.75		4.1
Argon	Ar	18	36	35.96755	0.337	Gas	6.0
			38	37.96272	0.063		0.8
			40	39.96238	99.60		0.61
Arsenic	As	33	75	74.9216	100	5.73	4.5
Beryllium	Be	4	9	9.01218	100	1.85	0.0095
Bismuth	Bi	83	209	208.9804	100	9.8	0.034
Boron	B	5	10	10.01294	19.78	2.3	3837
			11	11.00931	80.22		0.04
Cadmium	Cd	48	106	105.9070	1.22	8.65	2450
			108	107.9040	0.88		2.0
			110	109.9039	12.39		0.1
			111	110.9042	12.75		–
			112	111.9028	24.07		0.03
			113	112.9046	12.26		20,000
			114	113.9036	28.86		0.14
			116	115.9050	7.58		1.4
Carbon	C	6	12	12.0000	98.89	1.60	0.0034
			13	13.00335	1.11		0.0009
Cesium	Cs	55	133	132.9051	100	1.9	29.0
Cobalt	Co	27	59	59.93344	100	8.8	37.2
Copper	Cu	29	63	62.9298	60.09	8.96	4.5
			65	64.9278	30.91		2.2
Gadolinium	Gd	64	152	151.9195	0.20	7.95	< 180
			154	153.9207	2.15		100.0
			155	154.9226	14.73		61,000
			156	155.9221	20.47		11.5
			157	156.9339	15.68		254,000
			158	157.9241	24.87		3.5
Germanium	Ge	32	70	69.9243	20.52	5.36	3.68
			72	71.9217	27.43		0.98
			73	72.9234	7.76		14.0
			74	73.9212	36.54		0.45
			76	75.9214	7.76		0.2
Gold	Au	79	197	196.9666	100	19.32	98.8
Helium	He	2	3	3.01603	0.00013	Gas	5327
			4	4.00260	99.99987		–
Hydrogen	H	1	1	1.007825	99.985	Gas	0.332
			2	2.01410	0.015		0.0005
Indium	In	49	113	112.9043	4.28	7.31	11.1
			115	114.9041	95.72		193.2
Iodine	I	53	127	126.9044	100	4.93	6.2

continued

473

Atomic Masses and Other Properties of Isotopes[a]

Name	Symbol	Z	A	Isotopic Mass (u)[b]	Natural Abundance (%)	Density of Element (in 10^3 kg/m³)	σ_a(b) for (n, γ) (0.0253-eV neutrons)
Iron	Fe	26	54	53.9396	5.82	7.87	2.3
			56	55.9349	91.66		2.7
			57	56.9354	2.19		2.5
			58	57.9333	0.33		1.2
Lead	Pb	82	204	203.9730	1.48	11.34	0.17
			206	205.9745	23.6		0.0305
			207	206.9759	22.6		0.709
			208	207.9766	52.3		< 0.03
Lithium	Li	3	6	6.01512	7.42	0.53	245 (n, α)
			7	7.01600	92.58		0.037
Mercury	Hg	80	196	195.9650	0.146	13.55	3100.0
			198	197.9668	10.02		0.018
			199	198.9683	16.84		2500.0
			200	199.9683	23.13		50.0
			201	200.9703	13.22		50.0
			202	201.9706	29.80		4.5
			204	203.9735	6.85		0.4
Nickel	Ni	28	58	57.9353	67.88	8.90	4.4
			60	59.9308	26.23		2.6
			61	60.9310	1.19		2.0
			62	61.9283	3.66		14.2
			64	63.9280	1.08		1.5
Nitrogen	N	7	14	14.00307	99.63	Gas	1.81
			15	15.00011	0.37		0.00004
Oxygen	O	8	16	15.99491	99.759	Gas	0.000178
			17	16.99914	0.037		0.04
			18	17.99915	0.204		0.00016
Phosphorus	P	15	31	30.97376	100	1.82	0.19
Platinum	Pt	78	190	189.9600	0.0127	21.45	150
			192	191.9614	0.78		8.0
			194	193.9628	32.19		1.2
			195	194.9648	33.8		27.0
			196	195.9650	25.3		1.0
			198	197.9675	7.21		4.0
Rhodium	Rh	45	103	102.9048	100	12.41	150
Silicon	Si	14	28	27.97693	92.21	2.33	0.080
			29	28.97649	4.70		0.28
			30	29.97376	3.09		0.40
Silver	Ag	47	107	106.9041	51.82	10.49	37.0
			109	108.9047	45.18		92.0
Sodium	Na	11	23	22.98977	100	0.97	0.534
Uranium	U	92	234	234.0409	0.0057	19.1	95
			235	235.0439	0.710		678
			238	238.0508	99.284		2.73

[a] A complete list of isotope properties, for any isotope, can be retrieved from several websites such as http://periodic.lanl.gov/default.htm; http://www.lenntech.com/periodic/elements/pb.htm; http://www.chemicool.com/; http://www.nndc.bnl.gov/; http://atom.kaeri.re.kr/ton/

[b] Isotopic masses from *Nuclear Heat Transfer* by M. M. El-Wakil, International Textbook Co., N.Y. (1971).

[c] Same number gives density in g/cm³ or 10^3 kg/m³.

Appendix C

Alpha, Beta, and Gamma Sources Commonly Used

Table C1
Alpha Sources

Isotope	Half-Life	Alpha Energy (MeV)	Relative Intensity[a] (%)
^{210}Po	138.38 d	5.304	100
^{234}U	2.446×10^5 y	4.774	72
		4.723	28
^{233}U		4.824	84.3
		4.783	13.2
^{235}U	7.038×10^8 y	4.397	57
		4.367	18
^{238}U	4.468×10^9 y	4.196	77
		4.149	23
^{238}Pu	87.7 y	5.499	70.9
		5.546	29.0
^{239}Pu	2.413×10^4 y	5.155	73.3
		5.143	15.1
		5.105	11.5
^{241}Am	432.02 y	5.486	86
		5.443	12.7
^{252}Cf	2.646 y	6.118	81.6
		6.076	15.2
^{237}Np	2.14×10^6 y	4.788	47.6
		4.471	23.2
^{244}Cm	18.1 y	5.805	76.9
		5.762	23.1

[a] Only intensities greater than 10% are listed.

Table C2
Electron and Beta Sources

Isotope	Half-Life	Type of Particle	Energy (MeV)
^{3}H	12.33 y	β^-	0.0186
^{14}C	5730 y	β^-	0.1565
^{32}P	14.28 d	β^-	1.7104
^{35}S	87.4 d	β^-	0.1675
^{36}Cl	3.0×10^5 y	β^-	0.7095
^{89}Sr	50.55 d	β^-	1.463
^{90}Y	64 h	β^-	2.282
^{99}Tc	2.14×10^5 y	β^-	0.292
^{63}Ni	100 y	β^-	0.0659
^{113}Sn	115.1 d	IC electron	$E_K = 0.3625$
			$E_L = 0.3875$
^{137}Cs[a]	30.17 y	IC electron	$E_K = 0.626$
			$E_L = 0.656$
^{207}Bi	38 y	IC electron	$E_K = 0.4816, 0.5558$
			$E_L = 0.9754, 1.0496$
			$E_M = 1.060$

[a] ^{137}Cs is also a β^- emitter.

Table C3
Commonly Used Gamma Rays for Energy Calibration[a]

Source	Energy (keV)			Source	Energy (keV)		
^{241}Am	59.536	±	0.001	^{192}Ir	468.060	±	0.010
^{109}Cd	88.034	±	0.010	Annihilation	511.003	±	0.002
^{182}Ta	100.106	±	0.001	^{207}Bi	569.690	±	0.030
^{57}Co	122.046	±	0.020	^{208}Tl	583.139	±	0.023
^{144}Ce	133.503	±	0.020	^{192}Ir	604.378	±	0.020
^{57}Co	136.465	±	0.020	^{192}Ir	612.430	±	0.020
^{141}Ce	145.442	±	0.010	^{137}Cs	661.615	±	0.030
^{182}Ta	152.435	±	0.004	^{54}Mn	834.840	±	0.050
^{139}Ce	165.852	±	0.010	^{88}Y	898.023	±	0.065
^{182}Ta	179.393	±	0.000	^{207}Bi	1063.655	±	0.040
^{182}Ta	222.110	±	0.000	^{60}Co	1173.231	±	0.030
^{212}Pb	238.624	±	0.008	^{22}Na	1274.550	±	0.040
^{203}Hg	279.179	±	0.010	^{60}Co	1332.508	±	0.015
^{192}Ir	295.938	±	0.010	^{140}La	1596.200	±	0.040
^{192}Ir	308.440	±	0.010	^{124}Sb	1691.022	±	0.040
^{192}Ir	316.490	±	0.010	^{88}Y	1836.127	±	0.050
^{131}I	364.491	±	0.015	^{208}Tl	2614.708	±	0.050
^{198}Au	411.792	±	0.008	^{24}Na	2754.142	±	0.060

[a] Knoll, G. F., *Radiation Detection and Measurement*, 3rd ed., Wiley, 2000.

Table C4
Multiple Gamma Rays Emitted in the Decay of ^{152}Eu for Efficiency Calibration[a]

Energy (keV)	Relative Intensity		
121.8	141	±	4
244.7	36.6	±	1.1
344.3	127.2	±	1.3
367.8	4.19	±	0.04
411.1	10.71	±	0.11
444.0	15	±	0.15
488.7	1.984	±	0.023
586.3	2.24	±	0.05
678.6	2.296	±	0.028
688.7	4.12	±	0.04
788.9	6.6	±	0.6
867.4	20.54	±	0.21
964.0	70.4	±	0.7
1005.1	3.57	±	0.07
1085.8	48.7	±	0.5
1089.7	8.26	±	0.09
1112.1	65	±	0.7
1212.9	6.67	±	0.07
1299.1	7.76	±	0.08
1408.0	100	±	1
1457.6	2.52	±	0.09

[a] Knoll, G. F., *Radiation Detection and Measurement*, 3rd ed., Wiley, 2000.

Table C5
Gamma Rays Emitted by ^{226}Ra in Equilibrium with Its Daughters for Efficiency Calibration

Isotope	Gamma-Ray Energy (keV)			Relative Intensity		
^{226}Ra	186.211	±	0.010	9.00	±	0.10
^{214}Pb	241.981	±	0.008	16.06	±	0.19
^{214}Pb	295.213	±	0.008	42.01	±	0.53
^{214}Pb	351.921	±	0.008	80.42	±	0.81
^{214}Bi	609.312	±	0.007	100.00	±	0.92
^{214}Bi	768.356	±	0.010	10.90	±	0.15
^{214}Bi	934.061	±	0.012	6.93	±	0.10
^{214}Bi	1120.287	±	0.010	32.72	±	0.39
^{214}Bi	1238.11	±	0.012	12.94	±	0.17
^{214}Bi	1377.669	±	0.012	8.87	±	0.15
^{214}Bi	1509.228	±	0.015	4.78	±	0.09
^{214}Bi	1729.595	±	0.015	6.29	±	0.10
^{214}Bi	1764.494	±	0.014	34.23	±	0.44
^{214}Bi	1847.42	±	0.025	4.52	±	0.09
^{214}Bi	2118.551	±	0.030	2.53	±	0.05
^{214}Bi	2204.215	±	0.040	10.77	±	0.20
^{214}Bi	2447.86	±	0.100	3.32	±	0.08

Source: Knoll, G. F., *Radiation Detection and Measurement*, 3rd ed., Wiley, 2000.

Table C6
Commonly Used Gamma-Rays for Efficiency Calibration[a]

Isotope	Half-Life		Energy (keV)	Intensity
^{22}Na	2.6	y	1274.5	99.95
^{24}Na	15	h	1368.5	100.0
			2754	99.85
^{46}Sc	83.7	d	889.2	99.98
			1120.5	99.99
^{54}Mn	312.5	d	864.8	99.98
^{57}Co	272	d	14.4	9.6
			122.1	85.6
^{60}Co	5.27	y	1173.2	99.88
			1332.5	99.98
^{85}Sr	64.8	d	13.4	50.7
			514	99.28
^{88}Y	106.6	d	14.2	52.5
			1836.1	99.4
^{95}Nb	35.15	d	765.8	99.8
^{113}Sn	115.2	d	24.1	79.5
^{131}I	8.02	d	364.5	82.4
^{134}Cs	2.06	y	604.6	97.5
^{137}Cs	30	y	31.8/32.2	5.64

continued

Table C6 (continued)
Commonly Used Gamma-Rays for Efficiency Calibration[a]

Isotope	Half-Life		Energy (keV)	Intensity
			661.6	85.3
^{139}Ce	137.6	d	33.0./33.4	64.1
			165.8	80.0
^{141}Ce	32.5	d	35.6/36.0	12.6
			145.5	48.4
^{140}La	40.27	h	1596.6	95.6
^{198}Au	2.696	d	411.8	95.53
^{203}Hg	46.6	d	70.8/72.9	10.1
			279.2	81.3
^{241}Am	432	y	59.5	36.0

[a] Knoll, G. F., *Radiation Detection and Measurement*, 3rd ed., Wiley, 2000.

Appendix D

Tables of Photon Attenuation Coefficients*

Table D1
Total Mass Attenuation Coefficients in cm²/g[a] for Gamma Rays[b]

Photon Energy (MeV)	H	Be	C	N	O	Na
1.00 – 02[c]	3.85 – 01	5.36 – 01	2.17 + 00	3.57 + 00	5.58 + 00	1.51 + 01
1.50 – 02	3.76 – 01	2.68 – 01	7.22 – 01	1.09 + 00	1.62 + 00	4.37 + 00
2.00 – 02	3.69 – 01	2.06 – 01	3.88 – 01	5.41 – 01	7.54 – 01	1.88 + 00
3.00 – 02	3.57 – 01	1.71 – 01	2.30 – 01	2.76 – 01	3.35 – 01	6.39 – 01
4.00 – 02	3.46 – 01	1.59 – 01	1.93 – 01	2.12 – 01	2.36 – 01	3.55 – 01
5.00 – 02	3.35 – 01	1.52 – 01	1.79 – 01	1.87 – 01	1.99 – 01	2.54 – 01
6.00 – 02	3.26 – 01	1.47 – 01	1.70 – 01	1.74 – 01	1.81 – 01	2.09 – 01
8.00 – 02	3.09 – 01	1.39 – 01	1.58 – 01	1.60 – 01	1.62 – 01	1.70 – 01
1.00 – 01	2.94 – 01	1.32 – 01	1.50 – 01	1.50 – 01	1.52 – 01	1.52 – 01
1.50 – 01	2.65 – 01	1.19 – 01	1.34 – 01	1.34 – 01	1.34 – 01	1.31 – 01
2.00 – 01	2.43 – 01	1.09 – 01	1.23 – 01	1.23 – 01	1.23 – 01	1.18 – 01
3.00 – 01	2.11 – 01	9.45 – 02	1.07 – 01	1.06 – 01	1.07 – 01	1.02 – 01
4.00 – 01	1.89 – 01	8.47 – 02	9.55 – 02	9.54 – 02	9.54 – 02	9.14 – 02
5.00 – 01	1.73 – 01	7.73 – 02	8.72 – 02	8.71 – 02	8.71 – 02	8.34 – 02
6.00 – 01	1.60 – 01	7.15 – 02	8.07 – 02	8.05 – 02	8.06 – 02	7.72 – 02
8.00 – 01	1.40 – 01	6.29 – 02	7.09 – 02	7.08 – 02	7.08 – 02	6.78 – 02
1.00 + 00	1.26 – 01	5.65 – 02	6.37 – 02	6.36 – 02	6.37 – 02	6.09 – 02
1.50 + 00	1.03 – 01	4.60 – 02	5.19 – 02	5.18 – 02	5.18 – 02	4.97 – 02
2.00 + 00	8.75 – 02	3.94 – 02	4.45 – 02	4.45 – 02	4.46 – 02	4.28 – 02
3.00 + 00	6.91 – 02	3.14 – 02	3.57 – 02	3.58 – 02	3.60 – 02	3.49 – 02
4.00 + 00	5.81 – 02	2.66 – 02	3.05 – 02	3.07 – 02	3.10 – 02	3.04 – 02
5.00 + 00	5.05 – 02	2.35 – 02	2.71 – 02	2.74 – 02	2.78 – 02	2.76 – 02
6.00 + 00	4.50 – 02	2.12 – 02	2.47 – 02	2.51 – 02	2.55 – 02	2.56 – 02
8.00 + 00	3.75 – 02	1.82 – 02	2.16 – 02	2.21 – 02	2.26 – 02	2.32 – 02
1.00 + 01	3.25 – 02	1.63 – 02	1.96 – 02	2.02 – 02	2.09 – 02	2.18 – 02

	Al	Si	Fe	Cu	Pb	NaI	Ge
1.00 – 02	2.58 + 01	3.36 + 01	1.72 + 02	2.23 + 02	1.28 + 02	1.36 + 02	3.74 + 01
1.50 – 02	7.66 + 00	9.97 + 00	5.57 + 01	7.33 + 01	1.12 + 02	4.59 + 01	9.15 + 01
2.00 – 02	3.24 + 00	4.19 + 00	2.51 + 01	3.30 + 01	8.34 + 01	2.12 + 01	4.22 + 01
3.00 – 02	1.03 + 00	1.31 + 00	7.88 + 00	1.06 + 01	2.84 + 01	6.86 + 00	1.39 + 01
4.00 – 02	5.14 – 01	6.35 – 01	3.46 + 00	4.71 + 00	1.31 + 01	1.89 + 01	6.21 + 00
5.00 – 02	3.34 – 01	3.96 – 01	1.84 + 00	2.50 + 00	7.22 + 00	1.05 + 01	3.34 + 00
6.00 – 02	2.55 – 01	2.92 – 01	1.13 + 00	1.52 + 00	4.43 + 00	6.42 + 00	2.02 + 00
8.00 – 02	1.89 – 01	2.07 – 01	5.50 – 01	7.18 – 01	2.07 + 00	3.00 + 00	9.50 – 01
1.00 – 01	1.62 – 01	1.73 – 01	3.42 – 01	4.27 – 01	5.23 + 00	1.64 + 00	5.55 – 01
1.50 – 01	1.34 – 01	1.40 – 01	1.84 – 01	2.08 – 01	1.89 + 00	5.90 – 01	2.49 – 01
2.00 – 01	1.20 – 01	1.25 – 01	1.39 – 01	1.48 – 01	9.45 – 01	3.14 – 01	1.66 – 01
3.00 – 01	1.03 – 01	1.07 – 01	1.07 – 01	1.08 – 01	3.83 – 01	1.58 – 01	1.13 – 01

continued

* Mass X-ray and photon mass attenuation and mass energy absorption coefficients for Z = 1 to Z = 92 and 48 compounds can be retrieved from the National Institute and Standards Technology website http://physics.nist.gov/PhysRefData/XrayMassCoef/cover.html

Table D1 (continued)
Total Mass Attenuation Coefficients in cm²/gᵃ for Gamma Rays[b]

Photon Energy (MeV)	Al	Si	Fe	Cu	Pb	NaI	Ge
4.00 – 01	9.22 – 02	9.54 – 02	9.21 – 02	9.19 – 02	2.20 – 01	1.12 – 01	9.33 – 02
5.00 – 01	8.41 – 02	8.70 – 02	8.29 – 02	8.22 – 02	1.54 – 01	9.21 – 02	8.21 – 02
6.00 – 01	7.77 – 02	8.05 – 02	7.62 – 02	7.52 – 02	1.20 – 1	8.02 – 02	7.45 – 02
8.00 – 01	6.83 – 02	7.06 – 02	6.65 – 02	6.55 – 02	8.56 – 02	6.63 – 02	6.43 – 02
1.00 + 00	6.14 – 02	6.35 – 02	5.96 – 02	5.86 – 02	6.90 – 02	5.80 – 02	5.73 – 02
1.50 + 00	5.00 – 02	5.18 – 02	4.87 – 02	4.79 – 02	5.10 – 02	4.66 – 02	4.66 – 02
2.00 + 00	4.32 – 02	4.48 – 02	4.25 – 02	4.19 – 02	4.50 – 02	4.12 – 02	4.09 – 02
3.00 + 00	3.54 – 02	3.68 – 02	3.62 – 02	3.59 – 02	4.16 – 02	3.66 – 02	3.52 – 02
4.00 + 00	3.11 – 02	3.24 – 02	3.31 – 02	3.32 – 02	4.14 – 02	3.50 – 02	3.28 – 02
5.00 + 00	2.84 – 02	2.97 – 02	3.14 – 02	3.18 – 02	4.24 – 02	3.46 – 02	3.16 – 02
6.00 + 00	2.66 – 02	2.79 – 02	3.05 – 02	3.10 – 02	4.34 – 02	3.47 – 02	3.11 – 02
8.00 + 00	2.44 – 02	2.57 – 02	2.98 – 02	3.06 – 02	4.59 – 02	3.55 – 02	3.10 – 02
1.00 + 01	2.31 – 02	2.46 – 02	2.98 – 02	3.08 – 02	4.84 – 02	3.68 – 02	3.16 – 02

	Air	H₂O	Compact Bone	Muscle, Striated	Concrete	Pyrex Glass	Lucite $(C_5H_8O_2)_n$
1.00 – 02	4.82 + 00	4.99 + 00	2.00 + 01	5.09 + 00	2.65 + 01	1.67 + 01	3.11 + 00
1.50 – 02	1.45 + 00	1.48 + 00	6.15 + 00	1.53 + 00	8.01 + 00	4.95 + 00	9.82 – 01
2.00 – 02	6.91 – 01	7.11 – 01	2.68+ 00	7.31 – 01	3.45 + 00	2.13 + 00	5.03 – 01
3.00 – 02	3.18 – 01	3.38 – 01	9.07 – 01	3.42 – 01	1.12 + 00	7.24 – 01	2.74 – 01
4.00 – 02	2.29 – 01	2.48 – 01	4.78 – 01	2.49 – 01	5.59 – 01	3.94 – 01	2.19 – 01
5.00 – 02	1.96 – 01	2.14 – 01	3.27 – 01	2.14 – 01	3.61 – 01	2.77 – 01	1.98 – 01
6.00 – 02	1.79 – 01	1.97 – 01	2.58 – 01	1.97 – 01	2.73 – 01	2.25 – 01	1.86 – 01
8.00 – 02	1.62 – 01	1.79 – 01	2.00 – 01	1.78 – 01	2.00 – 01	1.80 – 01	1.72 – 01
1.00 – 01	1.51 – 01	1.68 – 01	1.74 – 01	1.66 – 01	1.70 – 01	1.60 – 01	1.62 – 01
1.50 – 01	1.34 – 01	1.49 – 01	1.47 – 01	1.48 – 01	1.40 – 01	1.36 – 01	1.45 – 01
2.00 – 01	1.23 – 01	1.36 – 01	1.32 – 01	1.35 – 01	1.25 – 01	1.23 – 01	1.32 – 01
3.00 – 01	1.06 – 01	1.18 – 01	1.14 – 01	1.17 – 01	1.07 – 01	1.06 – 01	1.15 – 01
4.00 – 01	9.53 – 02	1.06 – 01	1.01 – 01	1.05 – 01	9.58 – 02	9.50 – 02	1.03 – 01
5.00 – 01	8.70 – 02	9.67 – 02	9.25 – 02	9.59 – 02	8.73 – 02	8.67 – 02	9.41 – 02
6.00 – 01	8.05 – 02	8.95 – 02	8.56 – 02	8.87 – 02	8.07 – 02	8.02 – 02	8.70 – 02
8.00 – 01	7.07 – 02	7.86 – 02	7.51 – 02	7.79 – 02	7.09 – 02	7.04 – 02	7.65 – 02
1.00 + 00	6.36 – 02	7.07 – 02	6.75 – 02	7.00 – 02	6.37 – 02	6.33 – 02	6.87 – 02
1.50 + 00	5.18 – 02	5.75 – 02	5.50 – 02	5.70 – 02	5.19 – 02	5.16 – 02	5.59 – 02
2.00 + 00	4.45 – 02	4.94 – 02	4.73 – 02	4.89 – 02	4.48 – 02	4.44 – 02	4.80 – 02
3.00 + 00	3.58 – 02	3.97 – 02	3.83 – 02	3.93 – 02	3.65 – 02	3.61 – 02	3.85 – 02
4.00 + 00	3.08 – 02	3.40 – 02	3.31 – 02	3.37 – 02	3.19 – 02	3.14 – 02	3.29 – 02
5.00 + 00	2.75 – 02	3.03 – 02	2.97 – 02	3.00 – 02	2.90 – 02	2.84 – 02	2.92 – 02
6.00 + 00	2.52 – 02	2.77 – 02	2.74 – 02	2.74 – 02	2.70 – 02	2.63 – 02	2.66 – 02
8.00 + 00	2.23 – 02	2.43 – 02	2.44 – 02	2.40 – 02	2.45 – 02	2.37 – 02	2.32 – 02
1.00 + 01	2.04 – 02	2.22 – 02	2.26 – 02	2.19 – 02	2.31 – 02	2.22 – 02	2.11 – 02

ᵃ Multiply by 10^{-1} to obtain m²/kg.
ᵇ From J. H. Hubbell, "Photon Cross Sections, Attenuation Coefficients, and Energy Absorption Coefficients from 10 keV to 100 GeV," NSRDS-NBS 29 C (1969).
ᶜ Read × 10^{-2}.

Table D2
Mass Energy Absorption Coefficients in cm²/g[a] for Gamma Rays[b]

Photon Energy (MeV)	H	Be	C	N	O	Na
0.01	0.00986	0.368	1.97	3.38	5.39	14.9
0.015	0.0110	0.104	0.536	0.908	1.44	4.20
0.02	0.0135	0.0469	0.208	0.362	0.575	1.70
0.03	0.0185	0.0195	0.0594	0.105	0.165	0.475
0.04	0.0231	0.0146	0.0306	0.0493	0.0733	0.199
0.05	0.0271	0.0142	0.0233	0.0319	0.0437	0.106
0.06	0.0306	0.0147	0.0211	0.0256	0.0322	0.0668
0.08	0.0362	0.0166	0.0205	0.0223	0.0249	0.0382
0.10	0.0406	0.0184	0.0215	0.0224	0.0237	0.0297
0.15	0.0481	0.0216	0.0245	0.0247	0.0251	0.0260
0.2	0.0525	0.0235	0.0265	0.0267	0.0268	0.0264
0.3	0.0569	0.0255	0.0287	0.0287	0.0288	0.0277
0.4	0.0586	0.0262	0.0295	0.0295	0.0295	0.0284
0.5	0.0593	0.0265	0.0297	0.0296	0.0297	0.0285
0.6	0.0587	0.0263	0.0295	0.0295	0.0296	0.0284
0.8	0.0574	0.0256	0.0288	0.0289	0.0289	0.0275
1.0	0.0555	0.0248	0.0279	0.0279	0.0278	0.0266
1.5	0.0507	0.0227	0.0255	0.0255	0.0254	0.0243
2	0.0464	0.0208	0.0234	0.0234	0.0234	0.0225
3	0.0398	0.0180	0.0204	0.0205	0.0206	0.0199
4	0.0352	0.0161	0.0185	0.0186	0.0188	0.0184
5	0.0317	0.0148	0.0171	0.0173	0.0175	0.0174
6	0.0290	0.0138	0.0161	0.0163	0.0166	0.0161
8	0.0252	0.0123	0.0147	0.0151	0.0155	0.0159
10	0.0225	0.0114	0.0138	0.0143	0.0148	0.0155

	Al	Si	Fe	Cu	Pb	Air	Ge
0.01	25.5	33.3	142	160	127	4.61	35.64
0.015	7.47	9.75	49.3	59.4	91.7	1.27	62.56
0.02	3.06	4.01	22.8	28.2	69.1	0.511	31.78
0.03	0.868	1.14	7.28	9.50	24.6	0.148	11.26
0.04	0.357	0.472	3.17	4.24	11.8	0.0668	5.152
0.05	0.184	0.241	1.64	2.22	6.54	0.0406	2.759
0.06	0.111	0.144	0.961	1.32	4.08	0.0305	1.642
0.08	0.0562	0.0700	0.414	0.573	1.86	0.0243	0.7184
0.10	0.0386	0.0459	0.219	0.302	2.28	0.0234	0.3803
0.15	0.0285	0.0312	0.0814	0.106	1.15	0.0250	0.1288
0.2	0.0276	0.0292	0.0495	0.0597	0.629	0.0268	0.0686
0.3	0.0282	0.0294	0.0335	0.0370	0.259	0.0287	0.0389
0.4	0.0287	0.0298	0.0308	0.0318	0.143	0.0295	0.0319
0.5	0.0286	0.0298	0.0295	0.0298	0.0951	0.0296	0.0293
0.6	0.0286	0.0295	0.0286	0.0286	0.0710	0.0295	0.0279
0.8	0.0277	0.0288	0.0273	0.0271	0.0481	0.0289	0.0261
1.0	0.0269	0.0277	0.0262	0.0258	0.0377	0.0278	0.0248
1.5	0.0245	0.0253	0.0237	0.0233	0.0271	0.0254	0.0224
2	0.0226	0.0234	0.0220	0.0217	0.0240	0.0234	0.0209

continued

Table D2 (continued)
Mass Energy Absorption Coefficients in cm²/g^a for Gamma Rays^b

Photon Energy (MeV)	Al	Si	Fe	Cu	Pb	Air	Ge
3	0.0202	0.0210	0.0204	0.0202	0.0234	0.0205	0.0198
4	0.0188	0.0196	0.0199	0.0200	0.0245	0.0186	0.0196
5	0.0179	0.0187	0.0198	0.0200	0.0259	0.0174	0.0198
6	0.0172	0.0182	0.0199	0.0202	0.0272	0.0164	0.0203
8	0.0168	0.0177	0.0204	0.0209	0.0294	0.0152	0.0121
10	0.0165	0.0175	0.0209	0.0215	0.0310	0.0145	0.0221

	H_2O	Compact Bone	Muscle, Striated	Concrete	Pyrex Glass	Lucite $(C_5H_8O_2)_n$
0.01	4.79	19.2	4.87	25.5	16.5	2.91
0.015	1.28	5.84	1.32	7.66	4.75	0.783
0.02	0.512	2.46	0.533	3.22	1.94	0.310
0.03	0.149	0.720	0.154	0.936	0.554	0.0899
0.04	0.0677	0.304	0.0701	0.393	0.232	0.0437
0.05	0.0418	0.161	0.0431	0.204	0.122	0.0301
0.06	0.0320	0.0998	0.0328	0.124	0.0768	0.0254
0.08	0.0262	0.0537	0.0264	0.0625	0.0428	0.0232
0.10	0.0256	0.0387	0.0256	0.0424	0.0325	0.0238
0.15	0.0277	0.0305	0.0275	0.0290	0.0274	0.0266
0.2	0.0297	0.0301	0.0294	0.0290	0.0276	0.0287
0.3	0.0319	0.0310	0.0317	0.0295	0.0289	0.0310
0.4	0.0328	0.0315	0.0325	0.0298	0.0295	0.0318
0.5	0.0330	0.0317	0.0328	0.0300	0.0297	0.0322
0.6	0.0329	0.0314	0.0325	0.0297	0.0294	0.0319
0.8	0.0321	0.0306	0.0318	0.0289	0.0287	0.0311
1.0	0.0309	0.0295	0.0306	0.0279	0.0277	0.0301
1.5	0.0282	0.0270	0.0280	0.0254	0.0252	0.0275
2	0.0260	0.0249	0.0257	0.0235	0.0233	0.0253
3	0.0227	0.0219	0.0225	0.0209	0.0207	0.0220
4	0.0206	0.0200	0.0204	0.0193	0.0190	0.0199
5	0.0191	0.0187	0.0189	0.0182	0.0179	0.0184
6	0.0180	0.0178	0.0178	0.0176	0.0171	0.0173
8	0.0166	0.0167	0.0164	0.0168	0.0163	0.0158
10	0.0157	0.0159	0.0155	0.0163	0.0157	0.0148

^{a} Multiply by 10^{-1} to obtain m²/kg.
^{b} From J. H. Hubbell, "Photon Cross Sections, Attenuation Coefficients, and Energy Absorption Coefficients from 10 keV to 100 GeV," NSRDS-NBS 29 C (1969).

Appendix E

Table of Buildup Factor Constants*

Coefficients of the Berger Equation for Dose Buildup Factors

$$B = 1 + a\,\mu r\,\exp(b\,\mu r)^{a}$$

E (MeV)	Concrete[b] a	b	Air[c] a	b	Water[c] a	b	Iron a	b	Lead a	b
0.015	0.01	0.029	0.08	–0.034	0.09	–0.036	0.00	0.000	0.00	0.000
0.02	0.03	0.041	0.23	–0.032	0.26	–0.032	0.02	–0.032	0.00	0.000
0.03	0.10	–0.036	0.93	–0.009	1.01	–0.006	0.01	–0.036	0.00	0.000
0.04	0.26	0.035	2.40	0.018	2.58	0.024	0.02	–0.032	0.01	–0.066
0.05	0.52	–0.026	4.05	0.050	4.36	0.057	0.04	–0.034	0.01	–0.046
0.06	0.78	–0.008	5.27	0.075	5.59'	0.082	0.07	–0.039	0.01	–0.028
0.08	1.42	0.007	6.11	0.102	6.47	0.108	0.14	–.034	0.02	–0.029
0.1	1.83	0.028	5.93	0.113	6.11	0.120	0.24	–0.030	0.20	0.479
0.15	2.19	0.054	4.70	0.121	4.88	0.125	0.52	–0.015	0.21	–0.075
0.2	2.20	0.065	3.94	0.113	4.13	0.118	0.77	0.004	0.08	–0.054
0.3	2.03	0.067	3.10	0.094	3.18	0.096	1.06	0.022	0.08	–0.040
0.4	1.87	0.061	2.61	0.079	2.67	0.080	1.15	0.033	0.11	–0.033
0.5	1.73	0.055	2.29	0.067	2.32	0.068	1.16	0.036	0.15	–0.028
0.6	1.60	0.049	2.05	0.058	2.07	0.059	1.14	0.036	0.19	–0.024
0.8	1.41	0.040	1.71	0.045	1.74	0.045	1.09	0.032	0.25	–0.019
1	1.27	0.032	1.50	0.035	1.50	0.036	1.03	0.028	0.30	–0.015
1.5	1.02	0.021	1.16	0.021	1.16	0.021	0.88	0.020	0.36	–0.007
2	0.89	0.014	0.97	0.013	0.97	0.013	0.76	0.018	0.38	0.004
3	0.71	0.007	0.75	0.005	0.74	0.005	0.66	0.014	0.37	0.019
4	0.59	0.004	0.61	0.001	0.62	0.000	0.56	0.015	0.31	0.038
5	0.49	0.004	0.53	–0.002	0.52	–0.002	0.49	0.017	0.24	0.062
6	0.45	0.002	0.47	–0.004	0.47	–0.005	0.42	0.021	0.19	0.082
8	0.36	0.001	0.37	–0.004	0.38	–0.006	0.33	0.028	0.11	0.125
10	0.30	0.003	0.31	–0.004	0.31	–0.005	0.25	0.039	0.07	0.161
15	0.21	0.004	0.23	–0.006	0.23	–0.008	0.15	0.066	0.00	0.000

[a] Good up to 40 mean free paths, for point isotropic source in infinite medium.
[b] From A. B. Chilton, *Nuc. Set Eng.* **69**:436 (1979).
[c] From A. B. Chilton, C. M. Eisenhauer, and G. L. Simmons, *Nuc. Sci. Eng.* **73**:97 (1980).

---

* Shultis, J., and Faw, R. E., *Radiation Shielding*, Prentice Hall, 1996.

Index